Nano-Enabled Medical Applications

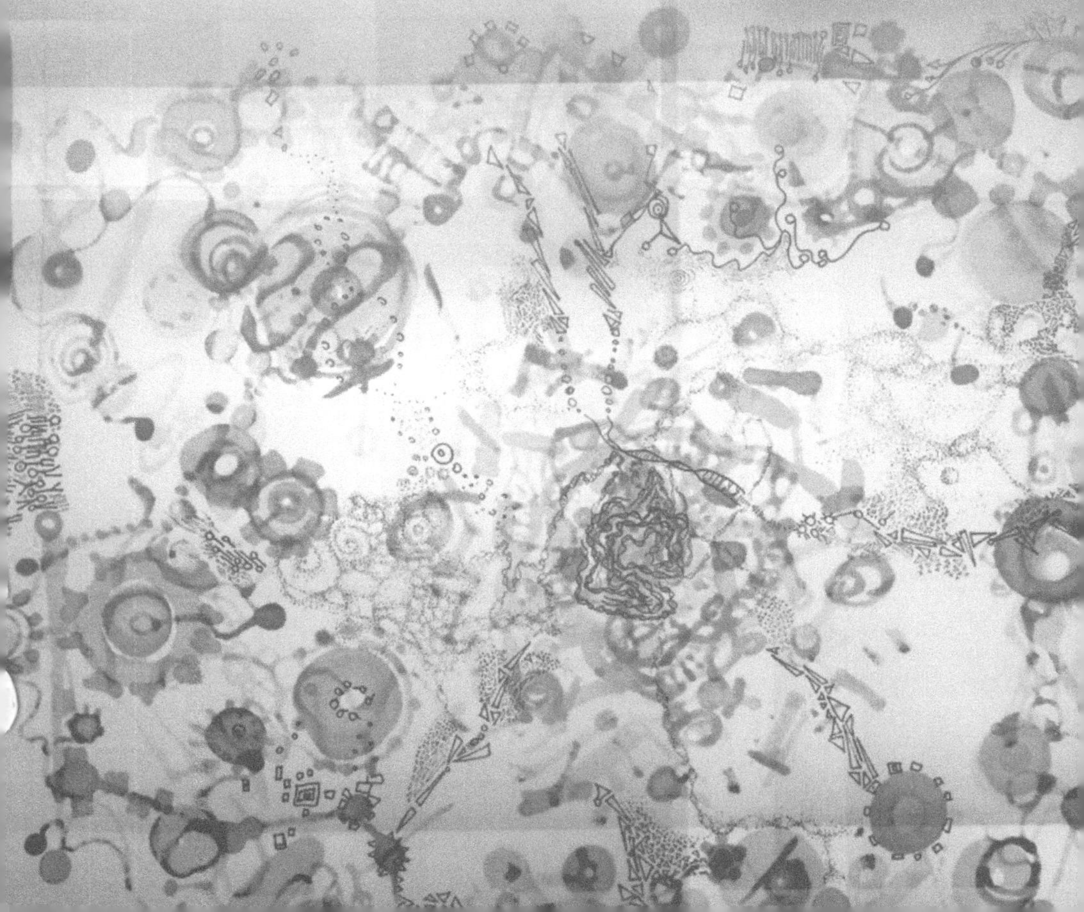

Nanomedicine's Most Cited

Lajos P. Balogh
Series Editor

Titles in the Series

Published

Vol. 1

Nanomedicine in Cancer

Lajos P. Balogh, ed.

2017

978-981-4745-80-2 (Hardcover)
978-1-315-11436-1 (eBook)

Vol. 2

Nano-Enabled Medical Applications

Lajos P. Balogh, ed.

2021

978-981-4800-32-7 (Hardcover)
978-0-429-39903-9 (eBook)

Forthcoming

Vol. 3

Functional Nanomaterials and Their Applications

Lajos P. Balogh, ed.

2021

Vol. 4

Theranostic Nanomedicine

Lajos P. Balogh, ed.

2022

Published by

Jenny Stanford Publishing Pte. Ltd.
Level 34, Centennial Tower
3 Temasek Avenue
Singapore 039190

Email: editorial@jennystanford.com
Web: www.jennystanford.com

British Library Cataloguing-in-Publication Data
A catalogue record for this book is available from the British Library.

Nano-Enabled Medical Applications

Cover image: Courtesy of Dr. Regina Valluzzi.

ISBN 978-981-4800-32-7 (Hardcover)
ISBN 978-0-429-39903-9 (eBook)

Nanomedicine's Most Cited – Volume 2

Nano-Enabled Medical Applications

edited by

Lajos P. Balogh

JENNY STANFORD
PUBLISHING

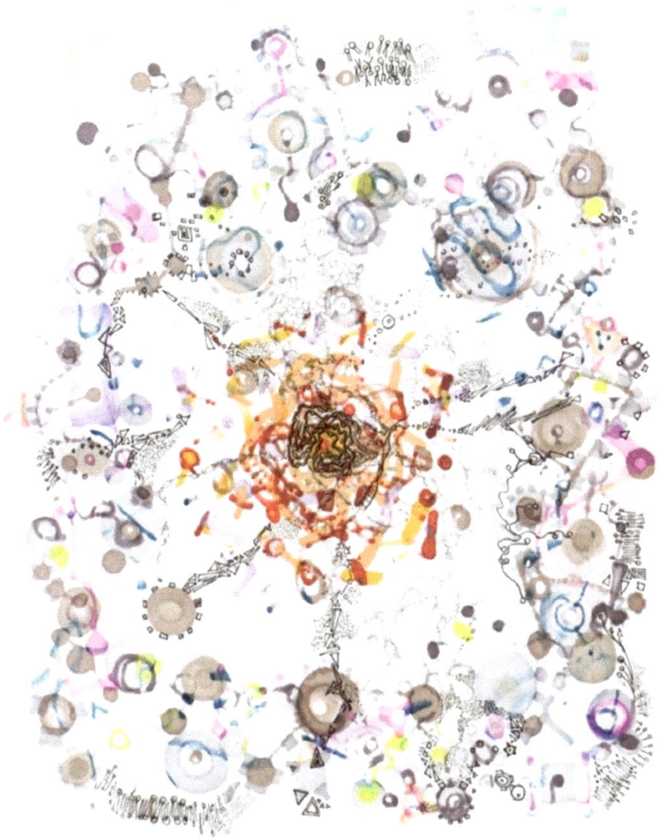

Scientific imagery is intriguing and beautiful, but also often impenetrable to the layperson. Powerful and beautiful ideas at the forefront of modern science become accessible when distilled through the lens of art. Dr. Regina Valluzzi achieves this distillation in a manner that respects the contexts of both art and science. A number of her pieces are on display in academic departmental offices and common spaces, corporate offices, and private collections worldwide. Her original work is available through her website, "The Nerdly Painter" http://www.NerdlyPainter.com. She also leases originals and prints through Turning Art (contact abby@turningart.com) and offers a range of prints and fun printed science-art items through https://regina-valluzzi.pixels.com/.

Contents

Preface

This unique collection of chapters in Volume 2 of the series provides the reader with an exceptional insight into some of the essential areas of nanomedicine with implications of their clinical applications.

The past 15 years of nanomedicine revolutionized the entire healthcare field and opened the medical and pharmaceutical mind-set to approaches that have not been considered feasible before. The period not only revealed new medicines and procedures but also advanced those to clinical translation. This book is a collection of the top-cited scientific articles selected from a variety of research areas that illustrate this movement in nanomedicine, enabling therapeutic, diagnostic, theranostic, and preventative technologies of the future.

Nanomedicine is a term that encompasses the use of nanotechnology in medicine and also to emphasize the significance of using sub-micron size biomaterials and interfaces for the treatment of diseases. The recognition and more detailed understanding that decreasing particle size fundamentally changes the properties and behavior of materials provided significant opportunities for the improvement of medicines.

These improvements are related to the utilization of novel biomaterials, superior controlled drug delivery, ultra-specific targeting, overcoming drug resistance, more patient-friendly drug administration methods, and earlier diagnosis. The excitement about the promising applications of novel biomaterials is seen from the results obtained with polymeric nanoparticles, liposomes, carbon nanotubes, nanodiamonds, quantum dots, and peptide- or RNA-based nanoparticles. Technological advances in the areas of highly controlled manufacturing methods using microfluidic techniques, development of nanomechanical analytical methods to distinguish healthy and cancer cells, and nano-electrophysiological devices to record intracellular activity are a few examples for the remarkable accomplishments in the field.

The fundamental knowledge about the fate of nanoparticles in the body was greatly expanded by studying the changes taking place on the surface of the nanoparticles after they are administered into the body. Regardless of the route of administration (injected, inhaled, topically applied into the eye or nose, or orally taken), all nanoparticles interact with biomolecules and attract some of these to their surface. This coating on the surface of nanoparticles (known as the "corona") greatly influences nanoparticle kinetics and cellular uptake, because in vivo it dynamically varies depending on the environment and changes the properties of the nanoparticles themselves thus affecting therapeutic efficacy.

Some of the most exciting developments in nanomedicine in the areas of cancer treatment, neuroscience, ophthalmology, and tissue engineering are presented here. The progression of innovative stages from designing nanoparticle systems with increasingly better effectiveness, specificities, and intelligent functions, understanding the basis of their biocompatibility to choose which one to manufacture, can be followed from the specially selected chapters. Most papers presented in this book have been cited several hundred times, and many have been cited over 1000 times, indicating the high level of activity, credibility, and importance to the field and researchers' accomplishments in these research areas. These chapters serve as inspiration for further research, development, and pursuit of solutions for therapeutic challenges.

Chapter 1

Intelligent Nanomaterials for Medicine: Carrier Platforms and Targeting Strategies—State of the Art

Georgette B. Salieb-Beugelaar,[a,b] Marc Wolf,[a] Roman Lehner,[c] Kegang Liu,[a] Stephan Marsch,[d] and Patrick Hunziker[a,b,d]

[a] *Nanomedicine Research Lab CLINAM, University of Basel, University Hospital Basel, Bernoullistrasse 20, CH-4056 Basel, Switzerland*
[b] *The European Foundation for Clinical Nanomedicine (CLINAM) Alemannengasse 12, CH-4016 Basel, Switzerland*
[c] *Adolphe Merkle Institute, University of Fribourg, Chemin des Verdiers 4, 1700 Fribourg, Switzerland*
[d] *Intensive Care Clinic, University Hospital Basel, Petersgraben 4, CH-4031 Basel, Switzerland*
beugelaar@swissnano.org

Nanoscience and nanotechnology have been significant drivers of medical knowledge generation and medical diagnostics and are bringing new therapeutic options to many medical fields. Intelligent nanomaterials have evolved from simpler nanosized materials such as liposomes and inorganic nanoparticles; they add environmental or remote sensing, switching, and stimuli-triggered activities to the carrier functionality of basic nanomaterials. This review examines properties and requirements for such nanodelivery platforms to deploy intelligent functionality, in particular for

Nano-Enabled Medical Applications
Edited by Lajos P. Balogh

ISBN 978-981-4800-32-7 (Hardcover), 978-0-429-39903-9 (eBook)
www.jennystanford.com

targeting receptors, cells, and tissues, for control of their nanobio interaction, and for their utility for clinical applications. Emphasis is placed on specific carrier platforms and targeting strategies required for real-world application, extending an earlier review with up-to-date information. Advantages and challenges to intelligent nanomaterials in a clinical context, including the important notion of particle-associated toxicity, e.g., due to their PEG content are discussed.

1.1 Introduction

During the past decade, the application of nanomaterials in medicine has extended to many new directions spanning from fundamental research (e.g., crossing biological barriers in drug delivery) to (commercially available) diagnostic tests, and nanotherapeutics. These developments have a profound impact on preclinical and clinical progress and shape the scientific field of nanomedicine. Currently, some of these developments are on the market, while many are being tested in clinical trials. Multi-functional, composite nanocarriers, such as nanotheranostics are a platform that combines the use of composite nanosystems for diagnosis and therapy. Such nanosystems often consist of (i) a carrier platform, (ii) a payload for imaging, sensing, or therapy, and (iii) optional targeting ligands. Beyond such basic systems, the large freedom in design allows to compose nanosystems with complex functionality that pave the way to intelligent and responsive behavior, potentially applicable in medicine. A typical example of such platforms are nanoparticles (NPs, diameter 1–200 nm) that have already been tested for medical applications as drug delivery platforms, e.g., for cancer therapy, but also as reagent in (rapid) diagnostic tests. NPs can carry different payloads such as small molecular drugs, imaging agents, proteins, nucleic acids, and other contents [1–4]. Nanocarriers are designed to improve efficacy and safety for drug delivery in general and for target specific non-viral drug delivery in particular [5, 6]. The design of such nanomaterials requires the ability to control particle size and shape, to assure

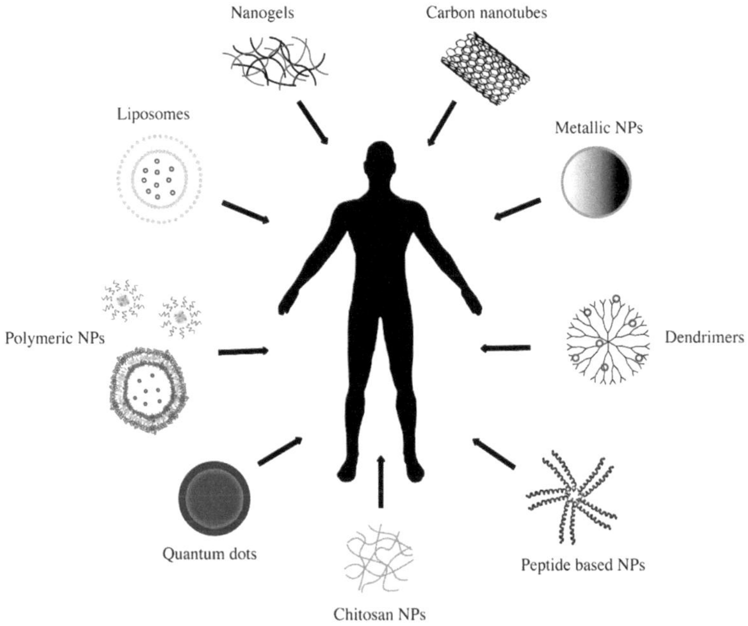

Figure 1.1 Schematic illustration, showing established therapeutic nanocarrier platforms (NPs) in (pre)clinical development.

biocompatibility and stealth properties, to optimize specificity, and to achieve controlled release and functionality [7, 10]. As basic materials, liposomes, dendrimers, polymers, carbon nanotubes, metallic NPs, silica NPs, organic NPs, quantum dots, nanogels, and peptidic NPs have been applied as possible nanotechnological carrier platforms (Fig. 1.1). With an emphasis on medical therapy, this review aims to shed light on design principles suited to create complex nanosystems by combining carrier platforms, engineering nanomaterial–cell interactions and enabling such systems to show stimuli responsiveness by equipping them with elements from the toolbox of switches on the nanoscale, considering the critical factors of such systems for success in clinical application with regard to complement activation and hypersensitivity reactions in particular against polyethylene glycol (PEG). This chapter extends and updates an earlier review on this topic [11].

1.2 Nanotechnology Carrier Platforms Suited for Switch Functionality

In recent years, major efforts have been devoted to develop suitable nanotechnological platforms to improve drug delivery to tumor tissue. For the development of such platforms, several challenges need to be mastered: (i) the control of the particle size, which can have influence on the NP distribution, clearance by kidney or liver and payload uptake; (ii) biocompatibility, to achieve an optimal benefit/risk relation; (iii) stealth properties, to escape immunological recognition and serum protein interactions; (iv) optimal blood circulation time for a specific application; (v) high target specificity for delivery of drugs or advanced functionality; (vi) controlled release mechanisms, e.g., endosomal escape; (vii) further functionality control through stimuli responsiveness. Multiple nanoscale platforms have been developed for this purpose, of which the most important will be discussed now.

1.2.1 Micelles and Liposomes

Lipid micelles are spherical nanosized structures characterized by a hydrophobic core and a hydrophilic coat (lipid bilayer) that may form spontaneously from amphiphilic molecules (phospholipids) in aqueous environments. They are suitable to encapsulate and transport hydrophobic molecules to (targeted) cells (e.g., hydrophobic drugs: taxanes [12, 13]). Liposomes can be prepared in various ways resulting in uni-laminar or multi-laminar liposomes. Uni-laminar liposomes range in size from approximately 50–250 nm have an aqueous core and are suitable for the loading of hydrophilic drugs, whereas multi-laminar liposomes usually have a diameter ranging 1–5 μm and are suitable for the loading of hydrophobic drugs due to the limited aqueous space [14, 15]. Under specific conditions, liposomes of ~100 nm in diameter have been successfully used to deliver chemotherapeutic agents to tumors [16]. Drug delivery of poorly soluble molecules can be achieved through micelles using lipid moieties as hydrophobic blocks linked to hydrophilic polymers [17]. Different lipids have different fatty acid chain lengths and head groups which is resulting in a broad range of achievable physicochemical characteristics like minimal micellar concentration or melting temperature, allowing the creation of environment-sensitive (e.g., temperature-, pH-sensitive, mechano-sensitive) liposomes by choosing the specific setup. Several liposome-based cancer drugs have entered the clinical domain, e.g., carrying the anthracyclines doxorubicin (Doxil, Myocet, Caelyx) for the treatment of Kaposi's sarcoma, treatment-refractory ovarian cancer, multiple myeloma, and metastatic breast cancer or daunorubicin (DaunoXome) for treatment of Kaposi's sarcoma [18–21]. Beyond approved agents, liposomal chemotherapeutics are finding their way into clinical trials [22–24]. One important direction is the development of receptor-specific targeting for cell-specific delivery, which may render liposomes suited for purposes such as siRNA delivery, as reported by Zang et al. delivering siRNA by using pH-sensitive liposomes targeted to breast cancer using an anti-EphA10 antibody [25].

Liposomes are interesting carrier candidates for the delivery of intelligent switches at the nanoscale because the inner aqueous core offers a "nanocompartment" where processes may take place that require protection from the surrounding body fluids when injected into an organism. Liposomes have already been converted into intelligent nanosystems by incorporating a wide variety of stimuli response functionalities such as temperature, light, pH, ultrasound, enzymatic response, and even drug delivery system for radiation-sensitive nanoparticles [26–35] highlighting that liposomes are simple, but effective carriers for multimodal nanoscale trigger and effector functionalities. The current advances in lipid-based nanostructures, such as lipid-polymer hybrid nanoparticles, nanostructured lipid carriers, and solid lipid nanoparticles, have gained interest due to their capability to load and release drugs that have (i) a low aqueous solubility and high intestinal permeability or (ii) low solubility and low permeability [36, 37].

1.2.2 Polymeric Systems

Polymeric and polymeric-biologic hybrid nanomaterials have gained increasing attention as modifiers of pharmacokinetics of "biologicals" (pharmaceutical products originating from biomacromolecules), as carriers of hydrophobic drugs, and, in particular, as non-viral vectors for nucleic acid delivery. A structurally simple approach is to couple a polymer (typically polyethylene glycol, PEG) with a protein, a strategy that allows the control of pharmacokinetics while maintaining the biological properties of the protein. Self-assembly to nanoparticles is not exploited here. Different types of polymers—biodegradable or non-degradable, synthetic and natural—are being used for the formation of polymeric micelles and polymersomes (polymeric vesicles) as drug delivery systems [18, 38–40]. One difference of certain synthetic polymers compared to natural polymers is their longer lifetime and therefore improved ability to concentrate at a disease location. This increases the possibility of prolonged drug release over several days to weeks. Polymeric micelles can be built by self-assembly of amphiphilic-block copolymers exhibiting a hydrophilic outer shell and a hydrophobic core, which can enable transport of water insoluble drugs to their specific target. Several important anticancer drugs like paclitaxel, tamoxifen, or campthotecin [41, 42] are highly water insoluble. The relatively small size of polymeric micelles (10–100 nm) compared to polymeric vesicles enhances their ability to accumulate in tumor tissues through the enhanced permeability and retention (EPR) effect [43]. Another important aim is the development of cationic polymers that are able to form complexes with nucleic acids with the goal achieving cell transfection and siRNA delivery, thereby promising new approaches to the therapy of severe diseases, including inherited metabolic disease or cancer. An important obstacle in cancer treatment is the multidrug resistance (MDR), for which the MDR-1 gene is responsible. Zhang et al. reported a triblock copolymer micelle that was designed for the delivery and reversal of this gene dependent multidrug resistance. They successfully demonstrated the effectiveness of this copolymer micelle with MDR reversal and with this novel treatment strategy [44]. Even though a number of cationic polymers used for encapsulating siRNAs have shown toxicity, nontoxic alternatives are also reported, for example, by Lehner et al. They developed novel biocompatible polyplexes by combining receptor-targeting, folic acid functionalized copolymers with cationic pentablock nucleic acid complexing polymers and achieved target specific delivery of siRNA with a knockdown efficiency of 31% [45]. Nevertheless, the positive charge of cationic polymers can facilitate cellular uptake and endolysosomal escape as reported by Park et al., thereby overcoming one major cellular barrier in target-specific drug delivery [46]. Nanoparticles with an almost neutral (between −10 and +10 mV) surface charge proved to affect less likely an immunological reaction in vivo. For example, the polyplexes of Lehner et al. [47] were near-neutral (+4 mV) [48, 49]. Other nucleic acid carrier systems include cationic polymers poly(L-lysine) (PLL), poly(ethyleneimine)(PEI), chitosan, and dendrimers [50]. Detailed information of cationic carriers for nucleic acid delivery is given by refs. [51–53]. Polymeric delivery platforms are excellent carrier platforms for the integration of switches at the nanoscale and other complex functionality, due to the large freedom of design, their improved stability compared to liposomes, and their ability to integrate functional biomolecules within the hydrophobic bilayer. Even though the experience in clinical application of composite polymeric nanosystems and vesicles is growing, the field is not yet as established as the liposome field, which is a factor to keep in mind when developing commercial products for clinical application that will require thorough insight into pharmacokinetics, metabolism and mass balance, toxicity for the regulatory process, and industrial scale-up and production.

1.2.3 Dendrimers

Dendrimers are large and complex molecules with very regular chemical structure, which were pioneered in the early 1980s [53]. They are nearly perfect monodisperse macromolecules with a regular and highly branched tree like architecture. Dendrimers are constructed through a repeating sequence of chemical reaction steps, leading to predictable alterations in their size determined by each generation [18, 54–56]. In typically used chemical syntheses, dendrimers are structures with a size of 1 to 10 nm and a hydrophobic interior, which enables drug delivery of hydrophobic compounds such as cancer drugs. Poly(amidoamine), PAMAM is the most well known dendrimer for biological applications [57–59]. Cationic PAMAM dendrimers show cytotoxic effects, which have been thought to result from the interaction with negatively charged molecules on the surface of the cell. To overcome this effect, Jia et al. reported the use of phosphorylcholine, a zwitterionic molecular segment found mainly at external surfaces of cell membranes and the end of some lipids [60]. Such zwitterionic PAMAM dendrimers encompassing a phosphorylcholine surface efficiently lower the cytotoxity compared with the native PAMAM dendrimers [60, 61]. The addition of a pyrrolidone prevents the activation of the pro-inflammatory signaling pathways in human monocytes [62]. As dendrimers can be designed such that they are multifunctional, biodegradable are capable in overcoming extra- and intracellular obstacles, they have also been explored clinically as outlined by refs. [63–66]. Dendrimers significantly differ in size from micelles, liposomes, and polymeric systems leading to differential distribution behavior after injection into an organism. Forming a single, covalently linked structure, they are mechanically more stable than liposomes. They offer some inner, hydrophobic cavity space for the encapsulation of payloads and switch functionality, but payload/carrier ratio, measured as weight/weight ratio, is lower compared to micelles and lipid/polymer vesicles, which represents a limitation when space-consuming smart sensors or nanoscale switches need to be integrated in the interior of a nanosystem. Malekzadeh et al. recently reported a novel multifunctional dendrimer, coated with magnetite nanoparticles suitable for both diagnosing cancer by the use of magnetic resonance imaging as cancer therapy [67].

1.2.4 Carbon Nanotubes

Carbon nanotubes are a distinct molecular form of carbon atoms, yielding a hexagonal arrangement. Carbon nanotubes exist as single-walled and multi-walled variants [68]. Their structure, formed from layered graphite sheets, gives them extreme physical strength, 10 times as strong as steel, and unusual heat and conductivity properties [69]. Recently carbon nanotubes have attracted attention due to their use in controlled drug release as well as delivery of nucleic acids, peptides, and antibodies. Their inner core and their outer surface allow the insertion of specific payload into the small inner core, while the outer surface can be modified to achieve the necessary biocompatibility within the body or to attach targeting ligands or drug payloads [70, 71]. Gu et al. published data concerning single-walled carbon nanotubes (SWNTs) coated with doxorubicin, exhibiting a pH-responsive drug release behavior [72]. They covalently attached hydrazino-benzoic acid to the SWNTs using hydrazine as linker molecule. Doxorubicin was bound to the modified SWNTs forming hydrazone bonds with hydrazinobenzoic acid. In vitro testing of these SWNTs showed drug release upon pH change. In an acidic tumor microenvironment, this application could provide a specific drug release system.

Osteosarcoma stem cells are critically associated with the progression of osteosarcoma. Miao and coworkers demonstrated that SWNTs have (i) the ability to specifically inhibit the process of transforming growth factor beta 1 (TGFb1) induced osteosarcoma cell dedifferentiation, (ii) prevent the stem cell phenotypes acquisition in osteosarcoma cells, and (iii) to reduce the osteosarcoma stem cell viability under conditions that mimic the osteosarcoma microenvironment [73].

Conflicting results have been reported for cytotoxicity from in vitro and in vivo studies dependent on size, length, and surface topology [74–76]. One challenge in toxicity testing is that due to their hydrophobicity, carbon nanotubes can only be brought into solution in physiologic liquids through detergents or surface modification. Although surface functionalization of carbon nanotubes by attachment of appropriate molecules showed dramatically lowering of their toxicity, more effort has to be put into understanding the cytotoxicity of carbon nanotubes [77–79]. Animal experiments resulted in specific targeting, tumor binding, and drug delivery from which one can envision new therapeutic approaches [80–82]. Hou et al. recently reported an in vitro and in vivo study dealing with phototherapy using the anticancer activity of SWNTs and compared this with graphene oxide and fullerene. All were modified with hyaluronic acid. Cytotoxicity studies revealed that the SWNTs had the lowest cell inhibitory efficacy and the highest tumor inhibition efficiency with 808 nm laser irradiation. It was concluded that both SWNTs and graphene oxide are suitable as photo thermal agents [83]. Clinically however carbon nanotubes have still not overcome phase I trials, as there are open questions regarding their elimination in vivo, and because the concern about potential undesired biologic effects of stiff, needlelike materials, although such concerns are probably highly variable between different lengths and types of nanotubes [84–86]. A number of research groups have reported the functionalization of carbon nanotubes to render them sensitive to environmental conditions such as pH, temperature, or glutathione. Most SWNTs have a diameter of close to 1 nm incorporating the graphene layer, forming an inner core of <1 nm. Compared to the achievable payload volume of other platforms such as liposomes and polymeric vesicles, the much smaller volume to weight ratio of carbon nanotubes is another limiting factor for their therapeutic use. To compete against micelles, liposomes, polymeric systems, and dendrimers as platform for clinical development of intelligent nanomaterials, a compelling argument based on their unique characteristic is probably required and not yet achieved.

1.2.5 Metallic Nanoparticles

Metallic nanoparticles such as iron oxide, gold, and silver have been developed and modified for use in drug delivery, magnetic separation, and diagnostic imaging [87–90]. Superparamagnetic nanoparticles (SPION) built from oxide nanoparticles, such as magnetite (Fe_3O_4) and maghemite (Fe_2O_3), exhibit particular features like ultrafine size, biocompatibility, and magnetic properties. The superparamagnetic properties become manifest when a magnetic moment is induced through the application of a magnetic field. The large magnetic moment yields a strong signal change in magnetic resonance imaging (MRI) allowing therefore sensitive detection at high resolution [91]. Another application of iron oxide nanoparticles is tumor treatment by magnetically induced hyperthermia [92]. Thanks to its chemical inertness and suited mechanical properties, gold has been used in medicine for teeth implants and is also in use in cancer radiotherapy [93]. Gold nanoparticles can be formed with core sizes ranging from 1 to 100 nm. The initial claim of absence of cytotoxicity has raised enthusiasm as an excellent drug delivery system although increasing recognition of size dependent cytotoxicity needs to be considered before their application [94, 95]. Gold nanoparticles are capable of delivering peptides, proteins nucleic acids, or small molecules. When functionalized with quaternary ammonium groups, they can bind negatively charged DNA or RNA and also protect the nucleic acids from enzymatic degradation [96].

Silver has been used for a long time as a staining agent for glass and as a bacteriostatic agent [97]. Elemental silver and silver salts are well known to be relatively nontoxic to human cells and have been applied in the treatment of burn wounds, diabetic skin ulcers, and newborn eye infections [98–100]. Silver nanoparticles have therefore been developed for antibacterial applications including coating of medical catheters to prevent contamination and infections [101]. Recent findings demonstrated such effects of silver nanoparticles are dependent of the shape of the silver nanoparticle [102].

The electromagnetic properties of various metallic nanoparticles render them interesting for electromagnetic control, in particular induced hyperthermia; plasmon effects on metallic nanoparticles are another physical effect that can be used to incorporate switches at the nanoscale [92, 103]. Metallic nanoparticles can be surface-functionalized with relative ease, allowing to equip the particles with a wide range of advanced functionalities, including pH sensitivity and redox responsiveness. A potential disadvantage is the weight/weight ratio between functional payload and inert platform, which tends to be low in metallic nanoparticles. The absence of an inner cavity is a disadvantage, and the question of long-term behavior (mass balance, bioaccumulation, elimination pathways) after injection into an organism, and of cell toxicity need to be carefully weighted when their application for clinical trials is considered.

A potential concern of metallic nanoparticles using heavy metals is their release into the environment in a potentially nonrecyclable form; a mass balance in the body and an estimation of the environmental mass balance of such materials seems prudent when industrial products are developed.

1.2.6 Other Organic Nanoparticles: Carbohydrate-Based NPs, e.g., Chitosan and Starch

Certain organic nanoparticles originating from natural materials feature good water solubility, low toxicity, biocompatibility, and biodegradability, rendering them interesting candidates for carriers. Chitosan is a polymer obtained by deacetylation of chitin. Chitin itself can be found in the exoskeleton of crustaceans such as shrimps and crabs but also in the cell walls of fungi [104]. Studies have shown the versatility of chitosan for drug delivery and stimuli responsiveness in pH- and temperature-triggered applications [105–108]. Starch can be found abundantly in nature in crops such as rice, corn, and potatoes. Chemical modification of starch allows a wide range of different applications e.g., biodegradable packaging material or as target-specific drug delivery system [109]. Xiao et al. showed preparation of folate-conjugated starch nanoparticles for tumor targeting. They encapsulated the anticancer drug doxorubicin with folate modified starch polymers. Successful cellular uptake followed by loss of tumor cell viability could be shown [110]. In another study, it was demonstrated that nanoparticles derived from green tea could be successfully loaded with doxorubicin via hydrophobic and electrostatic interactions and were capable to effectively deliver and potentiate the anticancer activity of doxorubicin in mouse tumors [111]. The biodegradability and the low potential of toxicity of organic materials render them a competitive nanomedical platform for clinical application. Nevertheless, issues regarding the dissolution of the polymer matrix at physiological pH as well as poor mechanical strength as seen for chitosan have been raised [107].

1.2.7 Quantum Dots and Carbon Quantum Dots

Quantum dots (QDs) are small sized (1–10 nm) semiconductor nanocrystals which were developed in the early 1980s by Alexei Ekimov and Louis E. Brus [112]. They are composed of an inorganic elemental core (e.g., Cd and Se) surrounded by a metallic shell (ZnS), which constitutes a barrier between the optically active core and the surrounding medium. QDs can be modified by ligand attachment or encapsulated with amphiphilic polymers to improve solubility, specificity, size, and visualization properties in tissue [113, 114]. They are widely used in biological research as fluorescence imaging tool due to their resistance to photobleaching and their excellent quantum yield compared to organic dyes [115]. QDs can also be used as drug carriers or simply as fluorescent labels for other drug carriers. Derfus et al. have demonstrated the use of QDs as beacons to track siRNA delivery by PEGylating QD cores, conjugating them with siRNA and tumor-homing peptides necessary for cellular uptake [116]. A second approach is the labeling of common drug carriers such as polymers, liposomes, and dendrimers with QDs. Zintchenko et al. have introduced a drug nanocarrier called "quantoplex." Quantoplexes are polyethylenimine nanocarriers incorporating plasmid DNA and QDs, allowing real-time nanoparticle tracking in living animals [117]. QDs are of particular interest when combined with other nanoplatforms, as one functional building block among other functionalities. For clinical application, their ability to combine molecular imaging and cancer therapy can open new doors in the field of cancer therapy [118–120], but the toxicity of certain materials used in QD synthesis is an important concern. Carbon quantum dots (CQDs) are a good alternative to QDs, due to their ease of synthesis, tunability, stability, strong photoresponsiveness, and high biocompatibility and are applicable in a similar manner. With respect to QDs, they have a much lower toxicity however each type needs to be characterized individually [121–123].

1.2.8 Nanogels

The field of nanogels is a new and fast developing domain of research in nanomedical drug delivery. Nanogels are a three-dimensional natural and/or synthetic polymeric network formed either chemically (covalent bonds) or physically (hydrogen bonds, Van der Waals, and electrostatic interactions) of cross-linked hydrophilic polymers with size of approximately 200 nm. Due to their hydrophilic properties, nanogels can swell and encapsulate a high volume of water when added to an aqueous solution. Due to their polymeric nature, a broad range of chemical modifications is possible. Such hydrogels can entrap through spontaneous processes a large amount of biological molecules as e.g., DNA, RNA, proteins, and drugs, rendering them well suited for drug delivery. A recent example is presented by Zhang et al., who demonstrated the development of a sequence-independent and size controllable DNA nanogel made of DNA fragments, Cisplatin and Doxorubicin, which potentially act as a drug carrier. In vitro studies showed that doxorubicin-loaded nanogel alters the drug distribution in cells without changing the cytotoxic potential. For site-specific tissue targeting within the body, the surface structure of nanogels can be chemically modified with different ligands. Nanogels proved their value as carrier platform for siRNA and for stimuli responsiveness (pH, temperature) [124–127]. Wu et al. have presented multifunctional core–shell nanogels combining magnetic regulation with biochemical sensing. They formed a core from magnetic Ni and fluorescent Ag surrounded by a pH-sensitive p(EG-MAA) nanogel shell. Combination of a pH-dependent magnetic response with fluorescent pH sensing allows the creation of an intelligent carrier system for diagnostic and therapeutic applications [128]. Nanogels exhibit high stability in vivo, controllability of size, biodegradability, surface functionality, and specific drug delivery and have found their way into clinical applications [129–131].

1.2.9 Peptide-Based Nanoparticles as Delivery System

Peptides can have lipophilic or hydrophilic properties based on their amino acid composition and can therefore be used to construct amphiphilic molecules that form nanostructures by self-assembly [132–134]. Design considerations of these biopolymers for drug carriers are similar to other biocompatible polymers, but should take into account that peptides may act as strong immunogens and that the body already contains several lines of defense against foreign proteins and peptidic structures like virus capsids. In addition, a variety of peptidases exist in the body; if a system is preclinically developed for later clinical use in man, species differences in peptidase expression needs to be carefully considered. Virus self-assembly can act as an inspiration to build hollow or solid peptidic nanostructures [135, 136]. Bawa et al. showed enhanced cellular delivery and activity of the anticancer drug ellipticine to human lung carcinoma A549 cells using self-assembling peptide-based nanoparticles [137]. Naskar et al. presented the formation of multivesicular structures from self-assembling peptides, depicting sensitivity upon exposure to calcium ions leading to vesicular disruption. This intelligent sensing/switching functionality, allows cargo release suited for medically relevant payloads [138]. However, a natural extension of peptide-based systems is the exploitation of biologic peptide functions like their use as receptor ligands or enzymatic activity, naturally leading to nanomaterials with complex or switchable functionalities. Clinically, peptidic systems have entered clinical trials dominantly as nanoplatforms for vaccines offering multivalency as a potent immune system stimulant.

1.3 Advanced Functionality Including Internal and External Stimuli-Responsiveness

Many different nanomaterial-based platforms have been developed to improve drug delivery to tumor tissue and expand diagnostic and enhance therapeutic efficacy while minimizing possible side-effects, but only a few have found their way into clinical application. An important direction of future research is the development of smarter, functional materials that turn nanosize delivery vehicles into "intelligent" nanoparticles through stimuli responsiveness. Insights into the complex biological properties of e.g., a diseased area may be exploited to enhance specific drug release within the diseased area or the cytoplasm of a key cell type involved in pathophysiology, by taking advantage of unique patterns or multimodal factors of the microenvironment. The design of such intelligent, stimuli responsive nanoplatforms also promises diagnostic opportunities with increased disease specificity.

Switchable nanoparticles can be classified based on the modality of stimulus into locally/internally triggered and externally switchable nanoparticles.

Internal stimuli such as pH, redox potential, enzymatic activity, or temperature are increasingly used for the design of intelligent, stimuli responsive nanoplatforms in preclinical studies [35, 139]. They have the advantages of high specificity due to the possible control by a molecular mechanism dependent on a certain disease. Internally switchable nanosystems are expected to exhibit less damage to healthy tissues compared to nonswitched materials; as their control signal is tissue specific, bystander organs are spared as well. Nanomaterials switched by internal stimuli have entered clinical trials [140, 141], but the widespread clinical benefit for the patient has not yet been achieved. An important challenge is the fact that not every disease shows a specific internal molecular trigger suitable for exploitation as a trigger for switchable nanosystems.

Externally controlled nanosystems can be divided into temperature-, light-, electromagnetically, radiation-, or ultrasound-sensitive delivery systems and have the advantage of external regulative application properties regarding the area of interest [35]. A limiting factor for externally controlled nanomaterials is the tissue penetration of the control signal and the potential for tissue damage between the energy source and the target organ. Tissue damage may arise upon interaction of certain forms of radiation (heat, light, electromagnetic fields) with biomacromolecules (e.g., membrane proteins), leading to tissue disruption and damage also in healthy bystander tissues. Some externally controlled systems have already proven clinical benefit as seen for iron oxide–based NanoTherm® nanoparticles developed by the company MagForce for the treatment of glioblastoma by magnetic hyperthermia [142] or the photodynamic therapy–based compounds Visudyne®, Photofrin®, and Levulan® Kerastick® for the treatment of agerelated macular degeneration, esophageal cancer, and actinic keratosis [30, 143]. In 2000, the U.S. Food and Drug Administration (FDA) approved the first antibody-linked drug named Mylotarg®, developed by the Celltech Group and American Home Products, for the treatment of acute myelogenous leukemia based upon redox-sensitivity [144]. Rejected by the European Medicines Agency (EMA) in 2008 due to failure of full remission and an inappropriate cost-benefit ratio for the patient, the FDA also decided to take Mytolarg® off the market.

1.4 Engineering Cell and Tissue Interaction at the Nanoscale

1.4.1 Targeting Strategies

Liposomes and more recently other nanoparticles have found broad application for delivery of diverse payloads. Their small size, biocompatibility, stability, and stealth properties render them capable of prolonged circulation half-life, targeting of tumor tissue and cellular uptake (Table 1.1). Successful targeting implies sufficient stability to avoid premature drug release into the bloodstream and to circumvent degradation of the payload in the circulation prior to target arrival. Of equal importance are adequate stealth properties, which allow nanoparticles to avoid interactions with plasma proteins or the immune system that could lead to early elimination of the material. Residence time in blood can be increased by conjugation with "self" markers, such as CD47 peptides to the NPs, which inhibits phagocytosis. This allows temporally extended accumulation in the targeted tissue [146]. When NPs come into contact with physiological fluids, a corona made of biomolecules, typically proteins, may form on their surface. Coronas influence many properties of the NPs, such as size, stability, and surface properties, and importantly thus entails them with a "biological identity," which defines the reactions of the organism towards them [147]. For example, a corona made of opsonins (complement system, antibodies, C-reactive protein) triggers clearance by MPS (mononuclear phagocyte system) cells, and a corona made of dysopsonins (e.g., albumin or apolipoproteins) imparts a stealth effect [148].

Cancer therapy is one of the most frequently pursued applications because of the severity of the disease and of the well-known severe side effects of many conventional therapies. Specific targeting of only the diseased tissue while sparing the rest of the body is therefore the holy grail for this clinical application. Targeting to tissues and cells can be either passive, through properties of cancer vasculature, or active, through receptor-specific ligands on the nanoparticle surface intended for cell binding. Intraarterial injection into disease-related arteries is a further variant of "targeted" delivery.

Table 1.1 Examples of nano-based platforms and their stages in clinical use

Nanoplatform	Size range	Drug/trade name	Application	Target	Status	References (or ClinicalTrials.gov Identifier)
Liposome	100–200 nm	Doxil/Caelyx	Cancer therapy	Kaposi's sarcoma, treatment	FDA approved (1995, 1999)	In market
		DaunoXome	Cancer therapy	refractory ovarian cancer	FDA approved (1996)	In market
Dendrimer	1–10 nm	VivaGel	Microbicide	Cervicovaginal	Phase III	NCT01577537
		IMDENDRIM	Cancer therapy	Liver cancers	Phase not provided	NCT03255343
Polymer	50–200 nm	Oncaspar	Cancer therapy	Acute lymphoblastic leukemia	FDA approved (1994)	In market
		Doxorubicin	Cancer therapy	Breast/Lung cancer	Phase II	[268]
		Paclitaxel	Cancer therapy	Breast cancer	Phase II	[269]
Micelle	10–100 nm	Genexol-PM	Cancer therapy	Metastatic breast cancer	Phase IV	NCT00912639
		Taxol	Cancer therapy	Psoriasis	Phase II	NCT00006276
Carbon nanotubes	1–25 nm diameter	Cisplatin	Cancer therapy	Head and neck	In vivo	[270]
		Doxorubicin	Cancer therapy	Breast cancer	In vitro	[271]

Metallic	1–150 nm	Feridex	MRI contrast agent	Liver	FDA approved (1996)	In market
		NanoTherm	Cancer therapy	Glioblastoma	EU approved	[272]
Organic nanoparticles	20–400 nm		Cancer therapy	Liver cancer	Phase II	[273]
Quantum dots	1–10 nm	Doxorubicin	Cancer therapy	Ovarian cancer	In vivo	[274]
			Cancer therapy	Breast, prostate cancer	In vivo	[275]
Nanogels	<200 nm	NB-001	Anti-viral	Herpes labialis	Phase III	NCT01324466
		MuGard	Mucositis	Head and neck	Phase IV	NCT01283906
CPP	~30aa residue long	Azurin	Cancer therapy	Refractory solid tumors	Phase I	[276] NCT00914914
	PepFect14 (PF14)	pDNA	Cancer therapy	Cancer	In vivo	[277]

While active targeting (e.g., via antibodies) may increase uptake of functionalized NPs by specific tumor cells, the additional manufacturing steps and the increased complexity of behavior are major hurdles for scale-up and clinical studies. Therefore, these constraints have to be balanced with potentially improved clinical outcome [149].

In general, there often is a translational gap between preclinical and (first-in-human) clinical trial results for nanodrugs. For example, translation of tumor chemotherapy with doxorubicin using liposome formulation from encouraging mouse experiments to clinical application did not show an increase in survival (14 clinical trials). The reasons for this are hypothesized to be due to enhanced EPR in mice due to rapid growth of tumor, lack of standardization of preclinical studies, use of tumor regression instead of survival rate in mice as an endpoint. However, the results of this study cannot be extrapolated to other nanocarrier systems [150].

1.4.1.1 Passive targeting

Passive targeting depends on the EPR effect, characterized by accumulation of nanoparticles in tumor tissues by diffusion from the blood circulation via concentration gradient [151–154]. Fast growing tumor tissue is characterized by rapid angiogenesis triggered by tissue anoxia and expression of growth factors like the vascular endothelial growth factor (VEGF). Newly formed blood vessels show a higher rate of vascular fenestration and therefore, permeability, in different tumors [155–157]. This means that the EPR effect is different for different tumors, i.e., some do not show EPR at all. Nanoparticles extravasate through the walls of such blood vessels and tend to show prolonged retention in the tumor tissue. Nanoparticle extravasation is size dependent [158–160]. In passive targeting strategies, a sufficiently long blood circulation time of the nanoparticles is important to allow a high contrast between tumor accumulation compared to that in liver, spleen, and kidney, where particles tend to accumulate in the absence of a suited tissue target. This goal can be achieved through optimal stealth properties, typically by incorporation of polymers like PEG into the shell of the nanomaterial. Theoretically, the optimal size of a candidate nanodrug needs to be large enough to avoid renal excretion but small enough to avoid liver uptake and sequestration. In addition, premature release of the payload from the nanocarriers (e.g., exchange of hydrophobic drugs with albumin) is an important issue, which needs to be considered. A frequently used animal model of human cancer, the xenograft mouse transplantation model, shows significant differences in vascular biology compared to human cancers [161], partly explaining why it is much easier to cure human cancer in mice than in humans. For this reason, the development of alternative models to study nanoparticle-tissue interaction before clinical application is desirable. For clinical settings, stratifying patients according to their probability of accumulating NP via EPR may improve treatment, i.e., patients with high tumoral EPR can be treated with nanotherapeutics using passive targeting [162].

1.4.1.2 Active targeting

Active targeting was conceptually introduced in 1906 by Paul Ehrlich, who coined the term "magic bullet" by describing the need for a system able to target specific drug delivery within the body [163–165]. For active targeting, nanocarriers need to first arrive in the target organ; therefore, initial design considerations are similar to those for passive targeting. Active targeting then exploits specific mechanisms that increase specific particle–target cell interaction through suited ligands. These ligands (either small molecules such as folic acid and carbohydrates or macromolecules, including aptamers, antibodies, proteins, peptides, or oligonucleotides) must be carefully chosen to exhibit maximal binding to the target while minimizing binding to healthy tissue [160, 166, 167]. Popular cancer targets include transferrin, folate [168, 169], and epidermal growth factor receptors due to their over-expression in tumor tissues [170, 171]. Targeted delivery requires highly specific target recognition and target binding with sufficient affinity, but as stealth properties imparted through polymers tend to compete with specific binding sites, careful titration of these effects, e.g., by optimizing the length of branches carrying a ligand and the number of ligands per nanocarriers is of key importance [172].

Targeting may be especially beneficial in small metastases treatment due to their low vascularization and low EPR properties. Ligand functionalized NP may lose their targeting capability when a corona forms on their surface [173]. Also, there was evidence that in vitro results differed significantly from in vivo results [174]. It has been shown that the tumor penetration depth of targeted NPs with high-affinity ligands is lower than that of NPs with low-affinity ligands, which is termed "binding site barrier" [175]. Targeting may increase intra-tumoral accumulation, cellular uptake, define intra-cellular behavior, all of which are important.

Genetic and epigenetic profiles of tumor endothelial cells are influenced by tumor cell signals, which means that vascular targeting may not be straightforward for different tumors. In an effort to overcome limitations of passive transvascular drug delivery, such as concentration gradient dependency, the active transvascular caveolar transport pathway (also termed "caveolae pumping"), which differs between healthy and tumor tissue regarding its proteomic signature, was exploited. With an antibody against annexin A1, a cell surface protein present in the caveolae of human tumor tissue, but not in normal organs, active transendothelial transport of antibodies against a concentration gradient with selectivity to tumor tissue could be achieved. Translation of these results to treatment of solid human tumors with nanomedicines seems promising, but efficacy and toxicity have to be evaluated in the future [176].

Tumor-penetrating peptides are an option to deal with poor tumor tissue penetration depths. They can open an active trans-cellular transport pathway, C-end Rule (CendR), that is faster than diffusion, with payloads up to NP sizes. The payload does not have to be coupled to the peptide, as the peptide activates a bulk transport system. Tumor-specific accumulation of various payloads was achieved in murine in vivo and human ex vivo studies [177].

The CRISPR-Cas9 system has received much attention in the past years since its discovery. Despite its versatility to modify genes in vitro, in vivo targeted delivery is still a major obstacle to overcome. In a recent work, a nucleus-targeting core–shell artificial virus was described. It was capable of targeted delivery to CD44 displaying cells (as demonstrated for human ovarian cancer cells in mice), accelerated endosomal escape, and nuclear permeation. Thereby it outperformed other viral (e.g., adenoviral, lentiviral) and non-viral (e.g., lipofectamine) carrier systems, in regard of cargo size, immunogenicity, and ease of manufacturing [178]. However, the CRISPR-Cas9 system is not fully understood yet, and there is an ongoing debate about off-target effects. Therefore, the way to clinical application is still long.

1.4.2 Endocytotic Cellular Uptake

Once a delivery nanodevice has arrived in a target tissue, several barriers to optimize pharmaceutical activity remain. Cellular uptake is a first step: different routes to enter a cell exist, including the physiological uptake mechanisms like passive diffusion through the cell membrane, channel, or transport protein-mediated uptake, or various vesicular uptake mechanisms. The endocytotic pathway is the major uptake mechanism of cells for nutrients, but also for therapeutic DNA, siRNA, and proteins. Endocytotic pathways between various cell types differ for the same nanoparticle [179]. There are several distinct endocytotic pathways (with different sizes of the resulting vesicles), including clathrin-mediated endocytosis (~120 nm, the main route for nanoparticles), phagocytosis (>250 nm), caveolar endocytosis (~60 nm) and macropinocytosis (0.5 to 10 μm, hyperactive in some tumors). Detailed information concerning these uptake pathways were reviewed recently [80]. Various ligands used in active targeting actually bind to membrane proteins that are associated with endocytosis and thus trigger endocytosis upon receptor binding. Prominent examples are the folate receptor and the transferrin receptor. After cellular uptake, molecules internalized via the plasma membrane will follow the intracellular endocytotic pathway involving early endosomes (with potential receptor recycling to the cell surface) or may progress to late endosomes and lysosomes. If the payload has its cellular target in the nucleus, the nuclear membrane is a second barrier, which is difficult to pass. Endocytotic nanoparticle uptake can be achieved using liposomes as a delivery platform or through surface modification using cell-penetrating peptides as discussed below.

Light-responsive polymeric NPs were created by integration of indocyanine green, selenium inserted copolymer, and doxorubicin. After uptake, irradiation of indocyanine green with NIR light generated ROS, which led (i) to the oxidation of selenium in the polymer entailing drug release of the NP, and (ii) to the disruption of the lysosomes and the evacuation of their content into the cytoplasm within minutes. The antitumor efficacy was corroborated with tumor bearing mice experiments [181].

1.4.3 Cellular Uptake by Cell-Penetrating Peptides

More than 20 years ago, Frankel and Pabo found that the trans-activating transcriptional activator (TAT) protein of HIV-1 is able to cross mammalian cell membranes and translocate into the nucleus [182]. A few years later, Prochiantz et al. reported the same effect for the Drosophila melanogaster Antennapedia homeodomain [183]. It was found that a short sequence of 10–16 amino acids was responsible for translocation. Based on this discovery, numerous cell-penetrating peptides (CPPs) have been developed for potential delivery of various biomolecules such as oligonucleotides, DNA, RNA, proteins, peptides, and drugs. A review on CPP and tumor-targeting peptides appeared recently [184]. CPPs are typically cationic (Tat, penetratin) or amphiphatic peptides of less than 30 amino acids showing lack of toxicity and can be grouped into two major classes comprising covalent linkage and non-covalent complexation with cargo molecules [185]. Cationic CPPs are generally composed of positively charged amino acids as arginine, lysine, and histidine, whereas amphiphatic CPPs are made up of lipophilic and hydrophilic parts [186]. Cellular uptake can occur through endocytotic (clathrin dependent, macropinocytosis, via caveola) or non-endocytotic pathways although mechanisms are not fully elucidated.

Tat peptide that carries a small-molecular-weight payload, such as fluorescein, enters cells in a manner that may involve direct penetration through the cell membrane, whereas Tat with a macromolecular or nanoparticle payload uses an endocytic pathway.

Polyarginine peptide (R9) is one of the most effective CPPs. It has been shown to achieve improved translocation ability (~20-fold) compared to TAT from HIV-1 virus [187].

Effective CPP-based delivery to target different diseases has been shown using covalently linked cargos such as Tat, transportan, penetratin, polyarginine, or VP22 [188–191]. Covalent linking the cargo to CPPs can be a limitation by altering the biological activity of the payload as seen by Juliano et al., who examined CPPs conjugated to siRNA [192]. In contrast to a covalently bound payload, amphiphatic peptide carriers such as MPG and Pep-1 show successful in vivo and in vitro cargo delivery of siRNA, plasmid DNA, or proteins without the need for chemical cross-linking [193–194].

In general, the lack of target specificity of CPP is a major challenge for in vivo applications. Inclusion of cell-targeting peptides (CTP) or tumor targeting peptides (TTP) can direct the CPP to specific cell types. In an effort to overcome the non-specificity of CPP, a multi-component system comprising a MMP-2/9 cleavage site (reacting with MMP-2/9 enzymes in tumor tissue), oligoarginine (R9) as cell-penetrating domain, a cell targeting peptide (specific for CD13), a protective PEG chain, and a nanoliposome loaded with doxorubicin was presented. It improved the antitumor efficacy in vitro and in vivo, with no significant systemic toxicity [195]. Another work presented a branched R9-cysteine/pDNA polyplex, which exhibited improved gene delivery effectiveness compared to other CPPs. The polyplex was bioreducible, which led to the release of the cargo by reduction with intracellular glutathione. Together with its high transfection efficacy, it showed a low toxicity [196].

1.4.4 Endosomal Escape

After internalization, NPs cargo must either diffuse to their target or be directed there by intracellular trafficking pathways. Many examples of endosomal escape by pathogens can be observed. In fact, certain bacteria (e.g., listeria) or viruses are able to escape lysosomal degradation through specific interactions with the vesicle membrane in a surprisingly successful manner and are an example where human ingenuity has not yet been able to mimic the capability of biologic organisms [197]. Endosomal escape belongs to the key challenges still to overcome to enable clinical application of therapeutic delivery of nucleic acids to target cells. In endosomes, mild acidification detaches typical ligands from their receptors. The latter are recycled to the cell surface, while the vesicle content progresses to late endosomes and lysosomes. Various strategies have been explored to achieve endosomal escape. Behr et al. have proposed a proton sponge effect in 1995, suggesting that specific agents become protonated at low pH in endosomes, leading to an influx of ions and water, causing swelling and rupture of the endosome with release of the payload. A prominent example for this effect is the use of polyethylenimine (PEI) for transfection of plasmid DNA or siRNA, which becomes protonated due to its amino groups at endosomal pH [198, 199]. Another approach for endosomal escape is incorporation of fusogenic peptides into the endosomal membrane. These peptides are derived from viruses and show conformational changes upon changes in pH, which allow them to fuse with the endosomal membrane by inducing membrane perturbations. This has been well documented for hemagglutinin, an influenza virus derived fusogenic peptide [200, 201]. A strategy called photochemical internalization has been described by Bert et al. Payload release was achieved through incorporation of photosensitizers into the endosomal or lysosomal membrane. The membranes are disrupted through chemical action of singlet oxygen, which is formed upon light exposure of the photosensitizers [202, 203]. Several photosensitizers as $TPPS_4$, $TPPS_{2a}$, or $AlPcS_{2a}$ have already been used to this end in different studies [198, 204–207].

Gold nanorods (AuNR; 47 $\times$ 14 nm^2) and two proteins (BSA and β-galactosidase) were coupled with complimentary ssDNA, which, via DNA hybridization conjugated the proteins to the AuNR. Uptake into fibroblasts was routed mainly through clathrin and caveolin pathways. Laser irradiation with a NIR wavelength of 780 nm led to the melting of the dsDNA and release of the protein cargo via photothermal effect exerted by plasmon optical resonance of the AuNR (the plasmonic carrier). Importantly, there was temporal control of the release of the proteins via different melting temperatures of the coupling dsDNA. NIR laser excitation also entailed the endosomal escape as validated by fluorescence

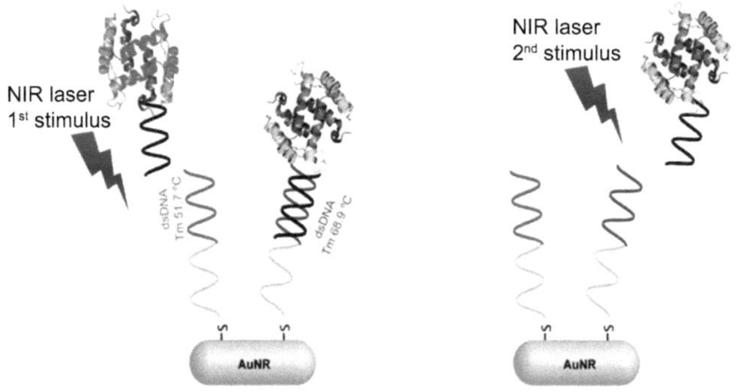

Figure 1.2 Two proteins (BSA, β-galactosidase) were attached to gold nanorods via hybridization with two different complimentary ssDNA strands. Sequential release of two proteins upon irradiation with near-infrared laser light was achieved. From [208].

colocalization and transmission electron microscopy (TEM) studies, presumably via generation of radical species (see Fig. 1.2 [208]). The intracellular routing with spatio-temporal control of release of more than one protein is innovative, but the in vivo cellular targeting still needs to be solved, e.g., via attachment of moieties such as cell targeting peptides. Site-specific (intra-tumoral) accumulation may also be achieved passively via the EPR effect, although the plasma clearance half-life should be sufficiently high in order to achieve this, which in this case may require further surface modification of the AuNRs to render them stealthy.

1.4.5 Exocytosis

In cell culture targeting studies using HeLa cell line, it was observed that following uptake of mesoporous silica NP (MSNP) with a diameter of 200 nm (2 h) a decrease in total intracellular NP concentration initiated, leading to values as low as ~15% of the peak values. After 6 h of incubation, a steady state was reached. In order to abrogate this mechanism, an exocytosis inhibitor was included in the payload of a nanovector system. As most exocytosis inhibitors are also endocytosis inhibitors, the exocytosis inhibitor must be encapsulated in the nanovector system, until released intracellularly, as otherwise also the uptake would be reduced. As a model system, doxorubicin was co-loaded with dimethilamyloride (DMA), an inhibitor of lysosomal exocytosis. As an indicator of exocytosis, temporal release of β-hexosaminidase (as lysosomal enzyme) into the culture medium was recorded. With DMA, its release was markedly reduced, and a gain in the intracellular dose of doxorubicin was up to sixfold. Moreover, an increased localization in the nucleus was observed, potentiating the cytotoxicity of doxorubicin, which is based on DNA intercalation. Increased cytotoxicity for the DMA-loaded nanovector systems was observed in cell culture and 3D microscale tissues. The exocytotic pathway may be a contributor to tumor chemoresistance [209].

1.5 Clinical Application of Nanomaterials

1.5.1 The Clinical Trial Landscape: An Emphasis on Cancer

On January 28, 2018, there were 271 clinical trials registered at http://www.clinicaltrials.gov and www.who.int/trialsearch mentioning the keyword "nanoparticle." Most of them are Phase I and Phase II trials in the status of "recruiting" or "active and not recruiting" (Fig. 1.3). The difference between the numbers found might be caused by the avoidance of the word "nano" in the trial details. Within the past 10 years, the number of clinical trials has increased and it is likely that this trend will continue in the future due to innovative research. In first generation nanodrugs like Doxil, improved safety rather than strongly increased efficacy contributed to their success, and for wider use, strategies that particularly look at enhanced disease specificity and enhanced therapeutic response without added toxicity are desired. Most nanomaterials that have entered clinical trial are currently devoid of switch functionality (Table 1.2). The majority of the mentioned trials are focusing on cancer treatment applying the cancer drug paclitaxel in a liposomal or nanoparticulate carrier, or its albumin bound formulation Abraxane®. According to the World Health Organization (WHO), cancer is a leading cause of death around the world, accounting for 8.8 million deaths in 2015. Even though cancer treatment has evolved significantly within the past few decades,

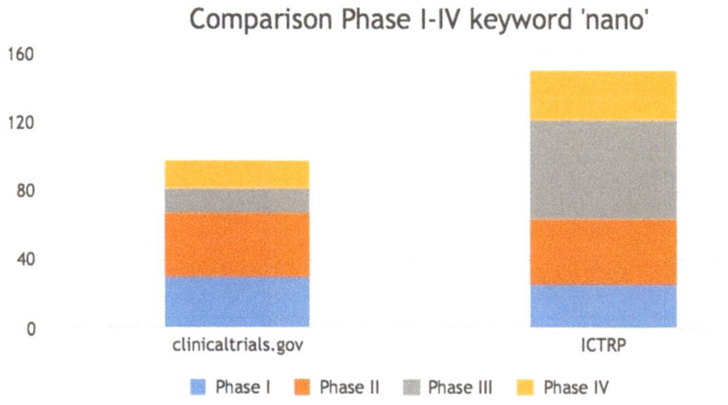

Figure 1.3 The numbers of ongoing clinical trials involving various nanoplatforms illustrating their application in cancer studies or global clinical application. Data adapted from www.clinicaltrials.gov and www.who.int/trialsearch (28.01.2018). ICTPR: International clinical trial registry platform.

surgery, radiation, and chemotherapy are still the main pillars for cancer therapy. While some spectacular successes have been achieved in specific malignant diseases (e.g., in certain leukemia types), most current chemotherapies for solid cancers are not curative and prolong life for a few months or years rather than for decades. Advances in cancer management with the goal of prolonging lifespan and eventually curing cancer depend on early cancer detection, prevention of recurrence and metastatic spread, and effective management of established metastases. All factors together play a key role, and this is where novel nanomaterial-based strategies could have the highest impact.

Early-stage cancer diagnosis is desirable to prevent metastasis but remains challenging because clinical symptoms usually become manifest only in advanced cancer stages. Nanoparticles have specific features that render them interesting for imaging in cancer as well as for therapy of solid cancer because of the propensity of nanoscale materials to distribute preferentially to vessels with altered permeability (EPR effect) and the additional possibility to apply active tissue targeting through targeting ligands or physical means. Conventional, small molecular chemotherapeutics often

Table 1.2 Nanoplatforms in clinical trials

	Clinical trials.gov		WHO ICTPR	
Platform	Cancer	Global	Cancer	Global
Liposome	1717	2074	415	936
Vesicle	125	597	17	207
Micelles	15	32	8	18
Polymer	3	7	0	6
Gold	2	7	0	10
Silver	2	7	0	37

Note: The numbers of ongoing clinical trials involving various nanoplatforms illustrating their application in cancer studies or global clinical applications. The keywords "gold" and "silver" were combined with "nano" and "nanoparticle" when searching the databases. Therefore, the presented numbers here might miss the trials that did not mention "nano" or "nanoparticle" in their keywords. Data adapted from www.clinicaltrials.gov and www.who.int/trialsearch (28.01.2018).

have a short circulation half-life due to rapid kidney filtration or liver metabolism, have a small therapeutic index limiting their dose and exhibit a relative lack of specificity for cancer tissue compared to other fast growing cells, which leads to the well known, undesirable side effects of cancer therapies. Many clinical cancer treatment protocols are dose-limited due to side effects and thus of suboptimal efficacy. The emergence of drug resistance is another challenge in conventional cancer chemotherapeutics. The combination of chemotherapy with nanoparticle-based drug delivery systems has shown to improve tolerability and there is some evidence for improved efficacy in specific contexts, but challenges for clinical applications still remain.

1.5.2 Potential Risk Factors Regarding Approved Nanoparticle-Based Therapies

A number of nanoparticle-based therapies have been approved by the U.S. Food and Drug Administration (FDA) for clinical use, highlighting the benefit that can be achieved through nanomedicine strategies (Table 1.3). Although all mentioned approved drugs have fulfilled the safety requirements of the FDA, the European Medicines Agency (EMA) or other agencies, clinical observation has led to heightened awareness of potential side effects. In particular,

Table 1.3 Clinically approved nanoparticle-based therapeutics and their benefits and risks

Nanoplatform	Constitution	Trade name	Application	Target	Benefits through platform implementation	General possible risks risks regarding platform
Liposomes	Liposomal amphotericin B	Abelcet	Antifungal	Fungal infections	Decrease in toxicity by factor 20 [210], lowering of nephrotoxicity of 58%, reducing risk of mortality by 28% [211]	Hypersensitivity reactions to liposomal drugs emerging from complement activation-related pseudoallergies (CARPA) as seen for Doxil, AmBisome, DaunoXome, can lead to life-threatening reactions in first time treated patients especially with a cardiovascular disease background [212–214]
		AmBisome	Antifungal	Fungal and protozoal infections	Lowering of nephrotoxicity of 30% [210, 215] and infusion related toxicity of 30–40% [210, 216], treatment success increase of 24% [217]	
		Amphotec	Antifungal	Fungal infections	Decrease of mortality of 10% and renal toxicity of 25% [218]	
		Amphocil	Antifungal	Fungal infections	Decrease in nephrotoxicity [219]	
	Liposomal daunorubicin	DaunoXome	Cancer therapy	HIV-related Kaposi's sarcoma, Leukemia, Non-Hodgkin lymphoma	Increase of drug concentration at the site of KS lesions [220], decrease in alopecia 30% and neuropathy 30% in Kaposi's sarcoma treatment [221], escape multi drug resistance [222]	
	Liposomal doxorubicin	Myocet	Cancer therapy	Combinational therapy with cyclophosphamide in metastatic breast cancer	Decrease of cardiotoxicity of 16% [223], total retention of activity [224]	
	Liposomal-PEG doxorubicin	Doxil, Caelyx	Cancer therapy	Metastatic breast and ovarian cancer, HIV-related Kaposi's sarcoma	Increase of circulation half-life of 74 h vs. 10 min, less frequent dosing schedule and reduction in cardiotoxicity of 20% for breast cancer [225]	

(*Continued*)

Table 1.3 *(Continued)*

Nanoplatform	Constitution	Trade name	Application	Target	Benefits through platform implementation	General possible risks risks regarding platform
	Liposomal verteporfin	Visudyne	Neovascularization	Age-related macular degeneration, pathologic myopia, ocular histoplasmosis	First and only drug therapy approved for treatment of macular degeneration [143]	Skin toxicity in areas of vitilgo related to pegylated liposomal doxorubicin [226]
	Liposomal vincristine	Onco TCS	Cancer therapy	Non-Hodgkin's lymphoma	Reduction of neurotoxicity [227], increase of circulation half-life of 72 h vs. several minutes, less frequent dosing schedule [228]	
	Liposomal vincristine	Marqibo	Cancer therapy	Hematologic malignancies and solid tutors	Vincristine sulfate liposome injection improves the pharmacokinetics and pharmacodynamics of vincristine [237, 238]	
	Liposomal cytarabine	DepoCyt	Cancer therapy	Neoplastic meningitis and lymphomatous meningitis	Less frequent dosing schedule and delaying neurological progression [229]	
	Liposomal IRIV vaccine	Epaxal/Inflexal V	Anti-viral	Hepatitis A, Influenza	Achievement of 100% seroprotection in infants and children, increase in immunogenicity of 10% [221]	
	Liposomal morphine	DepoDur	Postsurgical analgesia	Treatment of postsurgical analgesia	Increase in analgesia for hip arthroplasty up to 48 h [230]	
	Liposomal propofol	Diprivan	Anesthetic			
	Liposomal Irinotecan	Onivyde	Cancer therapy	Pancreatic cancer	Increased delivery to tumor site, lower systemic toxicity arising from side effects [241]	

Polymers	PEGylated formations	Oncaspar	Cancer therapy	Acute lymphoblastic leukemia	Increase of circulating half-life, less frequent dosing schedule [231], decrease of immunogenicity [232]	PEGylation is well tolerated and has not been associated with any adverse events regarding Pegasys, PegIntron, Neulasta, Somavert [256]
	Polymer-protein conjugate (PEGylated IFN beta 1A)	Plegridy	Immunomodulator	Multiple sclerosis	Improved stability of protein through PEGylation [246]	
	Polymer-protein conjugate (Pegylated factor VIII)	ADYNOVATE		Hemophilia	Improved stability of protein through PEGylation [247]	
		Somavert	Cancer therapy	Acromegaly	Increase in half-life of clearance [234]	
		Neulasta	Cancer therapy	Cancer chemotherapy associated with neutropenia	Increase of serum half-life of 40 h vs. 3 h [235]	PEG can generate complement activation products in human serum, leading to possible infusion-related reactions [239]
		Adagen	Immuno-suppression	Severe combined immune deficiency		
		Mircera	Anemia	Anemia treatment with associated chronic kidney disease	Increase in half-life, less frequent dosing schedule [235]	PEGylated proteins possibly trigger antibody formation against PEG, leading to accelerated clearance [233]
		Pegasys	Anti-viral	Hepatitis B, C	Longer half-life, reduced clearance [242]	

(Continued)

Table 1.3 (*Continued*)

Nanoplatform	Constitution	Trade name	Application	Target	Benefits through platform implementation	General possible risks regarding platform
		Macugen	Macula treatment	Age-related macular degeneration	First available ophthalmic anti-VEGF agent [248]	Detection of PEG antibodies of 22–25% in patients treated with PEG-asparaginase [243, 244]
		Cimzia	Inflammation	Crohn's disease	Increase in half-life, less frequent dosing [249]	
	Glu-Ala-Tyr copolymer	Copaxone	Immunomodulator	Multiple sclerosis		
	Poly(alylamine hydrochloride)	Renagel		Chronic kidney disease	Reduction of 20–35% in mortality for elderly people, decrease in bone disease, vascular calcification [250]	
	Colesevelam hydrochloride	Welchol	Anti-diabetic	Type 2 diabetes		
Iron nanoparticles	SPIONS	Feridex	Imaging	Liver		Dextran-coated superparamagnetic iron oxide nanoparticles (SPIO) as Feridex show interactions with plasma proteins, which can lead to complement activation and further clearance by the liver or spleen [251, 252]
		Gastromark	Imaging	Gastrointestinal		
		Lumirem	Imaging	Gastrointestinal		
		Resovist	Imaging	Liver		
	Iron oxide based	Venofer	Anemia	Chronic kidney disease		
		Cosmofer	Anemia			
		Ferrlecit	Anemia			
Other platforms	Albumin-bound paclitaxel	Abraxane	Cancer therapy	Metastatic breast cancer	Increase in overall response efficacy of 14%, reduction of risk for death by 28%, decrease of toxicity permitting administration of 50–70% higher doses [253, 254]	

Nanocrystalline	Emend	Antiemetic	Brain	Increase of bioavailability [255]
	Tricor	Anti-hyperlipidemic	Vascular endothelium	Increase of bioavailability [255]
	Triglide	Anti-hyperlipidemic	Vascular endothelium	Increase of bioavailability [264]
	Rapamune	Immuno-suppressant	Binding of mTOR complex 1	Increase of bioavailability 21% [255]
	Megace ES	Antianorectic	Pituitary	Increase of bioavailability [255]
	Avinza	Psychostimulant	Central nervous system and smooth muscle	Increased drug loading and bio-availability, extended release [260]
	Invega Sustena	Schizoaffective disorder	Dopamine D2 antagonist and Serotonin 5-HT2 receptor antagonist	Allows slow release of injectable low solubility drug [261]
	Ryanodex	Malignant hypothermia	Binding of ryanodine receptor 1	Faster administration at higher doses (No authors - 2015)
Emulsions	Estrasorb	Menopause treatment	Estrogen receptor	
	Elestrin	Menopause treatment	Estrogen receptor	

possible interactions involving complement activation and antibody formation against PEG that may lead to hypersensitivity reactions or preterm drug clearance by liver and kidney [233] are a concern, although PEGylation has been claimed to be safe in particular if applied to proteins [256]. The fact that such complement activation related pseudoallergic reactions (CARPA) [212, 213, 236–239, 257–259] have been observed only in a minority of patients, but in various liposomal drugs, including Doxil, AmBisome, DaunoXome, and others, points to individual predisposition. However, there is some controversy about mechanisms. This effect might be related to direct complement activation on the particle surface or to reactive antibodies against polymers such as PEG as triggers of complement activation, infusion reactions, and might also lead to altered particle clearance through opsonization. Garratty et al. observed a 22–25% occurrence of PEG antibodies in 350 healthy blood donors [243] whereas in 1984 a value of 0.2% had been described [263]. It has been hypothesized that such an increase in the occurrence of PEG antibodies in the last 30 years could be due to a higher appearance of PEGylated compounds within pharmaceuticals, cosmetics, and processed food products, as suggested by Armstrong et al. [244], though major methodological differences preclude firm conclusions. In addition to PEG, Meszaros et al. demonstrated the activation of the complement system by the ambisome, a charged-non-PEGylated lipose [245]. Recently, Zhang et al. reported that the IgE-independent mechanisms of CARPA may cause severe and even occasional lethal pseudoallergy by many drugs that are used in the clinic [266]. This remains an important unsolved public health issue and a further improvement of the screening method of components in medications is recommended.

1.6 Conclusions and Outlook

Medicine needs new solutions to achieve highly effective, side effect free, and cost-efficient solutions for major diseases [267]. The emerging field of intelligent nanomaterials for medical diagnosis, therapy, and their combination termed "theranostics" is based on a range of well studied carrier platforms, a number of targeting strategies each offering advantages and challenges and a "smart" payload. This combination has the goal of increasing specificity and efficacy in the diseased tissue, while abolishing toxicity largely. Although pioneering work has been done in oncology, both in preclinical development and clinical application, it is widely felt that the added capabilities of intelligent systems beyond "target cell killing" may render intelligent nanosystems a shaping force for a much broader range of diseases. To advance in such applications, a thorough knowhow about clinical challenges, material properties, physicochemical properties, nanobio interactions, toxicity, regulatory pathways, and clinical trials results is required.

Acknowledgment

Financial support of the Commission for Technology and Innovation of the Swiss Federation project 18994.1 is gratefully acknowledged.

References

1. Meng X, Liu Z, Coa Y, Dai W, Zhang K, Dong H, Feng X, Zhang X. Fabricating aptamer-conjugated PEGylated-MoS2/Cu1.8S theranostic nanoplatform for multiplexed imaging diagnosis and chemo-photothermal therapy of cancer. *Adv Funct Mat* 2017, **27**, 16, 1605592.
2. Deng K., Hou Z., Li X., Li C., Zhang Y., Deng X., Cheng Z., Lin J. Aptamer-mediated up-conversion core/MOF shell nanocomposites for targeted drug delivery and cell imaging. *Sci Rep* 2015, **5**, 7851.
3. Choi KY, Silvestre OF, Huang X, Min KH, Howard GP, Hida N, Jin AJ, Carvajal N, Lee SW, Hong JI, Chen X. Versatile RNA interference nanoplatform for systemic delivery of RNAs. *ACS Nano* 2014, **8**, 5, 4559–4570.
4. Xue P, Sun L, Li Q, Zhang L, Xu Z, Li CM, Kang Y. PEGylated magnetic Prussian blue nanoparticles as a multifunctional therapeutic agent for combined targeted photothermal ablation and pH-triggered chemotherapy of tumour cells. *J Coll Int Sci* 2018, **509**, 1, 384–394.
5. Pahle J, Walther W. Vectors and strategies for nonviral cancer gene therapy. *Exp Op Biol Ther* 2016, **16**, 4, 443–461.

6. Rekha MR, Sharma CP. Polymers for gene delivery: Current status and future perspectives. *Recent Patents on DNA and Gene Sequences* 2012, **6**, 2, 98–107.
7. Dai Y, Xu C, Sun X, Chen X. Nanoparticle design strategies for enhanced anticancer therapy by exploiting the tumour microenvironment. *Chem Soc Rev* 2017 12, 3839–3852.
8. Sunoqrot S, Hamed R, Abdel-Halim H, Tarawneh O. Synergistic interplay of medicinal chemistry and formulation strategies in nanotechnology: From drug discovery to nanocarrier design and development. *Curr Top Med Chem* 2017, **17**, 13, 1451–1468.
9. Marciello M, Pellico J., Fernandez-Barahona I, Herranz F, Ruiz-Cabello J, Filice M. Recent advances in the preparation and application of multifunctional iron oxide and liposome-based nanosystems for multimodal diagnosis and therapy. *Interface Focus* 2016, **6**, 6, 20160055.
10. Siafaka PI, Üstündag Okur N, Karavas E, Bikiaris DN. Surface modified multifunctional and stimuli responsive nanoparticles for drug targeting: Current status and uses. *Int J mol Sci* 2016, **17**, 9, 1440.
11. Lehner R, Wang X, Marsch, Hunziker P. Intelligent nanomaterials for medicine: Carrier platforms and targeting strategies in the context of clinical application. *Nanomed Nanotechnol Biol Med* 2013, **9**, 742–757.
12. Amirmahani N, Mahmoodi NO, Mohammadi Galangash M, Ghavidast A. Advances in nanomicelles for sustained drug delivery. *J Ind Eng Chem* 2017, **55**, 21–34.
13. Feng L, Mumper RJ. A critical review of lipid-based nanoparticles for taxane delivery. *Cancer Lett* 2013, **334**, 2, 157–175.
14. Fan Y, Zhang Q. Development of liposomal formulations: From concept to clinical investigations. *Asian J Pharm Sci* 2013, **8**, 2, 81–87.
15. Hwang JY, Li Z, Loh XJ. Small molecule therapeutic-loaded liposomes as therapeutic carriers: From development to clinical applications. *RSC Adv* 2016, **6**, 74, 70592–70615.
16. Ajorlou E, Khosroushahi AY. Trends on polymer- and lipid-based nanostructures for parenteral drug delivery to tumors. *Cancer Chemother Pharmacol* 2017, **79**, 2, 251–265.
17. Sawant RR, Torchilin VP. Multifunctionality of lipid-core micelles for drug delivery and tumour targeting. *Mol Membr Biol* 2010, **27**, 232–246.
18. Tang M-F, Lei L, Guo S-R, Huang W-L. Recent progress in nanotechnology for cancer therapy. *China J Cancer* 2010, **29**, 775–780.

19. Yingchoncharoen P, Kalinowski DS, Richardson DR. Lipid-based drug delivery systems in cancer therapy: What is available and what is yet to come. *Pharmacol Rev* 2016, **68**, 701–787.

20. Teixeira MC, Carbone C, Souto EB. Beyond liposomes: Recent advances on lipid based nanostructures for poorly soluble/poorly permeable drug delivery. *Prog Lipid Res* 2017, **68**, 1–11.

21. Barenholz YY. Doxil®–the first FDA-approved nano-drug: Lessons learned. *J Control Release* 2012, **160**, 117–134.

22. Hofheinz R-D, Gnad-Vogt SU, Beyer U, Hochhaus A. Liposomal encapsulated anti-cancer drugs. *Anticancer Drugs* 2005, **16**, 691–707.

23. Rochlitz C, Ruhstaller T, Lerch S, Spirig C, Huober J, Suter T, et al. Combination of bevacizumab and 2-weekly pegylated liposomal doxorubicin as first-line therapy for locally recurrent or metastatic breast cancer. A multicenter, single-arm phase II trial (SAKK 24/06). *Ann Oncol* 2011, **22**, 80–85.

24. Immordino ML, Dosio F, Cattel L. Stealth liposomes: Review of the basic science, rationale, and clinical applications, existing and potential. *Int J Nanomed* 2006, **1**, 297–315.

25. Zang X, Ding H, Zhao X, Li X, Du Z, Hu H, Qiao M, Chen D, Deng Y, Zhao X. Anti-EphA10 antibody-conjugated pH-sensitive liposomes for specific intracellular delivery of siRNA. *Int J Nanomed* 2016, **11**, 3951–3967.

26. Dabbagh A, Abdullah BJJ, Abdullah H, Hamdi M, Kasim NHA. Triggering mechanisms of thermosensitive nanoparticles under hyperthermia condition. *J Pharm Sci* 2015, **104**, 8, 2414–2428.

27. Miranda D, Lovell JF. Mechanisms of light-induced liposome permeabilization. *Bioeng Transl Med* 2016, **1**, 267–276.

28. Fomina N, Sankaranarayanan, Almutairi A. Photochemical mechanisms of light-triggered release from nanocarriers. *Adv Drug Deliv Rev* 2012, **64**, 11, 1005–1020.

29. Karimi M, Eslami M, Sahandi-Zangabad P, Mirab F, Farajisafiloo N, Shafaei Z, Ghosh D, Bozorgomid M, Dashkhaneh, Hamblin MR. pH-Sensitive stimulus-responsive nanocarriers for targeted delivery of therapeutic agents. *Wiley Interdiscip Rev Nanomed Nanobiotechnol* 2016, **8**, 5, 696–716.

30. Christie JG, Kompella UB. Ophthalmic light sensitive nanocarrier systems. *Drug Discov Today* 2008, **13**, 124–134.

31. Jain A, Tiwari A, Verma A, Jain SK. Ultrasound-based triggered drug delivery to tumors. *Drug Deliv Transl Res* 2018, **18**, 1, 150–164.

32. Wang Y, Yin T, Su Z, Qiu C, Wang Y, Zheng R, Chen M, Shuai X. Highly uniform ultrasound-sensitive nanospheres produced by a pH-induced micelle-to-vesicle transition for tumor-targeted drug delivery. *Nano Res* 2017, 1–12.
33. Scomparin A, Florindo HF, Tiram G, Ferguson EL, Satchi-Fainaro R. Two-step polymer- and liposome-enzyme prodrug therapies for cancer: PDEPT and PELT concepts and future perspectives. *Adv Drug Deliv Rev* 2017, **118**, 52–64.
34. Zhu C, Sempkowski M, Holleran T, Linz T, Bertalan T, Josefsson A, Bruchertseifer F, Morgenstern A, Sofou S. Alpha-particle radiotherapy: For large solid tumors diffusion trumps targeting. *Biomaterials* 2017, **130**, 67–75.
35. Lehner R, Wang X, Wolf M, Hunziker P. Designing switchable nanosystems for medical application. *J Control Release* 2012, **161**, 307–316.
36. Teixeira MC, Carbone C, Souto EB. Beyond liposomes: Recent advances on lipid based nanostructures for poorly soluble/poorly permeable drug delivery. *Prog Lipid Res* 2017, **68**, 1–11.
37. U.S. Department of Health and Human Services, Food and Drug Administration, Center for Drug Evaluation and Research (CDER). Bioavailability and bioequivalence studies for immediate-release solid oral dosage forms based on a biopharmaceutics classification system. Guidance for industry. December 2017, Biophamaceutics. https://www.fda.gov/downloads/Drugs/GuidanceComplianceRegulatoryInformation/Guidances/UCM070246.pdf
38. Almeida M, Magalhaes M, Veiga F, Figueiras A. Poloxamers, poloxamines and polymeric micelles: Definition, structure and therapeutic applications in cancer. *J Polym Res* 2018, **25**, 1, 31.
39. Mohamadi M, Ramezani M, Abnous K, Alibolandi M. Biocompatible polymersomes-based cancer theranostics: Towards multifunctional nanomedicine. *Int J Pharm* 2017, **517**, 1–2, 287–303.
40. Wang L, Zeng R, Li C, Qiao R. Self-assembled polypeptide-block-poly(vinylpyrrolidone) as prospective drug-delivery systems. *Colloids Surf B Biointerfaces* 2009, **74**, 284–292.
41. Lu J, Liong M, Zink JI, Tamanoi F. Mesoporous silica nanoparticles as a delivery system for hydrophobic anticancer drugs. *Small* 2007, **3**, 1341–1346.
42. Zhang X, Huang Y, Song L. Nanomicellar carriers for targeted delivery of anticancer agents. *Ther Deliv* 2014, **1**, 53–68.

43. Jhaveri AM, Torchilin VP. Multifunctional polymeric micelles for delivery of drugs and siRNA. *Front Pharmacol* 2014, **5**, 77.

44. Zhang CG, Yang SD, Zhu WJ, You BG, Liu Y, Yuan ZQ, Chen WL, Li JZ, Zhou XF, Liu C, Zhang XN. Distinctive polymer micelle designed for siRNA delivery and reversal of MDR1 gene-dependent multidrug resistance. *J Biomed Mater Res Part B* 2017, **105**, 2093–2106.

45. Jeong JH, Mok H, Oh Y-K, Park TG. siRNA conjugate delivery systems. *Bioconjug Chem* 2009, **20**, 5–14.

46. Park MR, Han KO, Han IK, Cho M-H, Nah JW, Choi YJ, et al. Degradable polyethylenimine–poly(ethylene glycol) copolymers as novel gene carriers. *J Control Release* 2005, **105**, 367–380.

47. Lehner R, Liu K, Wang X, Hunziker P. Efficient receptor mediated siRNA delivery in vitro by folic acid targeted pentablock copolymer-based micelleplexes. *Biomacromolecules* 2017, **8**, 2654–2662.

48. Davis ME. The first targeted delivery of siRNA in humans via a self-assembling, cyclodextrin polymer-based nanoparticle: from concept to clinic. *Mol Pharmaceutics* 2009, **6**, 659–668.

49. Salvador-Morales C, Zhang L, Langer R, Farokhzad OC. Immuno-compatibility properties of lipid-polymer hybrid nanoparticles with heterogeneous surface functional groups. *Biomaterials* 2009, **30**, 2231–2240.

50. He C-X, Tabata Y, Gao J-Q. Non-viral gene delivery carrier and its three-dimensional transfection system. *Int J Pharm* 2010, **386**, 232–242.

51. Oerlemans C, Bult W, Bos M, Storm G, Nijsen JFW, Hennink WE. Polymeric micelles in anticancer therapy: Targeting, imaging and triggered release. *Pharm Res* 2010, **27**, 2569–2589.

52. Navarro G, Pan J, Torchilin VP. Micelle-like nanoparticles as carriers for DNA and siRNA. *Mol Pharm* 2015, **12**, 2, 301–313.

53. Tomalia DA, Baker H, Dewald J, Hall M, Kallos G, Martin S, et al. A new class of polymers: Starburst-dendritic macromolecules. *Polym J* 1985, **17**, 117–132.

54. Paleos CM, Tsiourvas D, Sideratou Z, Tziveleka L-A. Drug delivery using multifunctional dendrimers and hyperbranched polymers. *Expert Opin Drug Deliv* 2010, **7**, 1387–1398.

55. Tomalia DA. In quest of a systematic framework for unifying and defining nanoscience. *J Nanopart Res* 2009, **11**, 1251–1310.

56. Farias ED, Bouchet LM, Brunetti V, Strumia MC. Dendrimers and dendronized materials as nanocarriers. *Nanostructures for Novel*

Therapy: Synthesis, Characterization and Applications. 2017, pp. 429–456.

57. Patil ML, Zhang M, Minko T. Multifunctional triblock nanocarrier (PAMAM-PEG-PLL) for the efficient intracellular siRNA delivery and gene silencing. *ACS Nano* 2011, **5**, 1877–1887.
58. Pavan GM, Posocco P, Tagliabue A, Maly M, Malek A, Danani A, et al. PAMAM dendrimers for siRNA delivery: Computational and experimental insights. *Chemistry* 2010, **16**, 7781–7795.
59. Navarro G. Tros de Ilarduya C. Activated and non-activated PAMAM dendrimers for gene delivery in vitro and in vivo. *Nanomedicine* 2009, **5**, 287–297.
60. Jia L, Xu J-P, Wang H, Ji J. Polyamidoamine dendrimers surface-engineered with biomimetic phosphorylcholine as potential drug delivery carriers. *Colloids Surf B Biointerfaces* 2011, **84**, 49–54.
61. Zhou W, Shao J, Jin Q, Wei Q, Tang J, Ji J. Zwitterionic phosphorylcholine as a better ligand for gold nanorods cell uptake and selective photothermal ablation of cancer cells. *Chem Commun* 2010, **46**, 1479.
62. Janaszewska A, Gorzkiewicz M, Ficker M, Petersen JF, Paolucci V, Christensen JB, Klajnert-Maculewicz B. Pyrrolidone modification prevents PAMAM dendrimers from activation of pro-inflammatory signaling pathways in human monocytes. *Mol Pharmaceutics* 2018, **5**, 1, 12–30.
63. McCarthy TD, Karellas P, Henderson SA, Giannis M, O'Keefe DF, Heery G, et al. Dendrimers as drugs: Discovery and preclinical and clinical development of dendrimer-based microbicides for HIV and STI prevention. *Mol Pharm* 2005, **2**, 312–318.
64. O'Loughlin J, Millwood IY, McDonald HM, Price CF, Kaldor JM, Paull JRA. Safety, tolerability, and pharmacokinetics of SPL7013 Gel (VivaGel (R)): A dose ranging, phase I study. *Sex Transm Dis* 2010, **37**, 100–104.
65. Price CF, Tyssen D, Sonza S, Davie A, Evans S, Lewis GR, et al. SPL7013 gel (VivaGel (R)) retains potent HIV-1 and HSV-2 inhibitory activity following vaginal administration in humans. *Plos One* 2011, **6**, e24095.
66. Luo K, He B, Wu Y, Shen Y, Gu Z. Functional and biodegradable dendritic macromolecules with controlled architectures as nontoxic and efficient nanoscale gene vectors. *Biotechnol Adv* 2014, **32**, 818–830.
67. Malekzadeh AM, Ramazani A, Rezaei SJ, Niknejad H. Design and construction of multifunctional hyperbranched polymers coated

magnetite nanoparticles for both targeting magnetic resonance imaging and cancer therapy. *J Colloid Interface Sci* 2017, **490**, 64–73.

68. Donaldson K. Carbon nanotubes: A review of their properties in relation to pulmonary toxicology and workplace safety. *Toxicol Sci* 2006, **92**, 5–22.
69. Kim KY. Nanotechnology platforms and physiological challenges for cancer therapeutics. *Nanomedicine* 2007, **3**, 103–110.
70. Beg S, Rizwan M, Sheikh AM, Hasnain MS, Anwer K, Kohli K. Advancement in carbon nanotubes: Basics, biomedical applications and toxicity. *J Pharm Pharmacol* 2010, **63**, 141–163.
71. Dinesh B, Bianco A, Ménard-Moyon C. Designing multimodal carbon nanotubes by covalent multi-functionalization. *Nanoscale* 2016, **8**, 18596.
72. Gu Y-J, Cheng J, Jin J, Cheng SH, Wong W-T. Development and evaluation of pH-responsive single-walled carbon nanotube-doxorubicin complexes in cancer cells. *Int J Nanomed* 2011, **6**, 2889–2898.
73. Miao Y, Zhang H, Pan Y, Red J, Ye M, Xia F, Huang R, Lin Z, Jiang S, Zhang Y, Songyang Z, Zhang Y. Single-walled carbon nanotube: One specific inhibitor of cancer stem cells in osteosarcoma upon downregulation of the TGFβ1 signaling. *Biomaterials* 2017, **149**, 29–40.
74. Lanone S, Boczkowski J. Biomedical applications and potential health risks of nanomaterials: Molecular mechanisms. *Curr Mol Med* 2006, **6**, 651–663.
75. Poland CA, Duffin R, Kinloch I, Maynard A, Wallace WAH, Seaton A, et al. Carbon nanotubes introduced into the abdominal cavity of mice show asbestos-like pathogenicity in a pilot study. *Nat Nanotechnol* 2008, **3**, 423–428.
76. Zhu L, Chang DW, Dai L, Hong Y. DNA damage induced by multiwalled carbon nanotubes in mouse embryonic stem cells. *Nano Lett* 2007, **7**, 3592–3597.
77. Ji S-R, Liu C, Zhang B, Yang F, Xu J, Long J, et al. Carbon nanotubes in cancer diagnosis and therapy. *Biochim Biophys Acta, Rev Cancer* 2010, **1806**, 29–35.
78. Sanginario A, Miccoli B, Demarchi D. Carbon nanotubes as an effective opportunity for cancer diagnosis and treatment. *Biosensors* 2017, **7**, 9.
79. Eldawud R, Wagner A, Dong C, Stueckle TA, Rojanasakul Y, Dinu CZ. Carbon nanotubes physicochemical properties influence the overall cellular behavior and fate. *NanoImpact* 2018, **9**, 72–84.

80. Villa CH, McDevitt MR, Escorcia FE, Rey DA, Bergkvist M, Batt CA, et al. Synthesis and biodistribution of oligonucleotide-functionalized, tumor-targetable carbon nanotubes. *Nano Lett* 2008, **8**, 4221–4228.
81. Martincic M, Tobias G. Filled carbon nanotubes in biomedical imaging and drug delivery. *Expert Opin Drug Deliv* 2015, **12**, 4, 563–581.
82. Ji Z, Lin G, Lu Q, Meng L, Shen X, Dong L, et al. Targeted therapy of SMMC-7721 liver cancer in vitro and in vivo with carbon nanotubes based drug delivery system. *J Colloid Interface Sci* 2012, **365**, 143–149.
83. Hou L, Yuan Y, Ren J, Zhang Y, Wang Y, Shan X, Liu Q, Zhang Z. In vitro and in vivo comparative study of the phototherapy anticancer activity of hyaluronic acid-modified single-walled carbon nanotubes, graphene oxide, and fullerene. *J Nanopart Res* 2017, **19**, 286.
84. Faraj AA, SWCNTs as novel theranostic nanocarriers for cancer diagnosis and therapy: towards safe translation to the clinics. *Nanomedicine* 2016, **11**, 11, 1431–1445.
85. Madani SY, Mandel A, Seifalian AM. A concise review of carbon nanotube's toxicology. *Nano Rev* 2013, **4**, 1–14.
86. Bussy C, Pinault M, Cambedouzou J, et al. Critical role of surface chemical modifications induced by length shortening on multi-walled carbon nanotubes-induced toxicity. *Part Fibre Toxicol* 2012, **9**, 46.
87. Alexiou C, Arnold W, Klein RJ, Parak FG, Hulin P, Bergemann C, et al. Locoregional cancer treatment with magnetic drug targeting. *Cancer Res* 2000, **60**, 6641–6648.
88. Frimpong RA, Hilt JZ. Magnetic nanoparticles in biomedicine: Synthesis, functionalization and applications. *Nanomedicine (Lond)* 2010, **5**, 1401–1414.
89. Hofmann-Amtenbrink M, Hofmann H, Montet X. Superparamagnetic nanoparticles – a tool for early diagnostics. *Swiss Med Wkly* 2010, **140**, w13081-13.
90. Salieb-Beugelaar GB, Hunziker P. Towards nano-diagnostics for rapid diagnosis of infectious diseases–current technological state. *Eur J Nanomed* 2014, **6**, 1, 11–28.
91. Bakhtiary Z, Saei AA, Hajipour MJ, Raoufi M, Vermesh O, Mahmoudi M. Targeted superparamagnetic iron oxide nanoparticles for early detection of cancer: Possibilities and challenges. *Nanomed Nanotechnol Biol Med* 2016, **12**, 2, 287–307.
92. Sohail A, Ahmad Z, Bég OA, Arshad S, Sherin L. A review on hyperthermia via nanoparticle-mediated therapy. *Bull Cancer* 2017, **104**, 5, 452–461.

93. Patra HK, Banerjee S, Chaudhuri U, Lahiri P, Dasgupta AK. Cell selective response to gold nanoparticles. *Nanomedicine* 2007, **3**, 111–119.
94. Schuemann J, Berbeca R, Chithrani D, Cho SH, Kumar R, McMahon SJ, Sridhar S, Krishnan S. Roadmap to clinical use of gold nanoparticles for radiation sensitization. *Int J Radiation Oncol Biol Phys* 2016, **94**, 1, 189–205.
95. Her S, Jaffray DA, Allen C. Gold nanoparticles for applications in cancer radiotherapy: Mechanisms and recent advancements. *Adv Drug Deliv Rev* 2017, **109**, 84–101.
96. Han G, Ghosh P, De M, Rotello VM. Drug and gene delivery using gold nanoparticles. *Nanobiotechnology* 2007, **3**, 40–45.
97. Sondi I, Salopek-Sondi B. Silver nanoparticles as antimicrobial agent: A case study on E. coli as a model for Gram-negative bacteria. *J Colloid Interface Sci* 2004, **275**, 177–182.
98. Silver S, Phung LT, Silver G. Silver as biocides in burn and wound dressings and bacterial resistance to silver compounds. *J Ind Microbiol Biotechnol* 2006, **33**, 627–634.
99. Aziz Z, Abu SF, Chong NJ. A systematic review of silver-containing dressings and topical silver agents (used with dressings) for burn wounds. *Burns* 2012, **38**, 307–318.
100. Storm-Versloot MN, Vos CG, Ubbink DT, Vermeulen H. Topical silver for preventing wound infection. *Cochrane Database Syst Rev* 2010, **17**, CD006478.
101. Beattie M, Taylor J. Silver alloy vs. uncoated urinary catheters: a systematic review of the literature. *J Clin Nurs* 2011, **20**, 2098–2108.
102. Acharya D, Singha KM; Pandey P, Mohanta B, Rajkumari J, Singha LP. Shape dependent physical mutilation and lethal effects of silver nanoparticles on bacteria. *Sci Rep* 2018, **8**, 1, 201.
103. Srinivasan S, Bhardwaj V, Nagasetti A, Fernandez-Fernandez A, McGoron AJ. Multifunctional surface-enhanced raman spectroscopy-detectable silver nanoparticles for combined photodynamic therapy and pH-triggered chemotherapy. *J Biomed Nanotechnol* 2016, **12**, 2202–2219.
104. Nagpal K, Singh SK, Mishra DN. Chitosan nanoparticles: A promising system in novel drug delivery. *Chem Pharm Bull* 2010, **58**, 1423–1430.
105. De Campos AM, Sánchez A, Alonso MJ. Chitosan nanoparticles: A new vehicle for the improvement of the delivery of drugs to the ocular surface. Application to cyclosporin A. *Int J Pharm* 2001, **224**, 159–168.

106. Han HD, Mora EM, Roh JW, Nishimura M, Lee SJ, Stone RL, et al. Chitosan hydrogel for localized gene silencing. *Cancer Biol Ther* 2011, **11**, 839–845.

107. Makhlof A, Tozuka Y, Takeuchi H. Design and evaluation of novel pH-sensitive chitosan nanoparticles for oral insulin delivery. *Eur J Pharm Sci* 2011, **42**, 445–451.

108. Sanoj Rejinold N, Sreerekha PR, Chennazhi KP, Nair SV, Jayakumar R. Biocompatible, biodegradable and thermo-sensitive chitosan-g-poly (N-isopropylacrylamide) nanocarrier for curcumin drug delivery. *Int J Biol Macromol* 2011, **49**, 161–172.

109. Simi CK, Emilia Abraham T. Hydrophobic grafted and cross-linked starch nanoparticles for drug delivery. *Bioprocess Biosyst Eng* 2007, **30**, 173–180.

110. Xiao S, Tong C, Liu X, Yu D, Liu Q, Xue C, Tang D, Zhao L. Preparation of folate-conjugated starch nanoparticles and its application to tumor-targeted drug delivery vector. *Chin Sci Bull* 2006, **51**, 1693–1697.

111. Wang YJ, Huang Y, Anreddy N, Zhang GN, Zhang YK, Xie M, Lin D, Yang DH, Zhang M, Chen ZS. Tea nano-particle, a safe and biocompatible nanocarrier, greatly potentiates the anticancer activity of doxorubicin. *Oncotarget* 2015, **7**, 5, 5877–5891.

112. Ekimov A. Quantum size effect in three-dimensional microscopic semiconductor crystals. *ZhETF Pis ma Redaktsiiu* 1981, **34**, 363.

113. Medintz IL, Mattoussi H, Clapp AR. Potential clinical applications of quantum dots. *Int J Nanomed* 2008, **3**, 151–167.

114. Reiss P, Protière M, Li L. Core/shell semiconductor nanocrystals. *Small* 2009, **5**, 154–168.

115. Resch-Genger U, Grabolle M, Cavaliere-Jaricot S, Nitschke R, Nann T. Quantum dots versus organic dyes as fluorescent labels. *Nat Methods* 2008, **5**, 763–775.

116. Derfus AM, Chen AA, Min D-H, Ruoslahti E, Bhatia SN. Targeted quantum dot conjugates for siRNA delivery. *Bioconjug Chem* 2007, **18**, 1391–1396.

117. Zintchenko A, Susha AS, Concia M, Feldmann J, Wagner E, Rogach AL, et al. Drug nanocarriers labeled with near-infrared-emitting quantum dots (quantoplexes): Imaging fast dynamics of distribution in living animals. *Mol Ther* 2009, **17**, 1849–1856.

118. Juzenas P, Chen W, Sun Y-P, Neto Coelho MA, Generalov R, Generalova N, et al. Quantum dots and nanoparticles for photodynamic and radiation therapies of cancer. *Adv Drug Deliv Rev* 2008, **60**, 1600–1614.

119. Luo G, Long J, Zhang B, Liu C, Ji S, Xu J, et al. Quantum dots in cancer therapy. *Expert Opin Drug Deliv* 2012, **9**, 47–58.

120. Hu R, Yong K-T, Roy I, Ding H, Law W-C, Cai H, et al. Functionalized near-infrared quantum dots for in vivo tumor vasculature imaging. *Nanotechnology* 2010, **21**, 145105.

121. Liu, WJ, Li C, Ren YJ, Sun XB, Pan W, Li YH, Wang JP, Wang W J. Carbon dots: Surface engineering and applications. *J Mater Chem B* 2016, **4**, 5772–5788.

122. Zhang, ZW, Duan Y, Yu Y, Yan ZY, Chen JQ. Carbon quantum dots: Synthesis, characterization, and assessment of cytocompatibility. *J Mater Sci: Mater Med* 2015, **26**, 213.

123. Hassan M, Gomes VG, Dehghani A, Ardekani SM. Engineering carbon quantum dots for photomediated theranostics. *Nano Res.* 2018, **11**, 1, 1–41.

124. Zhang Q, Zha L, Ma J, Liang B. A novel route to prepare pH- and temperature-sensitive nanogels via a semibatch process. *J Colloid Interface Sci* 2009, **330**, 330–336.

125. Mimi H, Ho KM, Siu YS, Wu A, Li P. Polyethyleneimine-based core-shell nanogels. A promising siRNA carrier for argininosuccinate synthetase mRNA knockdown in HeLa cells. *J Control Release* 2012, **28**, 158123–158130.

126. Neamtu I, Rusu AG, Diaconu A, Nita LE, Aurica P. Chiriac. Basic concepts and recent advances in nanogels as carriers for medical applications. *Drug Deliv* 2017, **24**, 1, 539–557.

127. Debele TA, Mekuria SL, Tsai HC. Polysaccharide based nanogels in the drug delivery system: Application as the carrier of pharmaceutical agents. *Mater Sci Eng C* 2016, **68**, 964–981.

128. Wu W, Shen J, Gai Z, Hong K, Banerjee P, Zhou S. Multi-functional core-shell hybrid nanogels for pH-dependent magnetic manipulation, fluorescent pH-sensing, and drug delivery. *Biomaterials* 2011, **32**, 9876–9887.

129. Dumville JC, O'Meara S, Deshpande S, Speak K. Hydrogel dressings for healing diabetic foot ulcers. *Cochrane Database Syst Rev* 2011, **9**, CD009101.

130. Skulason S, Holbrook WP, Thormar H, Gunnarsson GB, Kristmundsdottir T. A study of the clinical activity of a gel combining monocaprin and doxycycline: A novel treatment for herpes labialis. *J Oral Pathol Med* 2012, **41**, 61–67.

131. Soni KS, Desale SS, Bronich TK. Nanogels: An overview of properties, biomedical applications and obstacles to clinical translation. *J Control Release* 2016, **240**, 109–126.

132. Han S, Cao S, Wang Y, Wang J, Xia D, Xu H, et al. Self-assembly of short peptide amphiphiles: The cooperative effect of hydrophobic interaction and hydrogen bonding. *Chemistry* 2011, **17**, 13095–13102.

133. Meng Q, Kou Y, Ma X, Liang Y, Guo L, Ni C, et al. Tunable self-assembled peptide amphiphile nanostructures. *Langmuir* 2012, **28**, 5017–5022.

134. Ekiz, MS, Cinar, G, Khalily MA, Guler MO. Self-assembled peptide nanostructures for functional materials. *Nanotechnology* 2016, **27**, 40, 402002.

135. Fraysse-Ailhas C, Graff-Meyer A, Rigler P, Mittelhozer C, Raman S, Aebi U, et al. Peptide nanoparticles for drug delivery applications. *Eur Cell Mater* 2007, **14**, 115.

136. Raman S, Machaidze G, Lustig A, Aebi U, Burkhard P. Structure-based design of peptides that self-assemble into regular polyhedral nanoparticles. *Nanomedicine* 2006, **2**, 95–102.

137. Bawa R, Fung S-Y, Shiozaki A, Yang H, Zheng G, Keshavjee S, Liu M. Self-assembling peptide-based nanoparticles enhance cellular delivery of the hydrophobic anticancer drug ellipticine through caveolae-dependent endocytosis. *Nanomedicine* 2012, **8**, 647–654.

138. Naskar J, Roy S, Joardar A, Das S, Banerjee A. Self-assembling dipeptide-based nontoxic vesicles as carriers for drugs and other biologically important molecules. *Org Biomol Chem* 2011, **9**, 6610–6615.

139. Wang S, Huang P, Chen X. Hierarchical targeting strategy for enhanced tumor tissue accumulation/retention and cellular internalization. *Adv Mater* 2016, **28**, 34, 7340–7364.

140. Tagami T, Ernsting MJ, Li S-D. Efficient tumor regression by a single and low dose treatment with a novel and enhanced formulation of thermosensitive liposomal doxorubicin. *J Control Release* 2011, **152**, 303–309.

141. Lutz RJ, Whiteman KR. Antibody-maytansinoid conjugates for the treatment of myeloma. *MAbs* 2009, **1**, 548–551.

142. Maier-Hauff K, Ulrich F, Nestler D, Niehoff H, Wust P, Thiesen B, et al. Efficacy and safety of intratumoral thermotherapy using magnetic iron-oxide nanoparticles combined with external beam radiotherapy on patients with recurrent glioblastoma multiforme. *J Neurooncol* 2010, **103**, 317–324.

143. Keam SJ, Scott LJ, Curran MP. Verteporfin: A review of its use in the management of subfoveal choroidal neovascularisation. *Drugs* 2002, **63**, 2521–2554.

144. Coco Lo F, Ammatuna E, Noguera NL. Treatment of acute promyelocytic leukemia with gemtuzumab ozogamicin. *Clin Adv Hematol Oncol* 2005, **4**, 57–67.

145. Le UM, Shaker DS, Sloat BR, Cui Z. A thermo-sensitive polymeric gel containing a gadolinium (Gd) compound encapsulated into liposomes significantly extended the retention of the Gd in tumors. *Drug Dev Ind Pharm* 2008, **34**, 413–418.

146. Rodriguez PL, Harada T, Christian DA, Pantano DA, Tsai RK, Discher DE. Minimal "self" peptides that inhibit phagocytic clearance and enhance delivery of nanoparticles. *Science* 2013, **339**, 971–975.

147. Miceli E, Kar M, Caderon M. Interactions of organic nanoparticles with proteins in physiological conditions. *J Mater Chem B* 2017, **5**, 4393–4405.

148. Ritz S, Schottler S, Kotman N, Baier G, Musyanovych A, Kuharev J, Landfester K, Schild H, Jahn O, Tenzer S, Mailander V. Protein corona of nanoparticles: Distinct proteins regulate the cellular uptake. *Biomacromolecules* 2015, **16**, 1311–1321.

149. Cheng Z, Al Zaki A, Hui JZ, Muzykantov VR, Tsourkas A. Multifunctional nanoparticles: Cost versus benefit of adding targeting and imaging capabilities. *Science* 2012, **338**, 903–910.

150. Petersen GH, Alzghari SK, Chee W, Sankari SS, La-Beck NM. Meta-analysis of clinical and preclinical studies comparing the anticancer efficacy of liposomal versus conventional non-liposomal doxorubicin. *J Control Release* 2016, **232**, 255–264.

151. Pouponneau P, Leroux J-C, Soulez G, Gaboury L, Martel S. Co-encapsulation of magnetic nanoparticles and doxorubicin into biodegradable microcarriers for deep tissue targeting by vascular MRI navigation. *Biomaterials* 2011, **32**, 3481–3486.

152. Matsumura Y, Maeda H. A new concept for macromolecular therapeutics in cancer-chemotherapy – mechanism of tumoritropic accumulation of proteins and the antitumor agent Smancs. *Cancer Res* 1986, **46**, 6387–6392.

153. Maeda H. The enhanced permeability and retention (EPR) effect in tumor vasculature: The key role of tumor-selective macromolecular drug targeting. *Adv Enzyme Regul* 2001, **41**, 189–207.

154. Greish K. Enhanced permeability and retention of macromolecular drugs in solid tumors: A royal gate for targeted anticancer nanomedicines. *J Drug Target* 2007, **15**, 457–464.
155. Liberti PA, Rao CG, Terstappen LWMM. Optimization of ferrofluids and protocols for the enrichment of breast tumor cells in blood. *J Magn Magn Mater* 2001, **225**, 301–307.
156. Allen TM, Cullis PR. Drug delivery systems: Entering the mainstream. *Science* 2004, **303**, 1818–1822.
157. Bae YH. Drug targeting and tumor heterogeneity. *J Control Release* 2009, **133**, 2–3.
158. Paul F, Roath S, Melville D, Warhurst DC, Osisanya JO. Separation of malaria-infected erythrocytes from whole blood: Use of a selective high-gradient magnetic separation technique. *Lancet* 1981, **2**, 70–71.
159. Yuan F, Dellian M, Fukumura D, Leunig M, Berk DA, Torchilin VP, et al. Vascular permeability in a human tumor xenograft: Molecular size dependence and cutoff size. *Cancer Res* 1995, **55**, 3752–3756.
160. Torchilin V. Recent advances with liposomes as pharmaceutical carriers. *Nat Rev Drug Discov* 2005, **4**, 145–160.
161. Richmond A, Su Y. Mouse xenograft models vs GEM models for human cancer therapeutics. *Dis Model Mech* 2008, **1**, 78–82.
162. Shi JJ, Kantoff PW, Wooster R, Farokhzad OC. Cancer nanomedicine: Progress, challenges and opportunities. *Nat Rev Cancer* 2017, **17**, 20–37.
163. Elsherbini AAM, Saber M, Aggag M, El-Shahawy A, Shokier HAA. Magnetic nanoparticle-induced hyperthermia treatment under magnetic resonance imaging. *Magn Reson Imaging* 2011, **29**, 272–280.
164. Strebhardt KK, Ullrich AA. Paul Ehrlich's magic bullet concept: 100 years of progress. *Nat Rev Cancer* 2008, **8**, 473–480.
165. Kumar CSSR, Mohammad F. Magnetic nanomaterials for hyperthermia-based therapy and controlled drug delivery. *Adv Drug Deliv Rev* 2011, **63**, 789–808.
166. Shan Z, Li C, Zhang X, Oakes KD, Servos MR, Wu Q, et al. Temperature-dependent selective purification of plasmid DNA using magnetic nanoparticles in an RNase-free process. *Anal Biochem* 2011, **412**, 117–119.
167. Broz P, Benito SM, Saw C, Burger P, Heider H, Pfisterer M, et al. Cell targeting by a generic receptor-targeted polymer nanocontainer platform. *J Control Release* 2005, **102**, 475–488.

168. Low PS, Kularatne SA. Folate-targeted therapeutic and imaging agents for cancer. *Curr Opin Chem Biol* 2009, **13**, 256–262.

169. Leamon CP, Low PS. Delivery of macromolecules into living cells: A method that exploits folate receptor endocytosis. *Proc Natl Acad Sci U S A* 1991, **88**, 5572–5576.

170. Al-Deen FN, Ho J, Selomulya C, Ma C, Coppel R. Superparamagnetic nanoparticles for effective delivery of malaria DNA vaccine. *Langmuir* 2011, **27**, 3703–3712.

171. Danhier F, Feron O, Preat V. To exploit the tumor microenvironment: Passive and active tumor targeting of nanocarriers for anti-cancer drug delivery. *J Control Release* 2010, **148**, 135–146.

172. Ghaghada K, Saul J, Natarajan J, Bellamkonda R, Annapragada A. Folate targeting of drug carriers: A mathematical model. *J Control Release* 2005, **104**, 113–128.

173. Salvati A, Pitek AS, Monopoli MP, Prapainop K, Bombelli FB, Hristov DR, Kelly PM, Aberg C, Mahon E, Dawson KA. Transferrin-functionalized nanoparticles lose their targeting capabilities when a biomolecule corona adsorbs on the surface. *Nat Nanotechnol* 2013, **8**, 137–143.

174. Sakulkhu U, Maurizi L, Mahmoudi M, Motazacker M, Vries M, Gramoun A, Beuzelin MGO, Vallee JP, Rezaee F, Hofmann H. Ex situ evaluation of the composition of protein corona of intravenously injected superparamagnetic nanoparticles in rats. *Nanoscale* 2014, **6**, 11439–11450.

175. Peer D, Karp JM, Hong S, FaroKhzad OC, Margalit R, Langer R. Nanocarriers as an emerging platform for cancer therapy. *Nat Nanotechnol* 2007, **2**, 751–760.

176. Oh P, Testa JE, Borgstrom P, Witkiewicz H, Li Y, Schnitzer JE. In vivo proteomic imaging analysis of caveolae reveals pumping system to penetrate solid tumors. *Nat Med* 2014, **20**, 1062.

177. Ruoslahti E. Tumor penetrating peptides for improved drug delivery. *Adv Drug Deliv Rev* 2017, **110–111**, 3–12.

178. Li L, Song L, Liu X, Yang X, Li X, He T, Wang N, Yang S, Yu C, Yin T, Wen Y, He Z, Wei X, Su W, Wu Q, Yao S, Gong C, Wei Y. Artificial virus delivers CRISPR-Cas9 system for genome editing of cells in mice. *ACS Nano* 2017, **11**, 95–111.

179. Behzadi S, Serpooshan V, Tao W, Hamaly MA, Alkawareek MY, Dreaden EC, Brown D, Alkilany AM, Farokhzad OC, Mahmoudi M. Cellular uptake of nanoparticles: Journey inside the cell. *Chem Soc Rev* 2017, **46**, 4218–4244.

180. Yameen B, Choi WI, Vilos C, Swami A, Shi JJ, Farokhzad OC. Insight into nanoparticle cellular uptake and intracellular targeting. *J Control Release* 2014, **190**, 485–499.
181. Wang Y, Deng Y, Luo H, Zhu A, Ke H, Yang H, Chen H. Light-responsive nanoparticles for highly efficient cytoplasmic delivery of anticancer agents. *ACS Nano* 2017, **11**, 12134–12144.
182. Frankel AD, Pabo CO. Cellular uptake of the tat protein from human immunodeficiency virus. *Cell* 1988, **55**, 1189–1193.
183. Joliot A, Pernelle C, Deagostini-Bazin H, Prochiantz A. Antennapedia homeobox peptide regulates neural morphogenesis. *Proc Natl Acad Sci U S A* 1991, **88**, 1864–1868.
184. Dissanayake S, Denny WA, Gamage S, Sarojini V. Recent developments in anticancer drug delivery using cell penetrating and tumor targeting peptides. *J Control Release* 2017, **250**, 62–76.
185. Heitz F, Morris MC, Divita G. Twenty years of cell-penetrating peptides: From molecular mechanisms to therapeutics. *Br J Pharmacol* 2009, **157**, 195–206.
186. Bitler BG, Schroeder JA. Anti-cancer therapies that utilize cell penetrating peptides. *Recent Pat Anticancer Drug Discov* 2010, **5**, 99–108.
187. Wender PA, Mitchell DJ, Pattabiraman K, Pelkey ET, Steinman L, Rothbard JB. The design, synthesis, and evaluation of molecules that enable or enhance cellular uptake: Peptoid molecular transporters. *Proc Natl Acad Sci U S A* 2000, **97**, 13003–13008.
188. Dietz GPH, Bähr M. Delivery of bioactive molecules into the cell: The Trojan horse approach. *Mol Cell Neurosci* 2004, **27**, 85–131.
189. Snyder EL, Dowdy SF. Recent advances in the use of protein transduction domains for the delivery of peptides, proteins and nucleic acids in vivo. *Expert Opin Drug Deliv* 2005, **2**, 43–51.
190. Chen H-CG, Chiou S-T, Zheng J-Y, Yang S-H, Lai S-S, Kuo T-Y. The nuclear localization signal sequence of porcine circovirus type 2 ORF2 enhances intracellular delivery of plasmid DNA. *Arch Virol* 2011, **156**, 803–815.
191. Alves ID, Bechara C, Walrant A, Zaltsman Y, Jiao C-Y, Sagan S. Relationships between membrane binding. Affinity and cell internalization efficacy of a cell-penetrating peptide: Penetratin as a case study. *PLoS One* 2011, **6**, e24096.
192. Juliano R, Alam MR, Dixit V, Kang H. Mechanisms and strategies for effective delivery of antisense and siRNA oligonucleotides. *Nucleic*

Acids Res 2008, **36**, 4158–4171.

193. Morris MC, Depollier J, Mery J, Heitz F, Divita G. A peptide carrier for the delivery of biologically active proteins into mammalian cells. *Nat Biotechnol* 2001, **19**, 1173–1176.
194. Deshayes S, Morris M, Heitz F, Divita G. Delivery of proteins and nucleic acids using a non-covalent peptide-based strategy. *Adv Drug Deliv Rev* 2008, **60**, 537–547.
195. Shi N-Q, Li Y, Zhang Y, Shen N, Qi L, Wang S-R, Qi X-R. Intelligent "peptide-gathering mechanical arm" tames wild "trojan-horse" peptides for the controlled delivery of cancer nanotherapeutics. *ACS Appl Mater Inter* 2017, **9**, 41767–41781.
196. Yoo J, Lee D, Gujrati V, Rejinold NS, Lekshmi KM, Uthaman S, Jeong C, Park I-K, Jon S, Kim Y-C. Bioreducible branched poly(modified nona-arginine) cell-penetrating peptide as a novel gene delivery platform. *J Control Release* 2017, **246**, 142–154.
197. Shaughnessy LM, Hoppe AD, Christensen KA, Swanson JA. Membrane perforations inhibit lysosome fusion by altering pH and calcium in Listeria monocytogenes vacuoles. *Cell Microbiol* 2006, **8**, 781–792.
198. Salt C, Lennox A, Takagaki M, Maguire J, Hosmane N. Boron and gadolinium neutron capture therapy. *Russ Chem Bull* 2004, **53**, 1871–1888.
199. Günther M, Lipka J, Malek A, Gutsch D, Kreyling W, Aigner A. Polyethylenimines for RNAi-mediated gene targeting in vivo and siRNA delivery to the lung. *Eur J Pharm Biopharm* 2011, **77**, 438–449.
200. Walker SJ. Boron neutron capture therapy: Principles and prospects. *Radiography* 1998, **4**, 211–219.
201. Kakudo T, Chaki S, Futaki S, Nakase I, Akaji K, Kawakami T, et al. Transferrin-modified liposomes equipped with a pH-sensitive fusogenic peptide: An artificial viral-like delivery system. *Biochemistry* 2004, **43**, 5618–5628.
202. Barth RF, Coderre JA, Vicente MGH, Blue TE. Boron neutron capture therapy of cancer: Current status and future prospects. *Clin Cancer Res* 2005, **11**, 3987–4002.
203. Berg K, Selbo P, Prasmickaite L, Tjelle T, Sandvig K, Moan D, et al. Photochemical internalization: A novel technology for delivery of macromolecules into cytosol. *Cancer Res* 1999, **59**, 1180–1183.
204. Hwang KC, Lai PD, Chiang C-S, Wang P-J, Yuan C-J. Neutron capture nuclei-containing carbon nanoparticles for destruction of cancer cells. *Biomaterials* 2010, **31**, 8419–8425.

205. Strømhaug PE, Berg TO, Berg K, Seglen PO. A novel method for the study of autophagy: Destruction of hepatocytic lysosomes, but not autophagosomes, by the photosensitizing porphyrin tetra(4-sulphonatophenyl)porphine. *Biochem J* 1997, **321**, 217–225.

206. Prasmickaite L, Hogset A, Berg K. Evaluation of different photosensitizers for use in photochemical gene transfection. *Photochem Photobiol* 2001, **73**, 388–395.

207. Hawthorne MF. New horizons for therapy based on the boron neutron capture reaction. *Mol Med Today* 1998, **4**, 174–181.

208. Lino MM, Simoes S, Pinho S, Ferreira L. Intracellular delivery of more than one protein with spatio-temporal control. *Nanoscale* 2017, **9**, 18668–18680.

209. Corvaglia S, Guarnieri D, Pompa PP. Boosting the therapeutic efficiency of nanovectors: Exocytosis engineering. *Nanoscale* 2017, **9**, 3757–3765.

210. Johnson PC, Wheat LJ, Cloud GA, Goldman M, et al. Safety and efficacy of liposomal amphotericin B compared with conventional amphotericin B for induction therapy of histoplasmosis in patients with AIDS. *Ann Intern Med* 2002, **137**, 105–109.

211. Martino RR. Efficacy, safety and cost-effectiveness of amphotericin B lipid complex (ABLC): A review of the literature. *Curr Med Res Opin* 2004, **20**, 485–504.

212. Szebeni J, Baranyi L, Savay S, Lutz HU, Jelezarova E, Bünger R, et al. The role of complement activation in hypersensitivity to pegylated liposomal doxorubicin (Doxil®). *J Liposome Res* 1999, **10**, 467–481.

213. Szebeni J. Complement activation-related pseudoallergy: a new class of drug-induced acute immune toxicity. *Toxicology* 2005, **216**, 106–121.

214. Szebeni J, Muggia F, Barenholz Y. Activation of complement by therapeutic liposomes and other lipid excipient-based therapeutic products: Prediction and prevention. *Adv Drug Deliv Rev* 2011, **63**, 1020–1030.

215. Cagnoni PJ. Liposomal amphotericin B versus conventional amphotericin B in the empirical treatment of persistently febrile neutropenic patients. *J Antimicrob Chemother* 2001, **49**, 81–86.

216. Walsh TJ, Finberg RW, Arndt C, Hiemenz J, Schwartz C, Bodensteiner D, et al. Liposomal amphotericin B for empirical therapy in patients with persistent fever and neutropenia. National Institute of Allergy and Infectious Diseases Mycoses Study Group. *N Engl J Med* 1999, **340**, 764–771.

217. Moen MD, Lyseng-Williamson KA, Scott LJ. Liposomal amphotericin B: A review of its use as empirical therapy in febrile neutropenia and in the treatment of invasive fungal infections. *Drugs* 2009, **69**, 361–392.

218. Bowden R, Chandrasekar P, White M, Li X, Pietrelli L, Gurwith M, et al. A double-blind, randomized, controlled trial of amphotericin B colloidal dispersion versus amphotericin B for treatment of invasive aspergillosis in immunocompromised patients. *Clin Infect Dis* 2002, **35**, 359–366.

219. Working PK. Amphotericin B, colloidal dispersion. Pre-clinical review. *Chemotherapy* 1999, **45**, 15–26.

220. Petre CE, Dittmer DP. Liposomal daunorubicin as treatment for Kaposi's sarcoma. *Int J Nanomed* 2007, **2**, 277–288.

221. Chang H-I, Yeh M-K. Clinical development of liposome-based drugs: Formulation, characterization, and therapeutic efficacy. *Int J Nanomed* 2011, **7**, 49–60.

222. Sadava DD, Coleman AA, Kane SES. Liposomal daunorubicin overcomes drug resistance in human breast, ovarian and lung carcinoma cells. *J Liposome Res* 2002, **12**, 301–309.

223. Harris L, Batist G, Belt R, Rovira D, Navari R, Azarnia N, et al. Liposome-encapsulated doxorubicin compared with conventional doxorubicin in a randomized multicenter trial as first-line therapy of metastatic breast carcinoma. *Cancer* 2002, **94**, 25–36.

224. Batist G, Ramakrishnan G, Lee LW. Reduced cardiotoxicity and preserved antitumor efficacy of liposome-encapsulated doxorubicin and cyclophosphamide compared with conventional doxorubicin and cyclophosphamide in a randomized, multicenter trial of metastatic breast cancer. *J Clin Oncol* 2001, **19**, 1444–1454.

225. O'Brien M, Wigler N, Inbar M, Rosso R, Grischke E, Santoro A, et al. Reduced cardiotoxicity and comparable efficacy in a phase III trial of pegylated liposomal doxorubicin HCl (CAELYX (TM)/Doxil (R)) versus conventional doxorubicin for first-line treatment of metastatic breast cancer. *Ann Oncol* 2004, **15**, 440–449.

226. Yuan Y, Orlow S, Curtin J, Downey A, Muggia F. Pegylated liposomal doxorubicin (PLD): Enhanced skin toxicity in areas of vitiligo. *Ecancermedicalscience* 2008, **2**, 111.

227. Boman NLN, Masin DD, Bally MBM. Liposomal vincristine which exhibits increased drug retention and increased circulation longevity cures mice bearing P388 tumors. *Cancer Res* 1994, **54**, 2830–2833.

228. Mayer LDL, Masin DD, Bally MBM. Pharmacology of liposomal vincristine in mice bearing L1210 ascitic and B16/BL6 solid tumours. *Br J Cancer* 1995, **71**, 482–488.

229. Garcia-Marco JAJ, Panizo CC, Vela JAJG. Efficacy and safety of liposomal cytarabine in lymphoma patients with central nervous system involvement from lymphoma. *Cancer* 2009, **115**, 1892–1898.

230. Gambling D, Hughes T, Martin G, Horton W, Manvelian G. A comparison of Depodur (TM), a novel, single-dose extended-release epidural morphine, with standard epidural morphine for pain relief after lower abdominal surgery. *Anesth Analg* 2005, **100**, 1065–1074.

231. Zeidan A, Wang ES, Wetzler M. Pegasparaginase: Where do we stand? *Expert Opin Biol Ther* 2009, **9**, 111–119.

232. Pradhananga SS, Wilkinson II, Ross RJMR. Pegvisomant: Structure and function. *J Mol Endocrinol* 2002, **29**, 11–14.

233. Garay R, Labaune J. Immunogenicity of polyethylene glycol (PEG). *Open Conf Proc J* 2011, **2**, 104–107.

234. Molineux G. The design and development of pegfilgrastim (PEG-rmetHuG-CSF, neulasta®). *Curr Pharm Des* 2004, **10**, 1235–1244.

235. Sanchez-Fructuoso A, Guirado L, Ruiz JC, Torregrosa V, Gonzalez E, Suarez ML, et al. Anemia control in kidney transplant patients treated with methoxy polyethylene glycolepoetin beta (Mircera): The Anemiatrans Group. *Transplant Proc* 2010, **42**, 2931–2934.

236. Szebeni J. Hemocompatibility testing for nanomedicines and biologicals: Predictive assays for complement mediated infusion reactions. *Eur J Nanomed* 2012, **4**, 33–53.

237. Silverman JA, Deitcher SR. Marqibo® (vincristine sulfate liposome injection) improves the pharmacokinetics and pharmacodynamics of vincristine. *Cancer Chemother Pharmacol* 2013, **71**, 3, 555–564.

238. Shah NN, Merchant MS, Cole DE, Jayaprakash N, Bernstein D, Delbrook C, Richards K, Widemann BC, Wayne AS. Vincristine sulfate liposomes injection (VSLI, Marqibo®): Results from a phase I study in children, adolescents, and young adults with refractory solid tumors or leukemias. *Pediatr Blood Cancer* 2016, **63**, 6, 997–1005.

239. Hamad I, Hunter AC, Moghimi SM. Poly(ethylene glycol)s generate complement activation products in human serum through increased alternative pathway turnover and a MASP-2-dependent process. *Mol Immunol* 2008, **46**, 225–232.

240. Foster GR. Pegylated interferons for the treatment of chronic hepatitis C: Pharmacological and clinical differences between peginterferon-alpha-2a and peginterferon-alpha-2b. *Drugs* 2010, **70**, 147–165.

241. Wang-Gillam A, Li CP, Bodoky G, Dean A, Shan YS, Jameson G, Macarulla T, Lee KH, Cunningham D, Blanc JF, Hubner RA, Chiu CF, Schwartsmann G, Siveke JT, Braiteh F, Moyo V, Belanger B, Dhindsa N, Bayever E, Von Hoff DD, Chen LT, NAPOLI-1 Study Group. Nanoliposomal irinotecan with fluorouracil and folinic acid in metastatic pancreatic cancer after previous gemcitabine-based therapy (NAPOLI-1): A global, randomised, open-label, phase 3 trial. *Lancet* 2016, **387**, 10018, 545–557.

242. Zampros I, Praidou A, Brazitikos P, Ekonomidis P, Androudi S. Antivascular endothelial growth factor agents for neovascular age-related macular degeneration. *J Ophthalmol* 2012, **2012**, 319728.

243. Garratty G. Modulating the red cell membrane to produce universal/stealth donor red cells suitable for transfusion. *Vox Sang* 2007, **94**, 87–95.

244. Armstrong JK, Hempel G, Koling S, Chan LS, Fisher T, Meiselman HJ, et al. Antibody against poly(ethylene glycol) adversely affects PEG-asparaginase therapy in acute lymphoblastic leukemia patients. *Cancer* 2007, **110**, 103–111.

245. Mészáros T, Csincsi ÁI, Uzonyi B, Hebecker M, Fülöp TG, Erdei A, Szebeni J, Józsi M. Factor H inhibits complement activation induced by liposomal and micellar drugs and the therapeutic antibody rituximab in vitro. *Nanomedicine* 2016, **12**, 4, 1023–1031.

246. Hu X, Miller L, Richman S, Hitchman S, Glick G, Liu S, Zhu Y, Crossman M, Nestorov I, Gronke RS, Baker DP, Rogge M, Subramanyam M, Davar G. A novel PEGylated interferon beta-1a for multiple sclerosis: Safety, pharmacology, and biology. *J Clin Pharmacol* 2012, **52**, 6, 798–808.

247. Ing M, Gupta N, Teyssandier M, Maillere B, Pallardy M, Delignat S, Lacroix-Desmazes S, ABIRISK consortium. Immunogenicity of long-lasting recombinant factor VIII products. *Cell Immunol* 2016, **301**, 40–48.

248. Nesbitt A, Fossati G, Bergin M, Stephens P, Stephens S, Foulkes R, et al. Mechanism of action of certolizumab pegol (CDP870): In vitro comparison with other anti-tumor necrosis factor alpha agents. *Inflamm Bowel Dis* 2007, **13**, 1323–1332.

249. Veronese FM, Mero A, Pasut G. Protein PEGylation, basic science and biological applications. *PEGylated Protein Drugs: Basic Science and Clinical Applications. Milestones in Drug Therapy*. Switzerland: Birkhäuser Verlag, 2009, pp. 11–31.

250. Spaia S. Phosphate binders: Sevelamer in the prevention and treatment of hyperphosphataemia in chronic renal failure. *Hippokratia* 2010, **15**, 22–26.

251. Simberg D, Park J-H, Ruoslahti E. Differential proteomics analysis of the surface heterogeneity of dextran iron oxide nanoparticles and the implications for their in vivo clearance. *Biomaterials* 2009, **30**, 3926–3933.

252. Moore A, Weissleder R, Bogdanov A. Uptake of dextran-coated monocrystalline iron oxides in tumor cells and macrophages. *J Magn Reson Imaging* 1997, **7**, 1140–1145.

253. Gradishar WJW, Tjulandin SS, Davidson NN, Shaw HH, Desai NN, Bhar PP, et al. Phase III trial of nanoparticle albumin-bound paclitaxel compared with polyethylated castor oil-based paclitaxel in women with breast cancer. *J Clin Oncol* 2005, **23**, 7794–803.

254. Ibrahim NK, Desai N, Legha S, Soon-Shiong P, Theriault RL, Rivera E, et al. Phase I and pharmacokinetic study of ABI-007, a cremophor-free, protein-stabilized, nanoparticle formulation of paclitaxel. *Clin Cancer Res* 2002, **8**, 1038–144.

255. Junghanns J-UAH, Müller RH. Nanocrystal technology, drug delivery and clinical applications. *Int J Nanomed* 2008, **3**, 295–309.

256. Webster R, Elliott V, Park BK, Walker D, Hankin M, Taupin P. PEG and PEG conjugates toxicity: Towards an understanding of the toxicity of PEG and its relevance to PEGylated biologicals. *PEGylated Protein Drugs: Basic Science and Clinical Applications. Milestones in Drug Therapy*. Switzerland: Birkhäuser Verlag, 2009, pp. 127–146.

257. Szebeni J, Bedöcs P, Rozsnyay Z, Weiszhaár Z, Urbanics R, Rosivall L, et al. Liposome-induced complement activation and related cardiopulmonary distress in pigs: Factors promoting reactogenicity of Doxil and AmBisome. *Nanomedicine* 2012, **8**, 176–184.

258. Fülöp T, Nemes R, Mészáros T, Urbanics R, Kok RJ, Jackman JA, Cho NJ, Storm G, Szebeni J. Complement activation in vitro and reactogenicity of low-molecular weight dextran-coated SPIONs in the pig CARPA model: Correlation with physicochemical features and clinical information. *J Control Release* 2017, **270**, 268–274.

259. Chanan-Khan A, Szebeni J, Savay S, Liebes L, Rafique NM, Alving CR, et al. Complement activation following first exposure to pegylated liposomal doxorubicin (Doxil): Possible role in hypersensitivity reactions. *Ann Oncol* 2003, **14**, 1430–1437.

260. Hodgman M, Holland MG, Englich U, Wojcik SM, Grant WD, Leitner E. The influence of polyethylene glycol solution on the dissolution rate of sustained release morphine. *J Med Toxicol* 2016, **12**, 4, 391–395.

261. Nussbaum AM, Stroup TS. Paliperidone palmitate for schizophrenia. *Cochrane Database Syst Rev* 2012, **13**, 6, CD008296.

262. Ryanodex - a new dantrolene formulation for malignant hyperthermia. *Med Lett Drugs Ther* 2015, **57**, 1472, 100.

263. Richter AW, Akerblom E. Polyethylene-glycol reactive antibodies in man titer distribution in allergic patients treated with monomethoxy polyethylene-glycol modified allergens or placebo, and in healthy blood-donors. *Int Arch Allergy Appl Immunol* 1984, **74**, 36–39.

264. Tziomalos K, Athyros VG. Fenofibrate: A novel formulation (Triglide (TM)) in the treatment of lipid disorders: A review. *Int J Nanomed* 2006, **1**, 129–147.

265. Mészáros T, Csincsi ÁI, Uzonyi B, Hebecker M, Fülöp TG, Erdei A, Szebeni J, Józsi M. Factor H inhibits complement activation induced by liposomal and micellar drugs and the therapeutic antibody rituximab in vitro. *Nanomedicine* 2016, **12**, 4, 1023–1031.

266. Zhang B, Li Q, Shi C, Zhang X. Drug-induced pseudoallergy: A review of the causes and mechanisms. *Pharmacology* 2018, **101**, 104–110.

267. Hunziker P. "Knowledge-based (personalized) medicine" instead of 'evidence-based (cohort) medicine' – applying nanoscience and computational science to create an effective, safe, curative and affordable medicine of the future. *Eur J Nanomed* 2012, 1–2.

268. Bhirde AA, Patel V, Gavard J, Zhang G, Sousa AA, Masedunskas A, Leapman RD, Weigert R, Gutkind JS, Rusling JF. Targeted killing of cancer cells in vivo and in vitro with EGF-directed carbon nanotube-based drug delivery. *ACS Nano* 2009, **3**, 2, 307–316.

269. Pistone A, Iannazzo D, Ansari S, Milone C, Salamò M, Galvagno S, Cirmi S, Navarra M. Tunable doxorubicin release from polymer-gated multiwalled carbon nanotubes. *Int J Pharm* 2016, **515**, 1–2, 30–36.

270. Maier-Hauff K, Ulrich F, Nestler D, Niehoff H, Wust P, Thiesen B, Orawa H, Budach V, Jordan A. Efficacy and safety of intratumoral thermotherapy using magnetic iron-oxide nanoparticles combined with

external beam radiotherapy on patients with recurrent glioblastoma multiforme. *J Neurooncol* 2011, **103**, 2, 317–324.

271. Sohn JH, Choi HJ, Lee JT, Lee JD, Kim JH, Moon YM, Park K, Park KB, Kim E, Yoo NC. Phase II study of transarterial holmium-166-chitosan complex treatment in patients with a single, large hepatocellular carcinoma. *Oncology* 2009, **76**, 1, 1–9.

272. Veiman KL, Künnapuu K, Lehto T, Kiisholts K, Pärn K, Langel Ü, Kurrikoff K. PEG shielded MMP sensitive CPPs for efficient and tumor specific gene delivery in vivo. *J Control Release* 2015, **209**, 238–247.

Chapter 2

Nanocarriers as an Emerging Platform for Cancer Therapy

Dan Peer,[a,*] Jeffrey M. Karp,[b,c,*] Seungpyo Hong,[d,*] Omid C. Farokhzad,[e] Rimona Margalit,[f] and Robert Langer[c,d,*]

[a]*Laboratory of Precision NanoMedicine, School of Molecular Cell Biology and Biotechnology, Tel Aviv University, Tel Aviv 69978, Israel*
[b]*HST Center for Biomedical Engineering, Department of Medicine, Brigham and Women's Hospital, Harvard Medical School, Cambridge, Massachusetts 02139, USA*
[c]*Harvard-MIT Division of Health Sciences and Technology, Massachusetts Institute of Technology, Cambridge, Massachusetts 02139, USA*
[d]*Department of Chemical Engineering, Massachusetts Institute of Technology, Cambridge, Massachusetts 02139, USA*
[e]*Laboratory of Nanomedicine and Biomaterials and Department of Anesthesiology, Brigham and Women's Hospital, Harvard Medical School, Boston, Massachusetts 02115, USA*
[f]*Department of Biochemistry, George S. Wise Faculty of Life Sciences, and the Center for Nanoscience and Nanotechnology, Tel Aviv University, Tel Aviv 69978, Israel*
rlanger@mit.edu

Cancer remains one of the world's most devastating diseases, with more than 10 million new cases every year [1]. However, mortality has decreased in the past two years [2] owing to

---

*These authors contributed equally to this review.
Reproduced with permission from, *Nature Nanotechnology*, **2**, 751–760, 2007.

Nano-Enabled Medical Applications
Edited by Lajos P. Balogh

ISBN 978-981-4800-32-7 (Hardcover), 978-0-429-39903-9 (eBook)
www.jennystanford.com

better understanding of tumour biology and improved diagnostic devices and treatments. Current cancer treatments include surgical intervention, radiation and chemotherapeutic drugs, which often also kill healthy cells and cause toxicity to the patient. It would therefore be desirable to develop chemotherapeutics that can either passively or actively target cancerous cells. Passive targeting exploits the characteristic features of tumour biology that allow nanocarriers to accumulate in the tumour by the enhanced permeability and retention (EPR) effect [2]. Passively targeting nanocarriers first reached clinical trials in the mid-1980s, and the first products, based on liposomes and polymer–protein conjugates, were marketed in the mid-1990s. Later, therapeutic nanocarriers based on this strategy were approved for wider use (Table 2.1) and methods of further enhancing targeting of drugs to cancer cells were investigated. Active approaches achieve this by conjugating nanocarriers containing chemotherapeutics with molecules that bind to overexpressed antigens or receptors on the target cells. Recent reviews provide perspective on the use of nanotechnology as a fundamental tool in cancer research and nanomedicine [3, 4]. Here we focus on the potential of nanocarriers and molecules that can selectively target tumours, and highlight the challenges in translating some of the basic research to the clinic.

2.1 Passive and Active Targeting

Nanocarriers encounter numerous barriers en route to their target, such as mucosal barriers and non-specific uptake [5, 6]. To address the challenges of targeting tumours with nanotechnology, it is necessary to combine the rational design of nanocarriers with the fundamental understanding of tumour biology (Box 2.1).

General features of tumours include leaky blood vessels and poor lymphatic drainage. Whereas free drugs may diffuse non-specifically, a nanocarrier can extravasate (escape) into the tumour tissues via the leaky vessels by the EPR effect [7] (Fig. 2.1). The increased permeability of the blood vessels in tumours is characteristic of rapid and defective angiogenesis (formation of new blood vessels

Table 2.1 Representative examples of nanocarrier-based drugs on the market

Compound	Commercial name	Nanocarrier	Indications
Styrene maleic anhydride-neocarzinostatin (SMANCS)	Zinostatin/Stimalmer	Polymer-protein conjugate	Hepatocellular carcinoma
PEG-L-asparaginase	Oncaspar	Polymer-protein conjugate	Acute lymphoblastic leukemia
PEG-granulocyte colony-stimulating factor (G-CSF)	Neulasta/PEGfilgrastim	Polymer-protein conjugate	Prevention of chemotherapy-associated neutropenia
IL2 fused to diphtheria toxin	Ontak (Denilelukin diftitox)	Immunotoxin (fusion protein)	Cutaneous T-cel l lymphoma
Anti-CD33 antibody conjugated to calicheamicin	Mylotarg	Chemo-immunoconjugate	Acute myelogenous leukemia
Anti-CD20 conjugated to yttrium-90 or indium-111	Zevalin	Radio-immunoconjugate	Relapsed or refractory, low-grade, follicular, or transformed non-Hodgkin's lymphoma
Anti-CD20 conjugated to iodine-1 31	Bexxar	Radio-immunoconjugate	Relapsed or refractory, low-grade, follicular, or transformed non-Hodgkin's lymphoma
Daunorubicin	DaunoXome	Liposomes	Kaposi's sarcoma
Doxorubicin	Myocet	Liposomes	Combinational therapy of recurrent breast cancer, ovarian cancer, Kaposi's sarcoma
Doxorubicin	Doxil/Caelyx	PEG-liposomes	Refractory Kaposi's sarcoma, recurrent breast cancer, ovarian cancer
Vincristine	Onco TCS	Liposomes	Relapsed aggressive non-Hodgkin's lymphoma (NHL)
Paclitaxel	Abraxane	Albumin-bound paclitaxel nanoparticles	Metastatic breast cancer

Box 2.1 Rational design of nanocarriers for cancer therapy

Nanocarriers can offer many advantages over free drugs. They:

- protect the drug from premature degradation;
- prevent drugs from prematurely interacting with the biological environment;
- enhance absorption of the drugs into a selected tissue (for example, solid tumour);
- control the pharmacokinetic and drug tissue distribution profile;
- improve intracellular penetration.

For rapid and effective clinical translation, the nanocarrier should:

- be made from a material that is biocompatible, well characterized, and easily functionalized;
- exhibit high differential uptake efficiency in the target cells over normal cells (or tissue);
- be either soluble or colloidal under aqueous conditions for increased effectiveness;
- have an extended circulating half-life, a low rate of aggregation, and a long shelf life.

from existing ones). Furthermore, the dysfunctional lymphatic drainage in tumours retains the accumulated nanocarriers and allows them to release drugs into the vicinity of the tumour cells. Experiments using liposomes of different mean size suggest that the threshold vesicle size for extravasation into tumours is $\sim$400 nm [8], but other studies have shown that particles with diameters $<$200 nm are more effective [5, 8–10].

Although passive targeting approaches form the basis of clinical therapy, they suffer from several limitations. Ubiquitously targeting cells within a tumour is not always feasible because some drugs cannot diffuse efficiently and the random nature of the approach makes it difficult to control the process. This lack of control may induce multiple-drug resistance (MDR)—a situation where chemotherapy treatments fail patients owing to resistance of cancer

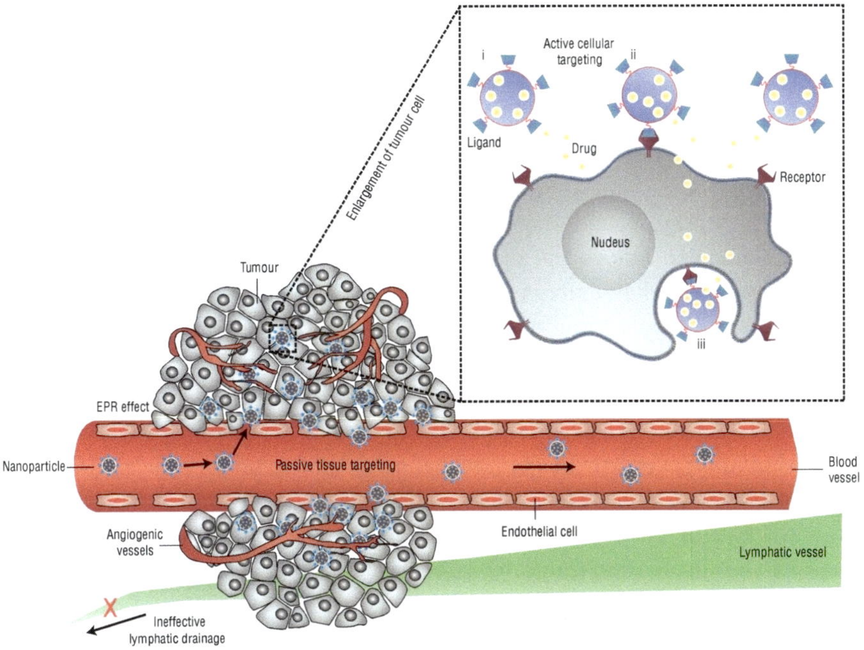

Figure 2.1 Schematic representation of different mechanisms by which nanocarriers can deliver drugs to tumours. Polymeric nanoparticles are shown as representative nanocarriers (circles). Passive tissue targeting is achieved by extravasation of nanoparticles through increased permeability of the tumour vasculature and ineffective lymphatic drainage (EPR effect). active cellular targeting (inset) can be achieved by functionalizing the surface of nanoparticles with ligands that promote cell-specific recognition and binding. the nanoparticles can (i) release their contents in close proximity to the target cells; (ii) attach to the membrane of the cell and act as an extracellular sustained-release drug depot; or (iii) internalize into the cell.

cells towards one or more drugs. MDR occurs because transporter proteins that expel drugs from cells are overexpressed on the surface of cancer cells [4, 11, 12]. Expelling drugs inevitably lowers the therapeutic effect and cancer cells soon develop resistance to a variety of drugs. The passive strategy is further limited because certain tumours do not exhibit the EPR effect, and the permeability of vessels may not be the same throughout a single tumour [13].

One way to overcome these limitations is to programme the nanocarriers so they actively bind to specific cells after extravasation. This binding may be achieved by attaching targeting agents such as ligands—molecules that bind to specific receptors on the cell surface—to the surface of the nanocarrier by a variety of conjugation chemistries [9]. Nanocarriers will recognize and bind to target cells through ligand–receptor interactions, and bound carriers are internalized before the drug is released inside the cell (Fig. 2.1). In general, when using a targeting agent to deliver nanocarriers to cancer cells, it is imperative that the agent binds with high selectivity to molecules that are uniquely expressed on the cell surface. Other important considerations are outlined below.

To maximize specificity, a surface marker (antigen or receptor) should be overexpressed on target cells relative to normal cells. For example, to efficiently deliver liposomes to B-cell receptors using the anti-CD19 monoclonal antibody (mAb), the density of receptors should be in the range of 10^4–10^5 copies per cell. Those with lower density are less effectively targeted [14]. In a breast cancer model, a receptor density of 10^5 copies of ErbB2 receptors per cell was necessary to improve the therapeutic efficacy of an anti-ErbB2-targeted liposomal doxorubicin relative to its non-targeted counterpart [15].

The binding of certain ligands to their receptors may cause receptor-mediated internalization, which is often necessary if nanocarriers are to release drugs inside the cell [16–18]. For example, a more significant therapeutic outcome was achieved when immunoliposomes targeted to human blood cancer (B-cell lymphoma) were labelled with an internalizing anti-CD19 ligand rather than a non-internalizing anti-CD20 ligand [19]. In contrast, targeting nanocarriers to non-internalizing receptors may sometimes be advantageous in solid tumours owing to the bystander effect, where cells lacking the target receptor can be killed through drug release at the surface of the neighbouring cells, where carriers can bind [20].

It is generally known that higher binding affinity increases targeting efficacy. However, for solid tumours, there is evidence that high binding affinity can decrease penetration of nanocarriers due to a 'binding-site barrier', where the nanocarrier binds to its target so strongly that penetration into the tissue is prevented [16, 21]. In addition to enhanced affinity, multivalent binding effects (or avidity) may also be used to improve targeting. The collective binding in a multivalent interaction is much stronger than monovalent binding. For example, dendrimer nanocarriers conjugated to 3–15 folate molecules showed a 2,500–170,000-fold enhancement in dissociation constants (K_D) over free folate when attaching to folate-binding proteins immobilized on a surface. This was attributed to the avidity of the multiple folic acid groups on the periphery of the dendrimers [22].

2.2 Types of Targeting Agents

Targeting agents can be broadly classified as proteins (mainly antibodies and their fragments), nucleic acids (aptamers), or other receptor ligands (peptides, vitamins, and carbohydrates).

Targeting cancer with a mAb was described by Milstein in 1981 [23]. Over the past two decades, the feasibility of antibody-based tissue targeting has been clinically demonstrated (reviewed in refs. [24, 25] with 17 different mAbs approved by the US Food and Drug Administration (FDA) [26]. The mAb rituximab (Rituxan) was approved in 1997 for treatment of patients with non-Hodgkin's lymphoma—a type of cancer that originates in lymphocytes [27]. A year later, Trastuzumab (Herceptin), an anti-HER2 mAb that binds to ErbB2 receptors, was approved for the treatment of breast cancer [28]. The first angiogenesis inhibitor for treating colorectal cancer, Bevacizumab (Avastin), an anti-VEGF mAb that inhibits the factor responsible for the growth of new blood vessels, was approved in 2004 [29]. Today, over 200 delivery systems based on antibodies or their fragments are in preclinical and clinical trials [16, 30]. Recent developments in the field of antibody engineering have resulted in the production of antibodies that contain animal and human origins such chimeric mAbs, humanized mAbs (those with a greater human contribution), and antibody fragments.

Antibodies may be used in their native state or as fragments for targeting (Fig. 2.2a). However, use of whole mAbs is advantageous because the presence of two binding sites (within a single antibody)

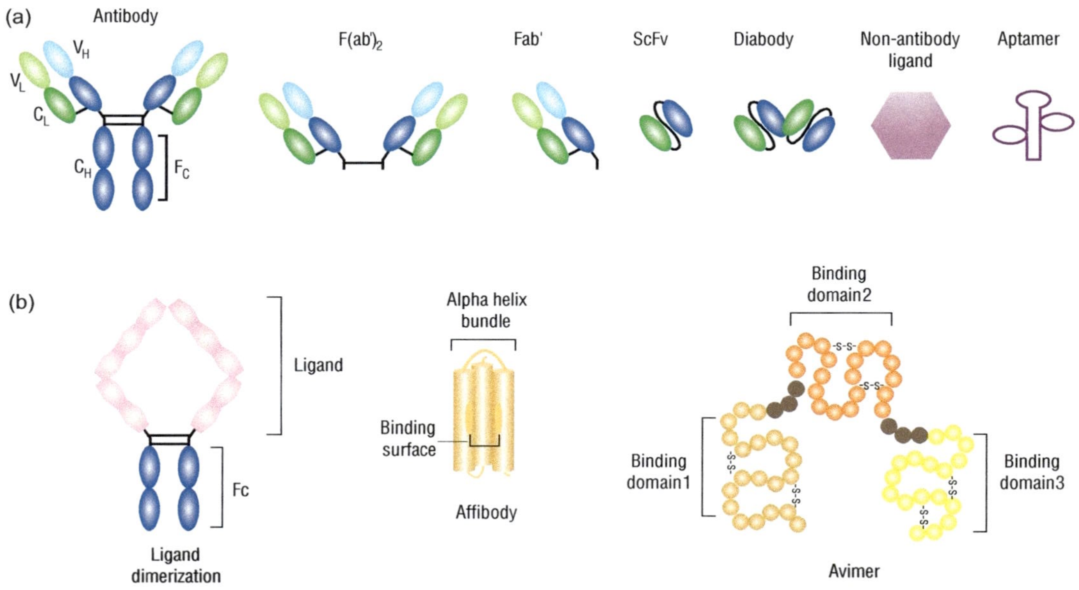

Figure 2.2 Common targeting agents and ways to improve their affinity and selectivity. (a) the panel shows a variety of targeting molecules such as a monoclonal antibody or antibodies' fragments, non-antibody ligands, and aptamers. the antibody fragments f(ab′)$_2$ and Fab are generated by enzymatic cleavage whereas the Fab, scfv, and bivalent scfv (diabody) fragments are created by molecular biology techniques. V_H: variable heavy chain; V_L: variable light chain; C_H: constant heavy chain; C_L: constant light chain. non-antibody ligands include vitamins, carbohydrates, peptides, and other proteins. aptamers can be composed of either DNA or RNA. (b) Affinity and selectivity can be increased through ligand dimerization or by screening for conformational-sensitive targeting agents such as affibodies, avimers and nanobodies, as well as intact antibodies and their fragments.

gives rise to a higher binding avidity. Furthermore, when immune cells bind to the Fc portion of the antibody, a signalling cascade is initiated to kill the cancer cells. However, the Fc domain of an intact mAb can also bind to the Fc receptors on normal cells, as occurs with macrophages. This may lead to increased immunogenicity—the ability to evoke an immune response—and liver and spleen uptake of the nanocarrier. An additional advantage of whole/intact antibodies is their ability to maintain stability during long-term storage. Although antibody fragments including antigen-binding fragments (Fab), dimers of antigen-binding fragments ($F(ab')_2$), single-chain fragment variables (scFv) and other engineered fragments are less stable than whole antibodies, they are considered safer when injected systemically owing to reduced non-specific binding [16, 30]. To rapidly select antibodies or their fragments that bind to and internalize within cancer cells, phage display libraries that involve a high throughput approach may be used [31, 32]. This method generates a multitude of potentially useful antibodies that bind to the same target cells but to different epitopes (a part of a macromolecule that is recognized by antibodies; one receptor may have several epitopes that will be recognized by multiple antibodies). For example, through a selective process, scFv antibodies have been identified for superior binding and internalization properties for prostate cancer cells [33].

It is possible to increase the efficacy of antibodies by conjugating a therapeutic agent directly to it for targeted delivery. For example, in 2000, the chemotherapeutic drug, calicheamicin, which is conjugated with the anti-CD33 antibody (marketed under the trade name Mylotarg), was the first clinically approved formulation that targets cancerous cells. Others include Zevalin and Bexxar, which use anti-CD20 antibodies to target radioisotopes to cancer cells (Table 2.1). Although the efficacy of these therapies has been proven, lethal side effects have been observed, likely due to non-specific binding [34] between the targeting agent and non-target moieties on the cell surface. Another reason could be the interaction of the targeting agent with its target expressed on non-cancerous cells. For example, BR96-doxorubicin—an immunoconjugate linked with doxorubicin and comprising an antibody that targets and binds to the Lewis-Y antigen (expressed on 75% of all breast cancers)—demonstrated significant anti-tumour activity in mouse tumour models. BR96-doxorubicin showed lower toxicity than that resulting from doxorubicin alone and it was efficacious in these animal models [35]. However, in dogs, an acute enteropathy (pathology of the intestine) was observed presumably due to binding of the conjugate to Lewis-Y-related antigens expressed by non-targeted gastrointestinal epithelial cells. In Phase II human clinical studies, BR96-doxorubicin immunoconjugates had limited anti-tumour activity and caused severe gastrointestinal toxicity, leading to termination of the study [36].

Although using genomics and proteomics technology to choose appropriate targets is an active area of research, to date no clinically effective targets have been identified. Creating new technologies to enhance selectivity and targeting efficacy with existing targets seem more promising. For example, fusion proteins can be created by combining two or more genes to produce a new protein with desired properties. Antibodies can be engineered so they bind to their target with high affinity, and using molecular biology techniques, it is possible to design protein-based ligand mimetics based on the structure of a receptor. Dimerization of proteins or peptides can increase ligand affinity through divalency—two simultaneous binding events, usually involving concurrent binding of a protein or a peptide to the two Fc domains of an antibody (Fig. 2.2b). For example, dimerization of a low-affinity scFv (also known as diabody) against the ErbB2, led to enhanced tumour localization in a mouse tumour model [37].

It is also possible to increase binding affinity and selectivity to cell surface targets by engineering proteins that detect a specific conformation of a target receptor. In a recent in vivo study using a fusion protein consisting of an scFv antibody fragment to target and deliver small interfering RNA (siRNA) to lymphocytes—a type of white blood cell—a 10,000-fold increased affinity for the target receptor, integrin LFA-1, was observed [18]. Integrin LFA-1 is usually present in a low-affinity non-adhesive form on naïve leukocytes (white blood cells that are not activated by cancer cells or pathogens that enter the body), but converts to the high-affinity adhesive form through conformational changes on activation of the immune system. Therefore, targeting the high-affinity form of LFA-1 enables drugs to be selectively delivered to the activated and adhesive leukocytes. New classes of targeting molecules can be engineered to target specific conformations. These include small protein domains, known as affibodies, that can be engineered to bind specifically to different target proteins in a conformational-sensitive manner. Other small proteins that act like antibodies—called avimers—are used to bind selectively to target receptors through multivalent effects. Nanobodies, which are heavy-chain antibodies engineered to one tenth of the size of an intact antibody with a missing light chain, have been used to bind to carcinoembryonic antigen (CEA), a protein used as a tumour marker [38–40] (Fig. 2.2b).

In addition to the rational design of antibodies, high-throughput approaches have been used to generate targeting agents such as aptamers, which are short single-stranded DNA or RNA oligonucleotides selected in vitro from a large number of random sequences ($\sim 10^{14}$–10^{15}). Aptamers are selected to bind to a wide variety of targets, including intracellular proteins, transmembrane proteins, soluble proteins, carbohydrates, and small molecule drugs. Several aptamers have also been developed to bind specifically to receptors on cancer cells, and thus may be suitable for nanoparticle-aptamer conjugate therapy [41]. For example, docetaxel (Dtxl)-encapsulated nanoparticles whose surface is modified with an aptamer that targets the antigen on the surface of prostate cancer cells, were delivered with high selectivity and efficacy in vivo [42].

Growth factor or vitamin interactions with cancer cells represent a commonly used targeting strategy, as cancer cells often overexpress the receptors for nutrition to maintain their fast-growing metabolism. Epidermal growth factor (EGF) has been shown to block and reduce tumour expression of the EGF receptor, which is overexpressed in a variety of tumour cells such as breast and tongue cancer [43]. Additionally, based on the same idea, the vitamin folic acid (folate) has also been used for cancer targeting because folate receptors (FRs) are frequently overexpressed in a range of tumour cells including ovarian, endometrial and kidney cancer [44]. Transferrin (Tf) interacts with Tf receptors (TfRs), which are overexpressed on a variety of tumour cells (including pancreatic, colon, lung, and bladder cancer) owing to increased metabolic rates [45]. Direct coupling of these targeting agents to nanocarriers containing chemotherapies such as drugs has improved intracellular delivery and therapeutic outcome in animal tumour models [46–48].

One challenge with targeting receptors whose expression correlates with metabolic rate, such as folate and Tf, is that these receptors are also expressed in fast-growing healthy cells such as fibroblasts, epithelial and endothelial cells. This could lead to non-specific targeting and subsequently decrease the effectiveness of the drug and increase toxicity [49].

The use of peptides as targeting agents—including arginine-glycine-aspartic acid (RGD), which is the ligand of the cell adhesion integrin $\alpha_v\beta_3$ on endothelial cells—results in increased intracellular drug delivery in different murine tumour models [50, 51]. However, RGD also binds to other integrins such as $\alpha_5\beta_1$ and $\alpha_4\beta_1$ and therefore is not specific to cancer cells, which may limit its use. In addition to cell surface antigens, extracellular matrices (ECMs) overexpressed in tumours, such as heparin sulphate, chondroitin sulphate, and hyaluronan (HA), may also serve as effective targets for specific ECM receptors [52, 53]. Coating liposomes with HA improves circulation time and enhances targeting to HA receptor-expressing tumours in vivo [54, 55].

2.3 The Arsenal of Nanocarriers

Nanocarriers are nanosized materials (diameter 1–100 nm) that can carry multiple drugs and/or imaging agents. Owing to their high surface-area-to-volume ratio, it is possible to achieve high ligand density on the surface for targeting purposes. Nanocarriers can also be used to increase local drug concentration by carrying the drug within and control-releasing it when bound to the targets. Currently, natural and synthetic polymers and lipids are typically used as drug delivery vectors; clinically approved formulations are listed in Table 2.1. The family of nanocarriers includes polymer conjugates, polymeric nanoparticles, lipid-based carriers such as liposomes and micelles, dendrimers, carbon nanotubes, and gold nanoparticles, including nanoshells and nanocages (Fig. 2.3a). These nanocarriers have been explored for a variety of applications such as drug delivery, imaging, photothermal ablation of tumours, radiation sensitizers, detection of apoptosis, and sentinel lymph-node mapping [3, 4, 56] (Table 2.2).

Table 2.2 Examples of nano-based platforms and their current stage of development for use in cancer therapy

Type of carrier and mean diameter (nm)	Drug entrapped or linked	Current stage of development	Type of cancer (for clinical trials)	References
Polymer-drug conjugates (6–15)	Doxorubicin, Paclitaxel, Camptothecin, Platinate, TNP-470	12 products under clinical trials (Phases I–III) and in vivo	Various tumours	Reviewed in 3, 61
Liposomes (both PEG and non-PEG coated) (85–100)	Lurtotecan, platinum compounds, Annamycin	Several products in clinical trials (Phases I–III) and in vivo	Solid tumours, renal cell carcinoma, mesothelioma, ovarian and acute lymphoblastic leukaemia	Reviewed in 9
Polymeric nanoparticles (50–200)	Doxorubicin, Paclitaxel, platinum-based drugs, Docetaxel	Several products are in clinical trials (Phases I–III) and in vivo	Adenocarcinoma of the oesophagus, metastatic breast cancer and acute lymphoblastic leukemia	5, 91, 100, 101
Polymersomes (~100)	Doxorubicin, Paclitaxel	In vivo		73, 74
Micelles (lipid based and polymeric) (5–100)	Doxorubicin	Clinical trials (Phase I)	Metastatic or recurrent solid tumours refractory to conventional chemotherapy	77, 92, 102
	Paclitaxel	Clinical trials (Phase I)	Pancreatic, bile duct, gastric and colonic cancers	
	Platinum-based drugs (carboplatin/cisplatin), Camptothecin, Tamoxifen, Epirubicin	In vivo and in vitro		Reviewed in 75

(*Continued*)

Table 2.2 (*Continued*)

Type of carrier and mean diameter (nm)	Drug entrapped or linked	Current stage of development	Type of cancer (for clinical trials)	References
Nanoshells (Gold-silica) (~130)	No drug (for photothermal therapy)	In vivo		37, 103
Gold nanoparticles (10–40)	No drug (for photothermal ablation)	In vivo		104
Nanocages (30–40)	No drug	Chemistry, structural analysis and in vitro		90, 105
Dendrimers (~5)	Methotrexate	In vitro/in vivo		46, 86
Immuno-PEG-liposomes (100)	Doxorubicin	Clinical trials (Phase I)	Metastatic stomach cancer	76
Immunoliposomes (100–150)	Doxorubicin, platinum-based drugs, Vinblastin, Vincristin, Topotecan, Paclitaxel	In vivo		Reviewed in 9, 106
Immunotoxins, Immunopolymers, and fusion proteins (3–15)	Various drugs, toxins	Clinical trials (Phases I–III)	Various types of cancer	Reviewed in 16, 17, 61

To date, at least 12 polymer–drug conjugates have entered Phase I and II clinical trials (Table 2.2 and Fig. 2.3a) and are especially useful for targeting blood vessels in tumours. Examples include anti-endothelial immunoconjugates, fusion proteins [57–59], and caplostatin, the first polymer-angiogenesis inhibitor conjugates [60]. Polymers that are chemically conjugated with drugs are often considered new chemical entities (NCEs) owing to a distinct pharmacokinetic profile from that of the parent drug. Despite the variety of novel drug targets and sophisticated chemistries available, only four drugs (doxorubicin, camptothecin, paclitaxel, and platinate) and four polymers (*N*-(2-hydroxylpropyl)methacrylamide (HPMA) copolymer, poly-*L*-glutamic acid, poly(ethylene glycol) (PEG), and Dextran) have been repeatedly used to develop polymer–drug conjugates [3, 61].

Polymers are the most commonly explored materials for constructing nanoparticle-based drug carriers. One of the earliest reports of their use for cancer therapy dates back to 1979 [62] when adsorption of anticancer drugs to polyalkylcyanoacrylate nanoparticles was described. Couvreur et al. revealed the release mechanism of the drugs from the polymer in calf serum, followed by tissue distribution and efficacy studies in a tumour model [63]. This work laid the foundation for the development of doxorubicin-loaded nanoparticles that were tested in clinical trials in the mid-1980s [64]. Polymeric nanoparticles can be made from synthetic polymers, including poly(lactic acid) (PLA) and poly(lactic co-glycolic acid) [65], or from natural polymers such as chitosan [66] and collagen [67] and may be used to encapsulate drugs without chemical modification. The drugs can be released in a controlled manner through surface or bulk erosion, diffusion through the polymer matrix, swelling followed by diffusion, or in response to the local environment. Several multifunctional polymeric nanoparticles are now in various stages of pre-clinical and clinical development [4, 56, 68, 69]. Concerns arising from the use of polymer-based nanocarriers include the inherent structural heterogeneity of polymers, reflected, for example, in a high polydispersity index (the ratio of the weight-and-number-average molecular weight (M_w/M_n)). There are, however, a few examples of polymeric nanoparticles that show near-homogenous size distribution [70].

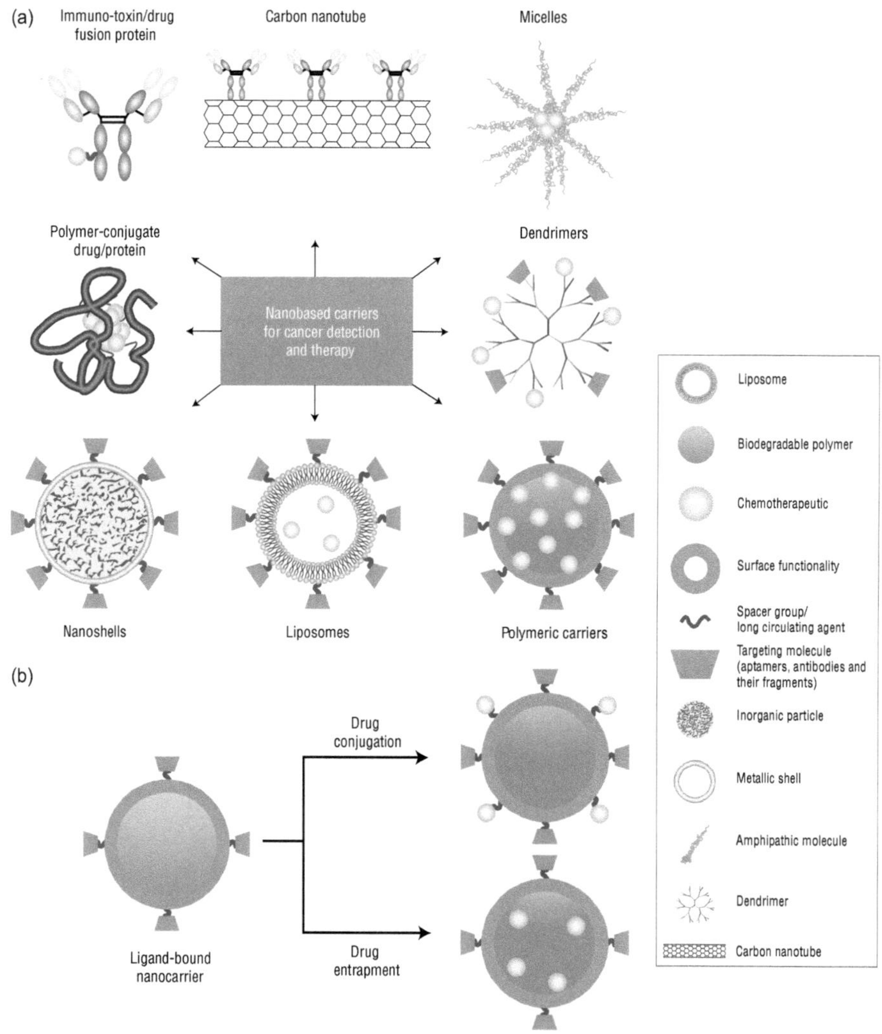

Figure 2.3 Examples of nanocarriers for targeting cancer. (a) a whole range of delivery agents are possible but the main components typically include a nanocarrier, a targeting moiety conjugated to the nanocarrier, and a cargo (such as the desired chemotherapeutic drugs). (b) Schematic diagram of the drug conjugation and entrapment processes. The chemotherapeutics could be bound to the nanocarrier, as in the use of polymer–drug conjugates, dendrimers and some particulate carriers, or they could be entrapped inside the nanocarrier.

Lipid-based carriers have attractive biological properties, including general biocompatibility, biodegradability, isolation of drugs from the surrounding environment, and the ability to entrap both hydrophilic and hydrophobic drugs. Through the addition of agents to the lipid membrane or by the alteration of the surface chemistry, properties of lipid-based carriers, such as their size, charge, and surface functionality, can easily be modified. Liposomes, polymersomes, and micelles represent a class of amphiphile-based particles. Liposomes are spherical, self-closed structures formed by one or several concentric lipid bilayers with inner aqueous phases. Today, liposomes are approved by regulatory agencies to carry a range of chemotherapeutics [26, 71, 72] (Table 2.1).

Polymersomes have an architecture similar to that of liposomes, but they are composed of synthetic polymer amphiphiles, including PLA-based copolymers [73, 74] (Table 2.2). However, as with polymer therapeutics, there are still no clinically approved strategies that use active cellular targeting for lipid-based carriers.

Micelles, which are self-assembling closed lipid monolayers with a hydrophobic core and hydrophilic shell, have been successfully used as pharmaceutical carriers for water-insoluble drugs (Table 2.2) [75]. They belong to a group of amphiphilic colloids that can be formed spontaneously under certain concentrations and temperatures from amphiphilic or surface-active agents (surfactants) (Fig. 2.3a). An example of a polymeric micelle under clinical evaluation is NK911, which is a block copolymer of PEG and poly(aspartic acid). NK911, which consists of a bound doxorubicin fraction ($\sim$45%) (Fig. 2.3b) and a free drug [76], was evaluated for metastatic pancreatic cancer treatment. Another carrier is NK105, a micelle containing paclitaxel, was evaluated for pancreatic, colonic and gastric tumour treatment [77].

Lipid-based carriers pose several challenges, which represent general issues in the use of other targeted nanocarriers such as polymeric nanoparticles. For example, upon intravenous injection, particles are rapidly cleared from the bloodstream by the reticuloendothelial defence mechanism, regardless of particle composition [78, 79]. Moreover, instability of the carrier and burst drug release, as well as non-specific uptake by the mononuclear phagocytic system (MPS), provides additional challenges for translating these carriers to the clinic.

Given their long history, liposome-based carriers serve as a classic example of the challenges encountered in the development of nanocarriers and the solutions that have been attempted. For example, PEG has been used to improve circulation time by stabilizing and protecting micelles and liposomes from opsonization—a plasma protein deposition process that signals Kupffer cells in the liver to remove the carriers from circulation [75, 80]. However, Daunosome and Myocet are examples of clinically used liposomes (80–90 nm in diameter) without PEG coating that have been reported to exhibit enhanced circulation times, although to a lesser degree than PEGylated liposomes such as Doxil/Caelyx (Table 2.1).

In addition to rapid clearance, another challenge is the fast burst release of the chemotherapeutic drugs from the liposomes. To overcome this phenomenon, doxorubicin, for example, may be encapsulated in the liposomal aqueous phase by an ammonium sulphate gradient [81]. This method achieves a stable drug entrapment with negligible drug leakage during circulation, even after prolonged residence in the blood stream [82]. In clinical practice, liposomal systems have shown preferential accumulation in tumours, via the EPR effect, and reduced toxicity of their cargo (Tables 2.1 and 2.2). However, long-circulating liposomes may lead to extravasation of the drug in unexpected sites. The most commonly experienced clinical toxic effect from the PEGylated liposomal doxorubicin is palmar-plantar erythrodysesthesia (PPE), also called the hand-foot syndrome. PPE—a dermatologic toxicity reaction seen with high doses of many types of chemotherapy—can be addressed by changing the dosing and scheduling of the treatment [83]. Other challenges facing the use of liposomes in the clinic include the high production cost, fast oxidation of some phospholipids, and lack of controlled-release properties of encapsulated drugs.

To achieve temporal release of two drugs, polymers and phospholipids can be combined as a single delivery agent (polymer core/lipid shell). After locating at a tumour site through the EPR effect, the outer phospholipid shell releases an anti-angiogenesis agent, and the inner polymeric nanoparticle subsequently releases a chemotherapy agent in response to local hypoxia—shortage of oxygen. This strategy led to reduced toxicity and enhanced anti-metastatic effects in two different mouse tumour models, emphasizing the advantages of a mechanism-based design for targeted nanocarriers [84].

Organic nanoparticles include dendrimers, viral capsids and nanostructures made from biological building blocks such as proteins. Abraxane is an albumin-bound paclitaxel nanoparticle formulation approved by the FDA in 2005 as a second-line treatment for metastatic breast cancer. Abraxane was designed to address insolubility problems encountered with paclitaxel. Its use eliminates the need for toxic solvents like Cremophor EL (polyoxyethylated castor oil), which has been shown to limit the dose of Taxol that can be administered [85].

Dendrimers are synthetic, branched macromolecules that form a tree-like structure whose synthesis represents a relatively new field in polymer chemistry. Polyamidoamine dendrimers have shown promise for biomedical applications because they (1) can be easily conjugated with targeting molecules, imaging agents, and drugs, (2) have high water solubility and well-defined chemical structures, (3) are biocompatible, and (4) are rapidly cleared from the blood through the kidneys, made possible by their small size (<5 nm), which eliminates the need for biodegradability. In vivo delivery of dendrimer–methotrexate conjugates using multivalent targeting results in a tenfold reduction in tumour size compared with that achieved with the same molar concentration of free systemic methotrexate [22, 46]. This work provided motivation for further pre-clinical development, and a variety of dendrimers are now under investigation for cancer treatment and are extensively reviewed elsewhere [86, 87]. Although promising, dendrimers are more expensive than other nanoparticles and require many repetitive steps for synthesis, posing a challenge for large-scale production.

Inorganic nanoparticles are primarily metal based and have the potential to be produced with near monodispersity. Inorganic materials have been extensively studied for magnetic resonance imaging and high-resolution superconducting quantum interference devices [88]. Inorganic particles may also be functionalized to introduce targeting molecules and drugs. Specific types of recently developed inorganic nanoparticles include nanoshells and gold nanoparticles.

Nanoshells (100–200 nm) may use the same carrier for both imaging and therapy (Table 2.2). They are composed of a silica core and a metallic outer layer. Nanoshells have optical resonances that can be adjusted to absorb or scatter essentially anywhere in the electromagnetic spectrum, including the near infrared region (NIR, 820 nm, 4 W cm^{-2}), where transmission of light through tissue is optimal. Absorbing nanoshells are suitable for hyperthermia-based therapeutics, where the nanoshells absorb radiation and heat up the surrounding cancer tissue. Scattering nanoshells, on the other hand, are desirable as contrast agents for imaging applications. Recently, a cancer therapy was developed based on absorption of NIR light by nanoshells, resulting in rapid localized heating to selectively kill tumours implanted in mice. Tissues heated above the thermal damage threshold displayed coagulation, cell shrinkage and loss of nuclear staining, which are indicators of irreversible thermal damage, whereas control tissues appeared undamaged [37, 89].

A similar approach involves gold nanocages which are smaller (<50 nm) than the nanoshells. These gold nanocages (Table 2.2) can be constructed to generate heat in response to NIR light and thus may also be useful in hyperthermia-based therapeutics [90]. Unlike nanoshells and nanocages, pure gold nanoparticles (Table 2.2) are relatively easy to synthesize and manipulate. Non-specific interactions that cause toxicity in healthy tissues may impede the use of many types of nanoparticles, but using inorganic particles for photo-ablation significantly limits non-specific toxicity because light is locally directed. However, inorganic particles may not provide advantages over other types of nanoparticles for systemic targeting of individual cancer cells because they are not biodegradable or small enough to be cleared easily, resulting in potential accumulation in the body, which may cause long-term toxicity.

2.4 The Challenges of Multidrug Resistance

The delivery of drugs through targeted nanocarriers that are internalized by cells provides an alternative route to diffusion of drugs into cells. This approach may allow targeted carriers to bypass the activity of integral membrane proteins, known as MDR transporters, which transport a variety of anticancer drugs out of the cancer cell and produce resistance against chemotherapy [11]. The molecular basis of cancer drug resistance is complex and has been correlated to elevated levels of enzymes that can neutralize chemotherapeutic drugs. More often, however, it is due to the overexpression of MDR transporters that actively pump chemotherapeutic drugs out of the cell and reduce the intracellular drug doses below lethal threshold levels. Because not all cancer cells express the MDR transporters, chemotherapy will kill only drug-sensitive cells that do not or only mildly express MDR transporters, while leaving behind a small population of drug-resistant cells that highly express MDR transporters. With tumour recurrence, chemotherapy may fail because residual drug-resistant cells dominate the tumour population.

Among the MDR transporters, the most widely investigated proteins are: P-glycoprotein (also referred to as MDR1 or ABCB1); the multidrug resistance associated proteins (MRPs), of which the most studied is the MRP1 (or ABCC1); and the breast cancer resistance protein (ABCG2). These proteins have different structures, but they share a similar function of expelling chemotherapy drugs from the cells [12]. Several studies have demonstrated the possibility of using nanocarriers to bypass the MDR transporters. SP1049C is a non-ionic (pluronic or also known as poloxamer) block copolymer composed of a hydrophobic core and hydrophilic tail that contains doxorubicin. SP1049C has been shown to circumvent p-glycoprotein-mediated drug resistance in a mouse model of leukaemia and is now under clinical evaluation [91, 92]. Folate receptor-mediated cell uptake of doxorubicin–loaded liposomes into an MDR cell line was shown to be unaffected by P-glycoprotein (Pgp)-mediated drug efflux, in contrast to the uptake of free doxorubicin [93]. In an attempt to reverse MDR, vincristine-loaded lipid nanoparticles conjugated to an anti-Pgp mAb (MRK-16), showed greater cytotoxicity in resistant human myelogenous leukaemia cell lines than control non-targeted particles—a response attributed to the inhibition of the Pgp-mediated efflux of vincristine by MRK-16 [94]. Additional reports have addressed the challenge of MDR using polymer therapeutics95, polymeric nanoparticles [96], lipid nanocapsules [97] and micelles [98] within cell lines or in mouse tumour models. Combination treatments with targeted nanocarriers for selective delivery of drugs and MDR pump inhibitors will likely address some of the problems posed by resistant tumours.

2.5 Into the Future

The choice of an appropriate nanocarrier is not obvious, and the few existing comparative studies are difficult to interpret because several factors may simultaneously affect biodistribution and targeting. In addition, developing suitable screening methodologies for determining optimal characteristics of nanocarriers remains elusive. Therefore, successful targeting strategies must be determined experimentally on a case-by-case basis, which is laborious. In addition, systemic therapies using nanocarriers require methods that can overcome non-specific uptake by mononuclear phagocytic cells and by non-targeted cells. It is also not clear to what extent this is possible without substantially increasing the complexity of the nanocarrier and without influencing commercial scale-up. Improved therapeutic efficacy of targeted nanocarriers has been established in multiple animal models of cancer, and currently more than 120 clinical trials are underway with various antibody-containing nanocarrier formulations [99]. For the clinician, in addition to enhancing confidence through the ability to image the type and location of the tumour, it is imperative to construct appropriate therapeutic regimens. When targeting cell surface markers presents a significant challenge, as in the case for solid tumours, targeting tumour vasculature or the extracellular matrix surrounding the tumour microenvironment may be necessary. In the case of circulating cancer cells, as in leukaemia and lymphoma, a therapy that targets surface antigens with high affinity and includes a carrier with a long circulating half-life may be the most efficacious. Similar to combination drug strategies that may be personalized to optimize treatment regimens, oncologists in the near future may be presented with the ability to choose specific nanocarrier/targeting molecule combinations which could lead to improved therapeutic outcomes and reduced costs.

Although we are still far from Nobel Prize winner Paul Ehrlich's 'magic bullet', many believe that we will soon enter an era in which nanocarrier-based approaches will represent an important modality within therapeutic and diagnostic oncology.

Acknowledgments

We would like to acknowledge Shiladitya Sengupta for critically reviewing the manuscript and Maeve Cullinane for helpful discussions. This work was supported by federal funds NIH/NCI CA119349, NIH/NIBIB EB 003647, and NIH R01-EB000244. The content is solely the responsibility of the authors and does not necessarily represent the official view of the NIH.

References

1. Stewart, B. W., and Kleihues, P. *World Cancer Report* (World Health Organization Press, Geneva, 2003).
2. *Cancer Facts & Figures 2007* (American Cancer Society, Atlanta, 2007).
3. Duncan, R. Polymer conjugates as anticancer nanomedicines. *Nat. Rev. Cancer* **6**, 688–701 (2006).
4. Ferrari, M. Cancer nanotechnology: opportunities and challenges. *Nat. Rev. Cancer* **5**, 161–171 (2005).
5. Couvreur, P., and Vauthier, C. Nanotechnology: Intelligent design to treat complex disease. *Pharm. Res.* **23**, 1417–1450 (2006).
6. Alonso, M. J. Nanomedicines for overcoming biological barriers. *Biomed. Pharmacother.* **58**, 168–172 (2004).
7. Matsumura, Y., and Maeda, H. A new concept for macromolecular therapeutics in cancer-chemotherapy—Mechanism of tumoritropic accumulation of proteins and the antitumor agent smancs. *Cancer Res.* **46**, 6387–6392 (1986).
8. Yuan, F., et al. Vascular-permeability in a human tumor xenograft—Molecular-size dependence and cutoff size. *Cancer Res.* **55**, 3752–3756 (1995).

9. Torchilin, V. P. Recent advances with liposomes as pharmaceutical carriers. *Nat. Rev. Drug Discov.* **4**, 145–160 (2005).
10. Hobbs, S. K., et al. Regulation of transport pathways in tumor vessels: role of tumor type and microenvironment. *Proc. Natl Acad. Sci. USA* **95**, 4607–4612 (1998).
11. Gottesman, M. M., Fojo, T., and Bates, S. E. Multidrug resistance in cancer: Role of ATP-dependent transporters. *Nat. Rev. Cancer* **2**, 48–58 (2002).
12. Peer, D., and Margalit, R. Fluoxetine and reversal of multidrug resistance. *Cancer Lett.* **237**, 180–187 (2006).
13. Jain, R. K. Barriers to drug-delivery in solid tumors. *Sci. Am.* **271**, 58–65 (1994).
14. de Menezes, D. E. L., Pilarski, L. M., and Allen, T. M. In vitro and in vivo targeting of immunoliposomal doxorubicin to human B-cell lymphoma. *Cancer Res.* **58**, 3320–3330 (1998).
15. Park, J. W., et al. Anti-HER2 immunoliposomes: enhanced efficacy attributable to targeted delivery. *Clin. Cancer Res.* **8**, 1172–1181 (2002).
16. Allen, T. M. Ligand-targeted therapeutics in anticancer therapy. *Nat. Rev. Cancer* **2**, 750–763 (2002).
17. Pastan, I., Hassan, R., FitzGerald, D. J., and Kreitman, R. J. Immunotoxin therapy of cancer. *Nat. Rev. Cancer* **6**, 559–565 (2006).
18. Peer, D., Zhu, P., Carman, C. V., Lieberman, J., and Shimaoka, M. Selective gene silencing in activated leukocytes by targeting siRNAs to the integrin lymphocyte function-associated antigen-1. *Proc. Natl Acad. Sci. USA* **104**, 4095–4100 (2007).
19. Sapra, P., and Allen, T. M. Internalizing antibodies are necessary for improved therapeutic efficacy of antibody-targeted liposomal drugs. *Cancer Res.* **62**, 7190–7194 (2002).
20. Allen, T. M. Long-circulating (sterically stabilized) liposomes for targeted drug-delivery. *Trends Pharmacol. Sci.* **15**, 215–220 (1994).
21. Adams, G. P., et al. High affinity restricts the localization and tumor penetration of single-chain Fv antibody molecules. *Cancer Res.* **61**, 4750–4755 (2001).
22. Hong, S., et al. The binding avidity of a nanoparticle-based multivalent targeted drug delivery platform. *Chem. Biol.* **14**, 107–115 (2007).
23. Warenius, H. M., Galfre, G., Bleehen, N. M., and Milstein, C. Attempted targeting of A monoclonal-antibody in a human-tumor xenograft system. *Eur. J. Cancer Clin. Oncology* **17**, 1009–1015 (1981).

24. von Mehren, A. G., Weiner L. M. Monoclonal antibody therapy for cancer. *Annu. Rev. Med.* **54**, 343–369 (2003).
25. Weiner, L. M., and Adams, G. P. New approaches to antibody therapy. *Oncogene* **19**, 6144–6151 (2000).
26. Gabizon, A. A. Pegylated liposomal doxorubicin: metamorphosis of an old drug into a new form of chemotherapy. *Cancer Invest.* **19**, 424–436 (2001).
27. James, J. S., and Dubs, G. FDA approves new kind of lymphoma treatment. *AIDS Treat. News* **284**, 2–3 (1997).
28. Albanell, J., and Baselga, J. Trastuzumab, a humanized anti-HER2 monoclonal antibody, for the treatment of breast cancer. *Drugs Today* **35**, 931–946 (1999).
29. Ferrara, N. VEGF as a therapeutic target in cancer. *Oncology* **69** (Suppl. 3), 11–16 (2005).
30. Carter, P. Improving the efficacy of antibody-based cancer therapies. *Nat. Rev. Cancer* **1**, 118–129 (2001).
31. Marks, J. D. Selection of internalizing antibodies for drug delivery. *Methods Mol. Biol.* **248**, 201–208 (2004).
32. Marks, J. D., et al. Human-antibody fragments specific for human blood-groups antigens from a phage display library. *Bio-Technol.* **11**, 1145–1149 (1993).
33. Liu, B., Conrad, F., Cooperberg, M. R., Kirpotin, D. B., and Marks, J. D. Mapping tumor epitope space by direct selection of single-chain Fv antibody libraries on prostate cancer cells. *Cancer Res.* **64**, 704–710 (2004).
34. Arnold, D. M., et al. Systematic review: efficacy and safety of rituximab for adults with idiopathic thrombocytopenic purpura. *Ann. Intern. Med.* **146**, 25–33 (2007).
35. Trail, P. A., et al. Cure of xenografted human carcinomas by Br96-doxorubicin immunoconjugates. *Science* **261**, 212–215 (1993).
36. Tolcher, A. W., et al. Randomized phase II study of BR96-doxorubicin conjugate in patients with metastatic breast cancer. *J. Clin. Oncology* **17**, 478–484 (1999).
37. Hirsch, L. R., et al. Nanoshell-mediated near-infrared thermal therapy of tumors under magnetic resonance guidance. *Proc. Natl Acad. Sci. USA* **100**, 13549–13554 (2003).
38. Silverman, J., et al. Multivalent avimer proteins evolved by exon shuffling of a family of human receptor domains. *Nat. Biotechnol.* **23**, 1556–1561 (2005).

39. Cortez-Retamozo, V., et al. Efficient cancer therapy with a nanobody-based conjugate. *Cancer Res.* **64**, 2853–2857 (2004).
40. Nord, K., et al. Binding proteins selected from combinatorial libraries of an alpha-helical bacterial receptor domain. *Nat. Biotechnol.* **15**, 772–777 (1997).
41. White, R. R., Sullenger, B. A., and Rusconi, C. P. Developing aptamers into therapeutics. *J. Clin. Invest.* **106**, 929–934 (2000).
42. Farokhzad, O. C., et al. Targeted nanoparticle-aptamer bioconjugates for cancer chemotherapy in vivo. *Proc. Natl Acad. Sci. USA* **103**, 6315–6320 (2006).
43. Sanfilippo, J. S., et al. Quantitative analyses of epidermal growth factor receptors, HER-2/neu oncoprotein and cathepsin D in nonmalignant and malignant uteri. *Cancer* **77**, 710–716 (1996).
44. Antony, A. C. The biological chemistry of folate receptors. *Blood* **79**, 2807–2820 (1992).
45. Prost, A. C., et al. Differential transferrin receptor density in human colorectal cancer: A potential probe for diagnosis and therapy. *Int. J. Oncol.* **13**, 871–875 (1998).
46. Kukowska-Latallo, J. F., et al. Nanoparticle targeting of anticancer drug improves therapeutic response in animal model of human epithelial cancer. *Cancer Res.* **65**, 5317–5324 (2005).
47. Iinuma, H., et al. Intracellular targeting therapy of cisplatin-encapsulated transferrin-polyethylene glycol liposome on peritoneal dissemination of gastric cancer. *Int. J. Cancer* **99**, 130–137 (2002).
48. Ishida, O., et al. Liposomes bearing polyethyleneglycol-coupled transferrin with intracellular targeting property to the solid tumors in vivo. *Pharm. Res.* **18**, 1042–1048 (2001).
49. Ekblom, P., Thesleff, I., Lehto, V. P., and Virtanen, I. Distribution of the transferrin receptor in normal human-fibroblasts and fibro-sarcoma cells. *Int. J. Cancer* **31**, 111–117 (1983).
50. Li, J., et al. Fusion protein from RGD peptide and Fc fragment of mouse immunoglobulin G inhibits angiogenesis in tumor. *Cancer Gene Ther.* **11**, 363–370 (2004).
51. Ruoslahti, E. Cell adhesion and tumor metastasis. *Princess Takamatsu Symp.* **24**, 99–105 (1994).
52. Peer, D., and Margalit, R. Tumor-targeted hyaluronan nanoliposomes increase the antitumor activity of liposomal Doxorubicin in syngeneic and human xenograft mouse tumor models. *Neoplasia* **6**, 343–353 (2004).

53. Hu, Z., Sun, Y., and Garen, A. Targeting tumor vasculature endothelial cells and tumor cells for immunotherapy of human melanoma in a mouse xenograft model. *Proc. Natl Acad. Sci. USA* **96**, 8161–8166 (1999).
54. Peer, D., and Margalit, R. Loading mitomycin C inside long circulating hyaluronan targeted nano-liposomes increases its antitumor activity in three mice tumor models. *Int. J. Cancer* **108**, 780–789 (2004).
55. Eliaz, R. E., and Szoka, F. C. Jr. Liposome-encapsulated doxorubicin targeted to CD44: a strategy to kill CD44-overexpressing tumor cells. *Cancer Res.* **61**, 2592–2601 (2001).
56. LaVan, D. A., McGuire, T., and Langer, R. Small-scale systems for in vivo drug delivery. *Nat. Biotechnol.* **21**, 1184–1191 (2003).
57. Arap, W., Pasqualini, R., and Ruoslahti, E. Cancer treatment by targeted drug delivery to tumor vasculature in a mouse model. *Science* **279**, 377–380 (1998).
58. Schraa, A. J., et al. Targeting of RGD-modified proteins to tumor vasculature: A pharmacokinetic and cellular distribution study. *Int. J. Cancer* **102**, 469–475 (2002).
59. Halin, C., et al. Enhancement of the antitumor activity of interleukin-12 by targeted delivery to neovasculature. *Nat. Biotechnol.* **20**, 264–269 (2002).
60. Satchi-Fainaro, R., et al. Targeting angiogenesis with a conjugate of HPMA copolymer and TNP-470. *Nat. Med.* **10**, 255–261 (2004).
61. Satchi-Fainaro, R., Duncan, R., and Barnes, C. M. in *Polymer Therapeutics II: Polymers as Drugs*, Conjugates and Gene Delivery Systems Vol. 193 (eds Satchi-Fainaro, R., and Duncan, R.) 1–65 (Springer-Verlag, Berlin, 2006).
62. Couvreur, P., Kante, B., Roland, M., and Speiser, P. Adsorption of antineoplastic drugs to polyalkylcyanoacrylate nanoparticles and their release in calf serum. *J. Pharm. Sci.* **68**, 1521–1524 (1979).
63. Couvreur, P., et al. Tissue distribution of anti-tumor drugs associated with polyalkylcyanoacrylate nanoparticles. *J. Pharm. Sci.* **69**, 199–202 (1980).
64. Couvreur, P., Kante, B., Grislain, L., Roland, M., and Speiser, P. Toxicity of polyalkylcyanoacrylate nanoparticles II: Doxorubicin-loaded nanoparticles. *J. Pharm. Sci.* **71**, 790–792 (1982).
65. Hrkach, J. S., Peracchia, M. T., Domb, A., Lotan, N., and Langer, R. Nanotechnology for biomaterials engineering: Structural characterization

of amphiphilic polymeric nanoparticles by H-1 NMR spectroscopy. *Biomaterials* **18**, 27–30 (1997).

66. Calvo, P., RemunanLopez, C., VilaJato, J. L., and Alonso, M. J. Chitosan and chitosan ethylene oxide propylene oxide block copolymer nanoparticles as novel carriers for proteins and vaccines. *Pharm. Res.* **14**, 1431–1436 (1997).
67. Elsamaligy, M. S., and Rohdewald, P. Reconstituted collagen nanoparticles, a novel drug carrier delivery system. *J. Pharm. Pharmacol.* **35**, 537–539 (1983).
68. Moses, M. A., Brem, H., and Langer, R. Advancing the field of drug delivery: taking aim at cancer. *Cancer Cell* **4**, 337–341 (2003).
69. Farokhzad, O. C., and Langer, R. Nanomedicine: Developing smarter therapeutic and diagnostic modalities. *Adv. Drug Deliv. Rev.* **58**, 1456–1459 (2006).
70. Guo, R., et al. Synthesis of alginic acid-poly[2-(diethylamino)ethyl methacrylate] monodispersed nanoparticles by a polymer-monomer pair reaction system. *Biomacromolecules* **8**, 843–850 (2007).
71. Gabizon, A. A. Stealth liposomes and tumor targeting: one step further in the quest for the magic bullet. *Clin. Cancer Res.* **7**, 223–225 (2001).
72. Safra, T., et al. Pegylated liposomal doxorubicin (doxil): reduced clinical cardiotoxicity in patients reaching or exceeding cumulative doses of 500 mg/m^2. *Ann. Oncol.* **11**, 1029–1033 (2000).
73. Ahmed, F., et al. Shrinkage of a rapidly growing tumor by drug-loaded polymersomes: pH-triggered release through copolymer degradation. *Mol. Pharm.* **3**, 340–350 (2006).
74. Discher, D. E., and Ahmed, F. Polymersomes. *Annu. Rev. Biomed. Eng.* **8**, 323–341 (2006).
75. Matsumura, Y., et al. Phase I clinical trial and pharmacokinetic evaluation of NK911, a micelle-encapsulated doxorubicin. *Brit. J. Cancer* **91**, 1775–1781 (2004).
76. Kato, K., et al. Phase I study of NK105, a paclitaxel-incorporating micellar nanoparticle, in patients with advanced cancer. *J. Clin. Oncol* **24** (suppl.), 2018 (2006).
77. Torchilin, V. P. Micellar nanocarriers: Pharmaceutical perspectives. *Pharm. Res.* **24**, 1–16 (2007).
78. Brigger, I., Dubernet, C., and Couvreur, P. Nanoparticles in cancer therapy and diagnosis. *Adv. Drug Deliv. Rev.* **54**, 631–651 (2002).

79. Kreuter, J., and Higuchi, T. Improved delivery of methoxsalen. *J. Pharm. Sci.* **68**, 451–454 (1979).

80. Papahadjopoulos, D., et al. Sterically stabilized liposomes - improvements in pharmacokinetics and antitumor therapeutic efficacy. *Proc. Natl Acad. Sci. USA* **88**, 11460–11464 (1991).

81. Haran, G., Cohen, R., Bar, L. K., and Barenholz, Y. Transmembrane ammonium-sulfate gradients in liposomes produce efficient and stable entrapment of amphipathic weak bases. *Biochim. Biophys. Acta* **1151**, 201–215 (1993).

82. Gabizon, A. A., Shmeeda, H., and Zalipsky, S. Pros and cons of the liposome platform in cancer drug targeting. *J. Liposome Res.* **16**, 175–183 (2006).

83. Lorusso, D., et al. Pegylated liposomal doxorubicin-related palmar-plantar erythrodysesthesia ('hand-foot' syndrome). *Ann. Oncol.* (2007).

84. Sengupta, S., et al. Temporal targeting of tumour cells and neovasculature with a nanoscale delivery system. *Nature* **436**, 568–572 (2005).

85. Damascelli, B., et al. Intraarterial chemotherapy with polyoxyethylated castor oil free paclitaxel, incorporated in albumin nanoparticles (ABI-007). *Cancer* **92**, 2592–2602 (2001).

86. Gillies, E. R., and Frechet, J. M. J. Dendrimers and dendritic polymers in drug delivery. *Drug Discov. Today* **10**, 35–43 (2005).

87. Malik, N., et al. Dendrimers: Relationship between structure and biocompatibility in vitro, and preliminary studies on the biodistribution of I-125-labelled polyamidoamine dendrimers in vivo. *J. Control. Release* **65**, 133–148 (2000).

88. Morawski, A. M., Lanza, G. A., and Wickline, S. A. Targeted contrast agents for magnetic resonance imaging and ultrasound. *Curr. Opin. Biotechnol.* **16**, 89–92 (2005).

89. Loo, C., Lowery, A., Halas, N., West, J., Drezek, R. Immunotargeted nanoshells for integrated cancer imaging and therapy. *Nano Lett.* **5**, 709–711 (2005).

90. Chen, J., et al. Gold nanocages: Bioconjugation and their potential use as optical imaging contrast agents. *Nano Lett.* **5**, 473–477 (2005).

91. Danson, S., et al. Phase I dose escalation and pharmacokinetic study of pluronic polymer-bound doxorubicin (SP 1049C) in patients with advanced cancer. *Brit. J. Cancer* **90**, 2085–2091 (2004).

92. Batrakova, E. V., et al. Anthracycline antibiotics non-covalently incorporated into the block copolymer micelles: In vivo evaluation of anticancer activity. *Brit. J. Cancer* **74**, 1545–1552 (1996).
93. Goren, D., et al. Nuclear delivery of doxorubicin via folate-targeted liposomes with bypass of multidrug-resistance efflux pump. *Clin. Cancer Res.* **6**, 1949–1957 (2000).
94. Matsuo, H., et al. Possibility of the reversal of multidrug resistance and the avoidance of side effects by liposomes modified with MRK-16, a monoclonal antibody to P-glycoprotein. *J. Control. Release* **77**, 77–86 (2001).
95. Duncan, R., Vicent, M. J., Greco, F., and Nicholson, R. I. Polymer-drug conjugates: towards a novel approach for the treatment of endrocine-related cancer. *Endocrine-Relat. Cancer* **12**, S189–S199 (2005).
96. Wong, H. L., et al. A new polymer-lipid hybrid nanoparticle system increases cytotoxicity of doxorubicin against multidrug-resistant human breast cancer cells. *Pharm. Res.* **23**, 1574–1585 (2006).
97. Garcion, E., et al. A new generation of anticancer, drug-loaded, colloidal vectors reverses multidrug resistance in glioma and reduces tumor progression in rats. *Mol. Cancer Ther.* **5**, 1710–1722 (2006).
98. Lee, E. S., Na, K., and Bae, Y. H. Doxorubicin loaded pH-sensitive polymeric micelles for reversal of resistant MCF-7 tumor. *J. Control. Release* **103**, 405–418 (2005).
99. Sapra, P., and Allen, T. M. Ligand-targeted liposomal anticancer drugs. *Prog. Lipid Res.* **42**, 439–462 (2003).
100. Moghimi, S. M. Recent developments in polymeric nanoparticle engineering and their applications in experimental and clinical oncology. *Anticancer Agents Med. Chem.* **6**, 553–561 (2006).
101. Lee, K. S., et al. Multicenter phase II trial of Genexol-PM, a Cremophor-free, polymeric micelle formulation of paclitaxel, in patients with metastatic breast cancer. *Breast Cancer Res. Treat.* (2007).
102. Nakanishi, T., et al. Development of the polymer micelle carrier system for doxorubicin. *J. Control. Release* **74**, 295–302 (2001).
103. Hirsch, L. R., et al. Metal nanoshells. *Ann. Biomed. Engin.* **34**, 15–22 (2006).
104. Sokolov, K., et al. Real-time vital optical imaging of precancer using anti-epidermal growth factor receptor antibodies conjugated to gold nanoparticles. *Cancer Res.* **63**, 1999–2004 (2003).

105. Chen, J. Y., et al. Facile synthesis of gold-silver nanocages with controllable pores on the surface. *J. Am. Chem. Soc.* **128**, 14776–14777 (2006).

106. Kontermann, R. E. Immunolliposomes for cancer therapy. *Curr. Opin. Mol. Ther.* **8**, 39–45 (2006).

Chapter 3

Microfluidic Technologies for Accelerating the Clinical Translation of Nanoparticles

Pedro M. Valencia,[a] Omid C. Farokhzad,[b,*] Rohit Karnik,[c,*] and Robert Langer[a,*]

[a]*Department of Chemical Engineering, Massachusetts Institute of Technology, Cambridge, Massachusetts 02139, USA*
[b]*Laboratory of Nanomedicine and Biomaterials and Department of Anaesthesiology, Brigham and Women's Hospital, Harvard Medical School, Boston, Massachusetts 02115, USA*
[c]*Department of Mechanical Engineering, Massachusetts Institute of Technology, Cambridge, Massachusetts 02139, USA*
ofarokhzad@zeus.bwh.harvard.edu, karnik@mit.edu, rlanger@mit.edu

Using nanoparticles for therapy and imaging holds tremendous promise for the treatment of major diseases such as cancer. However, their translation into the clinic has been slow because it remains difficult to produce nanoparticles that are consistent 'batch-to-batch', and in sufficient quantities for clinical research. Moreover, platforms for rapid screening of nanoparticles are still lacking. Recent microfluidic technologies can tackle some of these issues, and offer a way to accelerate the clinical translation of nanoparticles.

Reproduced with permission from *Nature Nanotechnology*, **7**, 623–629, 2012.

Nano-Enabled Medical Applications
Edited by Lajos P. Balogh

ISBN 978-981-4800-32-7 (Hardcover), 978-0-429-39903-9 (eBook)
www.jennystanford.com

In this chapter, we highlight the advances in microfluidic systems that can synthesize libraries of nanoparticles in a well-controlled, reproducible and high-throughput manner. We also discuss the use of microfluidics for rapidly evaluating nanoparticles in vitro under microenvironments that mimic the in vivo conditions. Furthermore, we highlight some systems that can manipulate small organisms, which could be used for evaluating the in vivo toxicity of nanoparticles or for drug screening. We conclude with a critical assessment of the near- and long-term impact of microfluidics in the field of nanomedicine.

Nanomedicine is the application of nanotechnology to medicine, specifically involving the use of engineered nanomaterials for therapy and diagnosis of major diseases such as cancer, cardiovascular and infectious diseases [1]. The first generation of nanoparticles with applications in medicine dates back to the 1970s, when drug-loaded nanoscale liposomes were developed to deliver their cargo to diseased cells in a 'Trojan horse' fashion [2]. Since then, a new generation of targeted drug delivery vehicles (for example, polymeric nanoparticles) [3], contrast agents (such as iron oxide nanoparticles) [4], diagnostic tools [5], and antennas for photothermal therapy (for example, gold nanoparticles) [6] have emerged. This is driven in part by further understanding of the biology of diseased states, and by technological advances in imaging techniques and synthesis of novel biocompatible and biodegradable materials [7]. Now, nanomedicine promises the precise delivery of drugs to disease sites (such as tumours and atherosclerotic plaques) without off-target toxicities, and the early detection of diseases using selective contrast agents and sensitive diagnostic tools [8].

Nanoparticles are attractive in medicine because their surfaces can be chemically modified for targeting specific disease tissues, or for in vivo stability. For therapy, drugs can be encapsulated inside nanoparticles and released in a controlled manner over time. For imaging, nanoparticles can provide higher contrast (for example, iron oxide nanoparticles for magnetic resonance imaging) or higher brightness (for example, quantum dots (QDs) for fluorescence imaging) than conventional small-molecule agents [9]. Despite these advantages and several decades of research, only a handful of nanoparticles have received approval from the US Food and Drug Administration (as of 2012). Examples include iron oxide nanoparticles for magnetic resonance imaging (for example, Feridex and Resovist), liposomes encapsulating the anticancer drug, doxorubicin, for chemotherapy (known as Doxil), and the protein-based nanoparticle encapsulating paclitaxel for chemotherapy (called Abraxane) [10].

In fact, translation of nanoparticles to the clinic has been slow compared with small-molecule drugs, with the majority of nanoparticles not even reaching the point of in vivo evaluation, and even fewer reaching clinical trials. This is due to a combination of factors. It remains difficult to reproducibly synthesize batches of nanoparticles that have identical properties and in sufficient quantities for clinical applications [11]. Moreover, knowledge on the fate of nanoparticles at the body-, organ- and cell-level remains limited [12]; this makes rational design of nanoparticles difficult and necessitates the use of screening-based approaches for synthesis. Furthermore, there are few platforms that can rapidly evaluate the biological behaviour of nanoparticles in vitro under conditions that can be correlated with their performance in vivo [11]. For example, there is a need for high-throughput methods for evaluating the binding and internalization of nanoparticles by cells, or the interaction of nanoparticles with plasma proteins and the complement system, among others. Finally, there is insufficient understanding of the biophysical and chemical interactions of nanoparticles with proteins, membranes, DNA and organelles. These interactions could have either beneficial or adverse outcomes [13]. It is expected that technologies tackling some of these challenges could significantly accelerate the discovery and clinical translation of nanomedicines.

Microfluidics — the science and technology of manipulating nanolitre volumes in microscale fluidic channels — has impacted a range of applications, including biological analysis, chemical synthesis, single-cell analysis and tissue engineering [14]. Building on its origins in semiconductor technology and chemical separations, the expansion of microfluidics has been driven by its ability to process small sample volumes and access biologically relevant length scales

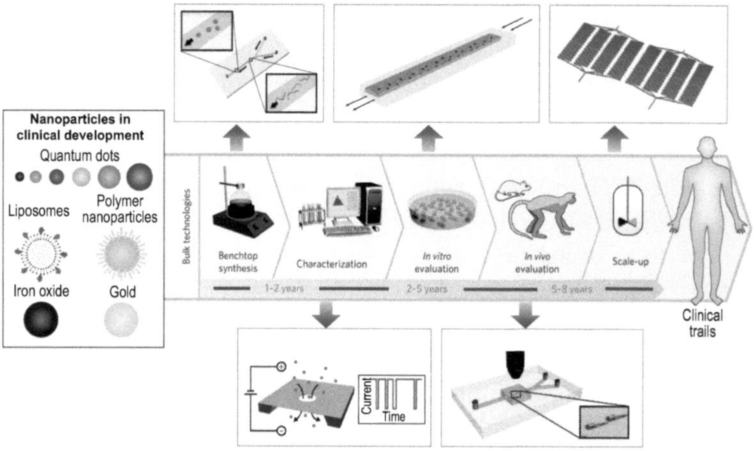

Figure 3.1 Nanoparticles in clinical development, steps for their translation (with average timescales) and microfluidic methods (green boxes) that could improve or complement current technologies. Synthesis is carried out in large reaction flasks, whereas microfluidic synthesis is carried out at micro and nano length scales that allow for improved control over reaction conditions. Characterization often involves taking a small sample of nanoparticles and measuring their properties offline, whereas nanopores embedded in microfluidic devices allow for real-time, in-line characterization. In vitro evaluation in plate wells produces a microenvironment far from that in vivo, whereas continuous flow in microfluidic systems result in conditions closer to those in vivo. In vivo evaluation in large animals is helpful for estimating the pharmacology of nanoparticles. To complement these studies microfluidic systems could enable real-time tracking of nanoparticles in large numbers of small organisms. Scale-up is generally carried out in reactor vessels several times larger than benchtop flasks, whereas parallelization of microfluidic channels can increase the production rate of nanoparticles with properties identical to the one at bench scale.

and microscale transport phenomena. This expansion has been largely facilitated by techniques, such as soft lithography, that enable rapid design and prototyping of microfluidic devices using a variety of materials [14]. Recent advances and innovations in microfluidics are expected to improve the synthesis of nanoparticles and accelerate their transition to clinical evaluation (Fig. 3.1). Although many of these microfluidic systems are still being developed, they have the potential to become widely adopted because they are economical, reproducible, amenable to modifications and can be integrated with other technologies [15]. In this chapter, we highlight some of these technologies and discuss their impact on accelerating the clinical translation of nanoparticles.

3.1 Well-Controlled Synthesis of Nanoparticles

Amphiphilic molecules such as block copolymers and lipids can self-assemble into nanoparticles when they experience a change in solvent quality (for example, from organic solvent to aqueous) (Fig. 3.2a). A common and flexible way to accomplish a change in solvent quality is by mixing the solvent with the anti-solvent, where mixing time directly influences the final size and size distribution of the nanoparticles formed [16]. If the mixing timescale, τ_{mix}, is longer than the characteristic timescale for chains to nucleate and grow (τ_{agg}, ~10–100 ms depending on the molecular weight of the chain), the nanoparticles begin to assemble under varying degrees of solvent quality. This heterogeneous environment prevents effective stabilization of the nanoparticles by the hydrophilic portion of the amphiphilic molecule and facilitates their aggregation, leading to the formation of larger, polydisperse nanoparticles. However, if $\tau_{mix} < \tau_{agg}$, particle self-assembly occurs primarily when the solvent change is complete. This homogenous solvent environment for nanoparticle assembly allows the hydrophilic portion of the molecule to stabilize the nanoparticles more effectively, and this yields smaller nanoparticles with uniform size [17]. Although conventional bulk mixing occurs at the timescale of seconds, in microfluidic devices the mixing time of solvents is controllable and tunable from the millisecond to microsecond scale (reaching $\tau_{mix} < \tau_{agg}$) [16, 18].

In recent years, several microfluidic systems that enable rapid mixing without the need of external actuators, such as stirrers or electric fields, have been developed [19]. The most widely used include flow-focusing mixers [20], droplet mixers [21] and those with micromixing structures embedded inside the channel [22]. Flow focusing squeezes the solvent stream between two anti-solvent

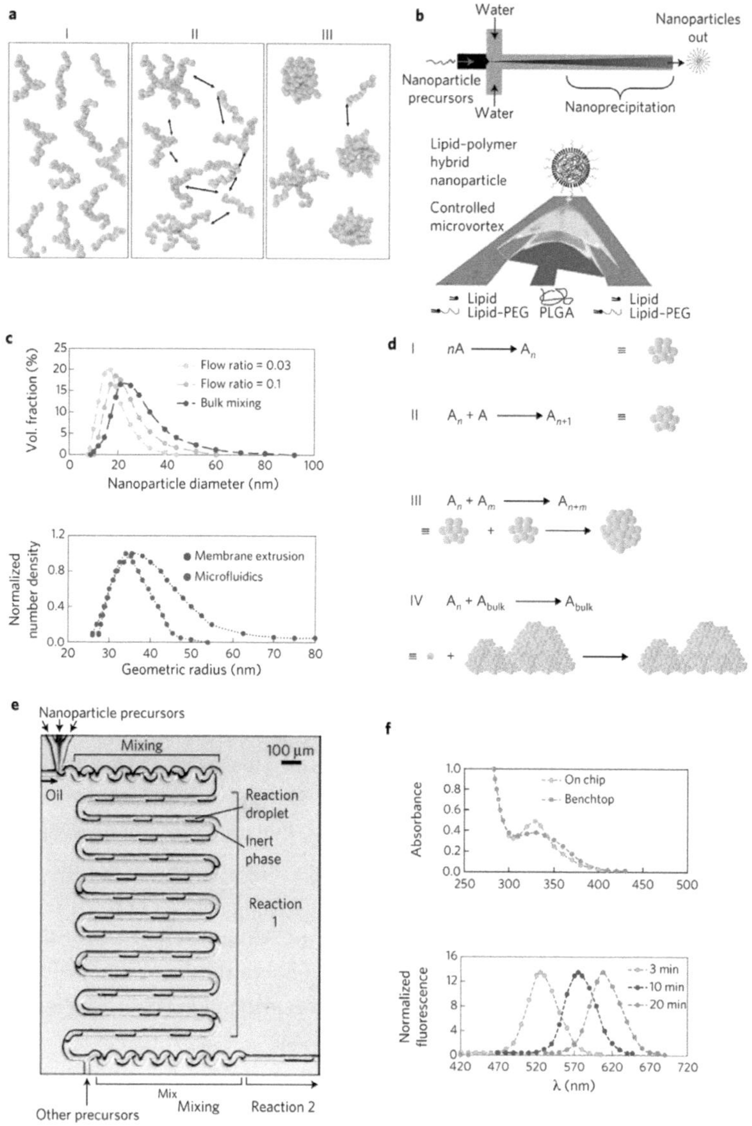

Figure 3.2 Microfluidic synthesis of nanoparticles. **a**, Schematic of the self-assembly mechanism of organic nanoparticles. On mixing with anti-solvent, polymers (or lipids) are brought to the vicinity of each other (I) then nucleate (II), subsequently aggregating into nanoparticles (III).

Figure 3.2 (*Continued*) **b**, Schematic of microfluidic synthesis of organic nanoparticles by rapid mixing through hydrodynamic flow focusing (top) and microvortices (bottom). Red and dark blue indicate organic and aqueous streams, respectively, while pink and light blue indicate their degree of mixing. PEG, polyethylene glycol; PLGA, poly(lactic-co-glycolic acid). **c**, Size distribution of polymeric nanoparticles (top) and liposomes (bottom) prepared in microfluidics compared with bulk synthesis. In both cases, narrower particle-size distributions are produced through microfluidics. **d**, Schematic of the self-assembly mechanism of inorganic nanoparticles. Individual molecules first nucleate (I and II), followed by aggregation of nuclei into nanoparticles (III). If the reaction is not quenched or stabilized, nanoparticles tend to agglomerate into bulk material (IV). A refers to individual molecules forming the nanoparticle, and A_n and A_m refer to nuclei formed of n and m number of A molecules, respectively. **e**, Microfluidic synthesis of inorganic nanoparticles by rapid mixing through two-phase flow where reagents are embedded in fluid droplets carried by an inert fluid. **f**, Top: sharp versus broad absorption maximum of QDs synthesized in microchannels and bulk, respectively. Bottom: control of the absorption spectra of QDs as function of reaction time. Figure reproduced with permission from: **a**, ref. 16, © 2003 APS; **b**, Top: ref. 18, © 2008 ACS; Bottom: ref. 35, © 2012 ACS; **c**, Top: ref. 18, © 2008 ACS; Bottom: ref. 23, © 2008 Springer; **d**, ref. 24, © 2005 ACS; **e**, ref. 27, © 2004 RSC; **f**, Top: ref. 28, © 2010 Wiley; Bottom: ref. 27, © 2004 RSC.

streams, resulting in rapid solvent exchange via diffusion (Fig. 3.2b). Droplets and three-dimensional microchannel geometries result in complex folding of fluid flows, which can completely mix two or more streams in milliseconds (Fig. 3.2b). The implementation of these mixing techniques for the formation of organic nanoparticles in continuous flow has resulted in polymeric and lipid nanoparticles with tunable nanoparticle size, narrower size distribution, higher drug loadings and greater batch-to-batch reproducibility relative to those made with conventional bulk techniques [23] (Fig. 3.2c).

Similarly, inorganic nanoparticles comprising transition metals such as gold, iron and cadmium, among others, undergo self-assembly where metal solutes nucleate, grow and agglomerate into nanoclusters (Fig. 3.2d) [24]. Obtaining narrow particle-size distribution requires rapid nucleation followed by growth of nanoparticles to the desired size in the absence of further nucleation, which can be accomplished by controlling the mixing time of reagents, reaction temperature and reaction time [25]. In bulk, these parameters are difficult to control, leading to uneven mixing, local temperature fluctuations and uncontrolled reaction times [25]. In contrast, microfluidic devices allow for control over the mixing time by varying solvent flow rates or channel geometry. Moreover, better heat transfer owing to large surface areas enables better temperature control, preventing the formation of large temperature gradients. Finally, as the channel length directly corresponds to the time taken by the reactants to flow through it in continuous flow synthesis, the reaction time can be controlled by tuning the channel length or by adding reagents at precise downstream locations during the particle formation process to quench the reaction [26].

Two-phase droplet mixers where reagents are encapsulated in droplets and separated by inert fluids are commonly used for the synthesis of inorganic nanoparticles (Fig. 3.2e) [27]. In this configuration, rapid mixing of the solutions occurs inside the droplets, which serve as identical microscale reactors providing homogeneous conditions for nanoparticle nucleation and growth. Both droplet-based and single-phase systems have been used to synthesize QDs that exhibit narrow size distributions, which translates into sharper absorption peaks and better luminescence qualities [27], with control over size to tune the absorption spectra [28] (Fig. 3.2f). Finally, similar systems have been implemented in the controlled synthesis of gold nanoparticles of defined size and shape, iron oxide nanoparticles with higher magnetization, and QDs with controlled size and biocompatible coatings [29].

Over the past four years, several examples showing the use of microfluidics for the synthesis of nanoparticles with different size, shape and surface compositions have emerged [30]. At present, there are microfluidic systems capable of characterizing the nanoparticle size and stability following synthesis in a single platform [31, 32]. Moreover, large numbers of distinct nanoparticles can be obtained through combinatorial synthesis [33], and production rates of identical nanoparticles can be increased through parallelization or re-design of the devices [34, 35]. Similar to high-throughput synthesis of libraries of small molecules, these advantages could potentially enable screening and optimization of libraries of nanoparticles with distinct properties. One of the challenges in gene therapy, for example, is finding the right formulation for delivering nucleic acids to specific sites in the body. By mixing different precursors at varying ratios, microfluidic systems have enabled a one-step combinatorial synthesis of libraries of polymeric and lipid nanoparticles that encapsulate DNA and small-interfering RNA, respectively. Screening these libraries of nanoparticles has helped identify superior formulations for gene transfection, compared with conventional transfection agents such as lipofectamine 2000. Furthermore, using this method, potent lipid-based small-interfering RNA formulations for in vivo delivery to the liver have also been discovered [33, 34].

3.2 Evaluation and Screening of Nanoparticles

Another challenge in nanoparticle development is the lack of in vitro models capable of predicting in vivo behaviour [36]. Conventionally, nanoparticles are evaluated in vitro using cells cultured in well plates, which does not capture the complexity of nanoparticle–cell interactions in vivo. For instance, a recent study showed that sedimentation of gold nanoparticles in well plates could lead to misinterpretation of results, such as increased nanoparticle uptake [37]. Microfluidics provides significant advantages over conventional methods for cell and tissue culture by displaying structures and networks at relevant physiological length scales, and by incorporating fluid flow and mechanical forces that bring the cell-based assays a step closer to mimicking the in vivo microenvironment [38] (Fig. 3.3a). This is especially advantageous, for instance, when investigating the toxic effects related to the cell uptake of nanoparticles. A recently developed microfluidic system for evaluating QD toxicity on mouse fibroblasts revealed increased cell viability under flow conditions compared with static incubation, possibly due to the absence of QD sedimentation [39].

Recently, researchers have focused on developing biomimetic microfluidic technologies capable of portraying organ-level functions on a chip, such as those observed in the lung, liver and kidneys, among others [38, 40]. For instance, microfluidic systems

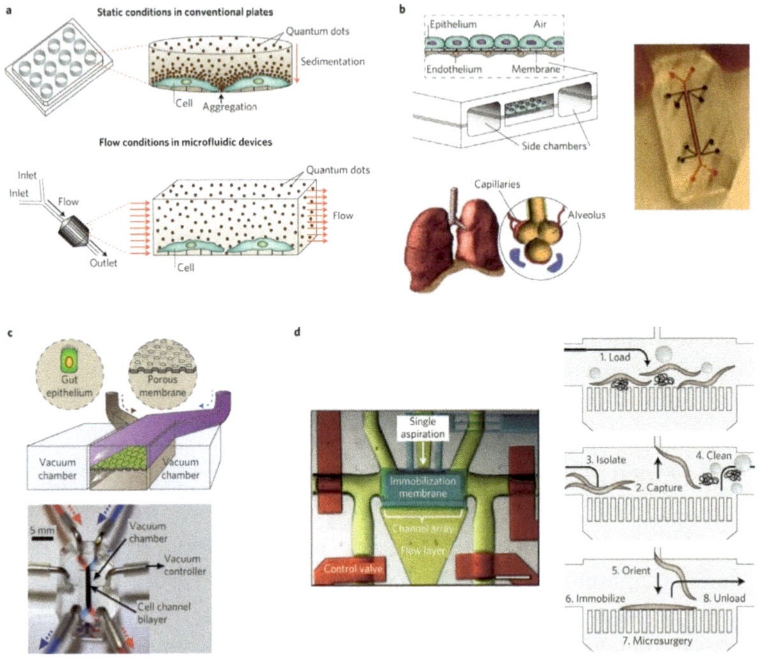

Figure 3.3 Microfluidic systems for in vitro evaluation and screening of nanoparticles. **a**, Schematic of nanoparticle sedimentation in conventional plates, which could result in misinterpretation of results. In contrast, flow conditions in microfluidics provide a more accurate method for evaluating nanoparticles in vitro. **b**, Left: schematic of the lung-on-a-chip that reconstitutes the critical functional alveolar-capillary interface of the human lung through a stretchable membrane containing an epithelium layer on one side and an endothelium layer on the other. Right: photograph of actual device. **c**, Top: schematic of the gut-on-a-chip made by flexible, porous, extracellular matrix-coated membrane lined by gut epithelial cells. The blue and brown arrows indicate two different streams of culture medium separated by a membrane, entering the channel from the top and bottom, respectively. Bottom: photograph of the gut-on-a-chip device made of polydimethylsiloxane elastomer. A syringe pump was used to perfuse dyes (red and blue) for channel visualization. **d**, Left: photograph of a dye-filled microfluidic system designed to handle *C. elegans* worms. Red, control valve layer; yellow, flow layer; blue, immobilization layer. Scale bar, 1 mm. Right: schematic showing load, capture, orient, immobilization and unload of the worm. Figure reproduced with permission from: **a**, ref. 39, © 2010 AIP; **b**, ref. 41, © 2010 AAAS; **c**, ref. 42, © 2012 RSC; **d**, ref. 48, © 2010 NAS.

that reconstitute the critical functional alveolar-capillary liquid/air interface of the human lung have been recently fabricated by growing alveolar epithelial cells and microvascular endothelial cells on different sides of a perforated silicone membrane. The membrane was pneumatically actuated to mimic the physiological expansion–contraction motion due to breathing. It was found that cyclic mechanical strain accentuates toxic and inflammatory responses in the lung when exposed to silica nanoparticles, which could not have been observed with other conventional in vitro systems [41] (Fig. 3.3b). Using a similar design approach, a 'gut-on-a-chip' was developed by coating both sides of a membrane separating two microfluidic devices with extracellular matrix and lined by human intestinal epithelial cells. It was demonstrated that by subjecting the membrane to flow and cyclic strains similar to those encountered in the gut, villi-like structures were formed and the co-culture of the intestinal microbes was made possible [42] (Fig. 3.3c).

Expansion of these technologies to other organs, for instance 'liver-on-a-chip' [43], could lead to platforms for evaluating and screening nanoparticle toxicity in organs where nanoparticles tend to accumulate and toxicity is likely to be a major concern (for example, liver, spleen and kidney). Although nanoparticles would still need to be evaluated in animals, such microfluidic systems could take in vitro nanoparticle screening to a new level of utility by selecting promising candidates with higher probabilities of success from a large pool of nanoparticle formulations, and eliminating those that would otherwise have failed in larger animal studies. Furthermore, coupling these technologies with microfluidic devices used for nanoparticle synthesis opens the possibility of rapid combinatorial screening of a large number of different nanoparticles under various conditions (for example, concentrations, pH values, under the presence of specific proteins and so on).

Nanoparticles exhibiting promising results in vitro are subsequently evaluated in vivo, which is considerably more expensive and resource intensive, especially in non-human primates. Although most of the parameters, such as pharmacokinetics, biodistribution and efficacy, are evaluated in mice and larger animals, tracking physiological effects of nanoparticles on animal development could potentially be obtained using a large number of smaller organisms.

The zebrafish and Caenorhabditis elegans worms are well-known models for studying fundamental mechanisms and progression of human diseases, and for drug screening [44]. For example, the zebrafish was recently used as an in vivo model to develop a hazard ranking for engineered nanoparticles based on their impact on mortality rate and morphological defects in zebrafish embryos exposed to these materials [45]. However, current methods for manipulating these organisms generally suffer from low throughput, low automation and imprecise delivery of external stimuli [46]. To solve these challenges, engineered microfluidic systems with dimensions comparable to small organisms and containing valves and suction points have been developed. These systems enable precise manipulation of these organisms with respect to placement and orientation for high-throughput screening [46, 47] (Fig. 3.3d). Other microfluidic systems are being developed that are capable of imaging dynamic cellular processes in small organisms, such as cell division and migration, degeneration, aging and regeneration [48]. With such technologies in place, it might be possible to use real-time microscopy to track physiological responses to fluorescently labelled nanotherapeutics and nano-imaging agents, as well as assess the distribution and efficacy of nanoparticles at both the organ and body level. Furthermore, real-time tracking of nanoparticle-induced toxicity at different concentrations and conditions in small organisms could enable rapid selection of nanoparticles (especially those made with novel synthetic materials) that are more likely to be non-toxic in larger animals.

3.3 Future Prospects

At present, the field of microfluidics applied to nanomedicine is still in its infancy. Although nanoparticles have a relatively small footprint in the pharmaceutical industry, it is anticipated that as these products bring in revenue, industry-led research and development efforts would probably adopt technologies, such as microfluidics, to accelerate their development. Nevertheless, microfluidic technologies, such as organ-on-a-chip and small-animal screening, are likely to be adopted first for the screening of small-molecule drug candidates, where the need for such tools is evident.

There are a few key directions at the intersection of microfluidics and nanomedicine that are likely to be pursued in the near future (Table 3.1). Although the quantity of nanoparticles synthesized by microfluidic devices is often in the micro- to milligram range, parallel and stackable microfluidic systems could continuously produce nanoparticles on the gram to kilogram scale with the same properties as those prepared at the bench scale. Similarly, the use of microfluidic platform technologies to reproducibly synthesize and screen libraries of nanoparticles with different chemical compositions and/or physical and chemical properties could potentially advance nanoparticle discovery analogously to how the high-throughput screening of small molecules in medicinal chemistry advanced small-molecule discovery. With respect to the design and development of novel nanoparticle constructs, the use of microfluidics could enable the synthesis of nanoparticles with properties not accessible by conventional synthesis, similar to what has already occurred for microparticle synthesis [49].

Another avenue of future research will be the integration of different steps of nanoparticle development into a single system (for example, nanoparticle synthesis characterization and evaluation), together with feedback control through a combination of microfluidics, robotics and automation, thus significantly cutting the time and cost of nanoparticle development. Finally, mass-produced microfluidic devices and well-defined nanoparticle precursors can aid in the synthesis of identical batches of nanoparticles with little to no variations introduced by user handling. This could lead to the use and commercialization of 'nanoparticle synthesis kits' composed of calibrated devices that can reproducibly synthesize a specific class of nanoparticle with well-defined properties for use as standards in conventional toxicological assays. Considering the large number of nanoparticles being made of novel synthetic materials or of unusual shapes, such standardization would be highly useful for regulatory purposes, among others.

Microfluidic technologies are capable of accelerating the discovery and translation of nanoparticles, and could serve as a tool for nanotherapeutics to reach a similar 'tipping point' reached

Table 3.1 Advantages, disadvantages/challenges, stage of development and potential impact of microfluidic systems on different steps in the clinical translation of nanoparticles

	Advantages	Disadvantages/challenges	Stage of development	Potential impact
Synthesis	• Tunable nanoparticle size • Narrower size distribution • Reproducible synthesis • Potential for high-throughput synthesis and optimization of nanoparticles	• Solvent and high-temperature incompatibility for low-cost polydimethylsiloxane microchannels • Higher costs and complexities in the fabrication of glass and silicon microdevices	*****	Rapid combinatorial, controlled and reproducible synthesis of libraries of distinct nanoparticles for a specific application, and/or reference nanoparticles for toxicology studies
Characterization	• Label-free characterization • Potential for feedback control and real-time nanoparticle optimization	• Current methods are not applicable to all classes of nanoparticles • Not all properties can be characterized, such as drug encapsulation and release, and signal-to-noise ratio	*	In-line rapid characterization and optimization of nanoparticles
In vitro	• Biological conditions closer to in vivo microenvironments • Potential for high-throughput screening of a large number of nanoparticles at different concentrations	• Higher costs and complexities in the fabrication and operation compared with well plates • Might not be reusable and if reusable, it would be difficult to keep sterile	****	High-throughput studies of nanoparticle toxicity, efficacy, tumour penetration and organ distribution, using 'organ-on-a-chip' systems

In vivo	• Large number of organisms could be used for a single measurement • High-throughput evaluation of toxicity for a large number of nanoparticles	• Lack of methods to translate data from small-scale organisms to other species • Pharmacokinetics or biodistribution cannot be determined	**	Real-time tracking of the distribution or toxicity of nanoparticles on small-scale organisms
Large-scale synthesis	• Continuous synthesis • Bench-scale to clinical-scale reproducibility • Parallelization allows for tuning scale of production	• Difficult to build systems at low-cost that are comparable to a batch reactor able to prepare grams or kilograms of nanoparticles	***	Synthesis of nanoparticles for human administration using stackable parallel microfluidic units

Rank: Most advanced in development (*****) to least advanced in development (*), based on the amount of research carried out on each category, as well as the potential ease of adoption by industry.

by genome sequencing in the past decade after high-throughput sequencing technologies were developed [50]. Among all the microfluidic technologies, those developed for synthesis and in vitro screening of nanoparticles have the highest probability of making an impact in the near future (Table 3.1). Specifically, microfluidic synthesis may be adopted as a second-generation manufacturing technology after the initial success of US Food and Drug Administration-approved nanoparticles in cases where the advantages of microfluidic synthesis are significant. Microfluidic synthesis may also be adopted as a screening tool to identify optimal nanoparticles in academic and industrial research laboratories. Alternatively, the impact of microfluidics might be observable in the medium- to long-term future for nanoparticle characterization and in vivo evaluation. For nanoparticle characterization, the use of microfluidics would probably increase once more-advanced technologies are developed to characterize several nanoparticle properties (for example, size, charge, surface composition and stability) in a single system. Similarly, in vivo evaluation of nanoparticles in microfluidics would probably mature once both easily adoptable microfluidic systems for manipulating small organisms and methods for translating data obtained from these organisms to larger animals are developed.

Overall, the use of microfluidic technologies in nanomedicine brings exciting opportunities to expand the body of knowledge in the field, advance the clinical translation of nano-based therapeutics and imaging agents, and demonstrate innovative ways to develop other classes of drugs.

Acknowledgements

This work was supported by the Koch-Prostate Cancer Foundation Award in Nanotherapeutics (R.L. and O.C.F.), the National Cancer Institute Center of Cancer Nanotechnology Excellence at MIT-Harvard (U54-CA151884, R.L. and O.C.F.), and the National Heart, Lung, and Blood Institute Programs of Excellence in Nanotechnology (HHSN268201000045C; R.L. and O.C.F.). P.M.V. acknowledges support by the National Science Foundation graduate research fellowship. We thank B. Timko and F. Karim for assistance in drafting Figs 3.1 and 3.2, respectively. We also thank A. Radovic-Moreno, C. Alabi and E. Pridgen for their insightful comments.

Additional Information

O.C.F. and R.L. disclose financial interest in Selecta Biosciences, Tarveda Therapeutics and Seer. R.L. discloses financial interest in Moderna.

References

1. Petros, R. A., and DeSimone, J. M. Strategies in the design of nanoparticles for therapeutic applications. *Nature Rev. Drug. Discov.* **9**, 615–627 (2010).
2. Gregoriadis, G. Drug entrapment in liposomes. *FEBS Lett.* **36**, 292–296 (1973).
3. Hrkach, J. et al. Preclinical development and clinical translation of a PSMA-targeted docetaxel nanoparticle with a differentiated pharmacological profile. *Sci. Transl. Med.* **4**, 128ra39 (2012). This article describes the translation of the first targeted polymeric nanoparticle for drug delivery from discovery to clinical trials.
4. Qiao, R., Yang, C., and Gao, M. Superparamagnetic iron oxide nanoparticles: from preparations to in vivo MRI applications. *J. Mater. Chem.* **19**, 6274–6293 (2009).
5. Haun, J. B. et al. Micro-NMR for rapid molecular analysis of human tumor samples. *Sci. Transl. Med.* **3**, 71ra16 (2011).
6. Kim, B. Y., Rutka, J. T., and Chan, W. C. Nanomedicine. *N. Engl. J. Med.* **363**, 2434–2443 (2010).
7. Kamaly, N., Xiao, Z., Valencia, P. M., Radovic-Moreno, A. F., and Farokhzad, O. C. Targeted polymeric therapeutic nanoparticles: design, development and clinical translation. *Chem. Soc. Rev.* **41**, 2971–3010 (2012).
8. Peer, D. et al. Nanocarriers as an emerging platform for cancer therapy. *Nature Nanotech.* **2**, 751–760 (2007).
9. Barreto, J. A. et al. Nanomaterials: applications in cancer imaging and therapy. *Adv. Mater.* **23**, H18–H40 (2011).

10. Shi, J., Xiao, Z., Kamaly, N., and Farokhzad, O. C. Self-assembled targeted nanoparticles: evolution of technologies and bench to bedside translation. *Acc. Chem. Res.* **44**, 1123–1134 (2011).
11. Murday, J. S., Siegel, R. W., Stein, J., and Wright, J. F. Translational nanomedicine: status assessment and opportunities. *Nanomedicine* **5**, 251–273 (2009).
12. Chou, L. Y., Ming, K., and Chan, W. C. Strategies for the intracellular delivery of nanoparticles. *Chem. Soc. Rev.* **40**, 233–245 (2011).
13. Nel, A. E. et al. Understanding biophysicochemical interactions at the nano–bio interface. *Nature Mater.* **8**, 543–557 (2009).
14. Whitesides, G. M. The origins and the future of microfluidics. *Nature* **442**, 368–373 (2006). An excellent classic review on the present and future of microfluidics by one of the fathers of the field, George Whitesides.
15. DeMello, A. J. Control and detection of chemical reactions in microfluidic systems. *Nature* **442**, 394–402 (2006).
16. Johnson, B. K., and Prud'homme, R. K. Mechanism for rapid self-assembly of block copolymer nanoparticles. *Phys. Rev. Lett.* **91**, 118302 (2003). This article describes the mechanism of nanoparticle self-assembly and explains how rapid mixing is key in controlling nanoparticle size.
17. Chen, T., Hynninen, A. P., Prud'homme, R. K., Kevrekidis, I. G., and Panagiotopoulos, A. Z. Coarse-grained simulations of rapid assembly kinetics for polystyrene-*b*-poly(ethylene oxide) copolymers in aqueous solutions. *J. Phys. Chem. B* **112**, 16357–16366 (2008).
18. Karnik, R. et al. Microfluidic platform for controlled synthesis of polymeric nanoparticles. *Nano Lett.* **8**, 2906–2912 (2008).
19. Capretto, L., Cheng, W., Hill, M., and Zhang, X. Micromixing within microfluidic devices. *Top. Curr. Chem.* **304**, 27–68 (2011).
20. Rhee, M. et al. Synthesis of size-tunable polymeric nanoparticles enabled by 3D hydrodynamic flow focusing in single-layer microchannels. *Adv. Mater.* **23**, H79–H83 (2011).
21. Liu, K. et al. A digital microfluidic droplet generator produces self-assembled supramolecular nanoparticles for targeted cell imaging. *Nanotechnology* **21**, 445603 (2010).
22. Valencia, P. M. et al. Single-step assembly of homogenous lipid-polymeric and lipid-quantum dot nanoparticles enabled by microfluidic rapid mixing. *ACS Nano* **4**, 1671–1679 (2010).

23. Jahn, A. et al. Preparation of nanoparticles by continuous-flow microfluidics. *J. Nanopart. Res.* **10**, 925–934 (2008).

24. Besson, C., Finney, E. E., and Finke, R. G. A mechanism for transition-metal nanoparticle self-assembly. *J. Am. Chem. Soc.* **127**, 8179–8184 (2005).

25. Song, Y., Hormes, J., and Kumar, C. S. Microfluidic synthesis of nanomaterials. *Small* **4**, 698–711 (2008).

26. Gu, F. X. et al. Targeted nanoparticles for cancer therapy. *Nano Today* **2**, 14–21 (2007).

27. Shestopalov, I., Tice, J. D., and Ismagilov, R. F. Multi-step synthesis of nanoparticles performed on millisecond time scale in a microfluidic droplet-based system. *Lab Chip* **4**, 316–321 (2004).

28. Kikkeri, R., Laurino, P., Odedra, A., and Seeberger, P. H. Synthesis of carbohydrate-functionalized quantum dots in microreactors. *Angew. Chem. Int. Ed.* **49**, 2054–2057 (2010).

29. Marre, S., and Jensen, K. F. Synthesis of micro and nanostructures in microfluidic systems. *Chem. Soc. Rev.* **39**, 1183–1202 (2010).

30. Zhao, C. X., He, L. Z., Qiao, S. Z., and Middelberg, A. P. J. Nanoparticle synthesis in microreactors. *Chem. Eng. Sci.* **66**, 1463–1479 (2011).

31. Fraikin, J. L., Teesalu, T., McKenney, C. M., Ruoslahti, E., and Cleland, A. N. A high-throughput label-free nanoparticle analyser. *Nature Nanotech.* **6**, 308–313 (2011).

32. Birnbaumer, G. et al. Rapid liposome quality assessment using a lab-on-a-chip. *Lab Chip* **11**, 2753–2762 (2011).

33. Wang, H. et al. A rapid pathway toward a superb gene delivery system: programming structural and functional diversity into a supramolecular nanoparticle library. *ACS Nano* **4**, 6235–6243 (2010). This article is one of the first examples that exploit microfluidic systems for rapid combinatorial synthesis of nanoparticles with a variety of physical and chemical properties.

34. Chen, D. et al. Rapid discovery of potent siRNA-lipid-nanoparticles enabled by controlled microfluidic formulation. *J. Am. Chem. Soc.* **134**, 6948–6951 (2012).

35. Kim, Y. et al. Mass production and size control of lipid-polymer hybrid nanoparticles through controlled microvortices. *Nano Lett.* **12**, 3587–3591 (2012).

36. Dobrovolskaia, M. A., Germolec, D. R., and Weaver, J. L. Evaluation of nanoparticle immunotoxicity. *Nature Nanotech.* **4**, 411–414 (2009).

37. Cho, E. C., Zhang, Q., and Xia, Y. The effect of sedimentation and diffusion on cellular uptake of gold nanoparticles. *Nature Nanotech.* **6**, 385–391 (2011).
38. Ziolkowska, K., Kwapiszewski, R., and Brzozka, Z. Microfluidic devices as tools for mimicking the in vivo environment. *New J. Chem.* **35**, 979–990 (2011).
39. Mahto, S. K., Yoon, T. H., and Rhee, S. W. A new perspective on in vitro assessment method for evaluating quantum dot toxicity by using microfluidics technology. *Biomicrofluidics* **4**, 034111 (2010).
40. Huh, D., Torisawa, Y. S., Hamilton, G. A., Kim, H. J., and Ingber, D. E. Microengineered physiological biomimicry: organs-on-chips. *Lab Chip* **12**, 2156–2164 (2012).
41. Huh, D. et al. Reconstituting organ-level lung functions on a chip. *Science* **328**, 1662–1668 (2010). (This article describes the design and assembly of a microfluidic system that recreates the alveolar-endothelial interface in lungs.)
42. Kim, H. J., Huh, D., Hamilton, G., and Ingber, D. E. Human gut-on-a-chip inhabited by microbial flora that experiences intestinal peristalsis-like motions and flow. *Lab Chip* **12**, 2165–2174 (2012).
43. Toh, Y. C. et al. A microfluidic 3D hepatocyte chip for drug toxicity testing. *Lab Chip* **9**, 2026–2035 (2009).
44. Crane, M. M., Chung, K., Stirman, J., and Lu, H. Microfluidics-enabled phenotyping, imaging, and screening of multicellular organisms. *Lab Chip* **10**, 1509–1517 (2010).
45. George, S. et al. Use of a high-throughput screening approach coupled with in vivo zebrafish embryo screening to develop hazard ranking for engineered nanomaterials. *ACS Nano* **5**, 1805–1817 (2011).
46. Shi, W., Wen, H., Lin, B., and Qin, J. Microfluidic platform for the study of Caenorhabditis elegans. *Top. Curr. Chem.* **304**, 323–338 (2011).
47. Baker, M. Screening: the age of fishes. *Nature Meth.* **8**, 47–51 (2011).
48. Samara, C. et al. Large-scale in vivo femtosecond laser neurosurgery screen reveals small-molecule enhancer of regeneration. *Proc. Natl Acad. Sci. USA* **107**, 18342–18347 (2010).
49. Dendukuri, D., and Doyle, P. S. The synthesis and assembly of polymeric microparticles using microfluidics. *Adv. Mater.* **21**, 4071–4086 (2009).
50. Zhao, J., and Grant, S. F. Advances in whole genome sequencing technology. *Curr. Pharm. Biotechnol.* **12**, 293–305 (2011).

Chapter 4

Theranostic Nanomedicine

Twan Lammers,[a,b,c] Silvio Aime,[d] Wim E. Hennink,[b] Gert Storm,[b,c] and Fabian Kiessling[a]

[a] *Department of Experimental Molecular Imaging, Helmholtz Institute for Biomedical Engineering, RWTH - Aachen University, Pauwelsstrasse 20, 52074 Aachen, Germany*
[b] *Department of Pharmaceutics, Utrecht Institute for Pharmaceutical Sciences, Utrecht University, Sorbonnelaan 16, 3584 CA Utrecht, The Netherlands*
[c] *Department of Innovative Cancer Diagnosis and Therapy, Clinical Cooperation Unit Radiation Oncology, German Cancer Research Center, Im Neuenheimer Feld 280, 69120 Heidelberg, Germany*
[d] *Department of Chemistry IFM and Molecular Imaging Center, University of Torino, Via P. Giuria 7, 10125 Torino, Italy*
tlammers@ukaachen.de

Nanomedicine formulations aim to improve the biodistribution and the target site accumulation of systemically administered (chemo)therapeutic agents. Many different types of nanomedicines have been evaluated over the years, including for instance liposomes, polymers, micelles and antibodies, and a significant amount of evidence has been obtained showing that these submicrometer-sized carrier materials are able to improve the balance between the efficacy and the toxicity of therapeutic interventions. Besides

---

Reproduced with permission from *Accounts of Chemical Research*, 44, 1029–1038, 2011.

Nano-Enabled Medical Applications
Edited by Lajos P. Balogh

ISBN 978-981-4800-32-7 (Hardcover), 978-0-429-39903-9 (eBook)
www.jennystanford.com

for therapeutic purposes, nanomedicine formulations have in recent years also been increasingly employed for imaging applications. Moreover, paralleled by advances in chemistry, biology, pharmacy, nanotechnology, medicine and imaging, several different systems have been developed in the last decade in which disease diagnosis and therapy are combined. These so-called (nano) theranostics contain both a drug and an imaging agent within a single formulation, and they can be used for various different purposes. In this Account, we summarize several exemplary efforts in this regard, and we show that theranostic nanomedicines are highly suitable systems for monitoring drug delivery, drug release and drug efficacy. The (pre)dinically most relevant applications of theranostic nanomedicines relate to their use for validating and optimizing the properties of drug delivery systems, and to their ability to be used for pre-screening patients and enabling personalized medicine. Regarding the former, the combination of diagnostic and therapeutic agents within a single formulation provides real-time feedback on the pharmacokinetics, the target site localization and the (off-target) healthy organ accumulation of nanomedicines. Various examples of this will be highlighted in this Account, illustrating that by non-invasively visualizing how well carrier materials are able to deliver pharmacologically active agents to the pathological site, and how well they are able to prevent them from accumulating in potentially endangered healthy tissues, important information can be obtained for optimizing the basic properties of drug delivery systems, as well as for improving the balance between the efficacy and the toxicity of targeted therapeutic interventions. Regarding personalized medicine, it can be reasoned that only in patients which show high levels of target site accumulation, and which respond well to the first couple of treatment cycles, targeted therapy should be continued, and that in those in which this is not the case, other therapeutic options should be considered. Based on these insights, we expect that ever more efforts will be invested in developing theranostic nanomedicines, and that these systems and strategies will contribute substantially to realizing the potential of personalized medicine.

4.1 Introduction

Nanomedicines are submicrometer-sized carrier materials designed to improve the biodistribution of systemically administered (chemo)therapeutic agents. Clinically relevant examples of nanomedicine formulations are liposomes, polymers, micelles, solid (lipid) nanoparticles, and antibodies. By delivering pharmacologically active agents more effectively and more selectively to the pathological site (site-specific drug delivery) and/or by guiding them away from potentially endangered healthy tissues (site-avoidance drug delivery), nanomedicines aim to improve the balance between the efficacy and the toxicity of systemic (chemo)therapeutic interventions [1–4].

Besides for drug targeting to pathological sites and for therapeutic purposes, nanomedicine formulations have also been more and more used for imaging applications as well as, in recent years, for theranostic approaches, that is, for systems and strategies in which disease diagnosis and therapy are combined [5–8]. To this end, on the one hand, "classical" drug delivery systems, such as liposomes, polymers, micelles, solid (lipid) nanoparticles, and antibodies, are being co-loaded both with drugs and with contrast agents (Fig. 4.1). On the other hand, also nanomaterials with an intrinsic ability to be used for imaging purposes, such as gold- and iron oxide-based nanoparticles, are increasingly being loaded with drugs and implemented for combining disease diagnosis and therapy.

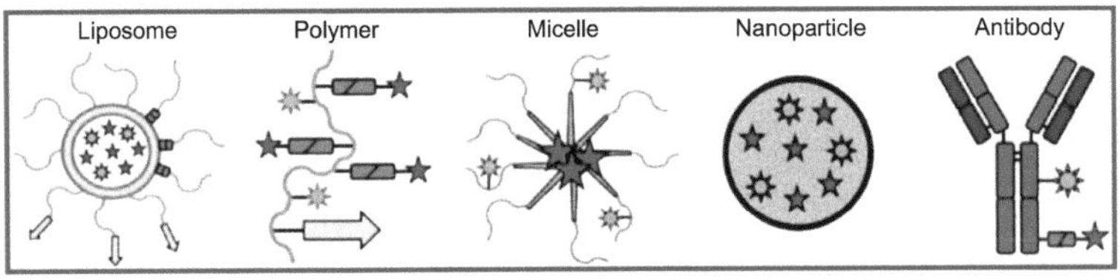

Figure 4.1 Theranostic nanomedicines. Liposomes and liposomal bilayers are shown in gray; polymers and polymer-coatings in green; solid (lipid) nanoparticle components in brown; antibodies in purple; linkers allowing for drug release and for sheddable stealth coatings in blue; targeting ligands in yellow; imaging agents in orange; and conjugated or entrapped pharmacologically active agents in red.

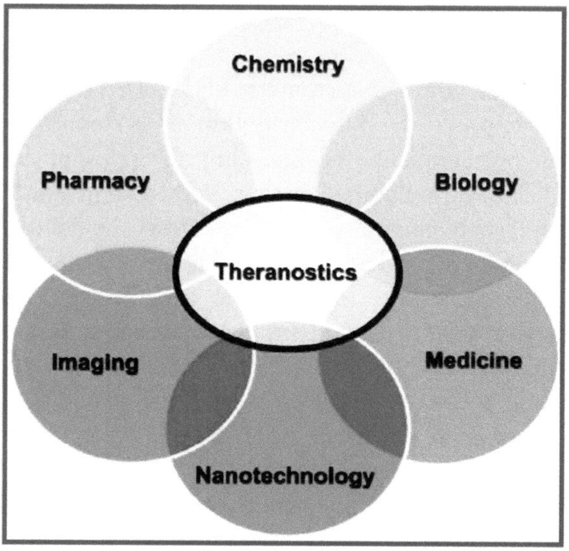

Figure 4.2 Theranostics. Schematic representation of the highly interdisciplinary field of (nano)theranostics. Theranostics are nanomedicine formulations which aim to combine disease diagnosis (in its broadest sense; including patient prescreening and therapy monitoring) and therapy, and which are developed and tested by researchers working at the intersection of several different scientific fields.

Based on the advances made in, and the close collaboration between, several different scientific disciplines, such as chemistry, biology, pharmacy, nanotechnology, medicine, and imaging, ever more of such theranostic nanomedicines have been designed and evaluated over the past few years, and ever more interesting applications are being envisioned for nanotheranostics. In the present Account, we summarize several exemplary efforts in this regard, and we show that theranostic nanomedicines are highly suitable systems for monitoring drug delivery, drug release, and drug efficacy.

4.2 Imaging Drug Delivery

As already briefly outlined above, nanomedicine formulations are primarily designed to improve the biodistribution and the target site accumulation of systemically administered (chemo)therapeutic agents. To facilitate pharmacokinetic and biodistributional analyses, and to thereby improve drug targeting to pathological sites, it would be highly useful if the circulation time and the organ accumulation of nanomedicine formulations could be visualized noninvasively in real-time. To achieve this goal, many different types of nanomedicines have been co-loaded both with drugs and with imaging agents.

In the majority of cases, radionuclides have been used for such purposes. Large numbers of radionuclide-labeled liposomes, polymers, micelles, nanoparticles, and antibodies have been subjected to biodistribution analyses over the years, both in animal models and in patients, and it has become clear that such studies substantially assist in improving our understanding of the drug delivery process, as well as in predicting the therapeutic potential of (tumor-) targeted nanomedicines.

An exemplary study in this regard showed that iodine-131-labeled HPMA (i.e., *N*-(2-hydroxypropyl)methacrylamide) copolymers can be used to visualize the biodistribution of long-circulating and passively tumor-targeted polymeric drug carriers in rats bearing subcutaneously transplanted Dunning ATI tumors. Using such radiolabeled copolymers, it has for instance been demonstrated that in line with the principles of enhanced permeability and retention (EPR)-mediated passive drug targeting [9], higher molecular weight HPMA copolymers circulate for significantly longer periods of time than do lower molecular weight copolymers and consequently accumulate in tumors significantly more effectively [10]. In a similar setup, it has been demonstrated that the physico-chemical modification of HPMA copolymers, e.g., with positively or negatively charged groups, as well their functionalization with drugs, with drug linkers and with targeting peptides, reduces the long-circulating properties of the copolymers, but that it does not negatively affect their tumor-targeting potential: both unfunctionalized and functionalized HPMA copolymers accumulated in tumors significantly more effectively than in 6–8 out of 9 healthy tissues [11].

In a pioneering clinical study, Seymour and colleagues imaged a radionuclide-labeled liver-targeted HPMA copolymer carrying doxorubicin (i.e., PK2). This theranostic nanomedicine formulation was targeted to the liver using galactosamine (which binds to the asialoglycoprotein receptor, which is overexpressed on hepatocytes), and effective organ targeting could be clearly visualized using 2-D gamma-scintigraphy [12]. When looking in more detail at the efficacy of drug targeting at the organ level, however, using anatomical CT imaging and functional SPECT imaging, it was found that the majority of PK2 ended up in healthy liver, rather than in the tumorous areas of the liver, thereby demonstrating that this formulation was not very effective in selectively delivering the conjugated drug to the pathological site.

In a comparable experimental setup, Harrington and colleagues evaluated the biodistribution and the target site accumulation of indium-Ill-labeled PEGylated liposomes in 17 patients suffering from various different types of malignancy (i.e., breast, head and neck, lung, brain, and cervical cancer) [13]. The authors showed that PEGylated liposomes circulate for prolonged periods of time, with an average $t_{1/2\beta}$ of 76 h, and with more than 50% of the injected dose still present in systemic circulation at 48 h post i.v. injection (p.i.). In addition, they showed that the radiolabeled liposomes effectively localized to tumors in 15 out of 17 cases, with the overall levels of liposome uptake varying from ~0.5 to ~3.5% of the injected dose at 72 h p.i., and with the degree of tumor accumulation closely correlating with the type of malignancy: the highest level of uptake was found for head and neck cancer (33 ± 16%ID/kg tumor, and the lowest level for breast cancer (5 ± 3%ID/kg tumor). Interestingly, in case of the latter, in spite of relatively low overall tumor concentrations, it was observed that the radiolabeled liposomes effectively accumulated in axillary lymph nodes, which are typically employed by breast cancers to metastasize. Significant accumulation of the radiolabeled liposomes was also observed in organs of the reticuloendothelial system (RES), like liver and spleen, which are known to be involved in the clearance of long-circulating nanomedicines from the circulation. Such image-guided insights into the efficacy of liposome-mediated drug targeting are considered to be highly useful for prescreening patients assigned to, e.g. Doxil (i.e., PEGylated liposomal doxorubicin), in order to identify which tumors are amenable to EPR-mediated drug targeting and which are not, and to thereby predict which patients are likely to respond to therapy and which are not.

An important downside of the pioneering work by Harrington and colleagues is that the liposomes they used did not contain a drug [13]. It is important to take into account in this regard that it is very likely that there are significant differences in the pharmacokinetics, the biodistribution, and the target site localization of drug-free "nanodiagnostics" versus drug-containing "nanotheranostics". This not only because the presence of a pharmacologically active agent can substantially affect the mode and/or the kinetics of clearance, but also because the incorporation of a drug can have major effects on the physicochemical properties of the formulation. Regarding the latter, it has, e.g. been shown that, upon conjugating doxorubicin and gemcitabine to iodine-131-labeled HPMA copolymers, because of altered physicochemical properties, the circulation time of the formulation decreases, while the levels localizing to kidney substantially increase [14]. Regarding the mode and/or the kinetics of clearance, the functionality of the RES needs to be taken into account. Without a drug, it is likely that the majority of radiolabeled liposomes or polymers eventually end up in macrophages. In the case of radiolabeled nanomedicines containing a drug (i.e., true nanotheranostics), on the other hand, it is likely that a significant portion of the circulating and/or tumor-associated macrophages are killed by the conjugated or entrapped chemotherapeutic drug, thereby substantially affecting the pharmacokinetics, the organ distribution, and the target site accumulation of the formulation. Because of the accelerated blood clearance (ABC) phenomenon, for instance, radiolabeled PEGylated liposomes are known to be cleared very rapidly from systemic circulation when administered repeatedly to the same animal (within a certain time frame; i.e., up to 3 weeks after the first i.v. injection) and to primarily accumulate in the liver [15]. In the case of doxorubicin-containing PEGylated liposomes, on the other hand, the ABC phenomenon turned out to be absent (because doxorubicin kills (some of) the macrophages responsible for rapid clearance), no decreases in circulation time were observed, and no increases in liver uptake were detected. This observation strongly suggests that by attenuating macrophage uptake and macrophage-mediated clearance, doxorubicin assists in providing long-circulating nanomedicines with proper passive targeting capabilities. Such insights are highly relevant for better understanding and optimizing the efficacy of targeted therapeutic interventions, and they underline the importance of using both drugs and imaging agents within a single formulation when intending to obtain really relevant results with regard to the noninvasive assessment of the biodistribution and the target site accumulation of nanomedicines.

4.3 Imaging Drug Release

To enable effective therapy, besides accumulating at the pathological site, the drug also needs to be efficiently released. Since drug release patterns greatly vary from formulation to formulation, and since there are large differences in the release patterns of, for example, liposomes versus polymers versus micelles, it is of the utmost importance to visualize and analyze drug release, not only under semiartificial in vitro conditions but also under physiologically relevant in vivo conditions. In vitro, drug release can generally be analyzed relatively easily, for example, using HPLC, but in vivo this is much more complicated: after harvesting the target tissue, for instance, the material generally needs to be homogenized, and the cells need to be lysed, in order to release the agents from certain intracellular compartments. During these processing steps, and especially during cell lysis (using detergents), many types of carrier materials are destabilized, and, for example, in the case of liposomes, it then is impossible to discriminate between the amount of drug that was still present within liposomes at the point of harvesting and the amount that was already released into the extra- and intracellular environment.

To overcome this shortcoming, and to enable noninvasive in vivo analyses on (the kinetics of) drug release, theranostic nanomedicine formulations have been developed in which drugs and imaging agents are co-loaded into the same delivery system. Since radionuclides render similar signals both in bound/entrapped form and in unbound/free form, these imaging agents are not suitable for visualizing drug release. Magnetic resonance (MR) contrast agents, on the other hand, such as gadolinium and manganese, depend on the interaction with surrounding water molecules to generate a signal, and since this interaction varies substantially when these agents are present within versus outside of water-impermeable vesicles, such as liposomes, MR probes are highly useful materials for monitoring drug release.

Pioneering work in this regard has been performed by Dewhirst and colleagues, who used manganese sulfate not only to load doxorubicin into liposomes (by means of a method comparable to ammonium sulfate-based pH-gradient loading) but also to noninvasively visualize drug release (through the generation of an increase in MR signal upon the simultaneous liberation of $MnSO_4$ and doxorubicin). They to this end prepared temperature-sensitive (TSL) and nontemperature-sensitive liposomes (NTSL), and they showed that the relaxivity, that is, the potential for MR signal enhancement, of TSL at temperatures below the transition temperature (T_g) was comparable to that of NTSL, but that it substantially increased upon heating to temperatures exceeding the T_g (i.e., $>39.5°C$), confirming Mn^{2+} release [16]. These findings were corroborated in a follow-up study, in which contrast agent release from TSL was correlated with doxorubicin release, and in which drug dose painting was performed on the basis of MRI [17], as well as by two recent studies in which doxorubicin release was correlated with ProHance (Gd-HP-DO3A) release [18, 19]. TSL rapidly released their contents upon i.v. administration to rats bearing preheated fibrosarcoma tumors, and they primarily did so in the well-vascularized and well-perfused periphery of the tumors.

Langereis, Gruell, and colleagues recently elegantly extended these efforts, developing liposomes containing two different contrast agents, that is, a chemical exchange saturation transfer (CEST) agent and a fluor-19-based probe [20]. As depicted schematically in Fig. 4.3A, the former only generates a signal when the liposomes are intact (and can consequently be used to track the biodistribution of the formulation), and the latter only upon the release of both contrast agents from the liposome (because of "quenching" of the ^{19}F-signal by the paramagnetic CEST agent). The suitability of

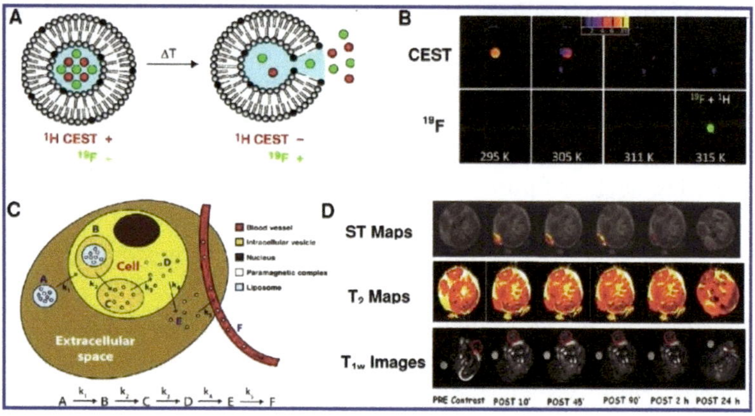

Figure 4.3 Noninvasive imaging of drug release. (A) Schematic depiction of a bimodal temperature-sensitive liposomal (TSL) MR contrast agent enabling the simultaneous visualization of target site accumulation (via a ^{1}H CEST agent; i.e., [Tm(HPDO3A)(H2O)]) and of hyperthermia-induced triggered drug release (via ^{19}F MRI). (B) ^{1}H CEST and ^{19}F MR images of TSL on a clinical 3.0 T MRI scanner. The CEST signal vanished close to T_g of the TSL (i.e., at 311 K; 38°C), while the fluorine signal appeared at 315 K (i.e., at 42°C). (C) Schematic representation of the kinetic model used for the analysis of the temporal evolution of three different MR signals upon the intratumoral administration of liposomes containing the T1 contrast agent Gd-HP-DO3A or the CEST agent Tm-DOTMA. The model is based on five consecutive steps, i.e., (i) cellular uptake of intact paramagnetic liposomes from the extracellular space (stage A) by endocytosis into intracellular vesicles (stage B); (ii) desintegration of the liposomal membrane and content release into endo/lysosomal compartments (stage C); (iii) release of the imaging agents from endo/lysosomes into the cytosol (stage D); (iv) cellular efflux of the contrast agents into the extracellular space (stage E); and finally (v) washout of the probes from the extracellular space in the tumor region via the bloodstream (stage F). (D) MR images (obtained at 7 T) illustrating the temporal evolution of the different MR contrast modes and of cellular trafficking of paramagnetic liposomes after intratumoral injection. Images adapted with permission from refs. [20, 22].

this agent for visualizing drug release was again demonstrated by preparing temperature-sensitive liposomes, with a T_g of 39°C. At physiological temperature, the liposomes maintained their integrity and produced a relatively strong CEST signal, while the fluorine signal could not be detected (Fig. 4.3B). At 42°C, on the other hand, the CEST signal had completely disappeared, and the fluorine signal had become apparent. Such conceptual advances are considered to be highly important for furthering the clinical development of MR-guided high-intensity focused ultrasound [21], which is currently only used to treat uterine fibroids but which, because of its ability to heat deep-seated tumors also holds significant potential for implementation in combination with chemotherapy-containing thermosensitive nanotheranostics.

A comparably elegant approach to visualize content release from liposomes has recently been reported by Delli Castelli and colleagues, who used two different types of paramagnetic liposomes, three different MR techniques, and mathematical modeling to kinetically analyze cellular uptake and cellular trafficking upon intratumoral injection [22]. As exemplified by Fig. 4.3C, they subdivided the fate of intratumorally injected paramagnetic liposomes into six different phases, that is, an initial extracellular localization (stage A), uptake into endocytic vesicles, such as endo- and lysosomes (stage B), release of the MR contrast agents in endo- and lyosomes (stage C), cytosolic entry of the contrast agents (stage D), efflux of the contrast agents out of tumor cells (stage E), and washout of the agents out of the tumor region via the vascular system (stage F). Using this methodology, they convincingly showed that it is not only possible to noninvasively assess content release from nanocarrier materials using multicontrast MRI but also to visualize and analyze (the kinetics of) cellular uptake and cellular trafficking.

Another interesting method to visualize drug release from theranostic nanomedicine formulations has recently been published by Onuki and colleagues, who prepared poly(lactic-*co*-glycolic acid) (PLGA)-based nanoparticles carrying the drug 5-fluorouracil and the contrast agents gadolinium-DTPA (Gd-DTPA) and superparamagnetic iron oxide (SPIO) [23]. They found that, immediately upon i.v. administration, the nanoparticles exhibited a strong T_2* contrast in the tumor area, as a result of the presence of the SPIO, which remained unaltered up until 2.5 h p.i. Conversely, TT mapping, which correlates to the release of Gd-DTPA from the SPIO-containing nanoparticles, only revealed a clear signal at 30 min p.i. This indicates that the release of Gd-DTPA, and likely also of co-loaded 5-

FU, occurs relatively early on after i.v. administration, that is, already to a significant extent within 30 min, and that after this, the released low-molecular-weight gadolinium complex diffuses away relatively fast, as evidenced by the fact that the Tl signal had vanished at 2.5 h p.i. Assuming that 5-FU is released from these particles with the same kinetics as is Gd-DTPA (which still needs to be validated), this setup thus enables the simultaneous and quantitative monitoring of drug delivery and drug release.

4.4 Imaging Drug Efficacy

Besides for noninvasively assessing drug delivery and drug release, theranostic nanomedicines are also highly useful for predicting and monitoring therapeutic responses. Concerning the prediction of therapeutic responses, as already briefly alluded to above, during phase I and phase II clinical trials, nanomedicine formulations could, for instance, be labeled with radionuclides, in order to obtain some initial noninvasive information with regard to target site accumulation. On the basis of this, rational predictions could then be made with regard to the potential effectiveness of nanomedicine-based therapeutic interventions. In the initial phase I trial focusing on HPMA-copolymer-based doxorubicin (i.e., PK1), for instance, a radiolabeled version of this passively tumor-targeted polymeric prodrug could have been used to pre-screen patients assigned to PK1, in order to identify which tumors are amenable to EPR-mediated drug targeting and which are not, and to thereby predict which patients are likely to respond to PK1 therapy and which are not. Similarly, in the phase I/II trial focusing on liver-targeted PK1 (i.e., PK2), the combination of functional SPECT imaging and anatomical CT imaging could have been used to visualize which HCC-patients show good tumor accumulation and which do not, and to on the basis of this decide in which patients PK2 treatment should be continued. Furthermore, by mixing in trace amounts of radiolabeled nanomedicines with (e.g., every second or third cycle of) drug treatment and by continuously subjecting patients to anatomical CT imaging in conjunction with 2D-scintigraphy or

3D-PET, it would have been possible to visualize the efficacy of the intervention in real-time, and to thereby provide important information for assisting in deciding whether or not to (dis)continue therapy, and whether or not to adjust drug doses. By doing so, theranostic nanomedicines might contribute to realizing the potential of "personalized medicine", that is, tailor-made therapy for individual patients, which, besides on the study of genetic polymorphisms and biomarkers, also relies on the development of imaging methods for predicting and measuring therapeutic responses.

Additional proof-of-principle for the use of theranostic nanomedicines for predicting and monitoring therapeutic responses is the accumulation of radiolabeled PEGylated liposomes in a patient suffering from AIDS-related Kaposi sarcoma [23]. Kaposi sarcomas (KS) are characterized by a dense and highly leaky vasculature, and long-circulating and passively tumor-targeted nanomedicines very effectively accumulate in these lesions by means of EPR. Consequently, KS patients generally respond very well to treatment with PEGylated liposomal doxorubicin (i.e., Doxil/Caelyx), and as opposed to other types of malignancy for which Doxil is approved, not only the tolerability of the intervention can be substantially improved, but also its efficacy: in a large phase III trial in which Doxil was compared to the formerly standard combination regimen ABV (i.e., adriamycin (doxorubicin), bleomycin, and vincristine), it was found that Doxil produced 1 complete response and 60 partial responses, as compared to only 31 partial responses for ABV [24]. This highly significant improvement in therapeutic efficacy can be explained by the fact that Doxil accumulates very well in KS lesions, whereas in other types of malignancies its target site accumulation tends to be less effective and varies substantially from patient to patient (see Fig. 4.4I). Following up on this and the above reasoning, if also KS and non-KS patients would be prescreened with radiolabeled Doxil, in order to identify which ones show high tumor accumulation and which ones do not, it would also in these cases be possible to decide in which patients Doxil treatment should be continued and in which ones other therapeutic options should be considered. The same holds true for adding in minute amounts of radiolabeled Doxil with (e.g., every second cycle of) drug treatment, to be able noninvasively visualize the efficacy of the intervention during follow-up, and to thereby assist in deciding whether or not to continue therapy and whether or not to adjust drug doses. It should be kept in mind in this regard, however, that the in vivo visualization of the biodistribution and the target site accumulation of theranostic nanomedicines is an add-on parameter for providing feedback on the potential therapeutic efficacy of the intervention, and that it by no means replaces established (PET-, MR- or CT-based) diagnostic procedures.

Theranostic nanomedicines could furthermore also be used to noninvasively assess treatment efficacy in the case of antiangiogenic and antimetastatic treatments. Regarding antiangiogenic therapy, for instance, it has already been demonstrated that, by conjugating the highly putative antiangiogenic agent TNP-470 to HPMA copolymers, the balance between the efficacy and the toxicity of this drug can be substantially improved [25]. On the other hand, gadolinium-labeled HPMA copolymers have on various occasions been shown to circulate for prolonged periods of time and, consequently, to hold significant potential for experimental MR angiography [26, 27], showing that a 25 kDa-sized gadolinium-labeled HPMA copolymer can be used to visualize blood vessels in subcutaneous rat tumors with relatively high resolution. When combining the two above approaches, that is, by synthesizing a theranostic polymer therapeutic containing both TNP-470 and gadolinium, it would be possible to induce and monitor antiangiogenesis simultaneously.

Regarding metastasis, theranostic nanomedicine formulations are also highly useful for noninvasively assessing the antimetastatic efficacy of targeted therapeutic interventions. The accumulation of gadolinium-labeled polypropylene diaminobutane-based dendrimers in a healthy and in a metastatic mouse liver, respectively, clearly demonstrates the suitability of these nanomedicine materials to visualize metastatic liver lesions [28]. Co-loading liposomes both with drugs and with imaging agents, and adding in trace amount of radiolabeled liposomes with (e.g., every second or third cycle of) of Doxil treatment, the antimetastatic efficacy of the intervention can be noninvasively visualized in real-time [13].

4.5 Conclusions

Theranostic nanomedicines can be used for various different purposes. By enabling a noninvasive assessment of the pharmacokinetics, the biodistribution and the target site localization of conjugated or entrapped pharmacologically active agents, nanotheranostics allow for the optimization of drug delivery systems. In addition, by combining information on overall target site localization with noninvasive imaging insights on the local distribution of the drug and/or the carrier material at the target site, nanotheranostics can also be used for predicting treatment responses. This is exemplified by the above-mentioned clinical studies on the tumor accumulation of radiolabeled PEGylated liposomes in Kaposi sarcoma patients, as well as by the liver versus tumor localization of galactosamine-targeted HPMA copolymers carrying doxorubicin. Furthermore, by noninvasively imaging (the kinetics of) drug release in vivo, some of the basic properties of drug delivery systems can be visualized and analyzed, and attempts can be made to correlate the in vitro characteristics of carrier materials with their in vivo capabilities. Related to this, by using contrast agents to monitor the release of pharmacologically active agents from stimuli-sensitive nanomedicines, the efficacy of triggerable drug delivery systems can be optimized, as exemplified by several studies on thermosensitive liposomes. And finally, by providing real-time feedback on the efficacy of targeted therapeutic interventions, theranostic nanomedicines can also be used to facilitate (pre-) clinical efficacy analysis, to prescreen patients, and to realize the potential of personalized medicine.

Acknowledgment

The authors gratefully acknowledge financial support by High-tech.NRW (ForSaTum) and by the European Commission's Framework Programme (FP6: "MediTrans—Targeted Delivery of Nanomedicines"; and FP7: "COST-Action TD1004: Theragnostics imaging and therapy—An action to develop novel nanosized systems for imaging-guided drug delivery").

References

1. Peer, D., Karp, J. M., Hong, S., Farokhzad, O., Margalit, R., Langer, R. Nanocarriers as an emerging platform for cancer therapy. *Nat. Nanotechnol.* 2007, **2**, 751–760.
2. Sanhai, W. R., Sakamoto, J. H., Canady, R., Ferrari, M. Seven challenges for nanomedicine. *Nat. Nanotechnol.* 2008, **3**, 242–244.
3. Davis, M. E., Chen, Z. G., Shin, D. M. Nanoparticle therapeutics: an emerging treatment modality for cancer. *Nat. Rev. Drug Discovery* 2008. **7**, 771–782.
4. Lammers, T., Hennink, W. E., Storm, G. Tumour-targeted nanomedicines: principles and practice. *Br. J. Cancer* 2008. **99**, 392–397.
5. Caruthers, S. D., Wickline, S. A., Lanza, G. M. Nanotechnological applications in medicine. *Curr. Opin. Biotechnol.* 2007, **18**, 26–30.
6. Xie, J., Lee, S., Chen, X. Nanoparticle-based theranostic agents. *Adv. Drug Delivery Rev.* 2010, **62**, 1064–1079.
7. Sun, D. Nanotheranostics: integration of imaging and targeted drug delivery. *Mol. Pharmaceutics* 2010, **7**, 1879.
8. Lammers, T., Kiessling, F., Hennink, W. E., Storm, G. Nanotheranostics and image-guided drug delivery: Current concepts and future directions. *Mol. Pharmaceutics* 2010, **7**, 1899–1912.
9. Maeda, H., Wu, J., Sawa, T., Matsumura, Y., Hori, K. Tumour vascular permeability and the EPR effect in macromolecular therapeutics: a review. *J. Controlled Release* 2000, **65**, 271–284.
10. Lammers, T., Subr, V., Ulbrich, K., Hennink, W. E., Storm, G., Kiessling, F. Polymeric nanomedicines for image-guided drug delivery and tumor-targeted combination therapy. *Nano Today* 2010, **5**, 197–212.
11. Lammers, T., Kuehnlein, R., Kissel, M., Subr, V., Etrych, T., Pola, R., Pechar, M., Ulbrich, K., Storm, G., Huber, P., Peschke, P. Effect of physicochemical modification on the biodistribution and tumor accumulation of HPMA copolymers. *J. Controlled Release* 2005, **110**, 103–118.
12. Seymour, L W., Ferry, D. R., Anderson, D., Hesslewood, S., Julyan, P. J., Poyner, R., Doran, J., Young, A. M., Burtles, S., Kerr, D. J. Hepatic drug targeting: phase I evaluation of polymer-bound doxorubicin. *J. Clin. Oncol.* 2002, **20**, 1668–1676.
13. Harrington, K. J., Mohammadtaghi, S., Uster, P. S., Glass, D., Peters, A. M., Vile, R. G., Stewart, J. S. Effective targeting of solid tumors in patients with locally advanced cancers by radiolabeled pegylated liposomes. *Clin. Cancer Res.* 2001, **7**, 243–254.

14. Lammers, T., Subr, V., Ulbrich, K., Peschke, P., Huber, P. E., Hennink, W. E., Storm, G. Simultaneous delivery of doxorubicin and gemcitabine to tumors in vivo using prototypic polymeric drug carriers. *Biomaterials* 2009, **30**, 3466–3475.
15. Dams, E. T., Laverman, P., Oyen, W. J., Storm, G., Scherphof, G. L., van Der Meer, J. W., Corstens, F. H., Boerman, O. C. Accelerated blood clearance and altered biodistribution of repeated injections of sterically stabilized liposomes. *J. Pharmacol. Exo. Ther.* 2000, **292**, 1071–1079.
16. Viglianti, B. L., Abraham, S. A., Michelich, C. R., Yarmolenko, P. S., MacFall, J. R., Bally, M. B., Dewhirst, M. W. In vivo monitoring of tissue pharmacokinetics of liposome/drug using MRI: illustration of targeted delivery. *Magn. Reson. Med.* 2004, **51**, 1153–1162.
17. Viglianti, B. L., Ponce, A. M., Michelich, C. R., Yu, D., Abraham, S. A., Sanders, L., Yarmolenko, P. S., Schroeder, T., MacFall, J. R., Barboriak, D. P., Colvin, O. M., Bally, M. B., Dewhirst, M. W. Chemodosimetry of in vivo tumor liposomal drug concentration using MRI. *Magn. Reson. Med.* 2006, **56**, 1011–1018.
18. De Smet, M., Heijman, E., Langereis, S., Hijnen, N. M., Griill, H. Magnetic resonance imaging of high intensity focused ultrasound mediated drug delivery from temperature-sensitive liposomes: An in vivo proof-of-concept study. *J. Controlled Release* 2011, **750**, 102–110.
19. Negussie, A. H., Yarmolenko, P. S., Partanen, A., Ranjan, A., Jacobs, G., Woods, D., Bryant, H., Thomasson, D., Dewhirst, M. W., Wood, B. J., Dreher, M. R. Formulation and characterisation of magnetic resonance imageable thermally sensitive liposomes for use with magnetic resonance-guided high intensity focused ultrasound. *Int. J. Hyperthermia* 2011, **27**, 140–155.
20. Langereis, S., Keupp, J., van Velthoven, J. L., de Roos, I. H., Burdinski, D., Pikkemaat, J. A., Gruell, H. A temperature-sensitive liposomal 1H CEST and 19F contrast agent for MR image-guided drug delivery. *J. Am. Chem. Soc.* 2009, **131**, 1380–1381.
21. Deckers, R., Moonen, C. T. Ultrasound triggered, image guided, local drug delivery. *J. Controlled Release.* 2010, **148**, 25–33.
22. Delli Castelli, D., Dastru, W., Terreno, E., Cittadino, E., Mainini, R., Torres, E., Spadaro, M., Aime, S. In vivo MRI multicontrast kinetic analysis of the uptake and intracellular trafficking of paramagnetically labeled liposomes. *J. Controlled Release* 2010, **144**, 271–279.
23. Onuki, Y., Jacobs, I., Artemov, D., Kato, Y. Noninvasive visualization of in vivo release and intratumoral distribution of surrogate MR contrast

agent using the dual MR contrast technique. *Biomaterials* 2010, **31**, 7132–7138.

24. Northfelt, D. W., Dezube, B. J., Thommes, J. A., Miller, B. J., Fischl, M. A., Friedman-Kien, A., Kaplan, L. D., Du Mond, C., Mamelok, R. D., Henry, D. H. Pegylated-liposomal doxorubicin versus doxorubicin, bleomycin, and vincristine in the treatment of AIDS-related Kaposi's sarcoma: results of a randomized phase III clinical trial. *J. Clin. Oncol.* 2005, **16**, 2445–2451.
25. Satchi-Fainaro, R., Puder, M., Davies, J. W., Iran, H. T., Sampson, D. A., Greene, A. K., Corfas, G., Folkman, J. Targeting angiogenesis with a conjugate of HPMA copolymer and TNP-470. *Nat. Med.* 2004, **10**, 255–261.
26. Kiessling, R., Heilmann, M., Lammers, T., Ulbrich, K., Subr, V., Peschke, P., Waengler, B., Mier, W., Schrenk, H. H., Bock, M., Schad, L., Semmler, W. Synthesis and characterization of HE-24.8: a polymeric contrast agent for magnetic resonance angiography. *Bioconjugate Chem.* 2006, **77**, 42–51.
27. Lammers, T., Subr, V., Peschke, P., Kuehnlein, R., Hennink, W. E., Ulbrich, K., Kiessling, R., Heilmann, M., Debus, J., Huber, P. E., Storm, G. Image-guided and passively tumour-targeted polymeric nanomedicines for radiochemotherapy. *Br. J. Cancer* 2008, **99**, 900–910.
28. Kobayashi, H., Brechbiel, M. W. Nano-sized MRI contrast agents with dendrimer cores. *Adv. Drug Delivery Rev.* 2005, **57**, 2271–2286.

Chapter 5

The Emerging Field of RNA Nanotechnology

Peixuan Guo
Center for RNA Nanobiotechnology and Nanomedicine;
Division of Pharmaceutics and Pharmacology, College of Pharmacy;
College of Medicine, Dorothy M. Davis Heart and Lung Research Institute and James Comprehensive Cancer Center; The Ohio State University, Columbus, Ohio 43210, USA
guo.1091@osu.edu

RNA can be designed and manipulated just like DNA while having different rules for base-pairing and displaying functions similar to proteins. The large variety of loops and motifs in RNA allow them to fold into numerous complicated structures. This diversity provides a platform for identifying viable building blocks for particle assemblies, substrate binding and manufacture engineering. RNA thermal stability allows production of multivalent nanostructures with defined stoichiometry. Here we review the unique qualities of RNA nanotechnology and their distinct properties inside the body. We describe techniques for constructing RNA nanoparticles from different building blocks and their applications in nanomedicine. Finally, we discuss challenges in predicting and synthesizing RNA and offer some perspectives on the yield and cost of RNA production.

Reproduced with permission from *Nature Nanotechnology*, **5**, 833–842, 2010.

Nano-Enabled Medical Applications
Edited by Lajos P. Balogh

ISBN 978-981-4800-32-7 (Hardcover), 978-0-429-39903-9 (eBook)
www.jennystanford.com

Nanotechnology addresses the creation and application of materials at the nanometer scale using either top-down approaches or bottom-up assembly. Macromolecules of DNA, RNA and proteins have intrinsically defined features at the nanoscale and may serve as powerful building blocks for bottom-up fabrication of nanostructures and nanodevices. A pioneering concept proposed by Ned Seeman 30 years ago has led to the development of the field known as DNA nanotechnology [1–3]. Peptides or proteins have also been studied extensively for applications in nanotechnology [4–7]. The concept of RNA nanotechnology has been around for more than

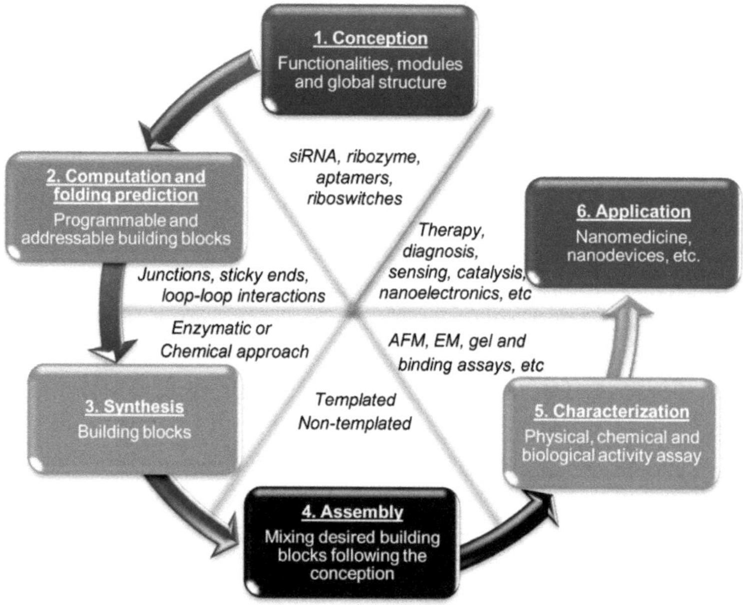

Figure 5.1 Approaches in RNA Nanotechnology. The construction of RNA nanoparticles involves several steps: Following the conception, a computational approach can be applied to predict the folding and structure of the building blocks as well as the consequences of inter-RNA interactions in RNA nanoparticle assembly. After the synthesis of monomeric building blocks, the individual subunits can be further assembled into quaternary architectures utilizing the spontaneous self-folding property of RNA. The assembled RNA nanostructures will be characterized to ensure proper folding with desired structural/functional capabilities. After thorough evaluation, the nanoparticles will be then be used for various applications.

a decade [8–13] (for review, see refs. [14–15]) and the first evidence for the construction of RNA nanoparticles through self-assembly of multiple reengineered natural RNA molecules was reported in 1998 [8]. However, RNA nanotechnology has only recently become increasingly popular due to the recognition of its potential in the treatment of cancer, viral infection, genetics diseases and other applications in nanomedicine (Fig. 5.1).

5.1 Significance and Uniqueness of RNA Nanotechnology

RNA can be designed and manipulated with a level of simplicity characteristic of DNA, meanwhile displaying flexibility in structure and diversity in function similar to that of proteins, including enzymatic activities. Although RNA nanotechnology can be regarded as a subdivision of DNA nanotechnology, the uniqueness of RNA properties distinct from those of DNA might warrant a new emerging discipline.

RNA is a polymer made up of four different nucleotides: adenine, cytosine, guanine and uridine. RNA possesses not only Watson–Crick base pairing but also noncanonical base pairing which promotes folding into rigid structural motifs distinct from the structure of single-stranded DNA. The noncanonical property facilitates loop-receptor interactions and allows the creation of synthetic ribozymes [16, 17]. It also allows the formation of special structural motifs [18–20]. Tertiary interactions mediate GNRA/receptor interactions [10], and formation of micrometer sized RNA filaments [10–12, 21, 24] (Fig. 5.2g–j). Currently, an RNA up to 80 nucleotides can be synthesized commercially. An 80-nucleotide RNA can display up to 10^{48} (4^{80}) different structures with unique shapes involving non-canonical interactions. Such a huge pool of rich structural conformations would ease the search for viable partners in particle assemblies, substrate binding, building architectures, and manufacture engineering.

Typically, RNA contains a large variety of single-stranded stem-loops for intra- and/or inter-molecular interactions. These loops can serve as mounting dovetails, and thus external linking dowels

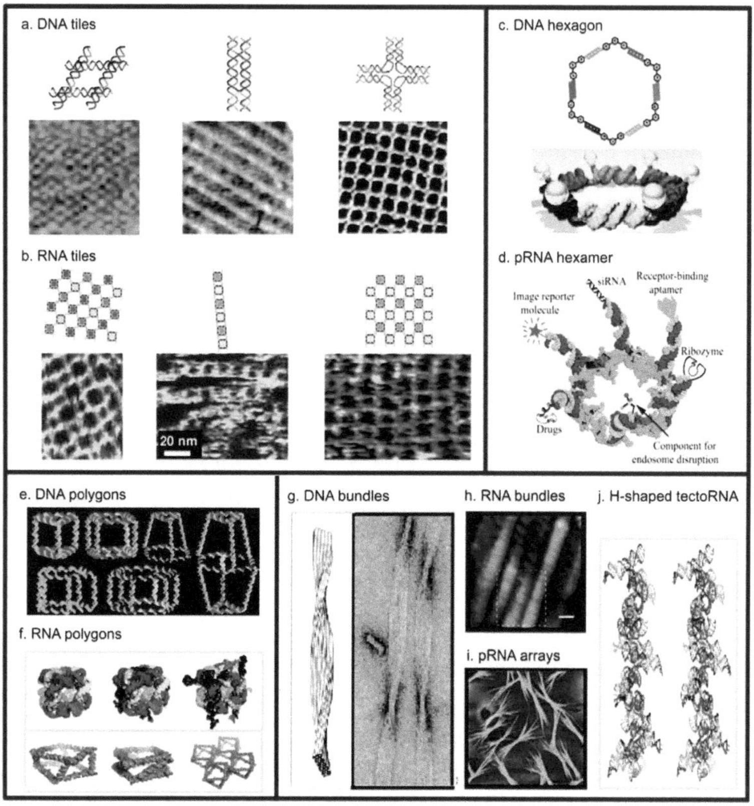

Figure 5.2 Comparison of self-assembled DNA (a,c,e,g), and RNA (b,d,f,h,I,j) nanoparticles. Images of DNA tiles **(a)** [1, 2] and RNA tiles via tectosquares **(b)** [13]; illustration of hexameric DNA gold nanoparticle **(c)** [1, 2], pRNA hexameric ring **(d)** [50, 87]; DNA 3D polygons **(e)** [2]; and RNA cubic scaffolds **(f)** [30, 77]; images of DNA bundles **(g)** [70], RNA bundles **(h)** [23], pRNA arrays **(i)** [12]; and 3D model of H-shaped tectoRNA **(j)** [21]. All images are taken by AFM except **(g)** and **(j)**, as well as the first two images of (a), which are TEM images. All images were adapted from the individual references with permission.

might not be needed for nanomachine fabrication and assembly. Loops and motifs also allow for construction of a more complicated secondary structure. Versatility and low-energy folding delivers a significant advantage. Furthermore, RNA molecules can possess special functionalities such as aptamer, riboswitch, ribozyme, and siRNA.

The RNA/RNA double helix is the most stable among the three helices: RNA/RNA, RNA/ DNA and DNA/DNA [25, 26]. RNA motifs and modules with special bends or stacks are particularly stable. The thermodynamic stability has been defined as the free energy required for complex formation, or in some cases, to unwind the helix ($\Delta G^0 = -G^0_{\text{helix}} = G^0_{\text{unwind}}$); thus, the lower the free energy ($-G^0_{\text{helix}}$) the complex holds, the more stable the complex. ΔG^0 is affected by neighboring sequences. Based on the nearest-neighbor model, $-G^0_{\text{helix}}$ has been calculated to be lower for RNA than DNA [25, 26]. The hydrogen bonding and phosphodiester backbone of RNA and DNA helix are very similar, and the difference in energy between DNA dA:dT and RNA A:U base pairs are trivial (Table 5.1). However, under physiological conditions, the RNA helix displays A-type configuration whereas the DNA helix is predominantly B-type. The 2′-OH in RNA ribose locks the ribose into a 3′-endo chair conformation which does not favor a B-helix. Base stacking is governed by van der Waals interaction, which contributes directly to the enthalpy. Though the difference in the stacking interaction is small between DNA and RNA, the sum over multiple base pairs can make a difference in the helix stability. Thus, RNA nanoparticles are more stable thermodynamically than their DNA counterparts. DNA tiles can assemble through sticky ends of four to six nucleotides. Surely, four to six nucleotides of RNA are sufficient to produce stable RNA helices in solution [12]. In certain cases, as few as two nucleotides can promote complex formations in RNA [27–31].

5.2 Distinct Attributes of RNA inside the Body

Cell endosome escape is an important consideration for in vivo delivery. Therapeutic particles are initially recognized by cell surface receptor(s) and then delivered through entry into the cellular endosome. The pH within the endosome ranges from 4.3 to 5.8 [32], where RNA is more stable than DNA (Table 5.1). In acidic environments [33], the protonation of DNA purine bases leads to depurination and the resulting apurinic DNA is susceptible to

Table 5.1 Differences between DNA and RNA

		DNA	RNA
Elements		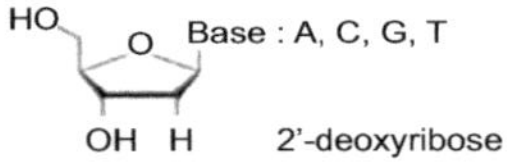	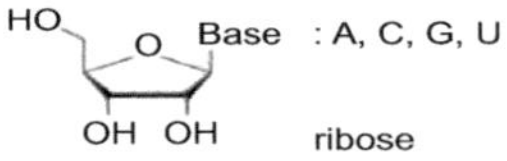
Base pairing		Canonical Watson–Crick (W-C)	Canonical and Non-canonical W-C
Acidic effect		Depurination: Apurine DNA sensitive to cleavage	Stable
Alkaline effect		Stable up to pH12	Sensitive to alkaline hydrolysis
Configuration		Predominantly B form: • Base pairs/turn of the helix: 10.5 • Pitch: 3.5 nm • Helix rise/bp: 0.314 nm • Humidity: Nucleotide:H_2O = 1:1	A form: • Base pairs/turn of the helix: 10.9 • Pitch: 2.5 nm • Helix rise/bp: 0.275 nm • Humidity: Nucleotide:H_2O =1:0.7
Chemical stability		Relatively stable but sensitive to DNase	Unstable, sensitive to RNase, but stable after chemical modification, e.g. 2′-F or2′-OMe modification
Thermal stability		G:C more stable than A:T	Thermally more stable than DNA, especially for RNA motifs and modules with particular bends or stacks
Free energy, $\cdot\Delta G^0$		-1.4^{-1} KJ.mol per bp stack [25]	–3.6 to –8.5 KJ.mol^{-1} per bp stack [25]
Helix formation		Need minimum 4 nucleotides	Need minimum 2 nucleotides [26, 27]
Intermolecular interactions		Cohesive ends, crossover motifs	Cohesive ends, crossover motifs, Kissing loops, interlocking loops
In vivo replication	Initiation	Origin of replication with primer	Promoter, exact nucleotide to start without primer
	Termination	No nature sequence for replication termination	Specific transcription terminators
In vitro synthesis	Enzymatic	DNA polymerase, PCR	T7/SP6 Transcription
	Chemical	Up to 160 nucleotides; low cost	Up to 117 nucleotide; high cost and low yield

Table 5.2 Techniques for constructing RNA nanoparticles

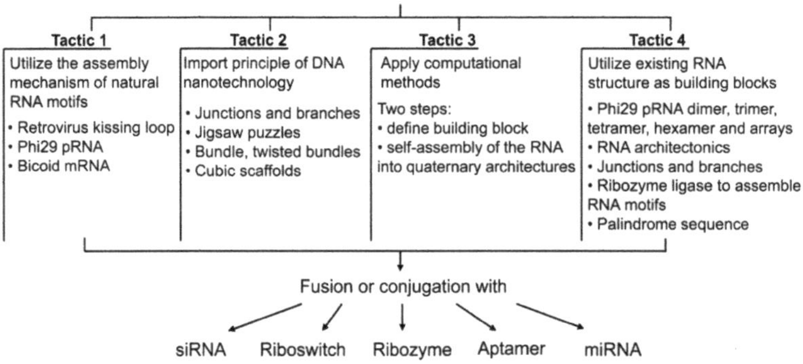

cleavage. The higher stability of RNA in acidic environments is especially useful in therapy since RNA will survive in the endosome after cell entry, and disperse throughout the cell during endocytosis.

Another intriguing property of RNA is the possibility to produce self-assembled RNA nanoparticles in vivo (Table 5.2). In contrast to DNA, small RNA molecules are transcribed in the cell using DNA as a template. By using an inducible promoter [34] and appropriate terminators for transcription, small RNA molecules can be controllably produced. RNA can be processed into the desired length by including delta ribozymes at both the upstream and downstream terminals for cis-cleavage [35]. Natural RNA nanoparticles such as dimers [36–39] and hexamers [8–9] have been discovered in cells. Sequences, such as pRNA [35] or tRNA [40, 41] for guiding the self-assembly of RNA nanoparticles with functionalities such as siRNA [42, 43], ribozyme [35], or aptamer [40] can be incorporated within the DNA template in vivo [35].

Small RNAs, such as riboswitch, with regulatory functions [44–47] within the cell may be viewed as Boolean networks based on logic operations [48, 49]. Input nodes can be viewed as RNA nanostructures and the output, e.g., the activation of a pathway, is based upon logic functions of input RNA concentrations. Multiple small RNA regulators can be used to regulate the in vivo products and functional pathways, with controls by induct ion or repression through the trans- and cis-actions. Varieties of small RNA can work cooperatively, synergistically, or antagonistically based on the design to produce computational logic circuits as conjunctive or disjunctive normal forms or other kind of logic operation. By designing the logic network of AND/NOT/OR different gates in the cell, an "RNA computer" can be theoretically implemented and applied to bacterial, yeast, and mammalian systems [48, 49].

5.3 Techniques for Constructing RNA Nanoparticles

Construction of nanoparticles requires the use of programmable, addressable and predictable building blocks. Self-assembly of RNA building blocks in a predefined manner to form larger two-, three-, and four-dimensional structures is a prominent bottom-up approach and represents an important means by which biological techniques and biomacromolecules can be successfully integrated into nanotechnology [12, 50, 51].

Within the realm of self-assembly there are two main subcategories: templated and non-templated assembly. Templated assembly involves the interaction of RNAs with one another under the influence of a specific external force, structure, or spatial constraint. RNA transcription, hybridization, replication, molding, and phi29 pRNA hexameric ring formation are all within this category. Non-templated assembly involves the formation of a larger structure by individual components without any external influence. Examples include, ligation, chemical conjugation, covalent linkages, loop/loop interactions of RNA such as the HIV kissing loop, and phi29 pRNA dimer or trimer formation [10, 12, 50, 52]. Various approaches available for RNA nanoparticle construction are discussed below.

The first approach utilizes the assembly mechanism of natural RNA nanoparticles that can form unique and intriguing multimers in vivo. For example, the retrovirus kissing loops facilitate genomic RNA dimerization [36, 37]. The pRNA of the bacteriophage phi29 DNA packaging motor assembles into dimers and hexamers via hand-in-hand interactions between two right and left interlocking loops [8, 12, 28, 39, 52, 53]. The bicoid mRNA of Drosophila embryos forms dimers via hand-in-arm interactions [38]. *E. coli* noncoding RNA DsrA assembles into stripe patterns via their built-in palindrome sequence [23]. The assemblies of RNA nanoparticles in vitro that mimic their natural counterparts were reported 12 years ago [8]. The novel HIV kissing loop mechanism has also inspired the design of tecto-RNA architectures [13, 29].

The Second tactic is to import the well-developed principle of DNA nanotechnology into the RNA field. While RNA is unique, certain common features between RNA and DNA in structure and chemistry make it compulsory for the developing RNA nanotechnology field to draw experiences from the well-developed DNA nanotechnology wealth and apply some, not all, of the approaches in DNA to the RNA field.

DNA nanotechnology utilizes the nature of DNA complementarity for the construction of nanomaterials via intermolecular interactions of DNA strands. A variety of elegant shapes have been created with precise control over their geometries, periodicities and topologies (Fig. 5.2) (see recent reviews [1–3] and the references therein). Various cross-over motifs have been designed via reciprocal exchange of DNA backbones [3]. Branched DNA tiles have been constructed using sticky ends and cross-over junction motifs, such as tensegrity triangles (rigid structures in periodic array form) [54], and algorithmic self-assembled Sierpinski triangles (aperiodic arrays of fractal patterns) [55]. The DNA tiles can further self-assemble into nanotubes, helix bundles [56] as well as complex DNA motifs and arrays for positioning nanoparticles, proteins or dyes with precise control, such as poly catenated DNA ladders [57]. Elegant 3D DNA networks using a minimal set of DNA strands with topologies such as cubes, polyhedrons, prisms, and buckyballs have also been fabricated based on junction flexibility and edge rigidity [3, 58]. A continuous growth of the tensegrity triangle within the periodic DNA module has resulted in the formation of DNA crystals diffracting to 4 Å resolution [59].

A striking marvel demonstrating the addressable and programmable property of DNA is Rothemund's DNA origami [60], where a long single-stranded viral DNA is used as a scaffold for binding shorter strands to generate well-defined 2D and 3D configurations. DNA origami was subsequently applied to build 3D boxes that can be locked and unlocked [61], nanoarrays for label-free detection of substrates [62] and for structure elucidation of organized proteins [63]. Rationally designed supramolecular DNA assemblies can be conjugated with organic and inorganic molecules, such as conjugation of porphyrins on parallel DNA helix bundles [64] nanomagnets [65] and elegant nanomachines [58, 66]. Replicable DNA architectures have been achieved to scale-up DNA nanostructure production for practical applications by using enzymatic rolling circle replication, bacterial cells infected with a viral vector [67], or chemical approaches for amplifying branched DNA arms [68].

Although the folding properties of RNA and DNA are not exactly the same, the fundamental principles in DNA nanotechnology are applicable to RNA nanotechnology. The use of 3WJ (three-way junction) and 4WJ [18, 29] to build novel and diverse RNA architectures are very similar to the branching approaches in DNA [1, 3] (Fig. 5.2a,b,e,f). The formation of Jigsaw puzzles was demonstrated in both RNA [13] and DNA [69] (Fig. 5.2a,b). Both RNA [12, 23, 30] and DNA [70] polymers can be developed into bundles by combining elongation and expansion in the x-y directions (Fig. 5.2g–j). The finding that insertion of bulges in RNA helix leads to the formation of twisted bundles [12] (Fig. 5.2i) was later demonstrated in DNA (Fig. 5.2g), revealing that insertions and deletions of bases can lead to the formation of twisted DNA bundles with handedness [70] thereby illustrating the same basic principle. However, RNA is more rigid in bulge structure due to non-canonical interactions, while in DNA, the twisting requires the interaction of two DNA helixes with four strands [70].

Recently, RNA cubic scaffolds [71] were constructed using several RNA sequences that do not fold on themselves but self-assemble with one another in a defined manner. This strategy is reminiscent of DNA nanotechnology, but in contrast to DNA strategies, RNA synthesis can be coupled to RNA self-assembly to generate fully assembled RNA cubes during in vitro transcription.

The third tactic is to apply computational methods in RNA nanoparticle construction. Computational approaches can be used to guide the design of novel RNA assemblies and to optimize sequence requirements for the production of nanoscale fabrics with controlled direction and geometry [37, 72–75]. In contrast to traditional methods in which raw materials are selected rather than designed for a given application, the next generation of building blocks can be designed a priori for programmed assembly and synthesis. There are two steps in building RNA nanoparticles. The first is a computational approach (e.g. using *Kinefold* [12]) utilizing the spontaneous self-folding properly of RNA into defined structures via base/base interactions based on their characteristic ΔG [76]. The second is the spontaneous assembly of the resulting RNA building blocks into larger assemblies based on the predicted architecture. This creates an effective computational pipeline for generating molecular models of RNA nanostructures. A recent example is the construction of cubic RNA-based scaffolds, whereby RNA sequence designs were optimized to avoid kinetic traps [77].

The fourth tactic is to utilize the existing RNA structure, or with known function, as building blocks in RNA nanoparticle construction. The structure of RNA motifs and mechanisms of RNA folding and sequence interactions have been investigated for many years. Varieties of mechanisms in RNA loop/loop interactions [8, 12, 31], tertiary architecture contacts [12, 15, 30], and formation of special motifs [12, 21, 29–31, 78–82] have been elucidated. Building-blocks are first synthesized after computing intra- and inter-molecular folding. Nanoparticles are built via spontaneous templated or nontemplated self-assembly as planned. A rich resource of well-developed databases can be utilized to extract known RNA structural units for construction of novel RNA nanoparticles with desired properties [37, 83, 84].

Several methods have been pursued for RNA nanoparticle construction borrowing RNA properties in loop/loop interactions. The first method is based on the structural features of the pRNA of the bacteriophage phi29 DNA packaging motor [8, 85] which uses a hexameric RNA ring to gear the machine [28, 86, 87]. The pRNA has been reengineered to form dimers, trimers, tetramers, hexamers and arrays via hand-in-hand or foot-to-foot interactions between two interlocking loops [12, 52] (Fig. 5.3). Dimers are formed using two building blocks with A/b′ (right and left hand, respectively) and B/a′ (Fig. 5.3). Trimers are formed using three building blocks with A/b′, B/c′, and C/a′ [12, 50, 51]. Dimers of an extended configuration (twins) can also be efficiently

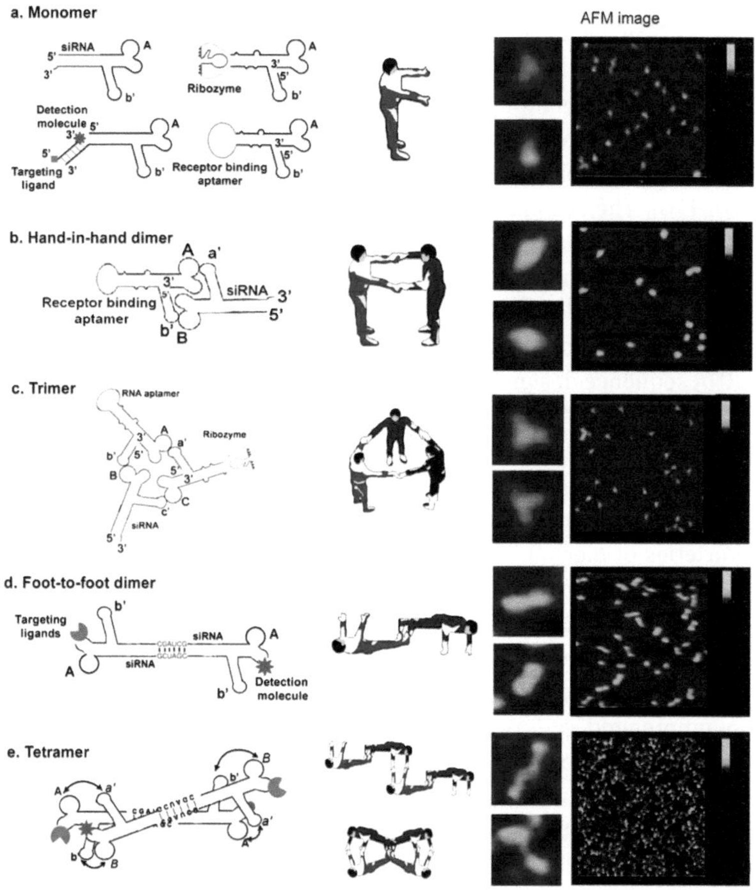

Figure 5.3 Applications of RNA nanotechnology. Therapeutic nanoparticles harboring siRNA, ribozyme, aptamer, and other moieties are constructed using bacteriophage phi29 pRNA left- and right-hand interlocking loops or palindrome sequence without template [8, 12, 39]. Uppercase and lowercase letters signify right and left hand **(a)**. Same letter pair, e.g., A–a′ indicates complementarity [28]. Dimers assemble via pRNA A–b′ and B–a′ **(b)**. Trimers form using pRNA A–b′, B–c′ and C–a′ **(c)**. Foot-to-foot dimers form via end Palindrome sequence **(d)**. Tetramers assemble by the combination of interlocking loops and palindrome mechanism **(e)**. The right panel depicts AFM images adapted from [12, 52] with permission.

self-assembled by introducing a palindrome sequence into the 3′-end of the pRNA [12]. These nanoparticles have been used successfully as polyvalent vehicles to deliver a variety of therapeutic molecules (Fig. 5.3f–j) (see Section 5.4) [12, 52]. The use of pRNA as building blocks for the construction of RNA arrays has also been achieved [12]. When three twins, Ab', Be' and Ca' are mixed, loop/loop interlocking makes the particles grow in three-dimensions.

The second method is the RNA "architectonics" [13], whereby structural modules specifying for bends or stack can be encoded within artificial RNA sequences for self-assembling higher order specific shapes of RNA. Examples include, RNA filaments [10, 21, 24] (Fig. 5.2j), molecular jigsaw puzzle units called tectosquares [13, 29] (Fig. 5.2f) and tRNA antiprisms [88].

The third method is the application of 3WJ (three way junction) and 4WJ that are selected from known RNA structures or motifs [18, 19] to serve as the cornerstone in nanoparticle construction (Fig. 5.2) [29, 76]. Some examples include: RNA-structural motif (from rRNA) to guide the tetramer assembly of L-shaped tecto RNAs; 3WJ-motif (from 23S rRNA) to construct T-shaped arrangement of three helices; and, tRNA motifs consisting of 4- and 5-WJ to fold L-shaped tertiary structures [29, 37].

The fourth method is to assemble non-natural functional RNAs with defined 3D structures using synthetic ribozyme ligase by employing molecular design of RNA based on the in vitro selection technique [16, 17]. Conformational switch of RNA nanostructures can also be constructed using a peptide-binding RNA structural motif [20].

The fifth method is the use of a palindrome sequence, which differs from the sticky end, at the 5′ or 3′ end of the RNA. The molecule will spontaneously assemble via self-annealing of the palindrome sequence immediately after in vitro transcription or chemical synthesis, before purification [12]. This method is useful for the creation of bundles, especially for designing 3D branches. Since each 11-nucleotides of the A-form RNA generates one helical turn of 360°, the angle or the direction for RNA fiber extension is controllable by varying the number of nucleotides in the helix containing the palindrome sequence.

5.4 Applications of RNA Nanotechnology

The versatility of RNA structure, the low free energy in RNA annealing, the amenability in sequence, the options in structure control, and the property of self-assembly make RNA an ideal material in nanotechnology applications. It is possible to adapt RNA to construct ordered, patterned, or preprogrammed arrays or superstructures (Fig. 5.2h,i). RNA sequences can mediate the growth of hexagonal palladium nanoparticles [89]; programmable self-assembling property of RNA ladders can direct the arrangement of cationic gold nanoparticles; periodically spaced RNA architectures can serve as a scaffold for nanocrowns [90]. Geometrically symmetrical shapes such as dimers, trimers, or polygons can be constructed from RNA [12, 13, 52]. As the symmetrical shapes facilitate the formation of crystals, RNA might serve as scaffolds for X-ray crystallography. In addition, self-assembly interaction between interlocking loops, self-linkages via a palindrome sequence, the continued growth into a hierarchical structure, and ease in conjugation and biocompatibility make RNA a good candidate for construction of scaffolds for tissue engineering [12, 21, 23]. Several laboratories have developed RNA aptamers as biosensors [91].

RNA's novel roles in nanomedicine application include cell recognition and binding for diagnosis [92]; targeted delivery via receptor mediated endocytosis [93]; intracellular control and computation via gene silencing and regulation [48, 49], nuclear membrane penetration, and brain blood barrier passing [94]. The most important therapeutic RNA moieties are discussed below.

Small interfering RNA [42, 44] *(siRNA)* is a helix with 20–25 nucleotides that interferes with gene expression through the cleavage of mRNA by a protein/RNA complex named RISC (RNA-induced silencing complex). The siRNA specifically suppresses the expression of a target protein whose mRNA includes a sequence identical to the sense strand of the siRNA. The discovery led to the award of the 2006 Nobel Prize to Andrew Fire and Craig Mello [42].

Ribozyme [95, 96] is an RNA molecule that has enzymatic activity. They have significant therapeutic potentials capable of regulating gene function by intercepting and cleaving RNA substrates, such as mRNA or the viral genome of RNA containing a sequence complementary to the catalytic center of the ribozyme. The discovery also led to the award of the 1989 Nobel Prize to Thomas Cech and Sydney Altman.

RNA aptamer [97, 98] is a family of oligonucleotides with functions similar to that of antibodies in their ability to recognize specific ligands (organic compounds, nucleotides, or peptides) through the formation of binding pockets [92]. SELEX [99] is the method to screen for the aptamers from randomized RNA pools developed in vitro by Ellington and Szostak [97], and Tuerk and Gold [98]. Using this technique, various aptamers have been selected for targeting markers relevant to diseases [92, 100, 101].

Riboswitches [102] are RNA components that bind small molecules and control gene expression in response to an organism's needs. As a biological control mechanism, riboswitches can recognize metabolites, induce premature termination of mRNA transcription, block ribosomes from translating mRNAs, cleave mRNAs, and even trigger mRNA destruction. Therefore, RNA switches can be reengineered to create a new generation of controllers regulated by drug-like molecules to tune the expression levels of targeted genes in vivo. Such RNA-based gene-control machines hold promise in future gene therapies by supplying nanoscale c/s-acting modulation [103, 104].

Various RNA moieties including siRNAs, ribozymes, antisense RNAs, aptamers, riboswitches, as well as other catalytic or editing RNAs can be easily fused or conjugated into RNA nanoparticles (Fig. 5.3f–j). The advantage of RNA nanomedicine includes: (1) self-assembly (see Section 5.2 for self-assembly and self-processing in vivo); (2) multi-valency; (3) targeted delivery; (4) protein-free; (5) nanoscale size; (6) controlled synthesis with defined structure and stoichiometry; (7) combining therapy and detection of therapy effects into one particle.

Bottom-up assembly of RNA can lead to multi-valency [51]. Each subunit may be separately functionalized to carry different therapeutic payloads, reporters and/or targeting ligands (Figs. 5.2d and 5.3f–j). Cell-type-specific delivery allows a lower concentration of the drug to be administered, thus reducing the side effects. The multivalent approach is similar to that of cocktail therapy, in which a mixture of drugs is used to produce a synergistic effect. The multi-valency offers an additional unique advantage in that therapy and detection of therapeutic effects may be combined into one nanoparticle conducted under a single administration [12, 50, 51].

Currently, a variety of other polyvalent nanoparticles have been developed; however, producing homologous particles and consistent reproduction of copy numbers within the population is challenging. Any uncertainty in structure and stoichiometry could cause unpredictable side effects or nonspecific toxicity. Using RNA nanotechnology, the production of homogeneous nanoparticles can be "manufactured" with high reproducibility, and defined structure and stoichiometry, thus facilitating quality and safety control.

The size of RNA particles in the nanometer-scale is another advantage. For effective delivery to diseased tissues, many studies suggest that particles ranging from 10–50 nm are optimal for a nonviral vector as they are large enough to be retained by the body yet small enough to pass through the cell membrane via the cell surface receptor mediated endocytosis [105]. Nanoparticle delivery has the potential to improve the pharmacokinetics, pharmacodynamics, biodistribution, and safety of this newly emerging modality.

The protein-free nature will avoid the induction of antibodies, thus allowing repeated administration for treatment of chronic diseases including cancers, viral infections, and genetic ailments. In addition, RNA nanoparticles are classified by the FDA as chemical rather than biological entities and this classification will speed-up the FDA approval.

The feasibility of RNA nanotechnology in disease therapy has been exemplified in the phi29 pRNA therapeutic system [14, 35, 50, 51, 106, 107]. Incubation of the synthetic polyvalent RNA nanoparticles containing receptor-binding aptamers or ligands resulted in cell binding and entry of the incorporated therapeutics, subsequently modulating apoptosis [50, 51]. The delivery efficiency and therapeutic effect were later confirmed in animal trials [50, 51]. The 3D design, circular permutation, folding energy alteration, and nucleotide modification of RNA were applied to generate RNase resistant RNA nanoparticles with low toxicity and to ensure processing of the chimeric RNA complexes into siRNA by Dicer after delivery.

5.5 Challenges and Perspectives

RNA nanoparticle construction involves conjugation of functionalities, crosslinking of modules, labelling subunits, and chemical modification of nucleotides. Methods in synthesizing RNA building blocks include both chemical and enzymatic approaches. While great progress has been made, improvements are in demand.

Prediction of RNA structure or folding for particle assembly remains a great challenge. Due to the unusual folding properties such as non-canonical base pairing, the rules that elucidate RNA folding are yet to be sorted out. Currently, using the RNA 2D prediction program by Zuker, typically only 70% of the 2D folding prediction is accurate based on experimental data [74, 75]. Clearly, predicting the RNA 3D and 4D structures is even more elusive.

Computer-aided programs in RNA structure prediction remain to be explored. Programs for the computation of inter- molecular interactions of RNA subunits for quaternary nanostructure formation are in imperative demand.

Natural RNA is sensitive to RNase and is especially unstable in serum or in the body. RNA's instability has long hindered its application as a construction material. Improvement of RNA stability is greatly important and is progressing rapidly, including chemical modification of the base (e.g. 5-Br-Ura and 5-I-Ura); phosphate linkage (e.g. phosphothioate, boranophosphate); and/or the C2′ (e.g. 2′-fluorine, 2′-O-methyl or 2′-amine) [108]. Other attempts include Peptide Nucleic Acids, Locked Nucleic Acids and their respective derivatives PolyCarbamate Nucleic Acids [109] or Locked Nucleic Acids with a bridge at different positions (2′–4′, 1′–3′) [110]. The 3′-end capping also improved the base pairing selectivity in duplex formation [111]. For all these methods, the 2′F-modification is the most appraisable since it has minimal detrimental effect on RNA folding and function.

Loop/loop interaction is one of the unique approaches to assemble quaternary RNA nanoparticles; however, dissociation of loops can occur when the concentration is reduced. Crosslinking agents, such as psoralen, nitrogen mustard derivatives, and transition metal compounds [112] can promote the formation of stable RNA complexes. Recent advancements include various bifunctional agents separated by linkers and phenolic derivatives [113 to increase the efficiency of crosslinking. Long-range (>9 Å) and short range (1.5 Å) photoaffinity crosslinking can be achieved using azidophenacyl derivatives and thionucleosides, such as 6-thioguanosine and 4-thiouridine, respectively.

For fluorescent labelling, single conjugation of fluorophores at the 5′- or 3′-end is preferable to prevent physical hindrance. End-labelling is not difficult with chemical synthesis of small RNA, however, it is challenging for long RNA requiring enzymatic methods. To meet this challenge, GMP or AMP derivatives that can only be used for transcription initiation, but not for chain elongation, have been used. Fluorescent RNA can also be easily synthesized in vitro with T7 RNA polymerase using a new agent tCTP [114].

The challenges of in vivo computation using RNA [48, 49] include scaling the logic operations with a large number of inputs, extending input signal types, and nonspecific actions resulting in targeting unexpected or undesired pathways.

The results of modification related to RNA folding and in vivo toxicity of the nucleotide derivatives remain to be explored. Due to the metabolism and biocompatibility issues, the most stable RNA might not necessarily be the most desirable; retention of particles within an appropriate time period is more attractive.

The most challenging aspect of RNA therapeutics is the yield and cost of RNA production. Commercial RNA chemical synthesis can only offer 40 (conservative) to 80 (with low yield) nucleotides. Acetalester 2′-OH protecting groups, such as pivaloyloxymethyl, have been reported to enhance chemical synthesis of RNA. RNase ligase II has been shown to be a good alternative over the traditional T4 DNA ligase to generate longer RNA by ligation of two shorter synthetic RNA fragments [115]. In enzymatic synthesis, heterogeneity of the 3′-end has been an issue [116]; this can be addressed by extending the transcribed sequence beyond the intended end and then cleaving the RNA at the desired site using ribozymes, DNAzymes, or RNase H [115–117]. Large scale RNA complexes produced in bacteria escorted by a tRNA vector have also been reported [40, 41]. Based on the rapid reduction of cost over the history of DNA synthesis, it is expected that the cost of RNA synthesis will gradually decrease with the development of industrial-scale RNA production technologies.

In conclusion, natural or synthetic RNA molecules can fold into pre-defined structures that can spontaneously assemble into nanoparticles with multiple functionalities. The field of RNA nanotechnology is emerging but will play an increasingly important role in medicine, biotechnology, synthetic biology and nanotechnology.

Acknowledgments

We thank John Rossi, Peter Stockley, Andrew Ellington, Shane Fimbel, Jason Lu, Farzin Haque, Anne Vonderheide, Randall Reif, Chaoping Chen, Mathieu Cinier and Feng Xiao for insightful comments; Chad Schwartz, Yi Shu, Jia Geng, for their assistance in preparation of this manuscript. The work in the author's laboratory is supported by NIH grants GM059944, EB003730, and NIH Nanomedicine Development Center entitled "Phi29 DNA Packaging Motor for Nanomedicine" (PN2 EY018 230) through the NIH Roadmap for Medical Research, as well as contract from Kylin Therapeutics, Inc., of which the author is a cofounder.

References

1. Lin C, Liu Y, Yan H. Designer DNA nanoarchitectures. *Biochemistry.* 2009, **48**, 1663–1674.
2. Aldaye FA, Palmer AL, Sleiman HF. Assembling materials with DNA as the guide. *Science*, 2008, **321**, 1795–1799.
3. Seeman NC. Nanomaterials based on DNA. *Annu Rev Biochem.* 2010, **79**, 65–87.
4. Moll D, et al. S-layer-streptavidin fusion proteins as template for nanopatterned D. molecular arrays. *Proc Natl Acad Sci USA*. 2002, 99, 14646–14651.

5. Cui H, Muraoka T, Cheetham AG, Stupp SI. Self-assembly of giant peptide nanobelts. *Nano Lett.* 2009, **9**, 945–951.
6. Adler-Abramovich L, et al. Self-assembled arrays of peptide nanotubes by vapour deposition. *Nat Nanotechnol.* 2009, **4**, 849–854.
7. Knowles TP, et al. Nanostructured films from hierarchical self-assembly of amyloidogenic proteins. *Nat Nanotechnol.* 2010, **5**, 204–207.
8. Guo P, et al. Inter-RNA interaction of phage phi29 pRNA to form a hexameric complex for viral DNA transportation. *Mol Cell.* 1998, **2**, 149–155.
9. Zhang F, et al. Function of hexameric RNA in packaging of bacteriophage phi29 DNA in vitro. *Mol Cell.* 1998, **2**, 141–147.
10. Jaeger L, Leontis NB. Tecto-RNA: one dimensional self-assembly through tertiary interactions. *Angew Chem Int Ed Engl.* 2000, **39**, 2521–2524.
11. Jaeger L, Westhof E, Leontis NB. TectoRNA: modular assembly units for the construction of RNA nano-objects. *Nucleic Acids Res.* 2001, **29**, 455–463.
12. Shu D, et al. Bottom-up assembly of RNA arrays and superstructures as potential parts in nanotechnology. *Nano Lett.* 2004, **4**, 1717–1723.
13. Chworos A, et al. Building programmable jigsaw puzzles with RNA. *Science.* 2004, **306**, 2068–2072.
14. Guo P. RNA nanotechnology: engineering, assembly and applications in detection, gene delivery and therapy. *J Nanosci Nanotechnol.* 2005, **5**(12), 1964–1982.
15. Jaeger L, Chworos A. The architectonics of programmable RNA and DNA nanostructures. *Curr Opin Struct Biol.* 2006, **16**, 531–543.
16. Ikawa Y, Tsuda K, Matsumura S, Inoue T. Denovo synthesis and development of an RNA enzyme. *Proc Natl Acad Sci U S A.* 2004, **101**, 13750–13755.
17. Matsumura S, et al. Coordinated control of a designed trans-acting ligase ribozyme by a loop-receptor interaction. *FEBS Lett.* 2009, **583**, 2819–2826.
18. Leontis NB, Lescoute A, Westhof E. The building blocks and motifs of RNA architecture. *Curr Opin Struct Biol.* 2006, **16**, 279–287.
19. Schroeder KT, McPhee SA, Ouellet J, Lilley DM. A structural database for k-turn motifs in RNA. *RNA.* 2010, **16**, 1463–1468.

20. Li X, Horiya S, Harada K. An efficient thermally induced RNA conformational switch as a framework for the functionalization of RNA nanostructures. *J Am Chem Soc.* 2006, **128**, 4035–4040.

21. Nasalean L, Baudrey S, Leontis NB, Jaeger L. Controlling RNA self-assembly to form filaments. *Nucleic Acids Res.* 2006, **34**, 1381–1392.

22. Liu B, Baudrey S, Jaeger L, Bazan GC. Characterization of tectoRNA assembly with cationic conjugated polymers. *J Am Chem Soc.* 2004, 126, 4076–4077.

23. Cayrol B, et al. A nanostructure made of a bacterial noncoding RNA. *J Am Chem Soc.* 2009, **131**, 17270–17276.

24. Geary C, Chworos A, Jaeger L. Promoting RNA helical stacking via A-minor junctions. *Nucleic Acids Res.* 2010, **39**(3), 1066–1080.

25. Sugimoto N, et al. Thermodynamic parameters to predict stability of RNA/DNA hybrid duplexes. *Biochemistry.* 1995, **34**, 11211–11216.

26. Searle MS, Williams DH. On the stability of nucleic acid structures in solution: enthalpy-entropy compensations, internal rotations and reversibility. *Nucleic Acids Res.* 1993, **21**, 2051–2056.

27. Kitamura A, et al. Analysis of intermolecular base pair formation of prohead RNA of the phage phi29 DNA packaging motor using NMR spectroscopy. *Nucleic Acids Res.* 2008, **36**, 839–848.

28. Chen C, Zhang C, Guo P. Sequence requirement for hand-in-hand interaction in formation of pRNA dimers and hexamers to gear phi29 DNA translocation motor. *RNA.* 1999, **5**, 805–818.

29. Severcan I, et al. Square-shaped RNA particles from different RNA folds. *Nano Lett.* 2009, **9**, 1270–1277.

30. Severcan I, et al. A polyhedron made of tRNAs. *Nat Chem.* 2010, **2**, 772–779.

31. Hansma HG, Oroudjev E, Baudrey S, Jaeger L. TectoRNA and 'kissing-loop' RNA: atomic force microscopy of self-assembling RNA structures. *J Microsc.* 2003, **212**, 273–279.

32. Lee RJ, Wang S, Low PS. Measurement of endosome pH following folate receptor-mediated endocytosis. *Biochim Biophys Acta.* 1996, **1312**, 237–242.

33. Pogocki D, Schoneich C. Chemical stability of nucleic acid-derived drugs. *J Pharm Sci.* 2000, **89**, 443–456.

34. Laurenti E, et al. Inducible gene and shRNA expression in resident hematopoietic stem cells in vivo. *Stem Cells.* 2010, **28**, 1390–1398.

35. Hoeprich S, et al. Bacterial virus phi29 pRNA as a hammerhead ribozyme escort to destroy hepatitis B virus. *Gene Ther.* 2003, **10**, 1258–1267.

36. Chang KY, Tinoco I Jr. Characterization of a "kissing" hairpin complex derived from the human immunodeficiency virus genome. *Proc Natl Acad Sci USA.* 1994, **91**(18), 8705–8709.

37. Bindewald E, et al. RNA Junction: a database of RNA junctions and kissing loops for three- dimensional structural analysis and nanodesign. *Nucleic Acids Res.* 2008, **36**, D392–D397.

38. Wagner C, Ehresmann C, Ehresmann B, Brunei C. Mechanism of dimerization of bicoid mRNA: initiation and stabilization. *J Biol Chem.* 2004, **279**, 4560–4569.

39. Chen C, Sheng S, Shao Z, Guo P. A dimer as a building block in assembling RNA: a hexamer that gears bacterial virus phi29 DNA-translocating machinery. *J Biol Chem.* 2000, **275**(23), 17510–17516.

40. Ponchon L, Beauvais G, Nonin-Lecomte S, Dardel F. A generic protocol for the expression and purification of recombinant RNA in Escherichia coli using a tRNA scaffold. *Nat Protoc.* 2009, **4**, 947–959.

41. Kuwabara T, et al. Formation of a catalytically active dimer by tRNA-driven short ribozymes. *Nat Biotechnol.* 1998, **16**, 961–965.

42. Fire A, et al. Potent and specific genetic interference by double-stranded RNA in Caenorhabditis elegans. *Nature.* 1998, **391**, 806–811.

43. Li H, Li WX, Ding SW. Induction and suppression of RNA silencing by an animal virus. *Science.* 2002, **296**, 1319–1321.

44. Breaker RR. Complex riboswitches. *Science.* 2008, **319**, 1795–1797.

45. Fabian MR, Sonenberg N, Filipowicz W. Regulation of mRNA translation and stability by microRNAs. *Annu Rev Biochem.* 2010, **79**, 351–379.

46. Zhang C. Novel functions for small RNA molecules. *Curr Opin Mol Ther.* 2009, **11**, 641–651.

47. Marvin MC, Engelke DR. Broadening the mission of an RNA enzyme. *J Cell Biochem.* 2009, **108**, 1244–1251.

48. Benenson Y. RNA-based computation in live cells. *Curr Opin Biotechnol.* 2009, **20**, 471–478.

49. Shlyakhtenko LS, et al. Silatrane-based surface chemistry for immobilization of DNA, protein-DNA complexes and other biological materials. *Ultramicroscopy.* 2003, **97**, 279–287.

50. Guo S, Tschammer N, Mohammed S, Guo P. Specific delivery of therapeutic RNAs to cancer cells via the dimerization mechanism of phi29 motor pRNA. *Hum Gene Ther.* 2005, **16**, 1097–1109.
51. Khaled A, Guo S, Li F, Guo P. Controllable self-assembly of nanoparticles for specific delivery of multiple therapeutic molecules to cancer cells using RNA nanotechnology. *Nano Lett.* 2005, **5**, 1797–1808.
52. Shu D, Huang L, Hoeprich S, Guo P. Construction of phi29 DNA-packaging RNA (pRNA) monomers, dimers and trimers with variable sizes and shapes as potential parts for nano-devices. *J Nanosci Nanotechnol.* 2003, **3**, 295–302.
53. Turner R, Tijan R. Leucine repeats and an adjacent DNA binding domain mediate the formation of functional c-Fos and c-Jun heterodimers. *Science.* 1989, **243**, 1689–1694.
54. Liu D, et al. Tensegrity: construction of rigid DNA triangles with flexible four-arm DNA junctions. *J Am Chem Soc.* 2004, **126**, 2324–2325.
55. Rothemund PW, Papadakis N, Winfree E. Algorithmic self-assembly of DNA Sierpinski triangles. *PLoS Biol.* 2004, **2**, e424.
56. Park SH, et al. Three-helix bundle DNA tiles self-assemble into 2D lattice or ID templates for silver nanowires. *Nano Lett.* 2005, **5**, 693–696.
57. Weizmann Y, et al. A polycatenated DNA scaffold for the one-step assembly of hierarchical nanostructures. *Proc Natl Acad Sci USA.* 2008, **105**, 5289–5294.
58. Aldaye FA, Sleiman HF. Modular access to structurally switchable 3D discrete DNA assemblies. *J Am Chem Soc.* 2007, **129**, 13376–13377.
59. Zheng J, et al. From molecular to macroscopic via the rational design of a self-assembled 3D DNA crystal. *Nature.* 2009, **461**, 74–77.
60. Rothemund PWK. Folding DNA to create nanoscale shapes and patterns. *Nature.* 2006, **440**, 297–302.
61. Andersen ES, et al. Self-assembly of a nanoscale DNA box with a controllable lid. *Nature.* 2009, **459**, 73–76.
62. Ke YG, et al. Self-assembled water-soluble nucleic acid probe tiles for label-free RNA hybridization assays. *Science.* 2008, **319**, 180–183.
63. Douglas SM, Chou JJ, Shih WM. DNA-nanotube-induced alignment of membrane proteins for NMR structure determination. *Proc Natl Acad Sci USA.* 2007, **104**, 6644–6648.
64. Endo M, Seeman NC, Majima T. DNA tube structures controlled by a four-way-branched DNA connector. *Angew Chem Int Ed Engl.* 2005, **44**, 6074–6077.

65. Tanaka K, et al. A discrete self-assembled metal array in artificial DNA. *Science.* 2003, **299**, 1212–1213.
66. Yurke B, et al. A DNA-fuelled molecular machine made of DNA. *Nature.* 2000, **406**, 605–608.
67. Lin C, et al. In vivo cloning of artificial DNA nanostructures. *Proc Natl Acad Sci USA.* 2008, **105**, 17626–17631.
68. Eckardt LH, et al. DNA nanotechnology: chemical copying of connectivity. *Nature.* 2002, **420**, 286.
69. Endo M, et al. Programmed-assembly system using DNA jigsaw pieces. *Chemistry.* 2010, **16**, 5362–5368.
70. Dietz H, Douglas SM, Shih WM. Folding DNA into twisted and curved nanoscale shapes. *Science.* 2009, **325**, 725–730.
71. Cherny DI, Eperon IC, Bagshaw CR. Probing complexes with single fluorophores: factors contributing to dispersion of FRET in DNA/RNA duplexes. *Eur Biophys J.* 2009, **38**, 395–405.
72. Shapiro BA. Computational Design Strategies for RNA Nanostructures. *J Biomol Str Dyn.* 2009, **26**, 820.
73. Yingling YG, Shapiro BA. Computational design of an RNA hexagonal nanoring and an RNA nanotube. *Nano Lett.* 2007, **7**, 2328–2334.
74. Zuker M. Mfold web server for nucleic acid folding and hybridization prediction. *Nucleic Acids Res.* 2003, **31**, 3406–3415.
75. Markham NR, Zuker M. UNAFold: software for nucleic acid folding and hybridization. *Methods Mol Biol.* 2008, **453**, 3–31.
76. Bindewald E, et al. Computational strategies for the automated design of RNA nanoscale structures from building blocks using NanoTiler. *J Mol Graphics Modelling.* 2008, **27**, 299–308.
77. Afonin KA, et al. In vitro assembly of cubic RNA-based scaffolds designed in silico. *Nat Nanotechnol.* 2010, **5**, 676–682.
78. Chakraborty S, Modi S, Krishnan Y. The RNA2-PNA2 hybrid i-motif-a novel RNA-based building block. *Chem Commun (Camb).* 2008, **0**, 70–72.
79. Afonin KA, Cieply DJ, Leontis NB. Specific RNA self-assembly with minimal paranemic motifs. *J Am Chem Soc.* 2008, **130**, 93–102.
80. Lescoute A, Westhof E. Topology of three-way junctions in folded RNAs. *RNA.* 2006, **12**, 83–93.
81. Ouellet J, et al. Structure of the three-way helical junction of the hepatitis C virus IRES element. *RNA.* 2010, **16**, 1597–1609.

82. dela PM, Dufour D, Gallego J. Three- way RNA junctions with remote tertiary contacts: a recurrent and highly versatile fold. *RNA*. 2009, **15**, 1949–1964.
83. Griffiths-Jones S, et al. Rfam: annotating non-coding RNAs in complete genomes. *Nucleic Acids Res*. 2005, **33**, D121–D124.
84. Abraham M, Dror O, Nussinov R, Wolfson HJ. Analysis and classification of RNA tertiary structures. *RNA*. 2008, **14**, 2274–2289.
85. Guo P, Erickson S, Anderson D. A small viral RNA is required for in vitro packaging of bacteriophage phi29 DNA. *Science*. 1987, **236**, 690–694.
86. Xiao F, Demeler B, Guo P. Assembly mechanism of the sixty-subunit nanoparticles via interaction of RNA with the reengineered protein connector of phi29 DNA-packaging motor. *ACS Nano*. 2010, **4**, 3293–3301.
87. Shu D, Zhang H, Jin J, Guo P. Counting of six p RNAs of phi29 DNA-packaging motor with customized single molecule dual-view system. *EMBO J*. 2007, **26**, 527–537.
88. Woodson SA. Compact intermediates in RNA folding. *Annu Rev Biophys*. 2010, **39**, 61–77.
89. Gugliotti LA, Feldheim DL, Eaton BE. RNA-mediated metal-metal bond formation in the synthesis of hexagonal palladium nanoparticles. *Science*. 2004, **304**, 850–852.
90. Koyfman AY, et al. Controlled spacing of cationic gold nanoparticles by nanocrown RNA. *J Am Chem Soc*. 2005, **127**, 11886–11887.
91. Oguro A, Ohtsu T, Nakamura Y. An aptamer-based biosensor for mammalian initiation factor eukaryotic initiation factor 4A. *Anal Biochem*. 2009, **388**, 102–107.
92. Mi J, et al. In vivo selection of tumor-targeting RNA motifs. *Nat Chem Biol*. 2010, **6**, 22–24.
93. Liu Y, et al. Targeting hypoxia-inducible factor-1 alpha with Tf-PEI-shRNA complex via transferrin receptor-mediated endocytosis inhibits melanoma growth. *Mol Ther*. 2009, **17**, 269–277.
94. Kumar P, et al. Transvascular delivery of small interfering RNA to the central nervous system. *Nature*. 2007, **448**, 39–43.
95. Kruger K, et al. Self-splicing RNA: autoexcision and autocyclization of the ribosomal RNA intervening sequence of Tetrahymena. *Cell*. 1982, **31**, 147–157.
96. Guerrier-Takada C, et al. The RNA moiety of ribonuclease P is the catalytic subunit of the enzyme. *Cell*. 1983, **35**, 849–857.

97. Ellington AD, Szostak JW. In vitro selection of RNA molecules that bind specific ligands. *Nature*. 1990, **346**, 818–822.
98. Tuerk C, Gold L. Systematic evolution of ligands by exponential enrichment: RNA ligands to bacteriophage T4 DNA ploymerase. *Science*. 1990, **249**, 505–510.
99. Ellington AD. Back to the future of nucleic acid self-amplification. *Nat Chem Biol*. 2009, **5**, 200–201.
100. Zhou J, Li H, Zaia J, Rossi JJ. Novel dual inhibitory function aptamer-siRNA delivery system for HIV-1 therapy. *Mol Ther*. 2008, **16**, 1481–1489.
101. Bunka DH, et al. Production and characterization of RNA aptamers specific for amyloid fibril epitopes. *J Biol Chem*. 2007, **282**, 34500–34509.
102. Sudarsan N, et al. Riboswitches in eubacteria sense the second messenger cyclic di-GMP. *Science*. 2008, **321**, 411–413.
103. Ogawa A, Maeda M. An artificial aptazyme-based riboswitch and its cascading system in *E. coli*. *Chembiochem*. 2008, **9**, 206–209.
104. Shahbabian K, Jamalli A, Zig L, Putzer H. RNase Y, a novel endoribonuclease, initiates riboswitch turnover in *Bacillus subtilis*. *EMBO J*. 2009, **28**, 3523–3533.
105. Prabha S, Zhou WZ, Panyam J, Labhasetwar V. Size-dependency of nanoparticle-mediated gene transfection: studies with fractionated nanoparticles. *Int J Pharm*. 2002, 244, 105–115.
106. Guo S, Huang F, Guo P. Construction of folate- conjugated pRNA of bacteriophage phi29 DNA packaging motor for delivery of chimeric siRNA to nasopharyngeal carcinoma cells. *Gene Ther*. 2006, **13**, 814–820.
107. Huifang M, Zhang, et al. Target delivery of anti-coxsachievirus siRNAs using ligand-conjugated packaging RNAs. *Antiviral Res*. 2009, **83**(3), 307–316.
108. Watts JK, Deleavey GF, Damha MJ. Chemically modified siRNA: tools and applications. *Drug Discov Today*. 2008, **13**, 842–855.
109. Madhuri V, Kumar VA. Design, synthesis and DNA/RNA binding studies of nucleic acids comprising stereoregular and acyclic polycarbamate backbone: polycarbamate nucleic acids (PCNA). *Org Biomol Chem*. 2010, **8**, 3734–3741.
110. Mathe C, Perigaud C. Recent approaches in the synthesis of conformationally restricted nucleoside analogues. *Eur J Org Chem*. 2008, **2008**(9), 1489–1505.

111. Patra A, Richert C. High fidelity base pairing at the 3′-terminus. J Am Chem Soc. 2009, 131, 12671–12681.

112. Efimov VA, Fediunin SV, Chakhmakhcheva OG. Cross-linked nucleic acids: formation, structure, and biological function. *Bioorg Khim*. 2010, **36**, 56–80.

113. Song Z, et al. Synthesis and oxidation-induced DNA cross-linking capabilities of bis(catechol) quaternary ammonium derivatives. *Chemistry*. 2008, **14**, 5751–5754.

114. Stengel G, Urban M, Purse BW, Kuchta RD. Incorporation of the fluorescent ribonucleotide analogue tCTP by T7 RNA polymerase. *Anal Chem*. 2010, **82**, 1082–1089.

115. Solomatin S, Herschlag D. Methods of site-specific labeling of RNA with fluorescent dyes. *Methods in Enzymology*, vol 469: *Biophysical, Chemical, and Functional Probes of RNA Structure, Interactions and Folding*, Pt B. 2009, **469**, 47–68.

116. Lavergne T, Bertrand JR, Vasseur JJ, Debart F. A base-labile group for 2′-OH protection of rib onucleosides: a major challenge for RNA synthesis. *Chemistry*. 2008, **14**, 9135–9138.

117. Hoeprich S, Guo P. Computer modeling of three-dimensional structure of DNA-packaging RNA(pRNA) monomer, dimer, and hexamer of phi29 DNA Packaging motor. *J Biol Chem*. 2002, **277**(23), 20794–20803.

Chapter 6

Drug Targeting to Tumors: Principles, Pitfalls and (Pre-) Clinical Progress

Twan Lammers,[a,b,c] Fabian Kiessling,[a] Wim E. Hennink,[b] and Gert Storm[b,c]

[a]*Department of Experimental Molecular Imaging, RWTH - Aachen University, Helmholtz Institute for Biomedical Engineering, Aachen, Germany*
[b]*Department of Pharmaceutics, Utrecht Institute of Pharmaceutical Sciences, Utrecht University, Utrecht, The Netherlands*
[c]*Department of Targeted Therapeutics, University of Twente, Enschede, The Netherlands*
tlammers@ukaachen.de

Many different systems and strategies have been evaluated for drug targeting to tumors over the years. Routinely used systems include liposomes, polymers, micelles, nanoparticles and antibodies, and examples of strategies are passive drug targeting, active drug targeting to cancer cells, active drug targeting to endothelial cells and triggered drug delivery. Significant progress has been made in this area of research both at the preclinical and at the clinical level, and a number of (primarily passively tumor-targeted) nanomedicine formulations have been approved for clinical use. Significant progress has also been made with regard to better understanding

---

Reproduced with permission from *Journal of Controlled Release*, **161**, 175–187, 2012.

Nano-Enabled Medical Applications
Edited by Lajos P. Balogh

ISBN 978-981-4800-32-7 (Hardcover), 978-0-429-39903-9 (eBook)
www.jennystanford.com

the (patho-) physiological principles of drug targeting to tumors. This has led to the identification of several important pitfalls in tumor-targeted drug delivery, including (I) overinterpretation of the EPR effect; (II) poor tumor and tissue penetration of nanomedicines; (III) misunderstanding of the potential usefulness of active drug targeting; (IV) irrational formulation design, based on materials which are too complex and not broadly applicable; (V) insufficient incorporation of nanomedicine formulations in clinically relevant combination regimens; (VI) negligence of the notion that the highest medical need relates to metastasis, and not to solid tumor treatment; (VII) insufficient integration of non-invasive imaging techniques and theranostics, which could be used to personalize nanomedicine-based therapeutic interventions; and (VIII) lack of (efficacy analyses in) proper animal models, which are physiologically more relevant and more predictive for the clinical situation. These insights strongly suggest that besides making ever more nanomedicine formulations, future efforts should also address some of the conceptual drawbacks of drug targeting to tumors, and that strategies should be developed to overcome these shortcomings.

6.1 Introduction

In the past few decades, significant progress has been made in understanding the molecular principles of many different diseases. In the case of cancer, these improved insights into the genetic and (patho-) physiological processes contributing to malignant transformation and tumorigenesis have resulted in the development of several novel (classes of) chemotherapeutic drugs. Such "molecularly targeted therapeutics," like the growth factor receptor inhibitor Herceptin, the proteasome inhibitor Velcade, the histone deacetylase inhibitor Vorinostat and the antiangiogenic agent Avastin, more selectively interfere with certain "hallmarks of cancer" [1, 2], like with the overexpression of growth factors and growth factor receptors, with the altered balance between apoptosis and anti-apo-ptosis, with the numerous genetic and epigenetic changes that are present in cancer cells, and with the development of a

Table 6.1 Barriers limiting the delivery of i.v. applied anticancer agents to tumors

Anatomical barriers	Physiological barriers	Chemical barriers	Clinical barriers
Vascular endothelium	Renal filtration	Low solubility	Low efficacy
Perivascular space	Hepatic degradation	Low stability	High toxicity
Cellular membrane	High tumor cell density	Low molecular weight	Need for hospitalization
Nuclear membrane	High interstitial fluid pressure	Large volume of distribution	Frequent administration
Blood–brain barrier	Drug efflux pumps	Charge interactions	Low cost-effectiveness

Note: Several barriers are inter-related, and that not all barriers apply to all types of (chemo-) therapeutic agents.
Source: Adapted, with permission, from [76].

dense vascular network, needed to provide tumors with oxygen and nutrients. By means of their pharmacologically and/or physiologically more optimal mechanism(s) of action, "molecularly targeted therapeutics" have been shown to be able to more preferentially kill cancer cells, both in vitro and in vivo, and to improve the balance between the efficacy and the toxicity of systemic anticancer therapy [3–5].

An important but often neglected property that such second-generation chemotherapeutics share with their first generation DNA-damaging counterparts is that upon intravenous administration, their pharmacokinetics and their tissue distribution often are far from optimal. Because of their low molecular weight, for instance, the majority of routinely used anticancer agents are rapidly cleared from systemic circulation (e.g. by means of renal filtration), and they do not accumulate well in tumors and in tumor cells. Also, because of their small size and their (generally) high hydrophobicity, drug molecules often have a large volume of distribution, and they tend to accumulate in and cause toxicity towards many different healthy tissues. In addition to this, as outlined in Table 6.1, a large number of other barriers need to be overcome before an i.v. administered anticancer agent can elicit antitumor efficacy, related e.g. to hepatic and enzymatic degradation, to the high interstitial fluid pressure that is typical of tumors, to cellular and nuclear membranes, and to the presence of drug efflux pumps.

To assist i.v. applied anticancer agents in overcoming such barriers, and to improve their pharmacokinetic profile and their target site accumulation, a large number of nanomedicine formulations have been designed and evaluated over the years. Nanomedicines are submicrometer-sized carrier materials which intend to improve the biodistribution of systemically administered (chemo-) therapeutic drugs. By delivering pharmacologically active agents more selectively to pathological sites ("site-specific drug delivery"), and/or by guiding them away from potentially endangered healthy tissues ("site-avoidance drug delivery"), nanomedicine formulations aim to improve the balance between the efficacy and the toxicity of systemic (chemo-) therapeutic interventions [6–11].

Various different concepts have been envisioned for nanomedicine-mediated drug targeting to tumors over the years, including e.g. passive drug targeting, active targeting to cancer cells, active targeting to endothelial cells and triggered drug delivery (using stimuli-responsive carrier materials). A huge number of preclinical and a significant number of clinical studies have provided insights into the validity of these approaches, and it has been realized that though they might hold true in several cases, there are also situations in which this is not the case. This is due to a series of misconceptions, misunderstandings and pitfalls, which together limit the bench-to-bedside translation of tumor-targeted nanomedicines, as well as their clinical benefit and their use in large numbers of patients.

In the present perspective, we briefly describe the basic principles of drug targeting to tumors, summarize the (pre-) clinical progress made with regard to tumor-targeted nanomedicines, address some of the pitfalls identified along the way, and propose strategies to overcome these shortcomings.

6.2 Principles of Drug Targeting to Tumors

6.2.1 Passive Targeting

Many different types of nanomedicines have been designed and evaluated for drug targeting to tumors. Prototypic examples of nanomedicine formulations are liposomes, polymers, micelles, nanoparticles and antibodies (Fig. 6.1A–E). The former four nanomedicines primarily aim to improve the circulation time of the conjugated or entrapped (chemo-) therapeutic drug and, by doing so, to enable it to exploit the pathophysiological fact that solid tumors tend to present with a tortuous and poorly differentiated vasculature, that in contrast to the vasculature in healthy tissues, allows for the extravasation of carrier materials with sizes of up to several hundreds of nanometers (Fig. 6.1G). Together with the fact that solid tumors tend to lack functional lymphatics, and therefore are unable to eliminate extravasated nanomaterials, this increase in vascular leakiness allows long-circulating nanomedicines to accumulate in tumors over time, by means of a mechanism known as the enhanced permeability and retention (EPR) effect [12, 13]. The exploitation of the EPR effect is arguably the most important strategy for improving the delivery of low-molecular-weight chemotherapeutic agents to tumors, and because of the fact that it essentially only relies on the pathophysiological properties of the target tissue, it is generally referred to as "passive drug targeting."

In spite of several conceptual drawbacks related to extravasation and penetration (see below; Sections 6.3.1 and 6.3.2), the vast majority of nanomedicines developed for drug targeting to tumors rely on the EPR effect. These primarily include long-circulating liposomes, polymers and micelles. Examples of passively targeted nanomedicines approved for clinical use are Myocet (non-PEGylated liposomal doxorubicin), Doxil (Caelyx in Europe; PEGylated liposomal doxorubicin), Daunoxome (non-PEGylated liposomal daunorubicin), Abraxane (albumin-based paclitaxel) and Genexol-PM (paclitaxel-containing polymeric micelles; pre-approved in Korea). Several additional passively tumor-targeted nanomedicines are currently in clinical trials (see Table 6.2), and a

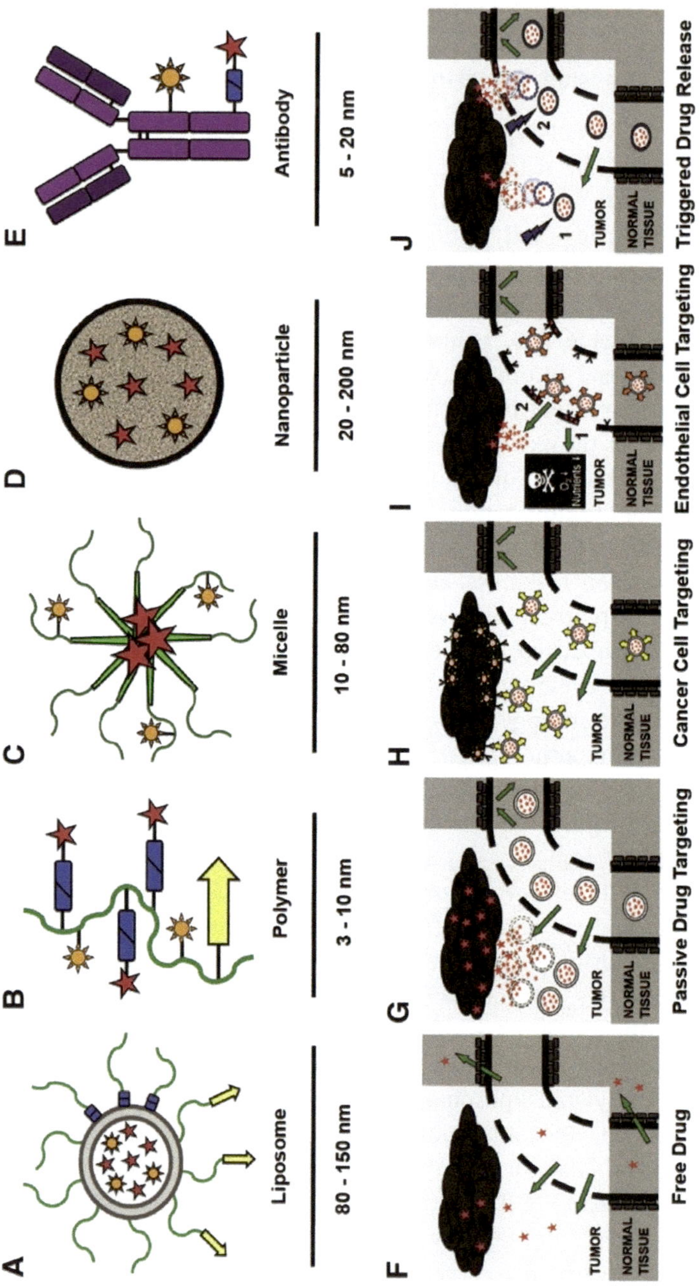
A
Liposome
80 - 150 nm
B
Polymer
3 - 10 nm
C
Micelle
10 - 80 nm
D
Nanoparticle
20 - 200 nm
E
Antibody
5 - 20 nm
F
TUMOR
NORMAL TISSUE
Free Drug
G
TUMOR
NORMAL TISSUE
Passive Drug Targeting
H
TUMOR
NORMAL TISSUE
Cancer Cell Targeting
I
TUMOR
NORMAL TISSUE
Endothelial Cell Targeting
J
TUMOR
NORMAL TISSUE
Triggered Drug Release

Figure 6.1 Systems and strategies used for drug targeting to tumors. A–E: Drug targeting systems. Liposomes and liposomal bilayers are depicted in gray, polymers and polymer-coatings in green, linkers allowing for drug release and for sheddable stealth coatings in blue (rectangles), targeting ligands in yellow (arrows), antibodies and antibody fragment in purple, imaging agents to monitor biodistribution and target site accumulation in orange (suns), and conjugated or entrapped (chemo-) therapeutic agents in red (stars). F–J: Drug targeting strategies. F: Upon the i.v. injection of a low-molecular-weight chemotherapeutic agent, which is often rapidly cleared from the blood, only low levels of the drug accumulate in tumors and in tumor cells, while its localization to normal organs and tissues can be relatively high. G: Upon the implementation of a passively targeted drug delivery system, by means of the EPR effect, the accumulation of the active agent in tumors and in tumor cells can be increased substantially, while its localization to healthy tissues can be attenuated. H: Active drug targeting to internalization-prone cell surface receptors (over-) expressed by cancer cells generally intends to improve the cellular uptake of nanomedicine formulations, and is particularly useful for the intracellular delivery of otherwise poorly internalized macromolecular drugs, such as DNA and siRNA. I: Active drug targeting to receptors (over-) expressed by angiogenic endothelial cells on the one hand aims to increase drug delivery to tumor endothelium, thereby eradicating tumor blood vessels and depriving tumor cells of oxygen and nutrients (I-1). On the other hand, reasoning that tumor endothelial cells are continuously exposed to long-circulating nanomedicines, endothelial cell targeting can likely also be employed to improve the overall accumulation of chemotherapeutic agents in tumors (I-2). J: Stimuli-sensitive nanomedicines, like Thermodox, can be activated (i.e. induced to release their contents) by externally applied physical triggers, like hyperthermia, ultrasound, magnetic fields and light. This can be done either after accumulation at the target site (J-l), or while circulating in the tumor vasculature (J-2). Images adapted, with permission, from [10, 42].

Table 6.2 Examples of clinically used tumor-targeted nanomedicines

	Name	Formulation	Indication	Status
Liposomes	Myocet	Liposomal doxorubicin	Breast cancer	Approved
	Daunoxome	Liposomal daunorubicin	Kaposi Sarcoma	Approved
	Depocyt	Liposomal cytarabine	Lymphomatus meningitis, leukemia, glioblastoma	Approved[a]
	Doxil/Caelyx	PEGylated liposomal doxorubicin	Breast cancer, ovarian cancer, multiple myeloma, Kaposi sarcoma	Approved
	Thermodox	Temperature-sensitive PEGylated liposomal doxorubicin	Liver cancer, breast cancer	Phase III
	NL CPT-11	PEGylated liposomal irinotecan	Glioma	Phase I
Polymers	Oncaspar	PEG-L-asparaginase	Leukemia	Approved
	Opaxio	PGA-paclitaxel	Lung cancer, ovarian cancer	Phase III
	PegAsys/PegIntron	PEG-IFNα2a/-IFNα2b	Melanoma, leukemia	Phase II
	PK1	PHPMA-doxorubicin	Breast cancer, lung cancer, colorectal cancer	Phase II
	ProLindac	PHPMA-oxaliplatin	Ovarian cancer	Phase II
	XMT-1001	Fleximer-bound CPT	Gastric cancer, lung cancer	Phase I
Micelles	Genexol-PM	Micellar paclitaxel	Breast cancer, lung cancer, ovarian cancer	Approved[b]
	Paclical	Micellar paclitaxel	Ovarian cancer	Phase III
	NK105	Micellar paclitaxel	Gastric cancer	Phase II
	NK911	Micellar doxorubicin	Various solid malignancies	Phase II
	NC-6004	Micellar cisplatin	Pancreatic cancer	Phase I/II
	NC-4016	Micellar oxaliplatin	Various solid malignancies	Phase I

Nanoparticles	Abraxane	Albumin-based nanoparticle containing paclitaxel	Breast cancer	Approved
	Docetaxel-PNP	Various solid malignancies	Various solid malignancies	Phase I
	CALAA-01	Cyclodextrin-based nanoparticle containing anti-RRM2 siRNA	Various solid malignancies	Phase I
	Atu027	Lipid-based nanoparticle containing anti-PKN3 siRNA	Various solid malignancies	Phase I
	C-VISA-BikDD	Lipid-based nanoparticle containing BikDD plasmid DNA	Pancreatic cancer	Phase I
	Rexin-G	Retrovirus-based nanoparticle containing dnG1 plasmid DNA	Various solid malignancies	Phase I
Antibodies	Ontak	CD25-targeted diphteria toxin-IL2 fusion protein	T-cell lymphoma	Approved
	Mylotarg	CD33-targeted ozogamycin-gemtuzumab	Leukemia	Approved[c]
	Zevalin	CD20-targeted yttrium-90-ibritumomab tiuxetan	Non-Hodgkin lymphoma	Approved
	Bexxar	CD20-targeted iodine- 131-tositumomab	Non-Hodgkin lymphoma	Approved
	SGN-35	CD30-targeted brentuximab vedotin	Non-Hodgkin lymphoma	Phase III
	AGS-5ME	SLC44A4-targeted monomethyl auristatin E	Prostate cancer, pancreatic cancer	Phase I

[a]Approved for lymphomatous meningitis; in Phase III for leukemia; and in Phase I/II for glioblastoma.
[b]Approved in Korea; in Phase II in US and Russia.
[c]Withdrawn in 2010.

large number of other ones are in early- and late-stage preclinical development [6–11].

6.2.2 Active Targeting to Cancer Cells

As opposed to passive drug targeting, "active drug targeting" relies on the use of targeting ligands, like antibodies and peptides, which specifically bind to receptor structures (over-) expressed at the target site [10, 14, 15]. Active drug targeting is generally implemented to improve target cell recognition and target cell uptake, and not to improve overall tumor accumulation. Examples of targeting ligands routinely used for actively targeting nanomedicine formulations to tumor cells are folate [16], transferrin [17] and galactosamine [18]. To date, however, in spite of significant advances made at the preclinical level with regard to active targeting, only antibody-based nanomedicines, such as Zevalin, Mylotarg, Ontak and Bexxar have been approved for clinical use (see Table 6.2).

For several reasons (as will be outlined in more detail below; see Section 6.3.3), no actively targeted liposomes, polymers, micelles and nanoparticles have thus far been approved for clinical use, and only very few are in clinical trials. Among those which are in clinical trials are formulations which really rely on improving cellular uptake for conferring therapeutic efficacy. A prototypic example of this is CALAA-01, i.e. a transferrin receptor-targeted cyclodextrin-based polymeric nanoparticle containing siRNA, which itself is unable to enter cells and which needs to be delivered into the cytoplasm of cancer cells to exert antitumor effects (see Table 6.2).

The observation that actively targeted "classical" nanomedicines, such as liposomes, polymers and micelles, have thus far largely failed to demonstrate benefit at the (pre-) clinical level can likely be mostly attributed to the fact that after leaving the highly leaky tumor vasculature, there are quite a number of anatomical and physiological barriers that need to be overcome before antibody- or peptide-targeted formulations can bind to (and enter) cancer cells. These include the presence of pericyte-, smooth muscle cell- and fibroblast-based cell layers between endothelial and tumor cells, the high cellular density within solid malignancies, and the high interstitial fluid pressure that is typical of tumors [11, 19].

Therefore, and also because of the binding-site barrier [20], which further limits the penetration of actively targeted nanomedicines into the tumor interstitium, actively targeted nanomedicines tend to have problems finding their target cells, and they generally fail to demonstrate an advantage over passively targeted formulations.

6.2.3 Active Targeting to Endothelial Cells

To overcome (some of) the abovementioned shortcomings with regard to active drug targeting—i.e. those related to the presence of several additional cell layers between endothelial and tumor cells, to the high tumor cellular density and to the high interstitial fluid pressure in tumors—a number of different endothelial cell-targeted nanomedicine formulations have been designed and evaluated over the years (Fig. 6.1H). Ligands used to target drugs and/or drug delivery systems to tumor blood vessels include the antibody fragment L19, which uses the EDB domain of the oncofetal protein fibronectin to home to angiogenic vasculature [21, 22], and several cyclic and linear derivatives of the oligopeptides RGD and NCR, which bind to angiogenic endothelium through the integrins $\alpha 2b\beta 3$, $\alpha v\beta 3$ and $\alpha 5\beta 1$, and through aminopeptidase-N (CD13), respectively [23–26]. Since tumor vasculature-targeted nanomedicines do not depend on extravasation and penetration across pericyte-, smooth muscle cell- and/or fibroblast-based cell layers, since they encounter their target receptors much more frequently than do cancer cell-targeted nanomedicines (especially if they circulate for prolonged periods of time), and since they do not suffer from the high tumor cell density and the high interstitial fluid pressure that unfavorably affect cancer cell-targeted nanomedicines, it is expected that in general (NB: excluding e.g. nucleic acid-containing formulations), endothelial cell-targeted nanomedicines possess significantly more potential for improving antitumor efficacy than do cancer cell-targeted nanomedicines. This is not only because they are more likely to find, bind and kill their target cells (i.e. endothelial cells; thereby depriving tumors of oxygen and nutrients; Fig. 6.1H-1), but also because they can be designed to release their contents within the tumor vasculature upon binding to tumor blood vessels, thereby enabling low-molecular-weight drugs to penetrate deep into the

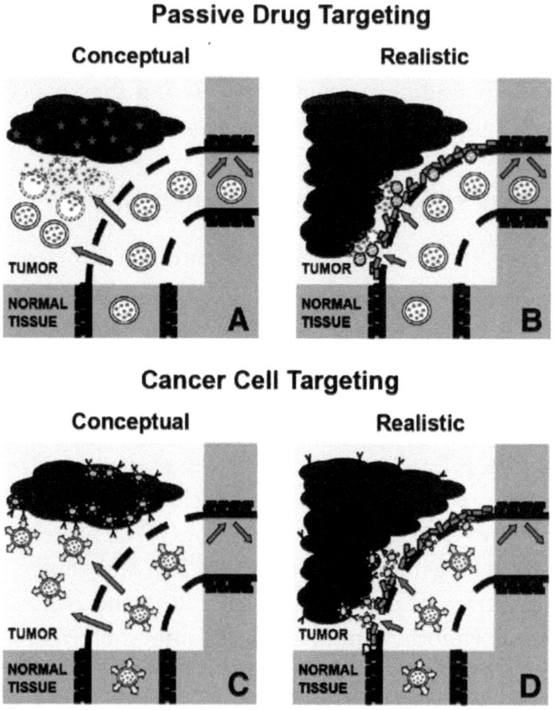

Figure 6.2 Conceptual and realistic models for passive and active drug targeting. A–B: In passive drug targeting, it is often mistakenly assumed that all tumor blood vessels are leaky, and that all tumors possess leaky blood vessels. This might be the case in rapidly growing tumor models in rodents, but is definitely not the case in humans, where substantially enhanced vascular leakiness is observed only in certain specific tumors (e.g. in Kaposi sarcoma), and where only certain parts of tumors are hyperpermeable. C–D: Active drug targeting to cancer cells if often mistakenly assumed to be able to increase overall tumor accumulation. This, however, cannot be the case, since nanomedicines enter the tumor interstitium via passive extravasation. After this, especially in physiologically relevant (i.e. slowly growing; comparable to the clinical situation) tumors, they need to cross several pericyte-, smooth muscle- and fibroblast-based cell layers before they are able to bind to cancer cells. Furthermore, even if they are able to reach cancer cell-containing compartments, their penetration deep into the tumor is limited by the binding-site barrier. Consequently, active cancer cell targeting is considered to be useful only for improving cellular uptake, as well as for targeting certain specific cell types within a solid tumor.

tumor interstitium (Fig. 6.1H-2). Based on these notions, on the significant amount of preclinical evidence that has been obtained with regard to active targeting to tumor blood vessels [21–26], and on the promising findings that have recently been reported in phase I trials for L19-targeted Interleukin-2 [27] and RGD-targeted adenoviruses [28, 29], it is expected that the future will see more and more endothelial cell-targeted nanomedicines entering and progressing through clinical trials.

6.2.4 Triggered Drug Delivery

Other nanomedicine formulations which are expected to gain more and more momentum in the years to come are systems which can be triggered to release their contents upon exposure to external stimuli, such as heat, light, ultrasound and magnetic fields. Such stimuli-responsive nanomedicines, like Thermodox (i.e. temperature-sensitive PEGylated liposomes containing doxorubicin), in principle hold significant clinical potential, since they are designed to only release the conjugated or entrapped chemotherapeutic drug upon applying locoregionally confined triggers, either upon EPR-mediated passive tumor accumulation (Fig. 6.1I-1), or already during circulation (Fig. 6.1I-2), thereby maximizing drug release at the pathological site, while preventing damage to potentially endangered healthy tissues. The downside of such formulations is that they are quite difficult to prepare, at least in such a way that they are really specific with regard to stimuli-responsive drug release; it is often observed that they either already release significant amounts of drug without actually being triggered, or they appear so stable that the triggering conditions required to induce drug release are so severe that they are hardly conformable with physiological processes (i.e. the stimuli themselves become toxic). To overcome these shortcomings, many different efforts are currently being undertaken, both at the academic and at the industrial level. These not only aim to improve the stimuli-responsiveness of tumor-targeted nanomedicines, but also to develop more suitable hardware tools to administer external stimuli more effectively and more selectively to the target tissue. Examples of the former for instance are liposomes and micelles with an optimized composition for heat-triggered drug release at temperatures close to 40°C Examples of the latter include fiberoptic catheters, which enable light-exposure in deep-seated tissues, and magnetic resonance-guided high-intensity focused ultrasound, which at the same generates, quantifies and auto-controls heating in deep-seated tissues [30–36].

6.3 Pitfalls in Drug Targeting to Tumors

6.3.1 Extravasation and the EPR Effect

One of the major pitfalls in the field of tumor-targeted drug delivery relates to the fact that the EPR effect is often overrated and/or misinterpreted. Drug delivery scientists generally have an incomplete knowledge of tumor biology and of the anatomical and (patho-) physiological properties of tumors, and therefore tend to neglect the fact that the EPR effect is a highly heterogeneous phenomenon, which varies substantially from tumor model to tumor model, as well as from patient to patient [11, 37]. Moreover, even within a single tumor, there are huge differences with regard to vascular permeability, and in many cases, there are parts in which particles as large as 200 nm are able to extravasate, whereas in (the vast majority of) other parts, not even molecules of the size of albumin, i.e. 3–4 nm, are able to enter the interstitium (see Fig. 6.2A,B). This is because in these areas, either the endothelial lining is intact, or because vascular leakiness is compromised by the presence of a dense peri-vascular lining, constituted e.g. of pericytes, smooth muscle cells and/or fibroblasts.

Apart from this intra- and inter-tumor- and intra- and inter-patient-variability, another important aspect to keep in mind in this regard is that research in animal models, especially in some of the most frequently used animal models (see Section 6.3.8), can lead to an over estimation of the potential usefulness of passively targeted nanomedicine formulations. This is due to the fact that the EPR effect is undoubtedly much larger in tumors in animal models than in patients, simply because most rodent tumors grow much faster. If a subcutaneously inoculated tumor in a mouse grows to 1 cm (~0.5 g) within 2–4 weeks, this would compare to a ~20 cm and

~1–2 kg tumor in humans, which would take years, and not weeks, to develop. Because of this rapid growth, blood vessels in mouse tumors generally do not develop properly, and they consequently tend to be much more leaky than their human counterparts. This can be exemplified by looking at the tumor accumulation of ~5 nm-sized radiolabeled HPMA copolymers shown in two different rat tumor models, i.e. in Dunning ATI and Dunning H prostate carcinomas [38]. The former grow to 1 cm in diameter within 2 weeks, their blood vessels (in red) therefore are poorly differentiated and hardly covered with pericytes (in green), and they consequently show a relatively high degree of EPR-mediated drug targeting (up to 0.5% of the injected dose per gram tumor tissue). The latter, on the other hand, take more than a year to grow to a size of 1 cm, their blood vessels therefore are properly differentiated and densely covered with pericytes and/or smooth muscle cells, and they consequently accumulate radiolabeled HPMA copolymers much less effectively (on average only 0.2% ID per gram tumor tissue).

In spite of the above notion, there are certain human tumors which are very leaky, e.g. because they express high levels of VEGF (which was initially identified as VPF, i.e. vascular permeability factor [39]), and simply also because they are very well-vascularized (thereby increasing the statistical chance of some of them being very leaky). A prototypic example of this is Kaposi sarcoma, which is known to contain many (and many leaky) blood vessels, and which therefore responds relatively well to treatment with passively tumor-targeted chemotherapeutic interventions. This is exemplified by the fact that Doxil works quite well in Kaposi sarcoma: in comparison to the formerly standard combination regimen ABV, i.e. adriamycin (doxorubicin), bleomycin and vincristine, Doxil monotherapy significantly improved response rates, from 25% to 46% [40].

There are several ways to address this shortcoming related to improper extravasation and highly heterogeneous EPR. These for instance include the use of contrast agent-labeled nanomedicines (i.e. theranostics) and appropriate imaging techniques to monitor tumor accumulation, and to thereby preselect patients [41–43]. The obvious aim of such strategies is to only treat patients who show a relatively high level of EPR-mediated drug targeting. In addition, pharmacologically active agents can be used to enhance extravasation (and penetration; see Section 6.3.2). Prominent examples of such agents are inflammatory mediators, such as tumor necrosis factor-α (TNF-α) and histamine, which have both been shown to be able to improve the extravasation and the penetration of nanomedicines. Both, however, tend to be relatively poorly tolerated when administered systemically, and therefore advanced locoregional setups, such as isolated-limb-perfusion, are needed to exploit their beneficial effects. In addition, they can be given at very low doses, just high enough to permeate tumor blood vessels, but not causing any systemic side effects. If such setups are available and achievable, the combination of extravasation-enhancing pretreatment with nanomedicine treatment can lead to large increases in therapeutic efficacy, as demonstrated e.g. by the pioneering work of Ten Hagen and colleagues [44, 45] and Seymour and colleagues [46].

6.3.2 Penetration

A second important pitfall relates to the (dis-) ability of nanomedicines to properly penetrate tumors. Upon leaving tumor blood vessels, extravasated nanoparticles and macromolecules need to penetrate into and distribute across the interstitium, to reach as many cancer cells as possible. Since they are much larger than conventional (chemo-) therapeutic drugs, however, their penetration is severely hampered, and due to the high tumor cell density and the high interstitial fluid pressure, they often do not cross more than one or two cell layers. This has recently been elegantly exemplified by Bawendi, Jain and colleagues, who used intravital microscopy to show that upon i.v. co-injecting three differently sized quantum dots (12, 60 and 125 nm), only the 12 nm-sized particles were able to properly penetrate, whereas the two larger particles clustered in peri-vascular regions [47]. Similar findings have been reported by Dreher and colleagues, who used five differently sized FITC-labeled dextrans (3.3, 10, 40, 70 and 2000 kDa), and who showed that both the extravasation and the penetration of macro-molecules with sizes of up to 70 kDa (i.e. a hydrodynamic diameter of ~6 nm) was reasonable, whereas that of 2000 kDa dextran (~50 nm) was very poor [48]. These insights strongly suggest that attempts should be made to try to tailor the size of nanomedicines to one that enables long-circulation properties, but that at the same time also allows for proper extravasation and penetration.

There are several ways to overcome this penetration barrier. These for instance include the use of inherently unstable or stimuli-sensitive nanomedicines (see above; Section 6.2.4), which either already in the tumor (micro-) circulation, or relatively fast upon extravasation, release their contents in the peri-vascular space, thereby setting free their low-molecular-weight payload, which can then penetrate deeply into the tumor. In addition, pharmacological treatments can be implemented, including besides the abovementioned extravasation- and penetration-enhancing inflammatory mediators TNF-α and histamine, e.g. also inhibitors of fibrosis and matrix-degrading enzymes. Regarding the former, Miyazono, Kataoka and colleagues have recently shown that a low dose of a transforming growth factor-β (TGF-β) inhibitor, which reduces fibrosis in the tumor microenvironment and the pericyte coverage of tumor blood vessels, can be used to substantially enhance both the extravasation and the penetration of both liposomal and micellar doxorubicin [49], as well as of iron oxide nanoparticles [50]. They demonstrated this in two xenograft models of pancreas cancer, which is notorious for its poor penetration, and also convincingly showed that co-treatment with the TGF-β inhibitor substantially improved the antitumor efficacy of micellar doxorubicin [49]. Regarding the latter, i.e. matrix-modifying agents, Jain and colleagues have shown that by pre-treating tumors with the enzyme collagenase (which degrades collagen; a major matrix component in tumors), caused a 2–3-fold increase in the penetration of antibodies and viral nanoparticles [51–53]. Similarly, pre-treatment of tumors with the hormone relaxin, which changes the structure of collagen, resulted in a 2–3-fold increase in the delivery of antibodies and macromolecules [51, 54]. An alternative non-pharmacological means to improve tumor penetration might be based on combinations with radiotherapy, which is known to be able to increase both the extravasation (via inducing the expression of VEGF and FGF) and the penetration (via lowering of the interstitial fluid pressure) of nanomedicines [38, 55–58]. A final elegant approach to improve the penetration and the intratumoral distribution of drug delivery systems relies on the development of ~100 nm-sized "multistage" nanoparticles, which initially profit from their relatively large size to ensure prolonged circulation times, but which upon extravasation are degraded to ~10 nm-sized "sub-particles" by tumor-associated proteases, such as matrix metalloproteinases, thereby enabling enhanced tumor penetration and improved intratumoral distribution [59].

6.3.3 Active Targeting

A third important pitfall relates to the overestimation and/or the misinterpretation of the potential usefulness of active drug targeting. In many papers in which new tumor-targeted nanomedicines are developed and tested, and in which the in vivo performance of the systems is not as good as initially envisioned, it is stated that in follow- up experiments, ligands will be attached to the surface of the carrier materials, to improve their biodistribution, tumor accumulation and therapeutic efficacy. As exemplified by Fig. 6.2C,D, however, it in many cases is incorrect to assume that active cancer cell targeting will improve target site accumulation. This is because nanomedicines do not accumulate in tumors to a higher extent upon incorporating ligands which bind to cancer cells, since their primary mode of tumor localization still relies on EPR-mediated passive extravasation. Upon leaving leaky tumor blood vessels and penetrating into the interstitium, they first have to reach the tumor cells, before being able to bind to them. Depending on the tumor model used and/or the human malignancy in question, and on how well they allow for extravasation and penetration, it can be more or less easy to find and bind cancer cells. If the endothelial lining is not very leaky, if there are a lot of pericytes or smooth muscle cells covering tumor blood vessels, if the tumor cell density is high, if there is a dense matrix hindering penetration, and/or if the interstitial fluid pressure is high, it is highly likely that active targeting to cancer cells will not at all lead to any benefit over passive targeting.

In case of rapidly growing tumors in animal models, where several of these (patho-) physiologically phenomena tend to be absent, cancer cells might be located directly behind tumor endothelial cells. In such cases, however, the initial (and overall) tumor accumulation is still based on passive extravasation, and an additional barrier comes into play, i.e. the binding-site barrier [20]. The binding-site barrier is based on the notion that ligand-modified nanomedicines will bind to the first receptors they encounter, and therefore will not penetrate very deeply into the tumor. Theoretically, the only benefit actively targeted nanomedicines may have over passively targeted formulations, is that they might be retained within tumors for a somewhat longer period of time, because their binding to and/or their uptake by cancer cells prevents them from rapidly re-entering systemic circulation. As an additional drawback, however, one should also take into account that the introduction of targeting moieties on the surface of nanocarriers often leads to an increase in immunogenicity and in protein adsorption. This generally has detrimental effects on their circulation time, thereby significantly lowering their ability to passively accumulate in tumors by means of EPR. These considerations explain why the vast majority of nanomedicines (NB: excluding antibody-based formulations; which in spite of their relatively small size and their high target specificity circulate relatively long) approved for clinical use and currently evaluated in clinical trials are passively, rather than actively, targeted.

There is one big advantage of actively (cancer cell-) targeted nanomedicines over passively targeted formulations, and this is that they are taken up by cancer cells much more efficiently. This can be exemplified by a study published by Kirpotin, Park and colleagues, who showed that in vitro, Her2-targeted liposomes are taken up >20 times more efficiently than are untargeted liposomes [60]. In vivo, on the other hand, in line with the above considerations, no difference was observed between the two formulations with regard to overall tumor accumulation (i.e. % injected dose per gram tumor; see Fig. 6.2C). Interestingly, the enhanced uptake by cancer cells did lead to a difference in the in vivo microdistribution of Her-2-targeted liposomes: a significantly higher portion of them was found in cancer cells within tumors, while untargeted liposomes primarily accumulated in stromal cells, and this led to a significant increase in antitumor efficacy [60]. Similar findings have been reported by

Davis and colleagues, who showed that overall, transferrin-targeted nanoparticles did not accumulate in tumors to a higher extent than did untargeted nanoparticles, but that they did improve the uptake of the particles by cancer cells in tumors [61].

The notion that active targeting to cancer cells improves cellular uptake within tumors implies that ligand-modified nanomedicines are highly useful systems for transporting agents which themselves are unable to enter cancer cells, such as nucleic acids. This can be exemplified by taking the recent clinical progress made with CALAA-01 into account (see Table 6.2). As opposed to the vast majority of other systems developed for siRNA delivery (which are mostly based on cationic polymer- or lipid-containing polyelectrolyte complexes, and which are not very stable, not very specific and not very biocompatible), this formulation is based on a charge-neutral and relatively biocompatible cyclodextrin-based polymeric backbone. Via adamantane inclusion complexes, siRNA (directed against ribonucleotide reductase M2), PEG (for stabilization purposes) and transferrin (for active targeting and uptake by cancer cells) are coupled to the carrier, and it has been convincingly shown that the therapeutic efficacy of this construct is highly dependent on the presence of transferrin [62, 63].

The above insights exemplify that in certain specific cases, active targeting to cancer cells is absolutely necessary to make formulations effective, whereas in (the majority of) other cases, it likely won't help at all, and might even have deleterious effects. This is not only because of an altered pharmacokinetic profile, but also because of drawbacks related to the complexity of the formulation, to upscaling and to industrial exploitation (see Section 6.3.4). Therefore, it is imperative to keep both the requirements of the drug, and the basic principles of passive and active drug targeting in mind, when intending to develop clinically relevant nanomedicine formulations.

6.3.4 Formulation

Another important pitfall relates to the rational design of the formulations to be developed and tested. To assure proper in vivo functionality, it not only is important to take size and stability into account, but also to try to understand (and incorporate) the complexity of tumor biology. This can be best exemplified by the release mechanism initially envisioned for passively tumor-targeted polymer therapeutics, which were designed to be absolutely stable during their transit in circulation, and to only release the conjugated chemo-therapeutic drug upon endocytosis. Several oligopeptide-based linkers were designed for this purpose, as they were found to be perfectly stable in blood, and to liberate active agents only upon incubation with lysosomal enzymes at low pH. A classical example of this is the tetrapeptide linker GFLG, which is cleaved by the lysosomal cysteine protease cathepsin B. PK1, i.e. Prague-Keele-1, the first passively tumor-targeted polymer-drug conjugate to enter clinical trials in 1994 [64], is based on this linker, as is Xyotax (i.e. Opaxio; paclitaxel polyglumex), which entered clinical trials in the mid 2000s. Both of these formulations worked quite well in animal models, but in patients, their therapeutic gain turned out to be very modest [65]. This in spite of reasonably effective passive tumor targeting in approximately 30% of patients treated with PK1 (analyzed using gamma-scintigraphy). In the case of Opaxio, data re-analysis and patient stratification showed that response rates correlated closely to hormone status, with proper responses in pre-menopausal women, but with hardly any responses in post-menopausal women and in men. This disparity was explained by the fact that the activity of cathepsin B closely correlates with estrogen levels, being higher in pre-menopausal women than in other patients. Such biological/physiological insights are highly important for optimal formulation design, especially when intending to treat large cohorts of patients.

A potential means to overcome such shortcomings is to use linkers which do not depend on enzyme activity for releasing conjugated drugs. Using the same type of carrier material, i.e. HPMA copolymers, Ulbrich and colleagues have in the past 10 years provided a significant amount of preclinical evidence demonstrating that this can be achieved using pH-responsive hydrazone linkers. As opposed to GFLG-based linkers, hydrazone bonds are not completely stable in blood at pH 7.4 (with 1–20% release within 24 h, depending on the exact chemical nature of the linker, its flanking groups and the drug), but release the conjugated active agent much faster at lower pH, with in the case of doxorubicin, ~50% release within 5 h at pH 5, and close to 100% release within 2 days. As a result of this less selective, more rapid and more complete release, hydrazone-based polymer-drug conjugates were found to be much more active in inhibiting tumor growth than were GFLG-based conjugates and strikingly, they were also found to be significantly less toxic [66, 67]. Therefore, such enzyme-independent and slightly less stable polymer therapeutics are considered to be particularly suitable formulations for treating large numbers of patients.

A final important consideration with regard to optimizing formulation design relates to the fact that many promising nanomedicines presented in the literature are quite complex, and therefore difficult to synthesize and scale-up by the pharmaceutical industry. Especially systems based on multiple and physicochemically very different components, such as polymers, lipids, antibodies, peptides, drugs and/or imaging agents, are difficult to scale-up, since their production involves many different synthetic and purification steps. This increases the cost, the complexity and the batch-to-batch variance of such formulations, and thereby decreases their commercial attractiveness and their clinical relevance. Only in cases in which multiple components are really necessary to achieve proper in vivo efficacy, as e.g. in the abovementioned case of delivery systems for siRNA (see Section 6.3.3), it seems to be justified to use multi-component systems. The CALAA-01 formulation, for instance, which is based on a cyclodextrin-based cationic polymer (as a backbone), siRNA (as a drug), adamantane-PEG (for stability and shielding against protein adsorption) and adamantane-PEG-transferrin (for enhancing cellular uptake) is probably the best example for such a complex multi-component system which still has significant clinical relevance [62, 63].

Besides for such nucleic acid-containing constructs, however, the most important dogma for forwarding nanomedicine formulations towards industrial exploitation and clinical translation is to "keep them simple." This might at first sound somewhat counter-intuitive, but the fact that the synthetic methods and the materials delivered for clinical testing need to be biocompatible, well-characterized and reproducible (i.e. low batch-to-batch variability), and relatively easy and cost-effective to prepare, underlines the notion that simple and straightforward systems might have much higher chances of success than highly elegant but also highly complex formulations. Taking these insights into account, it is advisable to stick to a phrase coined by Prestwich in the context of tissue engineering, stating that we should "embrace complexicity, engineer versatility, and deliver simplicity" [68]. This statement underlines the importance of I) realizing that biological systems are inherently very complex and variable, that II) carrier materials should be versatile, flexible and broadly applicable, and that III) pharmaceutical products should be as simple and straightforward as possible.

6.3.5 Efficacy vs. Toxicity

Many different nanomedicines have been evaluated and approved for clinical use in the past two decades. In the vast majority of cases, however, and especially in the case of cancer patients, they only turned out to be able to reduce the toxicity of systemic (chemo-) therapeutic interventions, rather than to improve their efficacy. Doxorubicin-containing liposomes, for instance, are well-known to be able to reduce the cardiotoxicity and/or the hematologic side effects associated with anthracycline therapy, but they generally fail to improve response rates and survival times. Only in certain specific cases, such as in patients suffering from Kaposi sarcoma and from cisplatin-responsive ovarian carcinoma, clear-cut improvements in response rates and overall survival times were observed, whereas in all other cases, such as in metastatic breast cancer and in multiple myeloma, only the incidence and/or intensity of side effects could be reduced [40, 69, 70]. Regarding Kaposi sarcoma, it should be noted that these tumors possess a dense and highly leaky vasculature, and that likely mostly because of this, Doxil turned out to be able to not only improve the toxicity but also the efficacy of the intervention: as compared to the formerly standard combination regimen ABV (i.e. adriamycin (doxorubicin), bleomycin and vincristine), which produced a partial response in 31 out of 125 patients (RR = 25%), Doxil achieved 1 complete response and 60 partial responses (RR = 46%) [40]. Regarding ovarian carcinoma, apart from reduced hematogical toxicity, no statistically significant difference could be observed between Doxil and topotecan-based standard treatment (RR = 20 vs. 17%; overall survival 60 vs. 57 weeks; both $p > 0.05$). When evaluating the efficacy of Doxil in cisplatin-responsive patients, however, a significant survival benefit was observed (108 vs. 71; $p = 0.008$) [69].

Comparable clinical observations have been made for polymers, micelles and nanoparticles. Two prototypic doxorubicin-containing polymer therapeutics, for instance, i.e. PK1 and PK2 (i.e. passively and actively liver-targeted HPMA copolymer-bound doxorubicin), have both been shown to be able to substantially reduce drug-related side effects, as exemplified by 5- and 2-fold increases in the maximum tolerated dose (MTD) that could be safely administered to patients, respectively, but they failed to achieve significant increases in response rates [18, 64, 65]. The same holds true for the mi cellar doxorubicin formulation NK911, which was tolerated reasonably well (i.e. same MTD as free Dox), but which only induced 1 partial response in 23 patients [71].

Abraxane (i.e. albumin-based nanoparticles containing paclitaxel), on the other hand, did present with quite promising therapeutic efficacy in a large phase III trial. In this trial, more than 400 breast cancer patients were treated either with the combination of Taxol (i.e. 175 mg/m^2 paclitaxel; administered via the castor oil-based solubilizer Cremophor, which is known to cause severe local irritation and inflammation) and corticosteroids (to inhibit Cremophor-associated inflammatory reactions), or with Abraxane alone (i.e. without corticosteroid co-medication; at a dose of 260 mg/m^2). As compared to Taxol, Abraxane significantly improved both the response rate (33 vs. 19%) and the progression-free survival time (23 vs. 17 weeks; $p = 0.024$) of the intervention, and it at the same time also attenuated the toxicity of the systemic taxane treatment: the incidence of grade 4 neutropenias was significantly lower for Abraxane (9 vs. 22%), despite the 50% higher dose, and no hypersensitivity reactions were observed, despite the absence of premedication [72].

To overcome the above shortcoming with regard to improving the balance between the efficacy and the toxicity of systemic nano-chemotherapeutic interventions, and to thereby broaden the clinical applicability of tumor-targeted nanomedicines, we and others have in the past 5 years developed several concepts for using nanomedicine formulations to improve the efficacy of combined modality anticancer therapy [73–78]. Convincing and clinically highly relevant evidence has for instance been obtained showing that nanomedicines are highly useful for improving the efficacy of radiochemotherapy and of chemotherapy combinations.

Regarding the former, we have shown that local external beam radiotherapy and polymeric nanomedicines interact synergistically, with radiotherapy improving the tumor accumulation of HPMA copolymers, and with the copolymers improving both the efficacy and the tolerability of radiochemotherapy [38, 78]. Using magnetic resonance imaging and γ-scintigraphy, we demonstrated in three different tumor models that pretreating tumors with radiotherapy increases their tumor accumulation (by 25–100%; depending on polymer size and on the tumor model used). These findings were explained by taking into account that radiotherapy increases the production of the permeability-enhancing growth factors VEGF and FGF, that it induces endothelial cell apoptosis, that it reduces the cell density in tumors, and that it lowers the interstitial fluid pressure [55, 56]. In addition to this, reasoning that I) the temporal and spatial interaction between i.v. applied weekly chemotherapy and clinically relevant daily radiotherapy is suboptimal, and that II) long-circulating and passively tumor-targeted nanomedicines are able to improve the temporal and spatial parameters of this interaction, we have shown that HPMA copolymers are able to improve both the efficacy and the toxicity of clinically relevant regimens of radiochemotherapy [78]. Both doxorubicin- and gemcitabine-containing copolymers were used for this purpose, and growth inhibition was achieved in an aggressively growing and radio- and chemo-resistant tumor model. These findings are in line with preclinical studies in which Doxil was combined with radiotherapy [57], as well as with the results of a phase I trial in which 12 patients with localized esophageal and gastric cancer were treated with the combination of poly(l-glutamic acid)-bound paclitaxel (i.e. Opaxio/Xyotax) and fractionated radiotherapy, and in which 4 complete responses and 7 partial responses (with reductions in tumor size of more than 50%) were achieved [79]. Together, these insights convincingly show that "carrier-based radiochemotherapy" might hold significant potential for improving the treatment of advanced solid malignancies.

Regarding chemotherapy combinations, following up on the pioneering efforts by Vicent, Duncan and colleagues [80], we have recently for the first time provided in vivo evidence showing that passively tumor-targeted polymeric drug carriers can be used to deliver two different drugs to tumors simultaneously. To this end, both doxorubicin and gemcitabine were co-conjugated to the same HPMA copolymer, and it was shown that this formulation—which we termed P-Gem-Dox—circulated for prolonged periods of time, that it localized to tumors both effectively and selectively, and that it increased the efficacy of the combination of doxorubicin plus gemcitabine without increasing its toxicity [81]. In addition to this, it was found that P-Gem-Dox more effectively induced apoptosis and reduced angiogenesis than did all relevant control regimens. These findings are in line with the results recently reported by Satchi-Fainaro and colleagues, who co-conjugated the anti-angiogenic agents aminobisphosphonate alendronate and TNP-470 to a single HPMA copolymer [82], as well as with those published by Tardi, Mayer and co-workers, who co-encapsulated optimal ("ratiometric") concentrations of doxorubicin and vincristine, of irinotecan and floxuridine and of daunorubicin and cytarabine into liposomes, and who are currently evaluating the potential of the latter two formulations in patients [83–86].

In addition to such targeted chemotherapy combination regimens, nanomedicines also perform quite well when combined with standard chemotherapy. Abraxane, for instance, has been shown to combine well with bevacizumab and with gemcitabine in patients suffering from metastatic breast cancer [87, 88], as did Genexol-PM (i.e. a polymeric micelle formulation of paclitaxel) with cislatin and with carboplatin [89, 90]. Similarly, Doxil has been shown to combine very well the microtubule-inhibitor docetaxel in metastatic breast cancer, with the alkylating agent canfosfamide in ovarian carcinoma and with the proteasome inhibitor bortezimob in multiple myeloma [91–93].

Collectively, the above insights and advances convincingly demonstrate that nanometer-sized carrier materials hold significant potential for improving the efficacy of combined modality anticancer therapy. Consequently, they strongly suggest that besides in developing novel and ever more advanced nanomedicine formulations, significant efforts should also be invested in establishing novel and more optimal combination regimens, in order to more optimally exploit the biocompatibility and the beneficial biodistribution of tumor-targeted nanomedicines.

6.3.6 Metastasis

The sixth important issue to take into account with regard to establishing effective, broadly applicable and clinically relevant carrier materials relates to the fact that virtually all anticancer nanomedicines developed to date are designed to target solid tumors. The vast majority of patients succumbing to cancer, however, eventually die from metastases, and not from locally confined tumors. Moreover, if tumors are locally confined, they are generally treated with surgery and/or with radiotherapy, and not with chemotherapy. Chemotherapy is often given in such cases, but generally not to be curative as such, but to pre-shrink tumors prior to surgery or radiotherapy (i.e. neo-adjuvant chemotherapy), or to prevent and/or treat metastasis after locally confined therapeutic interventions (i.e. adjuvant chemotherapy). Scientists working on nanomedicine formulations for cancer therapy are urged to keep this notion well in mind, and to try to come up with systems and strategies that enable more effective anti-metastatic treatments. This can e.g. be achieved by evaluating the ability of nanomedicines to deliver chemotherapeutic payloads to sentinel lymph nodes by involving the immune system, and by developing systems for compartmentalized chemotherapy (which might be particularly useful for treating locally confined ovarian carcinoma metastases; see below).

With regard to involving the immune system, Rihova and colleagues prepared polymeric nanomedicines which possess immunomodulatory properties [94]. They developed a human immunoglobulin- (HuIg-) modified version of PK1, and evaluated it in four end-stage patients in the Czech Republic. Improvements in disease parameters in blood were achieved in several cases, and in all four patients, evidence for an activation of lymphocyte activated killer (LAK) cells and nuclear killer (NK) cells could be observed [95]. At the preclinical level, they obtained similar immunostimulatory effects for a number of other antibody-based polymer therapeutics, e.g. for anti-Thyl.2-, anti-CD71- and B1-targeted versions of PK1, which were all found to be significantly more effective than antibody-free polymer-drug conjugates [96–98]. It is furthermore interesting to note that in addition to producing cures in up to 100% of mice, several of these antibody-targeted polymeric nanomedicines were shown to be able to induce antitumor immunity (i.e. relatively long-lasting immunoprotection), as exemplified by the fact that significant percentages of cured mice were found to be resistant to a re-challenge with a second (lethal) dose of cancer cells [99].

An alternative strategy to (try to) improve the treatment of metastasis using nanomedicine formulations is based on locally confined (i.e. compartmentalized) delivery. In the case of metastatic ovarian carcinoma, for instance, patients could be injected intraperitoneally with actively cancer cell-targeted polymers, liposomes, micelles or nanoparticles, which would then be present in the same compartment (i.e. in the peritoneal cavity) as are the metastases, and which within this compartment can be expected to be able to effectively target and treat metastatic tumor nodules. A similar approach could be envisioned for other abdominally localized and locally metastasized malignancies, such as liver, colorectal and pancreatic carcinomas. Related to this, for the treatment of metastatic liver carcinomas, also the "local" intra-hepatic artery administration of actively cancer cell-targeted nanomedicines could be considered. Since tumors in general, and liver tumors in particular, mainly rely on arterial blood (and oxygen and nutrient) supply, this would enable the delivery of large amounts of cancer cell-specific nanomedicines to primary liver tumors, as well as to liver metastases.

6.3.7 Personalization

Like for standard (chemo-) therapeutic drugs, it is increasingly being recognized that also for nanomedicine formulations, attempts should be made to personalize therapeutic interventions. Besides integrating knowledge on genetic polymorphisms, enzyme expression (see above: Section 6.3.4) and other biomarkers, personalized medicine also involves the establishment of visual methods for predicting and measuring therapeutic responses. Nanomedicines in principle are highly suitable systems for such purposes, since they can be relatively easily modified with imaging agents. By at the same time incorporating both drugs and imaging agents within a single formulation, they can be used to non-invasively monitor the biodistribution and the target site accumulation of the drug and/or carrier material, to visualize drug release, and to assess the therapeutic efficacy of the intervention in real-time [41–43].

Regarding the visualization of the biodistribution and the target site accumulation of such "theranostic" nanomedicines, it has already been convincingly shown that non-invasive imaging insights are highly useful for assessing the efficacy of drug targeting. In the majority of cases, radionuclides have been used to monitor biodistribution. Harrington and colleagues, for instance, radiolabeled PEGylated liposomes with indium-111, and evaluated their circulation time and their tumor accumulation in 17 patients suffering from various different types of locally advanced malignancy (breast, head and neck, lung, brain and cervical cancer) [100]. They were able to show that PEGylated liposomes circulate for very long periods of time, with an average distribution half-life time ($t1/2\alpha$) of 76 h, and with more than 50% of the injected dose still present in systemic circulation at 48 h p.i., thereby confirming the potential of these systems for passive drug targeting. In addition, they were able to show that radiolabeled liposomes relatively effectively localized to tumors in 15 out of 17 cases, and that tumor uptake depended significantly on tumor type: the highest levels were observed for head and neck cancer (33 ± 16%ID/kg tumor), and the lowest levels for breast cancer (5 ± 3%ID/kg tumor). Interestingly, however, in case of the latter, in spite of relatively low overall tumor accumulation, it was observed that the radiolabeled liposomes effectively accumulated in axillary lymph nodes, which are typically employed by breast cancers to spread throughout the body. Analogously, they also showed that indium-111-labeled PEGylated liposomes highly effectively accumulate in Kaposi sarcoma lesions, and that they do so both in primary tumors and in metastatic lesions, thereby explaining—at least in part—why Doxil is so effective in patients suffering from Kaposi sarcoma.

In a comparable setup, Seymour and colleagues visualized the biodistribution and the target site accumulation of liver-targeted HPMA copolymer-bound doxorubicin (PK2) [17]. In PK2, galactosamine, which binds to the asiolaglycoprotein-receptor overexpressed by hepatocytes, is used to direct the polymer-drug conjugates to hepatocellular carcinomas (HCC). Iodine-123-labeled PK2 highly specifically localized to the liver, and that therefore, organ-specific targeting could be considered highly effective. When looking in more detail at the efficacy of organ targeting, however, using anatomical CT imaging and functional SPECT imaging it was found that the majority of PK2 ended up in healthy liver tissue rather than in the tumor. Such image-guided insights show that PK2 was not very effective and selective in delivering the conjugated drug to the pathological site, and might explain why it was not very active in improving treatment efficacy in the majority of HCC patients included in this phase I/II trial.

The above examples show that image-guided insights can be used to pre-screen patients assigned to receive e.g. liposome- and polymer-based nanomedicine treatments, in order to identify which tumors are amenable to passive and active drug targeting and which are not, and to thereby predict which patients are likely to respond to such targeted therapeutic interventions and which are not. In addition to this, if minute amounts of radiolabeled nanomedicines could be added in during every (second or third) cycle of drug treatment, it would furthermore be possible to non-invasively visualize the efficacy of the intervention in real-time, e.g. by tracking the size of (and the accumulation of nanomedicines in) metastatic lesions. This could assist in deciding whether or not to (dis-) continue therapy, and whether or not to adjust drug doses. Therefore, theranostic nanomedicines and non-invasive imaging techniques might contribute substantially to realizing the potential of personalized medicine.

6.3.8 Translation

A final important pitfall with regard to realizing the potential of nanomedicines and drug targeting to tumors relates to the fact that the animal models routinely used in preclinical trials are far from being representative for the clinical situation. As already outlined above, tumors which grow to ~1 cm in diameter within 2–3 weeks in rodents are very different from tumors developing in patients, which generally take several years to grow to sizes beyond 1 cm in diameter; therefore, blood vessels in human tumors tend to be much less leaky than blood vessels in animal models (see Section 6.3.1). Also, tumors growing subcutaneously are anatomically and physiologically very different from tumors growing in their native (orthotopic) environment, and also very different from the metastatic lesions that (nanomedicine) formulations are often confronted with in early phase clinical trials. Furthermore, many in vivo efficacy experiments are merely performed with human xenograft tumors in immunodeficient nude mice, thereby excluding the possibility to assess the involvement of the immune system in positively or negatively affecting therapeutic outcome. Because of such biological, physiological and immunological differences, it can be argued that many agents which work well in animal models, might not at all work in patients.

To overcome this shortcoming, to better predict how well nanomedicines (and also standard chemotherapeutic drugs) will work in patients, and to thereby improve the time- and cost-effectiveness of clinical translation, it would be worthwhile to consider establishing a well-defined panel of animal models—comparable to the NCI-60 set of cancer cell lines—to test (and head-to-head compare) the efficacy of all formulations which are close to being evaluated in patients. Depending on the tumor type to be treated, such a panel could for instance include 1 representative subcutaneous xenograft model, 1 representative subcutaneous syngeneic model, 1 orthotopic model, 1 metastatic model and ideally also 1 transgenic model. By attempting to (pre-)

define which rodent tumor models are representative for the clinical situation, by validating these choices via standard chemotherapeutic interventions (i.e. those which are currently used clinically), and by evaluating how well novel (nanomedicine) formulations are able to improve therapeutic outcome in such models, it should in principle be possible to much better predict how well new therapeutic entities might eventually perform in patients, to more reliably compare novel to established treatments, and to thereby substantially facilitate the time- and cost-effectiveness of clinical translation.

6.4 (Pre-) Clinical Progress

In the past few decades, many different nanomedicines have been evaluated in (pre-) clinical trials. Initially, several types of doxorubicin-containing liposomes were tested, leading to the approval of Doxil in the US for the treatment of Kaposi sarcoma in 1995, and of Caelyx in Europe in 1997 (NB: both are similar, i.e. PEGylated liposomal doxorubicin). In the years that followed, Doxil/Caelyx was also approved for metastatic breast cancer, for ovarian cancer and for multiple myeloma, as was Myocet (i.e. non-PEGylated liposomal doxorubicin) for metastatic breast cancer, and DaunoXome (i.e. non-PEGylated liposomal daunorubicin) for Kaposi sarcoma. Furthermore, DepoCyt, i.e. non-PEGylated liposomal cytarabine, was approved in 1999 for the local intrathecal treatment of lymphomatous meningitis, i.e. a serious complication associated with brain cancer. In addition, DepoCyt is currently being tested in phase III trials for leukemia and in phase I/II clinical trials for glioblastoma (see Table 6.2). Other promising liposomal nanomedicines in clinical trials are Thermodox, a thermosensitive liposomal doxorubicin formulation which releases the drug at temperatures >39°C and which is currently being tested in phase III trials in hepatocellular carcinoma patients together with radiofrequency ablation; NL CPT-11, i.e. nanoliposomal irinotecan, which is in phase I trials for glioma; and several liposomal formulations carrying two different types of chemotherapeutics, such as cytarabine and daunorubicin, and irinotecan and floxuridine [84–

86]. At the preclinical level, many other liposomal nanomedicines are being evaluated, containing various different types of drugs, as are several image-guided formulations, used for instance to confirm EPR-mediated passive drug targeting to tumors, or to validate drug release from Thermodox-like triggerable formulations upon MR-guided high-intensity focused ultrasound [32–36].

Similarly, a large number of polymer-drug conjugates, micelles, nanoparticles and antibody-drug conjugates have been evaluated clinically. Several PEGylated proteins, such as Oncaspar for the treatment of leukemia, have already been approved, while other polymer therapeutics, such as Opaxio and ProLindac, are in late-stage clinical trials. Genexol-PM, a micellar formulation of paclitaxel, has been (pre-) approved in Korea for breast and lung cancer, and is currently undergoing phase II trials in the US and Russia for similar indications. Paclical, another micellar formulation of paclitaxel, has recently received FDA orphan drug designation, and is in phase III trials for ovarian carcinoma. In the nanoparticle field, most progress has thus far been made with Abraxane, which is a co-condensate of paclitaxel and albumin, and which has been shown to be able to improve the efficacy and reduce the toxicity of systemic taxane treatment. The majority of other nanoparticles in clinical trials, including the abovementioned multi-component CALAA-01 formulation, aim to deliver nucleic acid-based therapeutics, such as siRNA and plasmid DNA (see Table 6.2). Finally, a number of antibody-drug conjugates have been approved for clinical use, including e.g. Ontak, Zevalin and Bexxar (NB: Mylotarg was withdrawn in 2010, because of insufficient efficacy and unacceptable toxicity), carrying either bacterial toxins or radionuclides as therapeutic moieties.

The above advances clearly demonstrate that significant progress has been made in the targeted nanomedicine field, and that more progress is to be expected. Through the large number of clinical trials performed to date, however, as well as through several follow-up studies in which already approved formulations have been combined with other treatment modalities, such as with standard chemotherapy and/or with radiotherapy, it has become clear that tumor-targeted nanomedicines—as do standard chemotherapeutic drugs—perform particularly well when integrated in combined modality anticancer therapy. Therefore, in the years to come, besides making ever more carrier materials, and on attempting to better understand the biological and (patho-) physiological principles of drug targeting to tumors, efforts should also focus on establishing rational combination regimens, in order to more optimally exploit the biocompatibility and the beneficial biodistribution of nanomedicines.

6.5 Conclusion

Many different systems and strategies have been evaluated for drug targeting to tumors over the years. Several of these formulations have managed to gain FDA and/or EMA approval, and are routinely used in the clinic. The majority of them, however, have failed either in late-stage preclinical or in early-stage clinical trials. This might be due to several reasons, most prominently to the overinterpretation and/or the misunderstanding of some of the basic concepts used in tumor-targeted drug delivery, as well as to the fact that some of the formulations tested were not effective enough in monotherapy regimens, or too complex to be upscaled by the industry. Furthermore, also the lack of image-guided insights to personalize nanomedicine-based therapeutic interventions, and the notion that in many cases, inappropriate animal models were used at the preclinical level, might have hindered the clinical translation of tumor-targeted nanomedicines. Therefore, in future efforts, it is highly important to not only make ever more nanomedicine materials, but also to strive for a better understanding of the biological and the (patho-) physiological principles of drug targeting, to in-depth investigate the pitfalls in tumor-targeted drug delivery, and to come up with strategies to overcome these shortcomings.

Acknowledgment

The authors gratefully acknowledge financial support by the European Commission's Framework Programme 6 (MediTrans: Targeted Delivery of Nanomedicine) by the DFG and by Hightech.NRW (ForSaTum).

References

1. D. Hanahan, R. A. Weinberg, The hallmarks of cancer, *Cell* 100 (2000) 57–70.
2. D. Hanahan, R. A. Weinberg, Hallmarks of cancer: the next generation, *Cell* 144 (2011) 646–674.
3. D. S. Krause, R. A. Van Etten, Tyrosine kinases as targets for cancer therapy, *N. Engl. J. Med.* 353 (2005) 172–187.
4. R. M. Sharkey, D. M. Goldenberg, Targeted therapy of cancer: new prospects for antibodies and immunoconjugates, *CA Cancer J. Clin.* 56 (2006) 226–243.
5. I. Collins, P. Workman, New approaches to molecular cancer therapeutics, *Nat. Chem. Biol.* 2 (2006) 689–700.
6. V. P. Torchilin, Drug targeting, *Eur. J. Pharm. Sci.* 11 (2000) S81–S91.
7. T. M. Allen, P. R. Cullis, Drug delivery systems: entering the mainstream, *Science* 303 (2004) 1818–1822.
8. D. Peer, J. M. Karp, S. Hong, O. Farokhzad, R. Margalit, R. Langer, Nanocarriers as an emerging platform for cancer therapy, *Nat. Nanotechnol.* 2 (2007) 751–760.
9. M. E. Davis, Z. G. Chen, D. M. Shin, Nanoparticle therapeutics: an emerging treatment modality for cancer, *Nat. Rev. Drug Discov.* 7 (2008) 771–782.
10. T. Lammers, W. E. Hennink, G. Storm, Tumour-targeted nanomedicines: principles and practice, *Br. J. Cancer* 99 (2008) 392–397.
11. R. K. Jain, T. Stylianopoulos, Delivering nanomedicine to solid tumors, *Nat. Rev. Clin. Oncol.* 7 (2010) 653–664.
12. H. Maeda, Y. Matsumura, A new concept for macromolecular therapeutics in cancer chemotherapy: mechanism of tumoritropic accumulation of proteins and the antitumor agent Smancs, *Cancer Res.* 46 (1986) 6387–6392.
13. H. Maeda, J. Wu, T. Sawa, Y. Matsumura, K. Hori, Tumour vascular permeability and the EPR effect in macromolecular therapeutics: a review, *J. Control. Release* 65 (2000) 271–284.
14. T. M. Allen, Ligand-targeted therapeutics in anticancer therapy, *Nat. Rev. Cancer* 2 (2002) 750–763.
15. F. Marcucci, L. Lefoulon, Active targeting with particulate drug carriers in tumor therapy: fundamentals and recent progress, *Drug Discov. Today* 9 (2004) 219–228.

16. S. Wang, P. S. Low, Folate-mediated targeting of antineoplastic drugs, imaging agents, and nucleic acids to cancer cells, *J. Control. Release* 53 (1998) 39–48.
17. Z. M. Qian, H. Li, H. Sun, K. Ho, Targeted drug delivery via the transferrin receptor-mediated endocytosis pathway, *Pharmacol. Rev.* 54 (2002) 561–587.
18. L. W. Seymour, D. R. Ferry, D. Anderson, S. Hesslewood, P. J. Julyan, R. Poyner, J. Doran, A. M. Young, S. Burtles, D. J. Kerr, Cancer Research Campaign Phase I/II Clinical Trials Committee, hepatic drug targeting: phase I evaluation of polymer-bound doxorubicin, *J. Clin. Oncol.* 20 (2002) 1668–1676.
19. C. H. Heldin, K. Rubin, K. Pietras, A. Ostman, High interstitial fluid pressure—an obstacle in cancer therapy, *Nat. Rev. Cancer* 4 (2004) 806–813.
20. M. Juweid, R. Neumann, C. Paik, M. J. Perez-Bacete, J. Sato, W. van Osdol, J. N. Weinstein, Micropharmacology of monoclonal antibodies in solid tumors: direct experimental evidence for a binding site barrier, *Cancer Res.* 52 (1992) 5144–5153.
21. L. Borsi, E. Balza, M. Bestagno, P. Castellani, B. Carnemolla, A. Biro, A. Leprini, J. Sepulveda, O. Burrone, D. Neri, L. Zardi, Selective targeting of tumoral vasculature: comparison of different formats of an antibody (LI 9) to the ED-B domain of fibronectin, *Int. J. Cancer* 102 (2002) 75–85.
22. D. Neri, R. Bicknell, Tumour vascular targeting, *Nat. Rev. Cancer* 5 (2005) 436–446.
23. W. Arap, R. Pasqualini, E. Ruoslahti, Cancer treatment by targeted drug delivery to tumor vasculature in a mouse model, *Science* 279 (1998) 377–380.
24. R. Pasqualini, E. Koivunen, R. Kain, J. Lahdenranta, M. Sakamoto, A. Stryhn, R. A. Ashmun, L. H. Shapiro, W. Arap, E. Ruoslahti, Aminopeptidase N is a receptor for tumor-homing peptides and a target for inhibiting angiogenesis, *Cancer Res.* 60 (2000) 722–727.
25. K. Temming, R. M. Schiffelers, G. Molema, R. J. Kok, RGD-based strategies for selective delivery of therapeutics and imaging agents to the tumour vasculature, *Drug Resist. Updat.* 8 (2005)381–402.
26. A. Mitra, J. Mulholland, A. Nan, E. McNeill, H. Ghandehari, B. R. Line, Targeting tumor angiogenic vasculature using polymer-RGD conjugates, *J. Control. Release* 102 (2005) 191–201.

27. M. Johannsen, G. Spitaleri, G. Curigliano, J. Roigas, S. Weikert, C. Kempkensteffen, A. Roemer, C. Kloeters, P. Rogalla, G. Pecher, K. Miller, A. Berndt, H. Kosmehl, E. Trachsel, M. Kaspar, V. Lovato, R. Gonzalez-Iglesias, L. Giovannoni, H. D. Menssen, D. Neri, F. de Braud, The tumour-targeting human L19-IL2 immunocytokine: preclinical safety studies, phase I clinical trial in patients with solid tumours and expansion into patients with advanced renal cell carcinoma, *Eur. J. Cancer* 46 (2010) 2926–2935.
28. K. J. Kimball, M. A. Preuss, M. N. Barnes, M. Wang, G. P. Siegal, W. Wan, H. Kuo, S. Saddekni, C. R. Stockard, W. E. Grizzle, R. D. Harris, R. Aurigemma, D. T. Curiel, R. D. Alvarez, A phase I study of a tropism-modified conditionally replicative adenovirus for recurrent malignant gynecologic diseases, *Clin. Cancer Res.* 16 (2010) 5277–5287.
29. P. Nokisalmi, S. Pesonen, S. Escutenaire, M. Sarkioja, M. Raid, V. Cerullo, L. Laasonen, R. Alemany, J. Rojas, M. Cascallo, K. Guse, M. Rajecki, L. Kangasniemi, E. Haavisto, A. Karioja-Kallio, P. Hannuksela, M. Oksanen, A. Kanerva, T. Joensuu, L. Ahtiainen, A. Hemminki, Oncolytic adenovirus ICOVIR-7 in patients with advanced and refractory solid tumors, *Clin. Cancer Res.* 16 (2010) 3035–3043.
30. K. Kono, T. Ozawa, T. Yoshida, F. Ozaki, Y. Ishizaka, K. Maruyama, C. Kojima, A. Harada, S. Aoshima, Highly temperature-sensitive liposomes based on a thermosensitive block copolymer for tumor-specific chemotherapy, *Biomaterials* 31 (2010) 7096–7105.
31. C. Oerlemans, W. Bult, M. Bos, G. Storm, J. F. W. Nijsen, W. E. Hennink, Polymeric micelles in anticancer therapy: targeting, imaging and triggered release, *Pharm. Res.* 27 (2010) 2569–2589.
32. R. Deckers, C. T. Moonen, Ultrasound triggered, image guided, local drug delivery, *J. Control. Release* 148 (2010) 25–33.
33. T. Tagami, M. J. Ernsting, S. D. Li, Efficient tumor regression by a single and low dose treatment with a novel and enhanced formulation of thermosensitive liposomal doxorubicin, *J. Control. Release* 152 (2011) 303–309.
34. A. H. Negussie, P. S. Yarmolenko, A. Partanen, A. Ranjan, G. Jacobs, D. Woods, H. Bryant, D. Thomasson, M. W. Dewhirst, B. J. Wood, M. R. Dreher, Formulation and characterisation of magnetic resonance imageable thermally sensitive liposomes for use with magnetic resonance-guided high intensity focused ultrasound, *Int. J. Hyperthermia* 27(2011) 140–155.

35. M. de Smet, E. Heijman, S. Langereis, N. M. Hijnen, H. Grüll, Magnetic resonance imaging of high intensity focused ultrasound mediated drug delivery from temperature-sensitive liposomes: an in vivo proof-of-concept study, *J. Control. Release* 150 (2011) 102–110.
36. T. Tagami, M. J. Ernsting, S. D. Li, Optimization of a novel and improved thermosensitive liposome formulated with DPPC and a Brij surfactant using a robust in vitro system, *J. Control. Release* 154 (2011) 290–297.
37. Y. H. Bae, K. Park, Targeted drug delivery to tumors: Myths, reality and possibility, *J. Control. Release* 153 (2011) 198–205.
38. T. Lammers, P. Peschke, R. Kuhnlein, V. Subr, K. Ulbrich, J. Debus, P. E. Huber, W. E. Hennink, G. Storm, Effect of radiotherapy and hyperthermia on the tumor accumulation of HPMA copolymer-based drug delivery systems, *J. Control. Release* 117 (2007) 333–341.
39. H. F. Dvorak, L. F. Brown, M. Detmar, A. M. Dvorak, Vascular permeability factor/vascular endothelial growth factor, microvascular hyperpermeability, and angiogenesis, *Am. J. Pathol.* 146 (1995) 1029–1039.
40. D. W. Northfelt, B. J. Dezube, J. A. Thommes, B. J. Miller, M. A. Fischl, A. Friedman-Kien, et al., Pegylated-liposomal doxorubicin versus doxorubicin, bleomycin, and vincristine in the treatment of AIDS-related Kaposi's sarcoma: results of a randomized phase III clinical trial, *J. Clin. Oncol.* 16 (2005) 2445–2451.
41. D. Sun, Nanotheranostics: integration of imaging and targeted drug delivery, *Mol. Pharm.* 7 (2010) 1879.
42. T. Lammers, F. Kiessling, W. E. Hennink, G. Storm, Nanotheranostics and image-guided drug delivery: current concepts and future directions, *Mol. Pharm.* 7 (2010) 1899–1912.
43. T. Lammers, S. Aime, W. E. Hennink, G. Storm, F. Kiessling, Theranostic Nanomedicines, *Ace. Chem. Res.* 44 (2011) 1029–1038.
44. T. L. Ten Hagen, A. L. Seynhaeve, A. M. Eggermont, Tumor necrosis factor-mediated interactions between inflammatory response and tumor vascular bed, *Immunol. Rev.* 222 (2008) 299–315.
45. A. L. Seynhaeve, S. Hoving, D. Schipper, C. E. Vermeulen, G. de Wiel-Ambagtsheer, S. T. van Tiel, A. M. Eggermont, T. L. Ten Hagen, Tumor necrosis factor alpha mediates homogeneous distribution of liposomes in murine melanoma that contributes to a better tumor response, *Cancer Res.* 67 (2007) 9455–9462.

46. T. Seki, F. Carroll, S. Illingworth, R. Cawood, V. Subr, K. D. Fisher, L. W. Seymour, Tumour necrosis factor-alpha increases extravasation of virus particles into tumour tissue by activating the rho A/rho kinase pathway, *J. Control Release.* 156 (2011) 381–389.

47. Z. Popovic, W. Liu, V. P. Chauhan, J. Lee, C. Wong, A. B. Greytak, N. Insin, D. G. Nocera, D. Fukumura, R. K. Jain, M. G. Bawendi, A nanoparticle size series for in vivo fluorescence imaging, *Angew. Chem. Int. Ed Engl.* 49 (2010) 8649–8652.

48. M. R. Dreher, W. Liu, C. R. Michelich, M. W. Dewhirst, F. Yuan, A. Chilkoti, Tumor vascular permeability, accumulation, and penetration of macromolecular drug carriers, *J. Natl. Cancer Inst.* 98 (2006) 335–344.

49. M. R. Kano, Y. Bae, C. Iwata, Y. Morishita, M. Yashiro, M. Oka, T. Fujii, A. Komuro, K. Kiyono, M. Kaminishi, K. Hirakawa, Y. Ouchi, N. Nishiyama, K. Kataoka, K. Miyazono, Improvement of cancer-targeting therapy, using nanocarriers for intractable solid tumors by inhibition of TGF-beta signaling, *Proc. Natl. Acad. Sci. U. S. A.* 104 (2007) 3460–3465.

50. M. Kumagai, M. R. Kano, Y. Morishita, M. Ota, Y. Imai, N. Nishiyama, M. Sekino, S. Ueno, K. Miyazono, K. Kataoka, Enhanced magnetic resonance imaging of experimental pancreatic tumor in vivo by block copolymer-coated magnetite nanoparticles with TGF-beta inhibitor, *J. Control. Release* 140 (2009) 306–311.

51. E. Brown, T. McKee, E. diTomaso, A. Pluen, B. Seed, Y. Boucher, R. K. Jain, Dynamic imaging of collagen and its modulation in tumors in vivo using second-harmonic generation, *Nat. Med.* 9 (2003) 796–800.

52. J. Choi, K. Credit, K. Henderson, R. Deverkadra, Z. He, H. Wiig, H. Vanpelt, M. F. Flessner, Intraperitoneal immunotherapy for metastatic ovarian carcinoma: resistance of intratumoral collagen to antibody penetration, *Clin. Cancer Res.* 12 (2006) 1906–1912.

53. T. D. McKee, P. Grandi, W. Mok, G. Alexandrakis, N. Insin, J. P. Zimmer, M. G. Bawendi, Y. Boucher, X. O. Breakefield, R. K. Jain, Degradation of fibrillar collagen in a human melanoma xenograft improves the efficacy of an oncolytic herpes simplex virus vector, *Cancer Res.* 6 (2006) 2509–2513.

54. J. Y. Perentes, T. D. McKee, C. D. Ley, H. Mathiew, M. Dawson, T. P. Padera, L. L. Munn, R. K. Jain, Y. Boucher, In vivo imaging of extracellular matrix remodeling by tumor-associated fibroblasts, *Nat. Methods* 6 (2009) 143–145.

55. P. A. Netti, L. T. Baxter, Y. Boucher, R. Skalak, R. K. Jain, Time-dependent behavior of interstitial fluid pressure in solid tumors: implications for drug delivery, *Cancer Res.* 55 (1995) 5451–5458.

56. F. Samaniego, P. O. Markham, R. Gendelman, Y. Watanabe, V. Kao, K. Kowalski, J. A. Sonnabend, A. Pintus, R. C. Gallo, B. Ensoli, Vascular endothelial growth factor and basic fibroblast growth factor present in Kaposi's sarcoma (KS) are induced by inflammatory cytokines and synergized to promote vascular permeability and KS lesion development, *Am. J. Pathol.* 152 (1998) 1433–1443.

57. C. L. Davies, L. M. Lundstrom, J. Frengen, L. Eikenes, O. S. Bruland, O. Kaalhus, M. H. Hjelstuen, C. R. Brekken, Radiation improves the distribution and uptake of liposomal doxorubicin (caelyx) in human osteosarcoma xenografts, *Cancer Res.* 64 (2004) 547–553.

58. T. Lammers, V. Subr, K. Ulbrich, W. E. Hennink, G. Storm, F. Kiessling, Polymeric nanomedicines for image-guided drug delivery and tumor-targeted combination therapy, *Nano Today* 5 (2010) 197–212.

59. C. Wong, T. Stylianopoulos, J. Cui, J. Martin, V. P. Chauhan, W. Jiang, Z. Popovic, R. K. Jain, M. G. Bawendi, D. Fukumura, Multistage nanoparticle delivery system for deep penetration into tumor tissue, *Proc. Natl. Acad. Sci. U.S.A.* 108 (2011) 2426–2431.

60. D. B. Kirpotin, D. C. Drummond, Y. Shao, M. R. Shalaby, K. Hong, U. B. Nielsen, J. D. Marks, C. C. Benz, J. W. Park, Antibody targeting of long-circulating lipidic nanoparticles does not increase tumor localization but does increase internalization in animal models, *Cancer Res.* 66 (2006) 6732–6740.

61. C. H. Choi, C. A. Alabi, P. Webster, M. E. Davis, Mechanism of active targeting in solid tumors with transferrin-containing gold nanoparticles, *Proc. Natl. Acad. Sci. U. S. A.* 107 (2010) 1235–1240.

62. M. E. Davis, The first targeted delivery of siRNA in humans via a self-assembling, cyclodextrin polymer-based nanoparticle: from concept to clinic, *Mol. Pharm.* 6 (2009) 659–668.

63. M. E. Davis, J. E. Zuckerman, C. H. Choi, D. Seligson, A. Tolcher, C. A. Alabi, Y. Yen, J. D. Heidel, A. Ribas, Evidence of RNAi in humans from systemically administered siRNA via targeted nanoparticles, *Nature* 464 (2010) 1067–1070.

64. P. A. Vasey, S. B. Kaye, R. Morrison, C. Twelves, P. Wilson, R. Duncan, A. H. Thomson, L. S. Murray, T. E. Hilditch, T. Murray, S. Burtles, D. Fraier, E. Frigerio, J. Cassidy, Cancer Research Campaign Phase I/II Committee, Phase I clinical and pharmaco-kinetic study of PK1

[N-(2-hydroxypropyl)-methacrylamide copolymer doxorubicin]: first member of a new class of chemotherapeutic agents-drug-polymer conjugates, *Clin. Cancer Res.* 5 (1999) 83–94.

65. R. Duncan, Development of HPMA copolymer-anticancer conjugates: clinical experience and lessons learnt, *Adv. Drug Deliv. Rev.* 61 (2009) 1131–1148.
66. K. Ulbrich, T. Etrych, P. Chytil, M. Jelinkova, B. Rihova, HPMA copolymers with pH-controlled release of doxorubicin: in vitro cytotoxicity and in vivo antitumor activity, *J. Control. Release* 87 (2003) 33–47.
67. K. Ulbrich, V. Subr, Structural and chemical aspects of HPMA copolymers as drug carriers, *Adv. Drug Deliv. Rev.* 62 (2010) 150–166.
68. G. D. Prestwich, Engineering a clinically-useful matrix for cell therapy, *Organogenesis* 4 (2008) 42–47.
69. A. N. Gordon, J. T. Fleagle, D. Guthrie, D. E. Parkin, M. E. Gore, A. J. Lacave, Recurrent epithelial ovarian carcinoma: a randomized phase III study of pegylated liposomal doxorubicin versus topotecan, *J. Clin. Oncol.* 19 (2001) 3312–3322.
70. M. E. O'Brien, N. Wigler, M. Inbar, R. Rosso, E. Grischke, A. Santoro, R. Catane, D. G. Kieback, P. Tomczak, S. P. Ackland, F. Orlandi, L. Mellars, L. Alland, C. Tendler, CAELYX Breast Cancer Study Group, Reduced cardiotoxicity and comparable efficacy in a phase III trial of pegylated liposomal doxorubicin HC1 (CAELYX/Doxil) versus conventional doxorubicin for first-line treatment of metastatic breast cancer, *Ann. Oncol.* 15 (2004) 440–449.
71. Y. Matsumura, T. Hamaguchi, T. Ura, K. Muro, Y. Yamada, Y. Shimada, K. Shirao, T. Okusaka, H. Ueno, M. Ikeda, N. Watanabe, Phase I clinical trial and pharmacokinetic evaluation of NK911, a micelle-encapsulated doxorubicin, *Br. J. Cancer* 91 (2004) 1775–1781.
72. W. J. Gradishar, S. Tjulandin, N. Davidson, H. Shaw, N. Desai, P. Bhar, M. Hawkins, J. O'Shaughnessy, Phase III trial of nanoparticle albumin-bound paclitaxel compared with polyethylated castor oil-based paclitaxel in women with breast cancer, *J. Clin. Oncol.* 23 (2005) 7794–7803.
73. F. Greco, M. J. Vicent, Combination therapy: opportunities and challenges for polymer-drug conjugates as anticancer nanomedicines, *Adv. Drug Deliv. Rev.* 61 (2009) 1203–1213.
74. A. Dicko, L. D. Mayer, P. G. Tardi, Use of nanoscale delivery systems to maintain synergistic drug ratios in vivo, *Expert Opin. Drug Deliv.* 7 (2010) 1329–1341.

75. T. Lammers, G. Storm, F. Kiessling, Nanomedicine formulations for combination therapies, *Nano Rev.* 1 (2010) 5705.
76. T. Lammers, Improving the efficacy of combined modality anticancer therapy using HPMA copolymer-based nanomedicine formulations, *Adv. Drug Deliv. Rev.* 62 (2010) 203–230.
77. T. Lammers, V. Subr, K. Ulbrich, P. Peschke, P. E. Huber, W. E. Hennink, G. Storm, F. Kiessling, HPMA-based polymer therapeutics improve the efficacy of surgery, of radiotherapy and of chemotherapy combinations, *Nanomedicine (Lond.)* 5 (2010) 1501–1523.
78. T. Lammers, V. Subr, P. Peschke, R. Kuhnlein, W. E. Hennink, K. Ulbrich, M. Heilmann, F. Kiessling, P. E. Huber, G. Storm, Image-guided and passively tumour-targeted polymeric nanomedicines for radiochemotherapy, *Br. J. Cancer* 99 (2008) 900–910.
79. T. Dipetrillo, L. Milas, D. Evans, P. Akerman, T. Ng, T. Miner, D. Cruff, B. Chauhan, D. Iannitti, D. Harrington, H. Safran, Paclitaxel poliglumex (PPX-Xyotax) and concurrent radiation for esophageal and gastric cancer: a phase I study, *Am. J. Clin. Oncol.* 29 (2006) 376–379.
80. M. J. Vicent, F. Greco, R. I. Nicholson, A. Paul, P. C. Griffiths, R. Duncan, Polymer therapeutics designed for a combination therapy of hormone-dependent cancer, *Angew. Chem. Int. Ed.* 44 (2005) 4061–4066.
81. T. Lammers, V. Subr, K. Ulbrich, P. Peschke, P. E. Huber, W. E. Hennink, G. Storm, Simultaneous delivery of doxorubicin and gemcitabine to tumors in vivo using prototypic polymeric drug carriers, *Biomaterials* 30 (2009) 3466–3475.
82. K. Miller, R. Erez, E. Segal, D. Shabat, R. Satchi-Fainaro, Targeting bone metastases with a bispecific anticancer and antiangiogenic polymer-alendronate-taxane conjugate, *Angew. Chem. Int. Ed.* 48 (2009) 2949–2954.
83. S. A. Abraham, C. McKenzie, D. Masin, R. Ng, T. O. Harasym, L. D. Mayer, M. B. Bally, In vitro and in vivo characterization of doxorubicin and vincristine coencapsulated within liposomes through use of transition metal ion complexation and pH gradient loading, *Clin. Cancer Res.* 10 (2004) 728–738.
84. P. G. Tardi, R. C. Gallagher, S. Johnstone, N. Harasym, M. Webb, M. B. Bally, L. D. Mayer, Coencapsulation of irinotecan and floxuridine into low cholesterol-containing liposomes that coordinate drug release in vivo, *Biochim. Biophys. Acta* 1768 (2007) 678–687.
85. L. D. Mayer, T. O. Harasym, P. G. Tardi, N. L. Harasym, C. R. Shew, S. A. Johnstone, E. C. Ramsay, M. B. Bally, A. S. Janoff, Ratiometric dosing of

anticancer drug combinations: controlling drug ratios after systemic administration regulates therapeutic activity in tumor-bearing mice, *Mol. Cancer Ther.* 5 (2006) 1854–1863.

86. P. Tardi, S. Johnstone, N. Harasym, S. Xie, T. Harasym, P. Harvie, D. Bermudes, L. Mayer, In vivo maintenance of synergistic cytarabine: daunorubicin ratios greatly enhances therapeutic efficacy, *Leuk. Res.* 33 (2009) 129–139.
87. V. Roy, B. R. LaPlant, G. G. Gross, C. L. Bane, F. M. Palmieri, North Central Cancer Treatment Group, Phase II trial of weekly nab (nanoparticle albumin-bound)-paclitaxel (nab-paclitaxel) (Abraxane) in combination with gemcitabine in patients with metastatic breast cancer (N0531), *Ann. Oncol.* 20 (2009) 449–453.
88. C. Lobo, G. Lopes, O. Baez, A. Castrellon, A. Ferrell, C. Higgins, E. Hurley, J. Hurley, I. Reis, S. Richman, P. Seo, O. Silva, J. Slingerland, K. Tukia, C. Welsh, S. Gluck, Final results of a phase II study of nab-paclitaxel, bevacizumab, and gemcitabine as first-line therapy for patients with HER2-negative metastatic breast cancer, *Breast Cancer Res. Treat.* 123 (2010) 427–435.
89. D. W. Kim, S. Y. Kim, H. K. Kim, S. W. Kim, S. W. Shin, J. S. Kim, K. Park, M. Y. Lee, D. S. Heo, Multicenter phase II trial of Genexol-PM, a novel Cremophor-free, polymeric micelle formulation of paclitaxel, with cisplatin in patients with advanced non-small-cell lung cancer, *Ann. Oncol.* 18 (2007) 2009–2014.
90. H. J. Kim, K. H. Kim, J. Yun, S. H. Kim, H. J. Kim, S. C. Lee, S. B. Bae, C. K. Kim, N. S. Lee, K. T. Lee, D. J. Kim, S. K. Park, J. H. Won, D. S. Hong, H. S. Park, Phase II clinical trial of Genexol® (Paclitaxel) and Carboplatin for patients with advanced non-small cell lung cancer, *Cancer Res. Treat.* 43 (2011) 19–23.
91. J. A. Sparano, A. N. Makhson, V. F. Semiglazov, S. A. Tjulandin, O. I. Balashova, I. N. Bondarenko, N. V. Bogdanova, G. M. Manikhas, G. P. Oliynychenko, V. A. Chatikhine, S. H. Zhuang, L. Xiu, Z. Yuan, W. R. Rackoff, Pegylated liposomal doxorubicin plus docetaxel significantly improves time to progression without additive cardiotoxicity compared with docetaxel monotherapy in patients with advanced breast cancer previously treated with neoadjuvant-adjuvant anthracycline therapy: results from a randomized phase III study, *J. Clin. Oncol.* 27 (2009) 4522–4529.
92. I. Vergote, N. J. Finkler, J. B. Hall, O. Melnyk, R. P. Edwards, M. Jones, J. G. Keck, L. Meng, G. L. Brown, E. M. Rankin, J. J. Burke, R. V. Boccia,

C. D. Runowicz, P. G. Rose, Randomized phase III study of canfosfamide in combination with pegylated liposomal doxorubicin compared with pegylated liposomal doxorubicin alone in platinum-resistant ovarian cancer, *Int. J. Gynecol. Cancer* 20 (2010) 772–780.

93. R. Z. Orlowski, A. Nagler, P. Sonneveld, J. Blade, R. Hajek, A. Spencer, J. San Miguel, T. Robak, A. Dmoszynska, N. Horvath, I. Spicka, H. J. Sutherland, A. N. Suvorov, S. H. Zhuang, T. Parekh, L. Xiu, Z. Yuan, W. Rackoff, J. L. Harousseau, Randomized phase III study of pegylated liposomal doxorubicin plus bortezomib compared with bortezomib alone in relapsed or refractory multiple myeloma: combination therapy improves time to progression, *J. Clin. Oncol.* 25 (2007) 3892–3901.
94. B. Rihova, M. Kovar, Immunogenicity and immunomodulatory properties of HPMA-based polymers, *Adv. Drug Deliv. Rev.* 62 (2010) 184–191.
95. B. Rihova, J. Strohalm, J. Prausova, K. Kubackova, M. Jelinkova, L. Rozprimova, M. Sirova, D. Plocova, T. Etrych, V. Subr, T. Mrkvan, M. Kovar, K. Ulbrich, Cytostatic and immunomobilizing activities of polymer-bound drugs: experimental and first clinical data, *J. Control. Release* 91 (2003) 1–16.
96. B. Rihova, J. Strohalm, K. Kubackova, M. Jelinkova, O. Hovorka, M. Kovar, D. Plocova, M. Sirova, M. Stastny, L. Rozprimova, K. Ulbrich, Acquired and specific immunological mechanisms co-responsible for efficacy of polymer-bound drugs, *J. Control. Release* 78 (2002) 97–114.
97. M. Kovar, T. Mrkvan, J. Strohalm, T. Etrych, K. Ulbrich, M. Stastny, B. Rihova, HPMA copolymer-bound doxorubicin targeted to tumor-specific antigen of BCL1 mouse B cell leukemia, *J. Control. Release* 92 (2003) 315–330.
98. M. Kovar, J. Tomala, H. Chmelova, L. Kovar, T. Mrkvan, R. Joskova, Z. Zakostelska, T. Etrych, J. Strohalm, K. Ulbrich, M. Sirova, B. Rihova, Overcoming immunoescape mechanisms of BCL1 leukemia and induction of CD8+ T-cell-mediated BCL1-specific resistance in mice cured by targeted polymer-bound doxorubicin, *Cancer Res.* 68 (2008) 9875–9883.
99. M. Sirova, J. Strohalm, V. Subr, D. Plocova, P. Rossmann, T. Mrkvan, K. Ulbrich, B. Rihova, Treatment with HPMA copolymer-based doxorubicin conjugate containing human immunoglobulin induces long-lasting systemic anti-tumour immunity in mice, *Cancer Immunol. Immunother.* 56 (2007) 35–47.

100. K. J. Harrington, G. Rowlinson-Busza, K. N. Syrigos, R. G. Vile, P. G. Uster, A. M. Peters, J. S. Stewart, Pegylated liposome-encapsulated doxorubicin and cisplatin enhance the effect of radiotherapy in a tumor xenograft model, *Clin. Cancer Res.* 6 (2000) 4939–4949.

Chapter 7

Biomolecular Coronas Provide the Biological Identity of Nanosized Materials

Marco P. Monopoli,[a,b] Christoffer Aberg,[a] Anna Salvati,[a,b] and Kenneth A. Dawson[a]

[a] *Centre for BioNano Interactions, School of Chemistry and Chemical Biology, Belfield, Dublin 4, Ireland*

[b] *Conway Institute for Biomolecular and Biomedical Research, University College Dublin, Belfield, Dublin 4, Ireland*

kenneth.a.dawson@cbni.ucd.ie

The surfaces of all materials (including nanosized materials) have higher free energy than the bulk material itself. This means that surfaces of nanoparticles will progressively and selectively adsorb biomolecules when they come into contact with complex biological fluids [1–7], unless they have been designed to do otherwise. This 'corona' of biomolecules lowers the surface energy of the nanoparticle and promotes its dispersion. We propose that in many cases it is the biomolecular corona that is interacting with biological systems and thereby constitutes a major element of the biological identity of the nanoparticle [1, 4, 8–11] (Fig. 7.1a).

Reproduced with permission from *Nature Nanotechnology*, **7**, 779–786, 2012.

Nano-Enabled Medical Applications
Edited by Lajos P. Balogh

ISBN 978-981-4800-32-7 (Hardcover), 978-0-429-39903-9 (eBook)
www.jennystanford.com

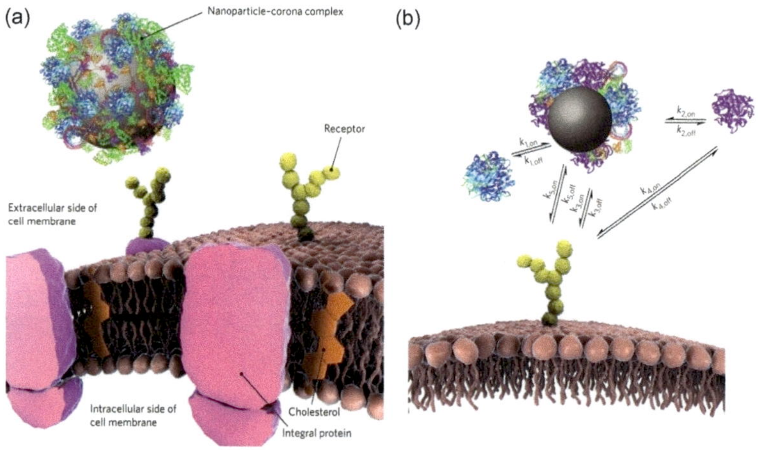

Figure 7.1 The nanoparticle-corona complex in a biological environment: (a) It is the nanoparticle-corona complex, rather than the bare nanoparticle, that interacts with biological machinery, here with a cell membrane receptor. (b), Relevant processes (arrows), in both directions (on/off), for a nanoparticle interacting with a receptor. Biomolecules in the environment adsorb strongly to the bare nanoparticle surface (k_1), forming a tightly bound layer of biomolecules, the 'hard' corona, in immediate contact with the nanoparticle. Other biomolecules, the 'soft' corona, have a residual affinity to the nanoparticle-hard-corona complex (primarily to the hard corona itself), but this is much lower, so those molecules are in rapid exchange with the environment (k_2). If sufficiently long-lived in the corona, a biomolecule may lead to recognition of the nanoparticle-corona complex as a whole by a cell- membrane receptor (k_3). The same biomolecule alone can also be recognized by the receptor (k_4). If present, the bare surface of the nanoparticle may also interact with cell surface receptors (k_5) or other constituents of the cell membrane. Proteins adapted from Protein Data Bank (http://www.pdb.org).

For several nanoparticles, such as gold [12], polystyrene [8, 13, 14], silica [5, 8, 14, 15], titania [15] and zinc oxide [15], a near-monolayer of biomolecules (called the 'hard' corona) binds tightly, but not completely irreversibly, to the nanoparticle surface. On top of this 'hard' corona is a 'soft' corona consisting of a more loosely associated and rapidly exchanging layer of biomolecules [8, 12–14, 16] (Fig. 7.1b). One expects this separation to be typical for nanoparticles that have similarly high energies of the bare surface.

Crucially, it is observed that only a few of the biomolecules available in typical biological environments are found in the hard corona. In the purely illustrative examples in Table 7.1, out of roughly 3,700 proteins in human blood plasma [17], only at most a few tens are ever abundant in the hard corona. Furthermore, these tens of proteins rarely correspond to the most abundant proteins in plasma, and are not necessarily those with the highest individual affinity to the nanoparticle surface [5, 6, 18–24].

As the hard corona is distinctively stable [5, 8, 12–14], any subsequent exposure of the nanoparticle to a new environment with different biomolecules may lead to only partial displacement of the original hard corona by new molecules [25, 26]. Biomolecules that are not replaced would serve as a corona 'memory' of the nanoparticle's previous environment. Therefore the corona composition could potentially depend not only on the current environment of the nanoparticle, but on all environments it has moved through [26, 27]. Figure 7.2 illustrates this concept for inhaled nanoparticles carrying typical lung surfactant proteins, and other membrane components, when they reach the bloodstream [28, 29].

The fact that the biomolecular corona defines the biological interactions of nanoparticles has been considered surprising, because we often classify substances according to the 'intrinsic properties of the bare material, rather than extrinsic' properties derived from the environment. However, the importance of extrinsic properties for the biological activity of small-molecule drugs biomaterial implants and cell scaffolds is already well known. For example, 99% of the anticoagulant drug warfarin is typically bound to plasma proteins in vivo, and this protein-bound drug fraction is retained much longer in the body than the more biologically active unbound fraction, thereby improving the overall circulation half life and effect of the drug [30]. Also, it is the biomolecules bound to the surface of biomaterial implants that are understood to interact with the surrounding tissue [31]; these biomolecules may mitigate material induced tissue damage (including late-stage thrombosis [31, 32]), or pro mote early inflammatory responses.

Despite the similarities, nanomaterials are different. Unlike macroscopic surfaces, nanoparticles can travel to almost every location within organisms [27–29] via endogenous transport pathways [33, 34] and potentially retain biomolecules from their previous environments in their corona, including those relevant to the transport process itself. Examples include apolipoproteins, which are abundant in the corona of very different nanomaterials (including polystyrene and other polymeric nanoparticles [1, 13, 35], silica [5] and quantum dots [36]) and are important for crossing biological barriers.

It is worth emphasizing that the biomolecular corona is also relevant to nanoparticles with grafted antibodies, proteins and pep tides that are specifically designed for targeting in nanomedicine [37, 38]. Although surface modifications (such as PEGylation [39]) reduce the binding of additional biomolecules, some association of biomolecules may still occur [35, 40]. Their presence may lead to more subtle and poorly understood biological consequences—issues possibly related to current struggles in achieving efficient targeting of nanomedicines in vivo. Indeed, in future, instead of chemical grafting, the corona itself might be controlled (for example, by surface templating approaches [41]) and used as a novel means to target diseases [21, 36].

7.1 Nanoscale Engagement with Biological Processes

Fundamentally, all interactions between the nanoparticle corona complex and the biological machinery are mediated by physical forces common to all molecular systems (for example, van der Waals, electrostatic and others [2]). A more useful classification is whether these interactions involve non-specific physical interactions [6, 42] or specific biological recognition [10]. Such recognition can, for example, be driven by either peptide sequences (epitopes) or more subtle properties, such as curvature of the complex [43]. Recognition is expected to lead to well-regulated biological processing and functional impacts. In contrast, strong non-specific interactions between particles and cells could lead to a range of quite different, but equally complex, outcomes. For example, in several cases, such as carbon nanotubes [6], silica [42] and graphene oxide [44], cell damage occurs in the absence of a protein corona (that is, cells exposed to particles in medium without proteins), but is mitigated in its presence. In such cases particles in the absence of protein may instead acquire a corona from biomolecules in the cell membrane, which suggests a mechanism for the damage [42]. Exceptional cases of direct entry into cytosol can also be related to a lack of corona [42, 45]. In nanoparticle-cell interactions, although the intended purpose in supplying cells with serum protein is to feed them, an important accidental role of serum is to limit direct contact of the nanoparticle high-energy surface with cells.

The entry of nanoparticles into cells and their translocation across biological barriers are important active processes [33, 34]. Small hydrophobic molecules, loosely speaking, partition in biological compartments according to physicochemical equilibrium principles [46], whereas larger, micrometre-sized objects engage significantly with macrophages and other elements of the front-line immune system [29]. In contrast, because of their size and surface, nanoparticle-corona complexes can engage with a wide range of endogenous cellular uptake and other processes, and potentially reach all cellular and organ compartments [27–29, 35], interact with them and initiate signalling processes [10]. Endocytic mechanisms of cellular uptake are known from the extensive biomolecular-transport literature [43, 47]. Whether these or novel mechanisms are being used in nanoparticle uptake [33, 48] will need to be clarified [49]. Almost certainly a variety of mechanisms is at work in the uptake of nanoparticles under different conditions. Several nanoparticles (for example, polystyrene [33, 46], silica [50] and others [49]) follow the endo-lysosomal pathway, leading finally to lysosomal accumulation.

However, there is now a need to revise nanoparticle-uptake questions in the context of the nanoparticle surface in situ. For example, multiple reports suggest that the rate of accumulation of nanoparticles in cells depends on the detailed nature of the biological milieu [4, 22, 51, 52]. In several cases of cell types and nanoparticles, including silica [42], carbon [53], iron oxide [54], polystyrene [55] and others [11], formation of a serum-derived corona leads to reduced cell-uptake. Other differences in uptake levels can arise from rather subtle differences in nanoparticle-dispersion preparation, such as serum heat inactivation [51], or even just different culture media, despite being supplemented with the same serum [22]. To rule out such changes as a source of differences, exact matching of the media will be required in future comparative experiments.

Similar implications apply for in vivo and in vitro comparative experiments, which are typically performed at different protein concentrations, thus resulting in different corona composition [14]. Furthermore, corona proteins from biological media derived from one species may not be recognized (or be recognized to a different degree) by receptors in different species, unless there is interspecies conservation. Thus, the role of serum in studies of cells from different species may need to be questioned, meaning that matching of the supplemented serum and organism may also be required in the future.

Lest it be forgotten, we emphasize that it is not only proteins that may be recognized by the biological machinery. For example, the lipid component, especially relevant for inhalation scenarios, also plays a crucial role for biorecognition [56]; thus carbon nanotubes covered by a phospholipid uptake signal, phosphatidylserine, are recognized and cleared by macrophages both in vitro and following inhalation in vivo [57].

In cases where the corona biomolecules have the propensity of being recognized by the biological machinery there may be a combination of two effects: (1) a non-specific lowering of the direct physical interactions between the bare particle and the cell surface; and (2) a specific interaction between adsorbed biomolecules and the biological machinery, resulting in the activation of more-specific regulated pathways. There have also been suggestions that surface biomolecules (including adsorbed endogenous proteins) could influence the crossing of biological barriers, such as the blood–brain barrier [35, 58, 59]. Mastering these phenomena will have deep implications for both nanosafety and nanotherapeutics [4].

Our hypothesis is that the biodistribution and biokinetics of many nanoparticles in vivo would largely be determined by the nanoparticle-corona complex rather than the bare material [2, 4, 11, 28, 29, 59, 60]. It is not yet know to what degree

Table 7.1 Quantitative abundance of proteins in the biomolecular corona of some example nanoparticles exposed to blood proteins (from previous literature)

Nanoparticle and protein sample	Relative abundance of top 20 proteins (%)	Total number of proteins identified
Polystyrene unmodified 200 nm in 10% plasma [14]	85.4	52
Polystyrene unmodified 200 nm in 55% plasma [14]	92.9	49
Silica 200 nm in 10% plasma [14]	94.9	47
Silica 200 nm in 55% plasma [14]	90.2	51
Carbon apatite in human serum [20]	73.5	100
Silica 55 nm (si125) in plasma [5]	71.2	125
Silica 16 nm (si20) in plasma [5]	72.1	125
Silica 10 nm (si8) in plasma [5]	70.4	125

Note: The 20 most-abundant proteins identified constitute the majority of the proteins in the 'average corona'. The abundance of the remaining proteins is, however, not negligible, and the number of such proteins is large. This gives an idea of the variance of protein abundance; specifically, it suggests that nanoparticles do not have identical coronas (for example, it seems clear that not all 125 proteins can be found on the 10 nm particle of the last sample, even if packed in a bilayer).

details of the preparation of nanoparticle dispersions and the biomolecules present in different animal species will affect the outcome. Individuals may express different molecules in their blood [17] and corona [61], and there have been propositions that the corona can identify cancer diagnostic markers [62]. These, and many others, are relevant questions that should be addressed.

A key question is whether it is the original corona at the point of entry (for example, blood, lung or other) or a corona modified by subsequent translocation that determines the biodistribution and effects of nanoparticles in vivo. At present, the detailed fate of the original corona, as it passes through membranes and barriers and interacts with the extracellular matrix, is unknown. The original corona may be replaced by new biomolecules (Fig. 7.2) or remain intact. Early results suggest that (part of) the original corona may

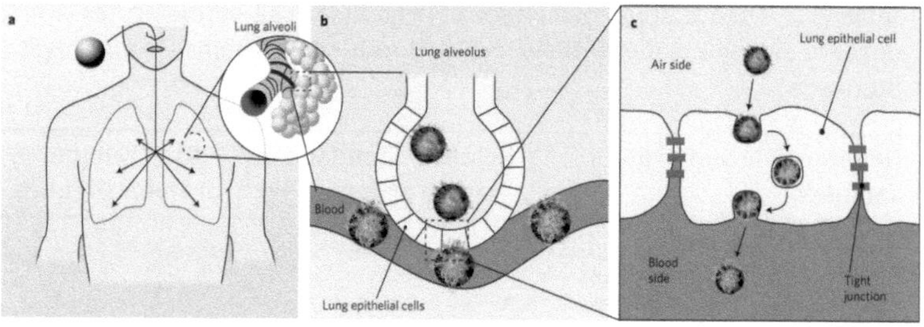

Figure 7.2 The evolution of the nanoparticle corona through the body. (a) An example for inhaled nanoparticles in the alveoli of the lungs. (b) An original corona (blue) forms when the nanoparticle comes in contact with the lung fluids in the alveoli. On subsequent translocation across the lung barrier (composed of a layer of epithelial cells), nanoparticles reach the bloodstream. In blood, some biomolecules from the original corona may be displaced by different biomolecules (purple) in this new compartment, forming a new corona. (c) A nanoparticle crossing the epithelial cells of the lung barrier, via regulated import, is further transported through different membrane compartments and is finally exported out from the cell. Partial displacement of the original corona (blue) by intracellular biomolecules (green) could potentially occur at any point along the pathway, or only once the nanoparticle is exposed to biomolecules in the blood (purple). Proteins adapted from Protein Data Bank (http://www.pdb.org).

be carried into the cell, rather than being stripped off by contact with cell-membrane proteins and lipids [63, 64]. This implies that the original corona, albeit modified by interactions at the cell membrane, may continue to hinder the direct contact between intracellular components and the bare nanomaterial surface. It is possible that the corona may be degraded entirely only if the particles end up in lysosomes or phagosomes [65]. For most of the early interactions with an organism, some variant of the corona would then constitute the biological identity of the nanoparticles, until final accumulation or clearance.

The role of the corona in signalling processes has also begun to emerge [66]. This highlights the fact that biological impacts might be driven by both the composition of the biomolecular corona and disturbances to the conformation of the proteins following adsorption on the nanoparticles [10, 67–72]. One example where the mechanism has been worked out involves nanoparticle-induced protein unfolding leading to initiation of the nuclear factor-KB (NF-KB) pathway and inflammation [10]. Protein unfolding due to nanoparticles may also lead to nucleation of protein fibrillation [73]. Certainly, interactions of nanoparticles and subsequent modulation of the immune system [40, 61, 74–77] require a systematic study in the context of the in situ nanoparticle surface in the relevant biological media.

7.2 Structure, Composition, and Dynamical and Kinetic Aspects

Owing to the complexity of the topic, we focus on two of the most immediate questions. Which aspects of the biomolecular corona are biologically relevant? And, what factors, besides the properties of the bare nanoparticle surface, determine these aspects of the corona?

First, even if the nanoparticles remain within the same biological compartment, the corona biomolecules must stay long enough on the surface of the nanoparticle to be biologically recognized at, for example, the outer plasma membrane of the cell [8]. Thus, besides the structure and composition of the corona, dynamical exchange processes will also determine what is biologically relevant (Fig. 7.1b). Biomolecules residing longer than the characteristic timescale of a given biological process may be relevant to that process [8], if appropriately oriented. For example, endocytosis of nanoparticles across the plasma membrane occurs within minutes [46, 50], and, consequently, a biomolecule must stay on the nanoparticle surface at least this long to be 'seen and processed by the endocytic machinery. For unmodified nanoparticles in biological media, several examples show that the tightly bound hard corona exchanges slowly, often over several hours [8, 12, 16]. In the case of the much more rapidly exchanging outer soft corona, we expect an exchange time of seconds or less [16]. This separation of timescales suggests that the key biological determinant for nanoparticles will often be the interface between the hard and soft corona. It may be that neither materials of the future nor some PEGylated particles will possess this simple separation of timescales, but the basic paradigm that the residence time determines the outcome will remain valid.

Second, it is important to connect the structure and composition of the hard corona to the nature of the original nanoparticle surface. This is not a simple task because the corona is not at thermodynamic equilibrium. Indeed, the molecular dynamical exchange times of the hard-corona biomolecules can be many hours [8, 12, 16]. This means that the composition and structure of the corona are not (and cannot be) equilibrated on a timescale of typical biological experiments, and the nearly stationary composition and structure is actually a consequence of the very slow exchange. Efforts to describe the hard corona by conventional methods of equilibrium physical sciences (for example, affinity constants and Langmuir isotherms) may therefore be incomplete. In other words, one should not expect the hard-corona structure and composition to reflect the basic interactions between the nanoparticle and biofluid in a simple manner.

For example, in the absence of rapid exchange, some biomolecules arriving early to the nanoparticle surface may subsequently either be displaced by those arriving later or, if they adhere strongly, may themselves play a role in determining the next set of proteins adsorbing (Fig. 7.3). As these processes are all affected by each biomolecule's abundance, rate of diffusion, affinity for the surface and/or other biomolecules already associated with the surface [72], nanoparticle-biomolecule and biomolecule-biomolecule interactions, as well as the conditions in which the corona was formed, are all relevant considerations. Still, practical approaches have been sought to predict the relationship between nanoparticle properties and the nature of the corona by the measurement of the binding affinities of a panel of small molecules to a nanoparticle surface [78]. It is too early to draw firm conclusions. As the biomolecular corona may not always be simply related to individual nanoparticle-biomolecule affinities, it may in the end be necessary to unravel the details of the sequential kinetics of the formation of the corona. Still, practically speaking, the corona, being relatively stationary in time, can be studied and characterized, allowing efforts to relate it to biological interactions.

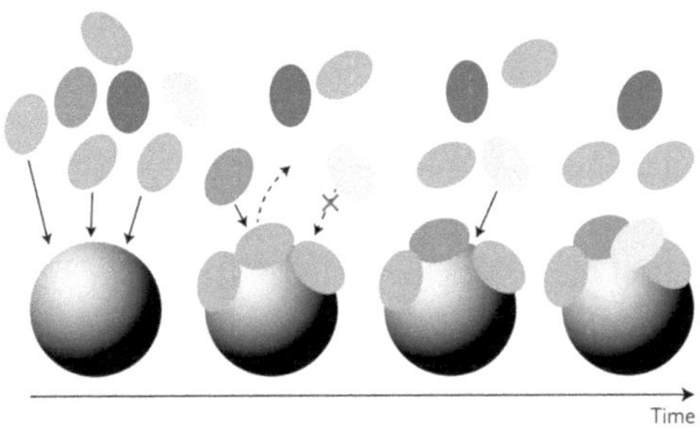

Figure 7.3 Formation of the biomolecular corona. From left to right: An initial corona forms from those biomolecules (in green) that arrive first to the surface (typically highly abundant proteins). An initially adsorbed molecule with low affinity (in green) is subsequently displaced by a different molecule with higher affinity (in blue) arriving later. A third molecule (in yellow), that initially had a low affinity for the bare surface, now adsorbs on the nanoparticle surface owing to favourable interactions with the adsorbed (green and blue) biomolecules. A different biomolecule (in red) does not adsorb at all.

Some factors that control the corona are now clear, and it will be useful if future studies are carried out as a function of these parameters. For example, as with any surf ace-coverage phenomenon, the ratio of nanoparticle surface area to protein concentration is an important factor [1, 14]. Still, there are strong indications that radius of curvature of the nanoparticle is also a key parameter for several reasons [1, 5, 13, 18, 19]. Thus the surface energy of a bare nanoparticle depends on the radius of curvature, because a highly curved surface causes local stresses on the microscopic organization of atoms at the surface. Indeed, the energy of a particle surface can be much larger than a flat surface of the same material [79].

The nanoparticle radius of curvature, if sufficiently high, will also affect the corona directly as large proteins seek to pack around small spaces [5, 13, 19]. For example, one expects radically different coronas in high-aspect-ratio materials (such as carbon nanotubes), where one of the lengths is typically smaller than that of the abundant proteins [80, 81]. In this case, biomolecules may align along the long axis to fit on the nanoparticle surface. Larger globular biomolecules, on the other hand, may simply not fit and hence be excluded from the corona.

All these considerations bring into question the concept of nanomaterial ontology (classification), which is under discussion for the purpose of biological interactions, as well as regulation. Thus, fine details of size, shape and bare-surface physicochemical properties (beyond the capacity of current nanoparticle characterization methods) can lead to different hard coronas. Therefore, properties of the nanoparticles themselves (in the absence of their biological milieu) may be insufficient to partition them into groups of similar particles in relation to, for example, their biological interactions. Fortunately, one may not have to base an ontology on particle properties; rather, knowledge of what is strongly bound to nanoparticles in canonical biological and other environments may be sufficient.

One can make progress in characterizing the corona, and connecting its properties to biological impacts. Systematic and good- quality information on the composition of the hard corona is now emerging, both in terms of identities [13, 15, 19, 20, 23, 35, 82–84] and, more recently, quantities [5, 14, 18, 20], of the different species. For animal-derived biological media, those corona compositions known at present are largely composed of proteins [5, 13, 14, 18–20, 23, 35, 82–84] lipids [56, 85] and sugars [86], but other components, derived from the overall composition of the environment, may be identified in future. Recent studies have gone further by application of tools from bioinformatics to correlate potential biological functions of the corona with nanoparticle properties [5].

It should be noted that most methods to determine the corona composition first separate all of the strongly adsorbed biomolecules from the particles into a single biomolecule sample, and then use mass spectrometry to identify the recovered proteins or other components. However, because the corona is not at thermo-dynamic equilibrium, there could be statistical fluctuations in its composition and organization from particle to particle within the same sample [8, 87]. Besides emphasizing that most of the hard corona is composed of few proteins, Table 7.1 suggests that there are many more proteins in the average corona than what would fit on one particle surface. If, as we believe, the sample of proteins extracted from the particle surface is not due to contamination by background proteins, this suggests that nanoparticles within the same sample may have slightly different hard coronas and, consequently, different biological actions. In future, different approaches are required to identify such corona subclasses, rather than averaging over them.

The hard-corona structure can also be studied by isolating the particle-corona complexes [8], and using physical characterization methods that are well known from colloid and interface science [88]. For example, dynamic light scattering [8] and fluorescent correlation spectroscopy [9, 16] measure the hydrodynamic radius, and differential centrifugal sedimentation [8, 14] sheds light on the distributions of such nanoparticle-corona complexes. From differential centrifugal sedimentation, the corona thickness can be estimated, and information on dynamical exchanges assessed by time-resolved experiments. It is also becoming common practice to use average-size and zeta-potential measurements to monitor the corona formation when nanoparticles are added to biological media and characterize the quality of the resulting dispersion of these complexes. Several techniques are now available to study adsorption kinetics of blood plasma proteins or enzymes on nanoparticles, including infrared and Raman spectroscopies [68, 69, 80, 89–91], fluorescence correlation spectroscopy [6, 9, 16, 91], surface plasmon resonance [32, 86, 92], small-angle X-ray scattering [67, 91, 93], quartz crystal microbalance [94], isothermal titration calorimetry [86, 93, 95] and ultraviolet-visible and fluorescence spectroscopies [68, 96]. Adsorption on the nanoparticle surface may result in protein conformational changes, and these have been studied by circular dichroism [6, 67, 68, 72, 80], infrared spectroscopy [69, 80], NMR [85, 97] and enzyme activity assays [67, 68, 98]. Morphological analysis of the corona-nanoparticle complexes has been carried out using transmission electron microscopy and atomic force microscopy [6, 67], and emerging computational studies seek to correlate with experimental results [6, 99–101].

So far, the focus has been on blood plasma-induced corona on nanoparticles. Studies of the corona of nanoparticles recovered from many other biologic fluids, such as urine, synovial fluid, cerebrospinal fluid and pleural effusion, are also emerging [20].

Very different coronas can also form when nanoparticles are dispersed in river water or similar aqueous media used for environmental studies [102, 103]. We should be aware that the environmental fluid first in contact with nanoparticles may also affect later interactions of the nanoparticles with organisms. As noted earlier, nanoparticles could derive their initial biomolecular corona from different sources—for example, lung entry leads to contact with lung fluids (Fig. 7.2) and early studies have focused on protein surfactants, such as surfactant protein A [60, 104] and related biomolecules. In a recent study, the selective adsorption of surfactant was demonstrated on single-walled carbon nanotubes recovered from broncheo-alveolar lavage of mice, and the presence of this surfactant lipid-protein corona was shown to influence the degree of macrophage uptake [56]. There is ample evidence showing that small nanoparticles pass the alveolar-capillary barrier in the lungs and enter the bloodstream, but it is not yet clear to what degree this is assisted by the early binding of lung-borne biomolecules [28, 29, 105].

7.3 Future Challenges

Although much progress has been made regarding biomolecular corona, there are several key problems that still need to be addressed [106]. The macroscopic composition of biomolecules that form the hard corona is valuable, but to fully link the properties of the corona to the biology, more detailed information on the composition, organization and dynamics of these biomolecules is needed. Techniques developed for protein characterization are likely to be valuable for identifying biologically functional peptide sequences (known as epitopes) at the interface of the hard and soft coronas (Fig. 7.4). Screening the biomolecules using antibody and protein microarrays will give information on the different types of binding targets and ultimately the epitopes exposed on corona complexes. Such methods also have the possibility to identify subclasses of particles with different bound biomolecules in different arrangements.

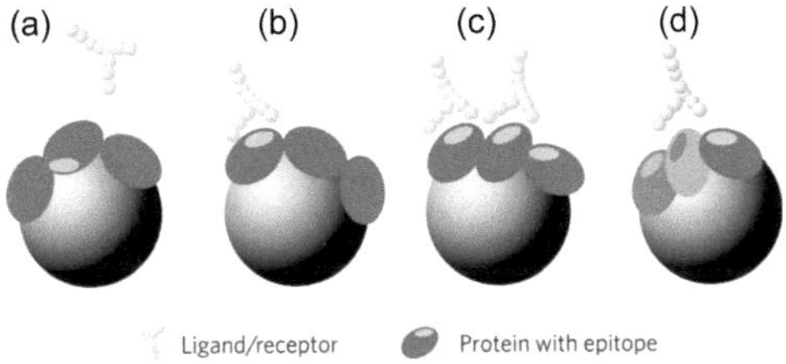

Figure 7.4 Epitope exposure on the nanoparticle/corona interface. (a) If the epitopes of proteins in the corona are not exposed, they are not recognized by the corresponding receptor (or other ligand). (b) The epitope on the corona is exposed and can be recognized. (c) Multiple (similar) epitopes are exposed, leading to a higher biological recognition (d) Several different epitopes of different proteins (shown in different colours) are exposed, leading potentially to anomalous or unusual combinations, which could be recognized by a different receptor (in green).

To determine the structure of the hard/soft corona interface in detail requires methods other than those used at present in the field. Several techniques from related areas, such as X-ray and neutron scattering, previously successful in studies of complex interfaces, have now started to be used for this purpose [7], and both spectroscopic and NMR approaches will be of help. It is clear that the combination of physicochemical and biological methodologies will continue, requiring scientists studying these problems to work across the traditional disciplines.

It is worth stressing that all of these approaches could support efforts to correlate and predict aspects of the biological interactions of new materials. Thus, quite different nanomaterials that share similar corona properties (such as size, shape, exchange dynamics and surface-expressed epitopes) will probably have similar early biological interactions, implying potentially similar biodistributions. Methods to fingerprint these (even without understanding them) will be of practical importance. Such approaches are considered important for nanosafety, because the range and complexity of relevant nanomaterials make their individual study very challenging.

All these questions are also relevant to nanoparticles made for medical applications. Although grafting targeting ligands onto nanoparticles can reduce non-specific binding of molecules from the environment, when studied in the presence of biological media, it is likely to emerge that the interface they form with biological tar- gets is much more complex than predicted, possibly even partly explaining the sometimes puzzling lack of success of some targeting strategies when applied in vivo [107].

It will be important to understand how the passage of nanoparticles inside cells, through biological barriers, and between compartments and organs in vivo is determined by and controls the evolving corona (Fig. 7.2). A key scientific question is the characterization and understanding, of the in situ biomolecular corona inside cells and organisms; particularly interesting is the detailed relationship 11 bears to trafficking to specific cellular and barrier targets [108]. Almost certainly this question has to be resolved for predicting biodistribution and biokinetics from nanomaterial design. It also impinges on broader questions, such as how nanoparticles that have passed through a variety of natural (say river) or production environments are changed from their pristine state [109, 110], via dissolution mechanisms as well as adsorption of molecules, and subsequently interact with organisms.

The general conceptual framework for interactions between nanoparticles and living systems is now falling into place, accompanied by an interdisciplinary community working on the question We may thus hope that a rational basis for biological identity at the nanoscale is a genuine possibility.

Acknowledgements

Funding has been generously provided by the INSPIRE (Integrated NanoScience Platform for Ireland) programme, funded by the Irish Government's Programme for Research in Third Level Institutions, Cycle 4, National Development Plan 2007–2013 (M.P.M. and A.S.), Science Foundation Ireland under Grant No. [09/RFP/MTR2425] (C. Å.) and the EU FP7 small collaborative project NanoTransKinetics under Grant No. NMP4-2010-EU-US-266737 (C.Å.). K.A.D. gratefully acknowledges the support of the European Commission Framework Program via several NMP programs (Nanolnteract, Neuronano, NanoTransKinetics) to the development of these ideas.

Competing Financial Interests

The authors declare no competing financial interests.

References

1. Cedervall, T. et al. Understanding the nanoparticle-protein corona using methods to quantify exchange rates and affinities of proteins for nanoparticles. *Proc. NatlAcad. Sci. USA* **104**, 2050–2055 (2007); introduction of the concept of 'nanoparticle-corona' and the linking of biomolecular exchange timescale to biological identity.
2. Nel, A. E. et al. Understanding biophysicochemical interactions at the nanobio interface. *Nature Mater.* **8**, 543–557 (2009).
3. Lynch, I., and Dawson, K. A. Protein-nanoparticle interactions. *Nano Today* **3**, 40–47 (2008).
4. Aggarwal, P., Hall, J. B., McLeland, C. B., Dobrovolskaia, M. A., and McNeil, S. E. Nanoparticle interaction with plasma proteins as it relates to particle biodistribution, biocompatibility and therapeutic efficacy. *Adv. Drug Deliv. Rev.* **61**, 428–437 (2009).
5. Tenzer, S. et al. Nanoparticle size is a critical physicochemical determinant of the human blood plasma corona: a comprehensive quantitative proteomic analysis. ACS *Nano* **5**, 7155–7167 (2011); extensive determination of protein corona composition, and bioinformatics analysis of protein function.
6. Ge, C. et al. Binding of blood proteins to carbon nanotubes reduces cytotoxicity. *Proc. Natl. Acad. Sci. USA* **108**, 16968–16973 (2011).
7. Ang, J. C., Lin, J-M., Yaron, P. N., and White, J. W. Protein trapping of silica nanoparticles. *Soft Matter* **6**, 383–390 (2010).
8. Walczyk, D., Baldelli Bombelli, E, Monopoli, M. P., Lynch, I., and Dawson, K. A. What the cell "sees" in bionanoscience. *J. Am. Chem. Soc.* **132**, 5761–5768 (2010).
9. Rocker, C., Potzl, M., Zhang, E, Parak, W. J., and Nienhaus, G. U. A quantitative fluorescence study of protein monolayer formation on colloidal nanoparticles. *Nature Nanotech.* **4**, 577–580 (2009). This is one of the first examples in which the corona lifetime has been determined.
10. Deng, Z. J., Liang, M., Monteiro, M., Toth, I., and Minchin, R. E Nanoparticle- induced unfolding of fibrinogen promotes Mac-1

receptor activation and inflammation. *Nature Nanotech.* **6**, 39–44 (2011); here it is demonstrated that the corona itself can include new (cryptic) epitopes due to unfolding on adsorption on nanoparticle surface, leading to activation of the NF-KB pathway.

11. Walkey, C. D., and Chan, W. C. W. Understanding and controlling the interaction of nanomaterials with proteins in a physiological environment. *Chem. Soc. Rev.* **41**, 2780–2799 (2012).
12. Casals, E., Pfaller, T., Duschl, A., Oostingh, G. J., and Puntes, V. Time evolution of the nanoparticle protein corona. ACS *Nano* **4**, 3623–3632 (2010); an early paper in which the kinetics of maturation of the protein hard corona is explored.
13. Lundqvist, M. et al. Nanoparticle size and surface properties determine the protein corona with possible implications for biological impacts. *Proc. Natl Acad. Sci. USA* **105**, 14265–14270 (2008).
14. Monopoli, M. P. et al. Physical-chemical aspects of protein corona: relevance to in vitro and in vivo biological impacts of nanoparticles. *J. Am. Chem. Soc.* **133**, 2525–2534 (2011); semi-quantitative determination of hard coronas. Also shows how the corona composition changes as the protein concentration is varied, for instance corresponding to typical in vitro and in vivo protein contents.
15. Deng, Z. J. et al. Differential plasma protein binding to metal oxide nanoparticles. *Nanotechnology* **20**, 455101 (2009).
16. Milani, S., Baldelli Bombelli, E, Pitek, A. S., Dawson, K. A., and Radler, J. Reversible versus irreversible binding of transferrin to polystyrene nanoparticles: soft and hard corona. ACS *Nano* **6**, 2532–2541 (2012).
17. Anderson, N. L., and Anderson, N. G. The human plasma proteome. *Mol. Cell. Proteomics* **1**, 845–867 (2002).
18. Zhang, H. et al. Quantitative proteomics analysis of adsorbed plasma proteins classifies nanoparticles with different surface properties and size. *Proteomics* **11**, 4569–4577(2011).
19. Dobrovolskaia, M. A. et al. Interaction of colloidal gold nanoparticles with human blood: effects on particle size and analysis of plasma protein binding profiles. *Nanomed. Nanotech. Biol. Med.* **5**, 106–117 (2009); here it is shown how particles of different sizes form different coronas in human blood and their effects on complement activation and coagulation.
20. Martel, J. et al. Comprehensive proteomic analysis of mineral nanoparticles derived from human body fluids and analyzed by liquid chromatography- tandem mass spectrometry. *Anal. Biochem.* **418**,

111–125 (2011); this is one of the first studies that includes protein corona composition in several human body fluids other than blood serum or plasma.

21. Dufort, S., Sancey, L., and Coll, J-L. Physico-chemical parameters that govern nanoparticles fate also dictate rules for their molecular evolution. *Adv. Drug Deliv. Rev.* **64**, 179–189 (2012).
22. Maiorano, G. et al. Effects of cell culture media on the dynamic formation of protein-nanoparticle complexes and influence on the cellular response. ACS *Nano* **4**, 7481–7491 (2010).
23. Simberg, D. et al. Differential proteomics analysis of the surface heterogeneity of dextran iron oxide nanoparticles and the implications for their in vivo clearance. *Biomaterials* **30**, 3926–3933 (2009).
24. Mahmoudi, M. et al. Protein-nanoparticle interactions: opportunities and challenges. *Chem. Rev.* **111**, 5610–5637 (2011).
25. Gasser, M. et al. The adsorption of biomolecules to multi-walled carbon nanotubes is influenced by both pulmonary surfactant lipids and surface chemistry. *J. Nanobiotechnol.* **8**, 31 (2010).
26. Lundqvist, M. et al. The evolution of the protein corona around nanoparticles: a test study. ACS *Nano* **5**, 7503–7509 (2011).
27. Schleh, C. et al. Size and surface charge of gold nanoparticles determine absorption across intestinal barriers and accumulation in secondary target organs after oral administration. *Nanotoxicology* **6**, 36–46 (2012).
28. Choi, H. S. et al. Rapid translocation of nanoparticles from the lung airspaces to the body. *Nature Biotechnol.* **28**, 1300–1303 (2010).
29. Oberdorster, G., Elder, A., and Rinderknecht, A. Nanoparticles and the brain: cause for concern? *J. Nanosci. Nanotechnol.* **9**, 4996–5007 (2009); early discussion of how secondary surface coating by molecules at portal of entry and during translocation may affect nanoparticle biodistribution in organisms, including to the brain. Results relevant to those reported in ref. 28.
30. Gaddum, J. *Gaddums Pharmacology* 9th edn (Oxford Univ. Press, 1985).
31. Ekdahl, K. N. et al. Innate immunity activation on biomaterial surfaces: a mechanistic model and coping strategies. *Adv. Drug Deliv. Rev.* **63**, 1042–1050 (2011).
32. Corbet, M. B., and Sefton, M. V. Biomaterial-associated thrombosis: roles of coagulation factors, complement, platelets and leukocytes. *Biomaterials* **25**, 5681–5703 (2004).

33. Rejman, J., Oberle, V., Zuhorn, I. S., and Hoekstra, D. Size-dependent internalization of particles via the pathways of clathrin- and caveolae-mediated endocytosis. *Biochem. J.* **377**, 159–169 (2004).
34. Chithrani, B. D., Ghazani, A. A., and Chan, W. C. W. Determining the size and shape dependence of gold nanoparticle uptake into mammalian cells. *Nano Lett.* **6**, 662–668 (2006).
35. Kim, H. R. et al. Analysis of plasma protein adsorption onto PEGylated nanoparticles by complementary methods: 2-DE, CE and protein lab-on-chip system. *Electrophoresis* **28**, 2252–2261 (2007).
36. Prapainop, K., Witter, D. P., and Wentworth, P. A chemical approach for cell- specific targeting of nanomaterials: small-molecule-initiated misfolding of nanoparticle corona proteins. *J. Am. Chem. Soc.* **134**, 4100–4103 (2012).
37. Ferrari, M. Cancer nanotechnology: opportunities and challenges. *Nature Rev. Cancer* **5**, 161–171 (2005).
38. Stephan, M. T, Moon, J. J., Urn, S. H., Bershteyn, A., and Irvine, D. J. Therapeutic cell engineering with surface-conjugated synthetic nanoparticles. *Nature Med.* **16**, 1035–1041 (2010).
39. Otsuka, H., Nagasaki, Y., and Kataoka, K. PEGylated nanoparticles for biological and pharmaceutical applications. *Adv. Drug Deliv. Rev.* **55**, 403–419 (2003).
40. Hamad, I. et al. Distinct polymer architecture mediates switching of complement activation pathways at the nanosphere-serum interface: implications for stealth nanoparticle engineering. ACS *Nano* **4**, 6629–6638 (2010).
41. Hoshino, Y. et al. Recognition, neutralization, and clearance of target peptides in the bloodstream of living mice by molecularly imprinted polymer nanoparticles: a plastic antibody. *J. Am. Chem. Soc.* **132**, 6644–6645 (2010); one of the first efforts to use molecular imprinting to control the protein corona on the nanoparticle surface.
42. Lesniak, A. et al. Effects of the presence or absence of a protein corona on silica nanoparticle uptake and impact on cells. ACS *Nano* **6**, 5845–5857 (2012).
43. Mercer, J., Schelhaas, M., and Helenius, A. Virus entry by endocytosis. *Annu. Rev. Biochem.* **79**, 803–833 (2010).
44. Hu, W. et al. Protein corona-mediated mitigation of cytotoxicity of graphene oxide. ACS *Nano* 5, 3693–3700 (2011).
45. Verma, A. et al. Surface-structure-regulated cell-membrane penetration by monolayer-protected nanoparticles. *Nature Mater.* **7**, 588–595 (2008).

46. Salvati, A. et al. Experimental and theoretical comparison of intracellular import of polymeric nanoparticles and small molecules: towards models of uptake kinetics. *Nanomed. Nanotechnol. Biol. Med.* **7**, 816–826 (2011).

47. Doherty, G. J., and McMahon, H. T. Mechanisms of endocytosis. *Annu. Rev. Biochem.* **78**, 857–902 (2009).

48. Dausend, J. et al. Uptake mechanism of oppositely charged fluorescent nanoparticles in HeLa cells. *Macromol. Biosci.* **8**, 1135–1143 (2008).

49. Iversen, T. G., Skotland, T., and Sandvig, K. Endocytosis and intracellular transport of nanoparticles: present knowledge and need for future studies. *Nano Today* **6**, 176–185 (2011).

50. Shapero, K. et al. Time and space resolved uptake study of silica nanoparticles by human cells. *Mol. Biosyst.* **7**, 371–378 (2011).

51. Lesniak, A. et al. Serum heat inactivation affects protein corona composition and nanoparticle uptake. *Biomaterials* **31**, 9511–9518 (2010).

52. Ehrenberg, M. S., Friedman, A. E., Finkelstein, J. N., Oberdorster, G., and McGrath, J. L. The influence of protein adsorption on nanoparticle association with cultured endothelial cells. *Biomaterials* **30**, 603–610 (2009).

53. Zhu, Y. et al. Effects of serum proteins on intracellular uptake and cytotoxicity of carbon nanoparticles. *Carbon* **47**, 1351–1358 (2009).

54. Bajaj, A., Samanta, B., Yan, H., Jerry, D. J., and Rotello, V. M. Stability, toxicity and differential cellular uptake of protein passivated-Fe_3O_4 nanoparticles. *J. Mater. Chem.* **19**, 6328–6331 (2009).

55. Lunov, O. et al. Differential uptake of functionalized polystyrene nanoparticles by human macrophages and a monocytic cell line. ACS *Nano* **5**, 1657–1669 (2011).

56. Kapralov, A. A. et al. Adsorption of surfactant lipids by single-walled carbon nanotubes in mouse lung upon pharyngeal aspiration. ACS *Nano* **6**, 4147–4156 (2012).

57. Konduru, N. V. et al. Phosphatidylserine targets single-walled carbon nanotubes to professional phagocytes in vitro and in vivo. *PLoS ONE* **4**, e4398 (2009).

58. Georgieva, J. V. et al. Surface characteristics of nanoparticles determine their intracellular fate in and processing by human blood-brain barrier endothelial cells in vitro. *Mol. Ther.* **19**, 318–325 (2011).

59. Kreuter, J. et al. Apolipoprotein-mediated transport of nanoparticle-bound drugs across the blood-brain barrier. *J. Drug Target.* **10**, 317–325 (2002).

60. Schleh, C., Rothen-Rutishauser, B., and Kreyling, W. G. The influence of pulmonary surfactant on nanoparticulate drug delivery systems. *Eur. J. Pharm. Biopharm.* **77**, 350–352 (2011).
61. Boraschi, D., Costantino, L., and Italiani, P. Interaction of nano-particles with immunocompetent cells: nanosafety considerations. *Nanomedicine.* **7**, 121–131 (2011).
62. Huo, Q. et al. A facile nanoparticle immunoassay for cancer biomarker discovery. *J. Nanobiotechnol.* **9**, 20 (2011).
63. Ghosh, P. et al. Intracellular delivery of a membrane-impermeable enzyme in active form using functionalized gold nanoparticles. *J. Am. Chem. Soc.* **132**, 2642–2645 (2010).
64. Doorley, G. W., and Payne, C. K. Cellular binding of nanoparticles in the presence of serum proteins. *Chem. Commun.* **47**, 466–468 (2011).
65. See, V. et al. Cathepsin L digestion of nanobioconjugates upon endocytosis. ACS *Nano* **3**, 2461–2468 (2009).
66. Marano, F, Hussain, S., Rodrigues-Lima, E, Baeza-Squiban, A., and Boland, S. Nanoparticles: molecular targets and cell signalling. *Arch. Toxicol.* **85**, 733–741 (2011).
67. Wang, J. et al. Soft interactions at nanoparticles alter protein function and conformation in a size dependent manner. *Nano Lett.* **11**, 4985–4991 (2011).
68. Gagner, J. E., Lopez, M. D., Dordick, J. S., and Siegel, R. W. Effect of gold nanoparticle morphology on adsorbed protein structure and function. *Biomaterials* 32, 7241–7252 (2011).
69. Roach, P., Farrar, D., and Perry, C. C. Surface tailoring for controlled protein adsorption: effect of topography at the nanometer scale and chemistry. *J. Am. Chem. Soc.* **128**, 3939–3945 (2006).
70. Mandal, H. S., and Kraatz, H. B. Effect of the surface curvature on the secondary structure of peptides adsorbed on nanoparticles. *J. Am. Chem. Soc.* **129**, 6356–6357 (2007).
71. Brown, D. M., Dickson, C., Duncan, P., Al-Attili, F., and Stone, V. Interaction between nanoparticles and cytokine proteins: impact on protein and particle functionality. *Nanotechnology* **21**, 215104 (2010).
72. Lacerda, S. H. et al. Interaction of gold nanoparticles with common human blood proteins. ACS *Nano* **4**, 365–379 (2010).
73. Linse, S. et al. Nucleation of protein fibrillation by nanoparticles. *Proc. Natl Acad. Sci. USA* **104**, 8691–8696 (2007).

74. Hubbell, J. A., Thomas, S. N., and Swartz, M. A. Materials engineering for immunomodulation. *Nature* **462**, 449–460 (2009).

75. Fadeel, B., and Garcia-Bennett, A. E. Better safe than sorry: understanding the toxicological properties of inorganic nanoparticles manufactured for biomedical applications. *Adv. Drug Deliv. Rev.* **62**, 362–374 (2010).

76. Karmali, P. P., and Simberg, D. Interactions of nanoparticles with plasma proteins: implication on clearance and toxicity of drug delivery systems. *Expert Opin. Drug Deliv.* **8**, 343–357 (2011).

77. Salvador-Morales, C. et al. Complement activation and protein adsorption by carbon nanotubes. *Mol. Immunol.* **43**, 193–201 (2006).

78. Xia, X. R., Monteiro-Riviere, N. A., and Riviere, J. E. An index for characterization of nanomaterials in biological systems. *Nature Nanotech.* **5**, 671–675 (2010).

79. Nanda, K. K., Maisels, A., Kruis, F. E., Fissan, H., and Stappert, S. Higher surface energy of free nanoparticles. *Phys. Rev. Lett.* **91**, 106102 (2003).

80. Chakraborty, S. et al. Contrasting effect of gold nanoparticles and nanorods with different surface modifications on the structure and activity of bovine serum albumin. *Langmuir* **27**, 7722–7731 (2011).

81. Dutta, D. et al. Adsorbed proteins influence the biological activity and molecular targeting of nanomaterials. *Toxicol. Sci.* **100**, 303–315 (2007).

82. Caracciolo, G. et al. Evolution of the protein corona of lipid gene vectors as a function of plasma concentration. *Langmuir* **27**, 15048–15053 (2011).

83. Gessner, A., Paulke, B. R., and Muller, R. H. Analysis of plasma protein adsorption onto polystyrene particles by two-dimensional electrophoresis: comparison of sample application and isoelectric focusing techniques. *Electrophoresis* **21**, 2438–2442 (2000).

84. Sund, J., Alenius, H., Vippola, M., Savolainen, K., and Puustinen, A. Proteomic characterization of engineered nanomaterial-protein interactions in relation to surface reactivity. ACS *Nano* **5**, 4300–4309 (2011).

85. Hellstrand, E. et al. Complete high-density lipoproteins in nanoparticle corona. *FEBSJ.* **276**, 3372–3381 (2009).

86. Zeng, Z. et al. Synthetic polymer nanoparticle-polysaccharide interaction: a systematic study. *J. Am. Chem. Soc.* **134**, 2681–2690 (2012).

87. Lartigue, L. et al. Nanomagnetic sensing of blood plasma protein interactions with iron oxide nanoparticles: impact on macrophage uptake. ACS *Nano* **6**, 2665–2678 (2012).
88. Hunter, R. J. *Foundations of Colloid Science* 2nd edn (Oxford Univ. Press, 2001).
89. Tsai, D. H. et al. Adsorption and conformation of serum albumin protein on gold nanoparticles investigated using dimensional measurements and in situ spectroscopic methods. *Langmuir* **27**, 2464–2477 (2011).
90. Turci, F. et al. An integrated approach to the study of the interaction between proteins and nanoparticles. *Langmuir* **26**, 8336–8346 (2010).
91. Sapsford, K. E., Tyner, K. M., Dair, B. J., Deschamps, J. R., and Medintz, I. L. Analyzing nanomaterial bioconjugates: a review of current and emerging purification and characterization techniques. *Anal. Chem.* **83**, 4453–4488 (2011).
92. Li, L., Mu, Q., Zhang, B., and Yan, B. Analytical strategies for detecting nanoparticle-protein interactions. *Analyst* **135**, 1519–1530 (2010).
93. Henzler, K. et al. Interaction strength between proteins and polyelectrolyte brushes: a small angle X-ray scattering study. *Phys. Chem. Chem. Phys.* **13**, 17599–17605 (2011).
94. Brewer, S. H., Glomm, W. R., Johnson, M. C., Knag, M. K., and Franzen, S. Probing BSA binding to citrate-coated gold nanoparticles and surfaces. *Langmuir* **21**, 9303–9307 (2005).
95. Liu, S. et al. Investigations on the interactions between plasma proteins and magnetic iron oxide nanoparticles with different surface modifications. *J. Phys. Chem.* C **114**, 21270–21276 (2010).
96. De Puig, H., Federici, S., Baxamusa, S. H., Bergese, P., and Hamad-Schifferli, K. Quantifying the nanomachinery of the nanoparticle-biomolecule interface. *Small* **7**, 2477–2484 (2011).
97. Giri, J. et al. Interactions of poly(amidoamine) dendrimers with human serum albumin: binding constants and mechanisms. ACS *Nano* **5**, 3456–3468 (2011).
98. Wu, Z. C., Zhang, B., and Yan, B. Regulation of enzyme activity through interactions with nanoparticles. *Int. J. Mol. Sci.* **10**, 4198–4209 (2009).
99. Obata, S., and Honda, K. Dynamic behavior of carbon nanotube and bio-/artificial surfactants complexes in an aqueous environment. *J. Phys. Chem. C* **115**, 19659–19667 (2011).
100. Hung, A. et al. Ordering surfaces on the nanoscale: implications for protein adsorption. *J. Am. Chem. Soc.* **133**, 1438–1450 (2011).

101. Dell'Orco, D., Lundqvist, M., Oslakovic, C., Cedervall, T., and Linse, S. Modeling the time evolution of the nanoparticle-protein corona in a body fluid. *PLoS* ONE **5**, e10949 (2010).
102. Gao, J., et al. Dispersion and toxicity of selected manufactured nanomaterials in natural river water samples: effects of water chemical composition. *Environ. Sci. Technol.* **43**, 3322–3328 (2009).
103. Keller, A. A. et al. Stability and aggregation of metal oxide nanoparticles in natural aqueous matrices. *Environ. Sci. Technol.* **44**, 1962–1967 (2010).
104. Schulze, C., Schaefer, U. R, Ruge, C. A., Wohlleben, W., and Lehr, C. M. Interaction of metal oxide nanoparticles with lung surfactant protein A. *Eur. J. Pharm. Biopharm.* **77**, 376–383 (2011).
105. Kreyling, W. G. et al. Size dependence of the translocation of inhaled iridium and carbon nanoparticle aggregates from the lung of rats to the blood and secondary target organs. *Inhal. Toxicol.* **21**, 55–60 (2009).
106. Faunce, T. A., White, J., and Matthaei, K. I. Integrated research into the nanoparticle-protein corona: a new focus for safe, sustainable and equitable development of nanomedicines. *Nanomedicine* **3**, 859–866 (2008).
107. Salvati, A. et al. Transferrin-functionalized nanoparticles lose their targeting capabilities when a biomolecule corona absorbs on the surface. *Nature Nanotech.* **8**(2), 137–143 (2013).
108. Sandin, P., Fitzpatrick, L. W., Simpson, J. C., and Dawson, K. A. High-speed imaging of Rab family small GTPases reveals rare events in nanoparticle trafficking in living cells. ACS *Nano* **6**, 1513–1521 (2012).
109. Lowry, G. V., Gregory, K. B., Apte, S. C., and Lead, J. R. Transformations of nanomaterials in the environment. *Environ. Sci. Technol.* **46**, 6893–6899 (2012).
110. Quik, J. T. et al. Effect of natural organic matter on cerium dioxide nanoparticles settling in model fresh water. *Chemosphere* **81**, 711–715 (2010).

Chapter 8

Protein–Nanoparticle Interactions

Iseult Lynch and Kenneth A. Dawson
UCD BioNano Institute, University College Dublin, Belfield, Dublin 4, Ireland
i.lynch@bham.ac.uk

The key role of protein–nanoparticle interactions in nanomedicine and nanotoxicity has begun to emerge recently with the development of the idea of the nanoparticle–protein "corona." This dynamic layer of proteins (and other biomolecules) adsorbs to nanoparticle surfaces immediately upon contact with living systems. While within the biomaterials field the role of adsorbed molecules in cellular responses is acknowledged, there are several new issues at stake where nanoparticles are concerned. We show here that highly selective protein adsorption, added to the fact that particles can reach subcellular locations, results in significant new potential impacts for nanoparticles on protein interactions and cellular behavior.

This chapter does not seek to provide a comprehensive review of all articles published on protein–nanoparticle interactions. Rather it seeks to provide a snapshot of the range of activities and to highlight some of the key research directions and paradigms that are emerging and evolving in this exciting arena. An important example

---

Reproduced with permission from *Nano Today*, **3**(1–2), 40–47, 2008.

Nano-Enabled Medical Applications
Edited by Lajos P. Balogh

ISBN 978-981-4800-32-7 (Hardcover), 978-0-429-39903-9 (eBook)
www.jennystanford.com

includes the recent discovery that nanoparticles can impact protein fibrillation processes.

Within the medical device community, it is now well accepted that material surfaces are modified by the adsorption of biomolecules such as proteins in a biological environment [1–10], and there is some consensus that cellular responses to materials in a biological medium reflect the adsorbed biomolecule layer, rather than the material itself [1, 3, 6]. An early study in the field of protein interactions with planar surfaces drew attention to the fact that distortion of the protein may occur upon adsorption [11]. However, the importance of the adsorbed protein layer in mediating interactions with living systems has been slower to emerge in the case of nanoparticle–protein interactions. While studies of protein adsorption to nanoparticles are beginning to Appear [12–14], the importance of the detailed structure of the adsorbed protein–solution interface (the outer surface of the adsorbed protein layer taking into account any changes in protein structure) has not yet been widely appreciated in the nanotoxicology literature [1, 15, 16], despite the fact that this is the primary surface in contact with cells.

Of course, as one changes from a flat surface to particles, and as the particles become smaller (eventually approaching the size of the proteins themselves), the composition and organization of the associated protein will change dramatically, departing from the simple limiting case of flat surfaces. We may expect this to lead to quite different biological consequences. Indeed, there is the potential that highly curved surfaces (very small nanoparticles) can suppress protein adsorption to the point where it no longer occurs, an effect likely to be selective to larger proteins, offering a route to differential control of protein adsorption [17]. In addition, flat surfaces can only affect biological process via cell surface receptors such as integrins, whereas nanoparticles can enter cells and thereby access a vast range of extra biological processes.

Interestingly, rather than complicating the story, recent studies suggest that nanomaterial surfaces, which have much larger surface area than flat ones, are more amenable to studies to determine the identity and residence times of adsorbed proteins [2, 18]. Indeed, direct determination of curvature effects on protein adsorption can be made, as with some material types the size of the particle can be increased until the curvature effects vanish, leaving an effectively flat surface. This offers several possibilities in terms of high-throughput or mass screening of nanoparticle–protein interactions, a concept that may have applicability as a new classification of nanoparticles based on their associated protein molecules, or "protein corona." We have recently introduced the concept of the "nanoparticle–protein corona" as the evolving collection of proteins that associate with nanoparticles in biological fluids, which is, in fact, the "biologically relevant entity" that interacts with cells [19].

Here we review some of the recent literature on nanoparticle–protein interactions in light of the fact that the biological impacts of nanoparticles are affected by the nature of the adsorbed protein layer, or the protein (biomolecule) corona. With the potential uses of nanoparticles in biological applications such as nanomedicine being well known, and the increasing importance of the emerging field of nanotoxicology, which aims to address the safety of engineered nanoparticles, the relevancy of protein–nanoparticle interactions cannot be overstated. While most of the knowledge regarding protein–nanoparticle interactions is from solution and in vitro studies, it is clear that future directions will require studies under competitive binding conditions such as occur in vivo.

8.1 The Nanoparticle–Protein Corona

It is a (near) universal rule of materials in biology that a material is always covered by proteins immediately upon contact with a physiological environment, and we believe that this phenomenon will also be key to understanding much of the bionanoscience world [19]. We have recently argued that the effective unit of interest in the cell-nanomaterial interaction is not the nanoparticle *per se*, but the particle and its "corona" of more or less strongly associated proteins from serum or other body fluids [1, 2]. It is important to understand, though, that it is not just the composition and organization of this protein layer, but the exchange times of the proteins on the nanoparticles that is "read" by living cells.

We conceive of the proteins associated with a particle as possessing a very wide range of affinities for the particle surface,

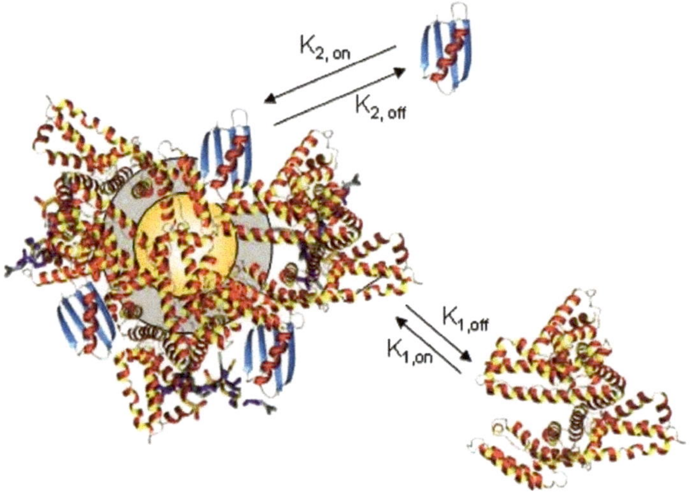

Figure 8.1 Schematic representation of the protein corona on a nanoparticle illustrating the exchange processes and equilibrium constants. The exchange rates are a complex function of the affinity for the surface, curvature effects from the surface, and changes in the surrounding milieu, and much work is needed to evaluate the equilibrium constants under different conditions.

resulting in a range of different residence times for proteins at a nanoparticle surface. In essence, we expect a huge range of equilibrium constants (one for each protein) representing the quite different (and competitive) binding mechanisms present. This means that we see the proteins associated with a particle as a "corona," rather than a solid fixed layer (Fig. 8.1). The composition of the protein corona at any given time will be determined by the concentrations of the over 3700 proteins in plasma [20], and the kinetic on and off rates (or equilibrium binding constants) of each protein for the particular nanoparticle. This corona may not immediately reach equilibrium when exposed to a biological fluid. Proteins with high concentrations and high association rate constants will initially occupy the nanoparticle surface, but may also dissociate quickly to be replaced by proteins of lower concentration, slower exchange, and higher affinity [18]. Thus the protein corona is the biological identity of a nanoparticle, as it is what the cell "sees" and interacts with. The exchange processes may also be important when particles redistribute from one compartment or organ to another, such as upon uptake into cells from the bloodstream, or upon transport from the cytosol to the nucleus.

Studies of protein adsorption bovine serum albumin (BSA), myoglobin (Mb), and cytochrome c (CytC)) onto self-assembled monolayers of mercaptoundecanoic acid (MUA) on Au nanoparticles using the quartz crystal microbalance (QCM) show that all three proteins form adsorption layers consisting of an irreversibly adsorbed fraction and a reversibly adsorbed fraction [21]. This corresponds very well with the notion of a nanoparticle–protein corona, which is proposed to have a hard corona composed of those proteins that have a strong adsorption to the nanoparticles and a long residence time and a softer corona composed of those proteins with shorter residence times and/or lower affinities.

A survey of the literature on nanoparticle–protein binding shows that the vast majority of nanoparticle types studied so far bind apolipoproteins [18]. At first sight this is a surprising result, and quite distinct from that for a flat surface. However, the fact that apolipoproteins are known to be involved in lipoprotein complexes, which themselves have sizes on the nanoscale (Fig. 8.2) ranging from 100 nm (chylomicron) to ~10 nm (high density lipoproteins), may mean that there are specific size-dependent interactions that drive the binding of apolipoproteins to nanoparticles. This is interesting from the point of view of nanoparticle interaction with cells, as lipoprotein complexes are involved in the general cellular processes of cholesterol metabolism [22, 23]. Thus, there are multiple receptors for apolipoprotein complexes at cell surfaces that nanoparticles with surface-adsorbed apolipoproteins can potentially exploit to enter cells [24].

If we consider again the issues of nanoparticle transport and fate in animals and humans, then it is also relevant that apolipoprotein E has been found to associate to some nanoparticles [18]. This has potentially significant consequences for neurotoxicity and the development of neurotherapies, as apolipoprotein E is known to be involved in trafficking to the brain [25, 26]. Thus, we hypothesize that it is the nanoparticle–protein corona (besides size and shape) that actually determines the final subcellular location of a specific

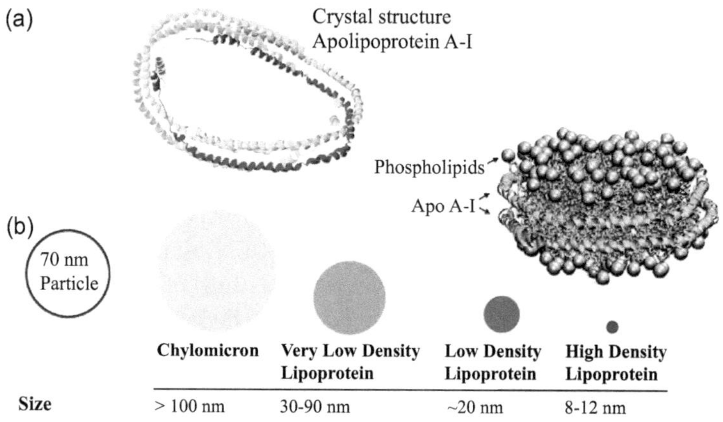

Figure 8.2 (a) Crystal structure of the protein apolipoprotein A-1, and schematic of apolipoprotein A-1 in a lipoprotein complex composed of phospholipids. (Courtesy of the Theoretical and Biophysical Computational Group, University of Illinois at Urbana-Champaign.) (b) Size comparison of a 70 nm nanoparticle with lipoprotein complexes—chylomicrons, very low, low, and high density lipoproteins. These lipoproteins routinely pass into cells, and it is clear that nanoparticles of a similar size coated with apolipoprotein A-1 could be recognized as lipoproteins.

nanoparticle upon interaction with a cell and, thereby, the range of disease processes that the nanoparticle can access [1]. Following on from this, we propose that, in the future, nanoparticles could be classified in terms of their biomolecule corona, which mediates their interaction with cellular machinery. This would represent a truly new paradigm in the field of nanoscale toxicology, and in the design of nanocarriers for nanomedicine.

The first reports of the direct biological influence of proteins adsorbed to nanoparticles are now emerging. Single-walled carbon nanotubes (SWNTs) and 10 nm amorphous silica coated with albumin have been shown to induce anti-inflammatory responses in macrophages, measured as inhibited induction of cyclooxygenase-2 (Cox-2) by lipopolysaccharide under serum-free conditions [27]. Blocking the adsorption of albumin by precoating the nanoparticles with a nonionic surfactant (Pluronic F127) also inhibits the anti-inflammatory properties of the nanoparticles. These observations suggest an important role for adsorbed proteins in modulating the uptake and toxicity of SWNTs and nanosized amorphous silica [27]. However, as these studies were conducted under serum-free conditions, it unclear whether the albumin would remain bound to nanoparticles under competitive binding conditions, such as occurs in plasma or in a cellular milieu.

8.2 Effects on Protein Conformation of Binding to Nanoparticles

Proteins are chains of amino acids, where the exact sequence of the amino acids determines the protein's shape, structure, and function. The principle units of protein secondary structure are α-helices and β-sheets, and the three-dimensional arrangement of these is the tertiary structure (α-helix, shown in red, and β-strand, blue, structures are illustrated in Fig. 8.1). The native conformation of a protein is tightly controlled by the shape complementarity of the hydrophobic residues that allow close packing of the cores [28]. Proteins are nevertheless marginally stable because the beneficial interactions that govern the native structure are counterbalanced by a large entropy loss associated with going from a large ensemble of states to a more restricted set of conformations, as well as by the repulsive electrostatic interactions present in the native state [29]. Thus, interaction with a surface can easily disrupt the native conformation and, therefore, the protein function. This has implications for the biological impact of nanoparticles.

The effect of the surface chemistry of biomaterials on the protein adsorption process has been a topic of great interest for many years, and much is known in this field [30]. Protein adsorption to various materials has been widely studied and it has been found that factors such as electrostatic interactions, hydrophobic interactions, and specific chemical interactions between the protein and the adsorbent play important roles. Selective adsorption of proteins on various synthetic adsorbents has been examined under different conditions (such as solution pH and protein concentration) and for many proteins the mechanism of selective adsorption has been attributed to electrostatic interactions [12].

More recently, many different nanoparticle and protein combinations have been studied using a range of different techniques. We have performed a detailed study of the interaction of human serum albumin (HSA) with polymeric nanoparticles of increasing hydrophobicity and size using isothermal titration calorimetry (ITC) [13], as well as studies of the interaction of complex mixtures of proteins with the same particles using techniques such as surface plasmon resonance (SPR), gel filtration, and mass spectrometry [2, 18]. In the case of polymer particles (described in detail elsewhere [2]), the interaction with HSA occurs with a release of heat (a discernible enthalpy change), whereas several other proteins that we have studied (including fibrinogen, lysosyme, ovalbumin, and human carbonic anhydrase II) bind with no enthalpy change. The binding of these proteins appears to be entropy driven as a result of the release of bound water from the surface of the nanoparticle [14]. Thus, the reduction in entropy of the protein is more than compensated for by the increased entropy of the water molecules. In the case of entropy-driven binding, the interaction does not result in a conformation change of the protein. This has been confirmed by circular dichroism (CD) spectroscopy and the binding has been confirmed by SPR measurements [14].

While there are many results in the literature, we have selected a few key papers to highlight here where the nanoparticles are of particular interest commercially, e.g. quantum dots (QDs) and Au nanoparticles, which are being developed for a range of in vivo imaging applications, or historically, e.g. asbestos, which is the only known case to date of particle-induced toxicity. We have chosen representative papers where very pronounced nanoscale effects are noted.

The interaction between human adult hemoglobin (Hb) and bare CdS QDs has been investigated by fluorescence, synchronous fluorescence, CD, and Raman spectroscopic techniques under physiological pH 7.43. CdS QDs dramatically alter the conformation of Hb, quenching the intrinsic fluorescence of Hb and decreasing the α-helix content of the secondary structure from 72.5% to 60.8%. Raman spectra results indicate that the sulfur atoms of the cysteine residues form direct chemical bonds on the surface of the CdS QDs [15].

Geoinspired synthetic chrysotile is an asbestos reference standard, and has been used to investigate homomolecular exchange of BSA between the adsorbed and dissolved state at the interface between asbestos fibers and the biological medium. Fourier transform infrared spectroscopy (FTIR) and CD spectroscopy show that, in the solid state, BSA modifications are driven by surface interaction with the substrate [16]. Once BSA is desorbed back into solution its structure rearranges, although some of the modifications with respect to the native species are irreversible [16]. Similar effects have been observed with polystyrene nanoparticles—adsorption and subsequent desorption from polystyrene particles causes irreversible changes in the stability and secondary structure of BSA [31]. The α-helix content is reduced, while the β-turn (a region of the protein involving four consecutive residues where the polypeptide chain folds back on itself by nearly 180°) [32] fraction is increased in the exchanged molecules. The irreversible surface-induced conformational change may be related to the aggregation of BSA molecules after exposure to a hydrophobic surface [31].

The effect of the curvature (angle) between the crystal faces of peptide-protected Au nanoparticles (which are polyhedral species) on the secondary structure of peptides is an unexplored but potentially critical issue. A recent structural investigation of a peptide containing 16 amino acids in both two- and three-dimensional self-assembled monolayers (SAMs) with increasing core diameters, and therefore decreasing curvature between the crystallographic faces, shows that the degree of surface curvature has a profound effect on the secondary structure of the peptide. In addition, a three-dimensional monolayer (on Au nanoparticles) does not always resemble the two-dimensional monolayer on a Au surface [33].

Functionalization of nanoparticle surfaces with peptides is increasingly being used to control the interaction of nanoparticles with proteins [34, 35]. Investigation of the effect of Au nanoparticles with positive, negative, and neutral ligands on attached CytC structure reveals that the protein retains its structure with neutral ligands but denatures in the presence of charged species [34]. Similar studies of the interaction of Au nanoparticles functionalized with L/D-leucine and/or L/D-phenylalanine residues with alpha-chymotrypsin (ChT) and CytC show that the chirality of the functionalized nanoparticle end-groups substantially affects the resultant complex stability, with up to 20-fold differences between particles of identical hydrophobicity. This demonstrates that structural information from the ligands can be used to control protein recognition [35].

8.3 Methods to Evaluate Protein Binding to Nanoparticles

Many methods are commonly used to study nanoparticle–protein interactions, several of which have already been mentioned in the sections above, such as FTIR, CD spectroscopy, ITC, SPR, mass spectrometry, and fluorescence spectroscopy [2, 13, 15, 16, 18, 31].

We have recently reported the use of size-exclusion chromatography to study nanoparticle–protein interactions, with specific emphasis on the determination of the residence times of proteins on nanoparticles in order to identify those proteins that may be relevant on the biological timescale [2]. This method is based on the fact that nanoparticles are too big to enter the pores of the size-exclusion matrix, and thus pass through the column in the void volume, whereas proteins are small enough to enter the pores, and the time they spend inside the pores depends on their molecular weight. Thus, each protein elutes with a characteristic volume. However, in the presence of nanoparticles, the elution volume of proteins is shifted toward earlier elution times depending on the duration of the interaction between the protein and the nanoparticle—that is, the protein hitches a ride through the column on the nanoparticle, and thus the nanoparticle-associated proteins are separated from the nonassociated proteins (Fig. 8.3). Using this technique, we have been able to identify those proteins that bind, as well as their association times [2].

The techniques reported here are a mixture of the easily accessible (fluorescence and CD) and the highly specialized (SPR and mass spectrometry). This distinction is an important one in terms of the mass screening of protein–nanoparticle interactions to characterize nanoparticles in terms of their protein corona. In order for this to become feasible for the screening/characterization of the

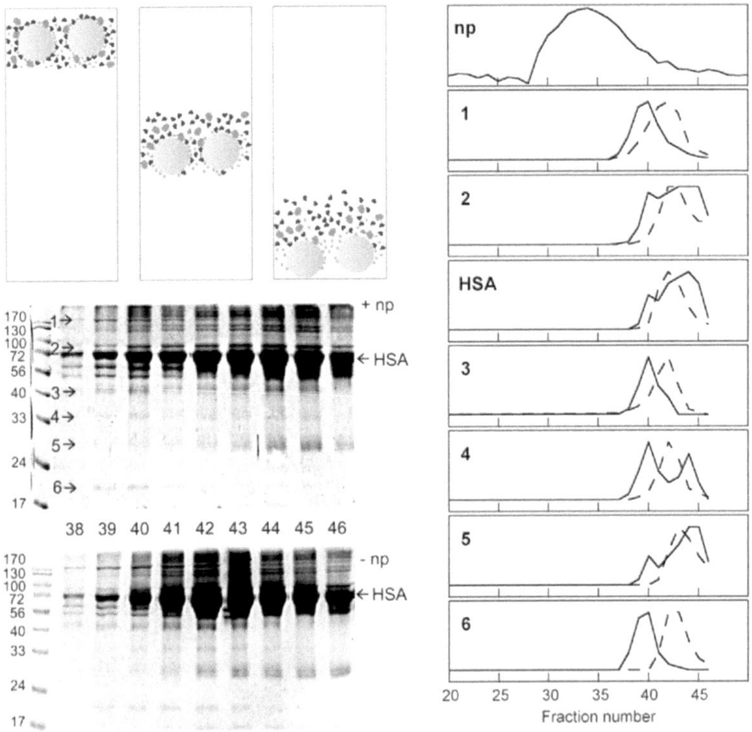

Figure 8.3 Size-exclusion chromatography study of nanoparticle–protein interactions. The elution time of proteins is shifted depending on their affinity for the nanoparticle surface, the longer the protein is associated with the nanoparticle the earlier the protein elutes from the column [2]. Proteins that have sufficiently long residence times elute in the void volume with the nanoparticles. It is clear that each fraction collected from the size-exclusion column contains many different proteins, which can be further separated by gel electrophoresis using denaturing acrylamide gels as shown on the right. The different gel bands can be cut out and the proteins identified by mass spectrometry. Reproduced with permission from [2]. © 2007 National Academy of Sciences.

30 000 nanoparticles that are claimed to be in the industrial pipeline worldwide, the experimental techniques need to be robust, easily accessible, and high throughput (where possible). We highlight here another approach that could potentially contribute to large-scale screening of nanoparticle–protein interactions, namely zeta potential.

The interaction of negatively charged BSA and positively charged lysozyme with a range of metal oxide particles (alumina, silica, titania, and zirconia particles with diameters 73–271 nm) has been studied by zeta potential [36]. The adsorbed proteins change the zeta potentials and the isoelectric points (IEP) of the oxide particles. The amount of adsorbed protein on the alumina, silica, and titania (but not on the zirconia) particle surfaces correlate with the zeta potential. For the slightly less hydrophilic zirconia particles, significant amounts of protein adsorption are observed even under repulsive electrostatic conditions, perhaps on account of the hydrophobic effect playing a more important role for zirconia than the electrostatic interactions [36]. While this data alone does not provide much insight into the effects of the interaction on protein conformation, it may be possible to correlate the zeta potential of nanoparticle–protein complexes with the nature of the principle proteins adsorbed and, in the longer term, to correlate this to nanoparticle uptake by cells.

8.4 Increased Protein Stability/Activity upon Binding to Nanoparticles

While, in general, the loss of secondary structure and consequent changes in the activity of proteins upon binding to nanoparticles can be seen as a drawback or a potential source of nanoparticle toxicity, there is a potential positive outcome too. Promising uses of nanoparticles include increasing protein stability toward enzyme degradation and increasing the activity of enzymes via immobilization at surfaces. Enzymes such as Candida rugosa lipase (CRL) and Pseudomonas cepacia lipase (PCL) have been adsorbed to nanostructured polystyrene (PS) and polymethylmethacrylate (PMMA) by simple addition of the lipase solution to the polymeric nanoparticles under protein-friendly conditions (pH 7.6) [37]. Adsorption leads to improved performance in terms of activity and selectivity with respect to that shown by lipases adsorbed on the same non-nanostructured carriers, as well as increased enantioselectivity and pH and thermal stability [37].

The highly curved surface of C_{60} fullerenes have also been shown to enhance enzyme stability in strongly denaturing environments to a greater extent than flat supports. The half life of a model enzyme, soybean peroxidase, adsorbed onto fullerenes at 95°C is ~2.5-fold higher than that of the enzyme adsorbed onto graphite flakes and ~13-fold higher than that of the native enzyme [38]. Similar observations have been found with other nanoscale supports including silica and Au nanoparticles. The ability to enhance protein stability by interfacing them with nanomaterials may impact numerous fields ranging from the design of diagnostics, sensors, and nanocomposites to drug delivery [38].

8.5 Potential Role of Nanoparticles in Protein Fibrillation

Amyloidogenic proteins are a group of proteins that aggregate under certain conditions to form highly insoluble structures (fibrils), which precipitate to form plaques. These aggregates are involved in a range of serious and irreversibly progressive pathological conditions (protein-misfolding diseases), such as Alzheimer's disease, Parkinson's disease, and dialysis-related amyloidosis. At least 20 such diseases are known.

We have reported recently that a range of different nanoparticles, including polymer particles, cerium oxide, carbon nanotubes, and poly(ethylene glycol) (PEG)-coated QDs, enhance the rate of fibrillation of the amyloidogenic protein β-2-microglobulin under conditions where the protein is in a slightly molten, globular state at pH 2.5 [39]. As the fibrils imaged by transmission electron microscopy (TEM) do not appear to grow out of the nanoparticles (Fig. 8.4), we have suggested a mechanism based on the locally increased concentration of the protein in the vicinity of the nanoparticle surface. This, we believe, increases the probability of the formation of a critical oligomer that, once formed, returns to the solution phase. Multiple layers of protein bound to the particles and interaction with the particles does not appear to have much effect on the protein conformation as determined by fluorescence of the

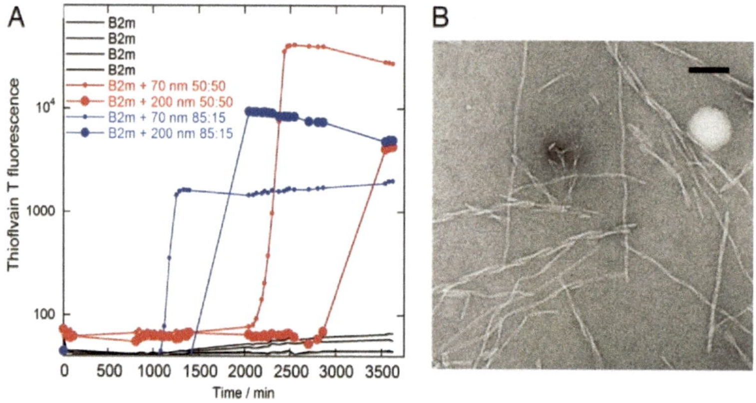

Figure 8.4 Nanoparticles have been shown to increase the rate of fibrillation of amyloidogenic proteins using assays based on the binding of thioflavin-T to protein fibrils [39]. The presence of 70 nm and 200 nm polymeric particles results in a reduced fibrillation time for β-2-microglobulin (B2m), the protein involved in dialysis-related amyloidosis. (a) Thioflavin-T assays in the absence (black) and presence of nanoparticles of different size and composition. As the thioflavin-T only fluoresces when it is bound to fibrils, the onset of fluorescence correlates with the onset of fibrillation. (b) TEM of the protein fibrils in the presence of nanoparticles showing that the fibrils do not grow out from the nanoparticles. Scale bar: 100 nm. Reproduced with permission from [39]. © 2007 National Academy of Sciences.

tryptophan residues, with slight effects observed for the innermost protein layer, and no observed effects for subsequent layers [39].

A more recent report from Bellezza et al. [40] suggests that the interaction of Mb with phosphate-grafted zirconia nanoparticles induces significant rearrangements in the Mb structure, particularly loss of the secondary structure (β-helices). The amount of bound Mb implies a monolayer of adsorbed molecules. Atomic force microscopy (AFM) measurements indicate that the interaction also affects the morphology of the bound protein, inducing the nucleation of prefibrillar-like aggregates at pH 4.7, whose appearance and height (2–4 nm) are consistent with the prefibrillar-like assemblies previously seen for Mb [40]. It is well known that Mb can nucleate fibrillar structures only under destabilizing conditions, where it is has at least a partially unfolded conformation [41]. In this case, the authors state that the prefibrillar-like aggregates are always observed next to the ZrO_2–P nanoparticles, suggesting that the prefibrillar-like structures develop from the bound protein.

Other nanoscale structures have also been recently reported to induce protein fibrillation: dendrimers (synthetic, symmetrically branched polymers that can be manufactured to a high degree of definition and, therefore, present themselves as monodisperse entities) interact with and perturb polypeptide conformations, with particular efficiency toward amyloid structures. Various cationic dendrimers have been shown to be potently disaggregative toward prion aggregates, and to be able to remove prion molecules in the infectious state (PrPSc) effectively from both PrPSc-infected neuroblastoma cells and from PrPSc-containing brain homogenates [42]. However, the same dendrimers have variable effects on the stability of different proteins, suggesting that they do not act as generic denaturants, but rather exert their effects via specific interactions with individual parts of each protein.

Studies in animals have shown that C_{60} hydrated fullerene may have anti-amyloidogenic capacity resulting from inhibition of the fibrillation of amyloid-beta 25–35 peptide [43]. A single intracerebroventricular injection of a C_{60} hydrated fullerene at a dose of 7.2 nmol/ventricle significantly improves the performance of a cognitive task in control rats. TEM studies have confirmed that C_{60} hydrated fullerene inhibits the fibrillation of amyloid-beta 25–35 peptide. This suggests a potential role for nanoparticles in the development of therapies against amyloidogenic diseases. There is currently no cure or reliable diagnosis until postmortem in many cases (e.g. Alzheimer's disease), so advances such as these are very exciting.

8.6 Use of Nanoparticles to Probe Protein–Protein Interactions

In order to fulfill their biological roles, proteins must interact with other molecules. For instance, enzymes, receptors, and transcription factors have to bind their substrates, ligands, and target DNA elements, respectively, to execute their function. Thus, small changes in protein conformation, as could be induced by interaction with the surface of an engineered nanoparticle, can have significant impacts on a protein's function and on its interaction with other proteins (protein–protein interactions). In addition to identifying the proteins directly involved in a nanoparticle's corona, it is also important to understand the effect of incorporating a protein into a nanoparticle corona on the protein–protein interactions, in order to ascertain the biological impacts induced by nanoparticles.

Nanoparticles and nanostructured surfaces offer a new route to study protein interactions, both protein-ligand and protein–protein. For example, we have studied the binding of HSA to a series of polymeric nanoparticles of increasing hydrophobicity in the presence and absence of oleic acid [13], which is one of the key ligands that HSA binds during its normal functioning [44]. Very different interaction patterns are observed with and without oleic acid – HSA in complex with oleic acid gives an endothermic signal and lower stoichiometry compared with the exothermic signal from apo-HSA (in the absence of oleic acid), as shown in Fig. 8.5. A recent study using semiconductor nanocrystals (QDs of average diameter less than 2 nm) directly conjugated to HSA has shown that these conjugates can be used as fluorescent biological labels. Using measurements of fluorescence resonance energy transfer (FRET) from the amino acid tryptophan (Trp214) to the QD, it is possible to follow the local and global changes in the HSA structure during thermal unfolding and refolding processes [45].

Recently, a dual-functionalized nanoparticle has been described that combines histidine-tagged protein purification with site-specific fluorophore labelling [46]. The particles are tetramethylrhodamine-doped nickel chelate silica nanoparticles surface modified with nitrilotriacetic acid. When exposed to a bacterial lysate containing the estrogen receptor alpha ligand binding domain (ERalpha) as a minor component, these beads show very high specificity binding, enabling protein purification in one step. The particles also exhibit good activity for ligand binding and ligand-induced binding to co-activators in solution FRET experiments and protein microarray fluorometric and FRET assays [46].

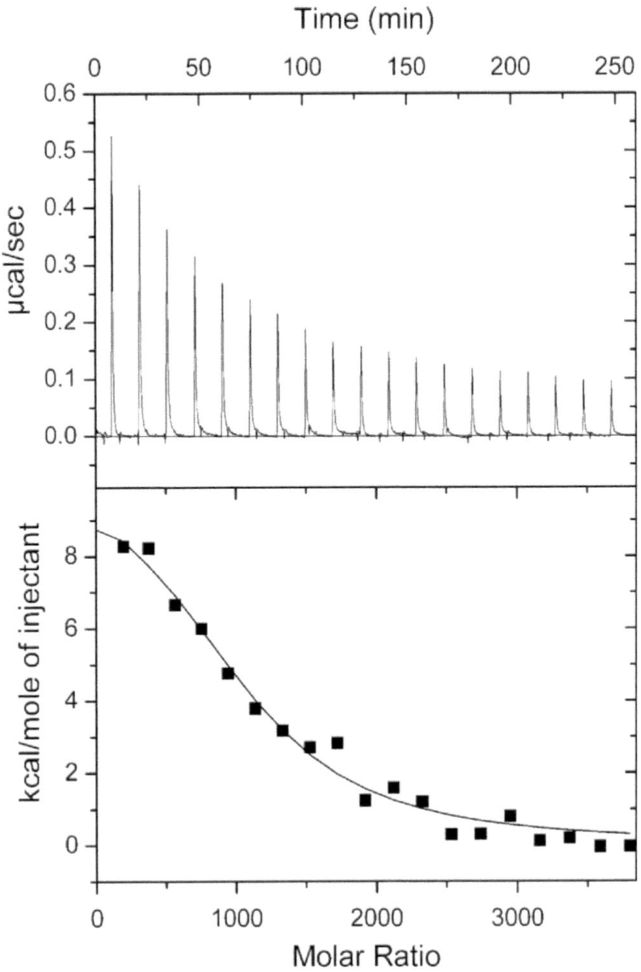

Figure 8.5 ITC has been used to study nanoparticle-protein and nanoparticle–protein–ligand interactions [13]. The graphs show the binding of HSA in the absence (a) and presence (b) of oleic acid (one of the major ligands that binds to HSA) to 70 nm hydrophobic nanoparticles. The binding isotherms also give information about the number of molecules bound and the energetics of binding. In these studies, 1 mg/ml of HSA was titrated into 10 mg/ml particles. Reproduced with permission from [13]. © 2007 American Chemical Society.

From the range of studies presented here, it is clear that research into nanoparticle–protein interactions is a dynamic area but there is much still to be done. There also appears to be considerable contradiction in terms of the findings, with for example, increased protein stability upon binding reported for some protein–nanoparticle pairs and protein destabilization reported for other protein–nanoparticle pairs. Also, the effects of nanoparticles on protein fibrillation seem to be nanoparticle–protein pair specific. Thus, we are some way from general paradigms as yet, although the nanoparticle–protein corona as the biological identity of nanoparticles does appear to be an overarching effect, with the potential to unify various findings into a general approach to characterizing nanoparticle–protein interactions.

8.7 Conclusion

Our core hypothesis is that, rather than the simple nanoparticle itself, it is the dynamical corona of associated biomolecules that defines the biological identity of the nanoparticle. It is therefore this corona that might lead to a more clear classification system of nanosafety, and it is also this corona that could be used to engineer nanomedical outcomes. We therefore believe that very considerable efforts will increasingly be directed at this challenge by the scientific community.

We also note the surprising feature that nanoparticles are sometimes able to induce dramatic effects on protein interactions, such as in the case of protein fibrillation. This observation, when combined with the potential of nanoparticles to be transported to the brain, implies the need to study this whole arena in more depth in the future.

Scientists in the 20th century developed surface physical chemistry and related fields to a high point of achievement. It is an interesting observation that once more, albeit in quite a different context, high surface area adsorption of biomolecules has become a key arena for the development of bionanointeractions.

Acknowledgments

Funding from ESF EPITOPEMAP, EU NanoInteract, EU CIPSNAC, EI, and SFI BioNanoInteract. The authors wish to thank Sara Linse, Tommy Cedervall, Celia Cabailero-Lago, and Anna Salvati for their contributions to the work reviewed here.

References

1. Lynch, I., et al., *Science STKE* (2006) **327**, pe14.
2. Cedervall, T., et al., *Proc. Natl. Acad. Sci. USA* (2007) **104**, 2050.
3. Norde, W., and Lyklema, J., *J. Biomater. Sci. Polym. Ed.* (1991) **2**, 183.
4. Norde, W., and Gage, D., *Langmuir* (2004) **20**, 4162.
5. Allen, L. T., et al., *Biomaterials* (2006) **27**, 3096.
6. Gray, J. J., *Curr. Opin. Struct. Biol.* (2004) **14**, 110.
7. Wilson, C. J., et al., *Tissue Eng.* (2005) **11**, 1.
8. Engel, M. F. M., et al., *Proc. Natl. Acad. Sci. USA* (2004) **101**, 11316.
9. Shen, M., et al., *J. Biomed. Mater. Res. A* (2004) **70**, 533.
10. Chen, C. S., et al., *Methods Mol. Biol.* (2000) **139**, 209.
11. Klein, J., *J. Colloid Interface Sci.* (1986) **111**, 305.
12. Arai, T., and Norde, W., *Colloids Surf.* (1990) **51**, 17.
13. Lindman, S., et al., *Nano Lett.* (2007) **7**, 914.
14. Dawson, K. A., et al., (2008), unpublished data.
15. Shen, X. C., et al., *J. Colloid Interface Sci.* (2007) **311**, 400.
16. Sabatino, P., et al., *J. Colloid Interface Sci.* (2007) **314**, 389.
17. Klein, J., *Proc. Natl. Acad. Sci. USA* (2007) **104**, 2029.
18. Cedervall, T., et al., *Angew. Chem. Int. Ed.* (2007) **46**, 5754.
19. Lynch, I., et al., *Adv. Colloid Interface Sci.* (2007) **134–135**, 167.
20. Muthusamy, B., et al., *Proteomics* (2005) **5**, 3531.
21. Kaufman, E. D., et al., *Langmuir* (2007) **23**, 6053.
22. Dagher, G., et al., *J. Lipid Res.* (2003) **44**, 1811.
23. Zhang, Y., et al., *Mol. Membr. Biol.* (2007) **24**, 442.
24. Kim, H. R., et al., *Cell. Mol. Life Sci.* (2007) **64**, 356.
25. Kim, H. R., et al., *Biomacromolecules* (2007) **8**, 793.

26. Michaelis, K., et al., *J. Pharmacol. Exp. Therap.* (2006) **317**, 1246.
27. Dutta, D., et al., *Toxicol. Sci.* (2007) **100**, 303.
28. Word, J. M., et al., *J. Mol. Biol.* (1999) **285**, 1711.
29. Kumar, S., and Nussinov, R., *Cell. Mol. Life Sci.* (2001) **58**, 1216.
30. Sigal, G. B., et al., *J. Am. Chem. Soc.* (1998) **120**, 3464.
31. Norde, W., and Giacomelli, C. E., *J. Biotechnol.* (2000) **79**, 259.
32. Lewis, P. N., et al., *Proc. Natl. Acad. Sci. USA* (1971) **68**, 2293.
33. Mandal, H. S., and Kraatz, H. B., *J. Am. Chem. Soc.* (2007) **129**, 6356.
34. Aubin-Tam, M.-E., and Hamad-Schifferli, K., *Langmuir* (2005) **21**, 12080.
35. You, C. C., et al., *Chemistry* (2008) **14**, 143.
36. Rezwan, K., et al., *J. Phys. Chem. B* (2005) **109**, 14469.
37. Palocci, C., et al., *Biomacromolecules* (2007) **8**, 3047.
38. Asuri, P., et al., *J. Nanosci. Nanotechnol.* (2007) **7**, 1675.
39. Linse, S., et al., *Proc. Natl. Acad. Sci. USA* (2007) **104**, 8691.
40. Bellezza, F., et al., *Langmuir* (2007) **23**, 13007.
41. Fändrich, M., et al., *Nature* (2001) **410**, 165.
42. Heegaard, P. M. H., et al., *Macromol. Biosci.* (2007) **7**, 1047.
43. Podolski, I. Ya., et al., *J. Nanosci. Nanotechnol.* (2007) **7**, 1479.
44. Sarver, R. W., et al., *Anal. Biochem.* (2005) **347**, 297.
45. Sarkar, R., et al., *J. Phys. Chem. B* (2007) **111**, 12294.
46. Kim, S. H., et al., *J. Am. Chem. Soc.* (2007) **129**, 13254.

Chapter 9

Rapid Formation of Plasma Protein Corona Critically Affects Nanoparticle Pathophysiology

Stefan Tenzer,[a,*] Dominic Docter,[b,*] Jörg Kuharev,[a] Anna Musyanovych,[c,g] Verena Fetz,[b] Rouven Hecht,[d] Florian Schlenk,[e] Dagmar Fischer,[e] Klytaimnistra Kiouptsi,[f] Christoph Reinhardt,[f] Katharina Landfester,[c] Hansjörg Schild,[a] Michael Maskos,[g] Shirley K. Knauer,[d] and Roland H. Stauber[b]

[a] *Institute for Immunology, University Medical Center of Mainz, Langenbeckstrasse 1, 55101 Mainz, Germany*
[b] *Molecular and Cellular Oncology/Mainz Screening Center, University Medical Center of Mainz, Langenbeckstrasse 1, 55101 Mainz, Germany*
[c] *Max Planck Institute for Polymer Research, Ackermannweg 10, 55128 Mainz, Germany*
[d] *Institute for Molecular Biology, University Duisburg-Essen, Universitätsstrasse 5, 45117 Essen, Germany*
[e] *Department of Pharmaceutical Technology, Friedrich-Schiller University Jena, Otto-Schott-Strasse 41, 07745 Jena, Germany*
[f] *Center for Thrombosis and Hemostasis, University Medical Center of Mainz, Langenbeckstrasse 1, 55101 Mainz, Germany*
[g] *Institute of Microtechnology Mainz GmbH, Carl-Zeiss-Strasse 18-20, 55129 Mainz, Germany*
roland.stauber@unimedizin-mainz.de, tenzer@uni-mainz.de

In biological fluids, proteins bind to the surface of nanoparticles to form a coating known as the protein corona, which can critically

---

*These authors contributed equally.
Reproduced with permission from *Nature Nanotechnology*, **8**, 772–781, 2013.

Nano-Enabled Medical Applications
Edited by Lajos P. Balogh

ISBN 978-981-4800-32-7 (Hardcover), 978-0-429-39903-9 (eBook)
www.jennystanford.com

affect the interaction of the nanoparticles with living systems. As physiological systems are highly dynamic, it is important to obtain a time-resolved knowledge of protein-corona formation, development and biological relevancy. Here we show that label-free snapshot proteomics can be used to obtain quantitative time-resolved profiles of human plasma coronas formed on silica and polystyrene nanoparticles of various size and surface functionalization. Complex time- and nanoparticle-specific coronas, which comprise almost 300 different proteins, were found to form rapidly (<0.5 min) and, over time, to change significantly in terms of the amount of bound protein, but not in composition. Rapid corona formation is found to affect haemolysis, thrombocyte activation, nanoparticle uptake and endothelial cell death at an early exposure time.

The use of nanomaterials in biomedical and biotechnological applications is growing, and there is also an increasing probability that such materials will be released into the environment as consumer products [1–6]. Owing to their high surface free energy, nanomaterials adsorb biomolecules on contact with biological fluids [7, 8]. In particular, proteins bind to the surface of nanoparticles to form a biological coating around the nanoparticle, known as the protein corona. This corona affects the biological identity of the nanoparticle and may thus affect biomedical applications or modulate nanotoxicological effects, including ecotoxicology [6, 7, 9–11]. Therefore, in the development of nanomaterials for any kind of biological or biomedical applications [8, 10, 12–16], it is essential to understand the formation and kinetic evolution of the protein corona.

It has been reported that the physicochemical properties of nanoparticles, and their exposure time to biological environments, can influence the protein corona, which affects the biodistribution of the nanoparticles and their therapeutic and pathophysiological effects [7, 8, 17, 18]. The current hypothesis suggests that the 'hard protein corona', which consists of less than 100 different proteins, evolves slowly and is highly dynamic with a composition that changes over time [7, 14, 19]. There have been several investigations into corona formation on prolonged nanomaterial exposure to complex biological environments [7, 13–16]. However, these studies, in general, failed to recognize that physiological systems are often highly dynamic and might need to react instantly to external stimuli. In particular, in the blood system, flow velocities in the ascending aorta can reach up to 60 cm s^{-1}, and processes that control haemostasis and thrombosis can be triggered within minutes [17]. Here we show that a time-resolved knowledge of nanoparticle-specific protein fingerprints, their evolution and impact on biological processes at early exposure times is required to understand fully the interaction of nanoparticles with biological systems.

Protein coronas and their composition are established rapidly. Currently, silica and polystyrene nanoparticles (PsNPs) are tested widely for numerous applications in nanobiomedicine, and are already used in consumer products [7, 15, 20, 21]. Hence, we investigated negatively or positively charged polystyrene nanoparticles (nPsNPs and pPsNPs, respectively) and silica nanoparticles of various sizes (Ø $\approx$ 35 nm, 120 nm, 140 nm), charge and surface modifications (amine, carboxylate, unmodified) (Table 9.1 and Supplementary Table S1). Nanomaterials were characterized thoroughly by transmission electron microscopy (TEM), dynamic light scattering (DLS) and zeta potential measurements (Table 9.1 and Supplementary Table S1 and Fig. S1). AmSil30 amorphous silica nanoparticles are an industrial product manufactured in large quantities and thus released into the environment. In contrast, the other silica nanoparticles examined are produced exclusively for research purposes on an academic scale, and have been used previously [7, 22]. Thus, the selected model systems are relevant for nanobiomedicine, nanotoxicology and also nanoecology [23].

To allow a snapshot resolution of the corona evolution, particles were incubated with human plasma for different time periods and centrifuged through a sucrose cushion, which allowed the rapid separation of unbound plasma as well as recovery of nanoparticle–corona complexes (Fig. 9.1a and see Supplementary Fig. S2 and Table S2 for controls). Thereby, in contrast to other methods such as mere centrifugation [14, 16, 24–26], we could minimize the nanoparticle–protein interaction time to <0.5 min. Protein binding visualized by gel electrophoresis demonstrated that a complex protein corona was established even at the earliest exposure time point (0.5 min) for all nanoparticles examined (Fig. 9.1a and Sup-

Table 9.1 Nanoparticle characterization

Nanoparticle	TEM diameter + s.d. (nm)	DLS hydrodynamic diameter + s.d. (nm) (PDI)		Zeta potential + s.d. (mV)	
	In dry state	In buffer-A	Plasma*	In buffer-A	Plasma
AmSil30	31.4 ± 3.8	35.2 ± 0.2 (0.0001)	48 ± 3 (0.004)	−25 ± 6	−14 ± 2
SiNP30-NH_2	30.2 ± 6.8	31 ± 10 (0.104)	72 ± 21 (0.085)	−10 ± 0.5	-9 ± 1
SiNP30-COOH	27.2 ± 3.8	28 ± 7 (0.063)	69 ± 18 (0.068)	−18 ± 2	−14 ± 2
SiNP30	31.6 ± 5.8	33 ± 7 (0.045)	75 ± 17 (0.051)	−15 ± 2	-9 ± 2
SiNP30F	30.6 ± 6.8	33.6 ± 8 (0.057)	80 ± 19 (0.056)	−15 ± 1	-9 ± 2
SiNP140-NH_2	144.8 ± 8	148 ± 10 (0.004)	172 ± 16 (0.009)	−18 ± 2	-8 ± 1
SiNP140-COOH	129 ± 4	133 ± 10 (0.006)	180 ± 15 (0.007)	−28 ± 3	-8 ± 2
SiNP140	141 ± 6	141 ± 6 (0.002)	158 ± 13 (0.007)	−21 ± 3	-6 ± 1
SiNP140F	140.8 ± 8	142.4 ± 6 (0.002)	164 ± 9 (0.003)	−20 ± 3	-7 ± 1
nPsNP	112 ± 21	124 ± 12 (0.009)	126 ± 13 (0.011)	−55 ± 5	-24 ± 5
pPsNP	115 ± 6	117 ± 12 (0.011)	156 ± 16 (0.011)	+51± 6	-18 ± 9

Note: The average size of the different nanoparticles was determined in the dry state (TEM) as well as in buffer-A by DLS. Plasma measurements were performed 10 min post-exposure to plasma, followed by plasma separation and subsequent washing as described in Supplementary Fig. S2. Zeta potentials were determined with a Zetasizer [15]. Values are mean ± s.d. from three independent experiments. PDI was calculated as the square of the standard deviation over the diameter [21]. Particle nomenclature: commercial silica nanoparticles (AmSil30), laboratory-synthesized silica nanoparticles (SiNP), with fluorescent label (F). *Plasma measurements 10 min post-exposure.

plementary Fig. S3). Distinct nanoparticle-specific kinetic protein-binding profiles were observed, including increased or decreased protein adsorption over time (Fig. 9.1b and Supplementary Fig. S3).

To identify and quantify biologically relevant corona components reliably, we next employed label-free quantitative liquid chromatography mass spectrometry (LC-MS). Quantitative time-resolved data are essential to dissect the evolution of the corona and have not been available so far [7]. Previous studies suggested that the protein corona consists of only a few tens of proteins, even when nanomaterials are introduced into a highly complex environment such as the human blood plasma [7, 14, 16, 19]. Surprisingly, we detected and quantified 166 different plasma corona proteins at the earliest exposure time point (0.5 min) for PsNPs (Ø $\approx$ 120 nm), as well as for silica nanoparticles (AmSil30, Ø $\approx$ 35 nm) (Fig. 9.1c,d and Supplementary Fig. S2). To correlate further the impact of the nanoparticles' physicochemical properties on the complexity and evolution of the plasma protein corona, additionally we investigated silica particles of various sizes (Ø $\approx$ 35 nm and 140 nm) and surface modifications (amine, carboxylate and unmodified, see Table 9.1), and extended the analysis to prolonged plasma exposure time points (0.5–480 min). Employing a latest-generation quadrupole time-of-flight instrument combined with a data-independent ion-mobility enhanced data-acquisition workflow, we were able to quantify almost 300 different corona proteins in this independent analysis (Fig. 9.1e) for all exposure time points.

Not only was the coronas' rapid formation unexpected, but also, importantly, so was their complexity. The complete data sets, including the calculated isoelectric point, molecular mass, localization and associated biological processes of all identified proteins, are provided in Supplementary Tables S3 and S4. Table 9.2 and Supplementary Table S5 further list the 20 most-abundant corona proteins for each nanoparticle at the earliest exposure time point.

Classifying bound proteins according to predicted molecular mass not only revealed an impact of plasma exposure time, but also that proteins $>$60 kDa constitute the majority of all corona components (Fig. 9.1c). High molecular mass proteins ($>$150 kDa) were preferentially bound by silica nanoparticles (Fig. 9.1c).

Table 9.2 Top 20 most-abundant corona proteins (ppm) detectable after 0.5 min of plasma exposure

No.	AmSil30	nPsNP	pPsNP
1	Serum albumin	Serum albumin	Serum albumin
2	Apolipoprotein A-I	Complement C3	Apolipoprotein A-I
3	Complement C3	Complement factor H	Ig γ-1 chain C region
4	Ig γ-1 chain C region	β2-glycoprotein 1	Inter-α-trypsin inhibitor heavy chain H4
5	Complement factor H	Kininogen-1	Ig μ chain C region
6	Kininogen-1	Inter- α-trypsin inhibitor heavy chain H4	Ig γ-3 chain C region
7	Complement C4-A	Ig γ-1 chain C region	Ig κ chain C region
8	Ig κ chain C region	Vitronectin	Vitronectin
9	Serotransferrin	Complement C1r subcomponent	Complement C3
10	Gelsolin	Ig γ-3 chain C region	Complement C1r subcomponent
11	Ceruloplasmin	Lipopolysaccharide-binding protein	Complement C4-A
12	Ig γ-3 chain C region	Gelsolin	Complement C1s subcomponent
13	α-2-macroglobulin	Complement C5	α-2-macroglobulin
14	Haemopexin	Complement C1s subcomponent	Serotransferrin
15	Ig γ-4 chain C region	Apolipoprotein A-I	Apolipoprotein A-IV
16	β-2-glycoprotein 1	Ig μ chain C region	α-1-antitrypsin
17	Complement C1r subcomponent	Complement C4-B	Keratin, type II cytoskeletal 1
18	Ig α-1 chain C region	Ig κ chain C region	Kininogen-1
19	Inter-α-trypsin inhibitor heavy chain H4	C4b-binding protein α chain	Clusterin
20	Haptoglobin	Histidine-rich glycoprotein	Ig α-1 chain C region

Proteins highlighted in gray are among the top 20 for all indicated nanoparticles

Further, bioinformatic analysis showed that proteins displaying a negative charge overall (isoelectric point (p*I*) <7) at physiological pH 7.3 constitute the majority of the corona components for all nanoparticles, irrespective of the particles' negative or positive surface functionalization or plasma exposure time (Fig. 9.1d). This is also reflected by zeta potential measurements of corona-coated nanoparticles, which all display an overall negative charge (Table 9.1 and Supplementary Table S1). Although it was suggested that negatively charged particles attract primarily positively charged proteins and *vice versa* [7, 14, 25], our data indicate that biological environments face predominantly negatively charged corona-covered nanoparticles. Hence the hypothesis that positively charged nanoparticles preferentially interact with the negatively charged cell membrane to result in facilitated cellular uptake seems too simplified [27]. Whether and how supramolecular interactions with charged domains on nanoparticles, corona proteins and/or cellular structures occur and if these are functionally relevant remain to be resolved.

To identify parameters that dictate the composition and evolution of the protein corona, we investigated factors including nanoparticle size, surface characteristics and plasma exposure time. In contrast to previous studies [7, 16, 19], unsupervised hierarchical clustering demonstrated clearly that plasma exposure time was, indeed, a significant influential factor of nanoparticle-bound protein abundance, as all nanoparticles were correctly kinetically classified (Fig. 9.1f and Supplementary Fig. S4). This unbiased analysis revealed that even small variations in the exposure time of <2 min had a significant impact on the corona composition, and significant changes occurred mostly at early rather than late exposure time points for all nanoparticles examined.

The dendrograms illustrate further that surface functionalization was dominant over exposure time, although silica nanoparticle size was dominant over surface functionalization in affecting the protein corona (Supplementary Fig. S4). Although functionalization did impact the formation and evolution of the protein corona, zeta potential measurements showed that modification with amine groups was not sufficient to compensate completely all the nega-

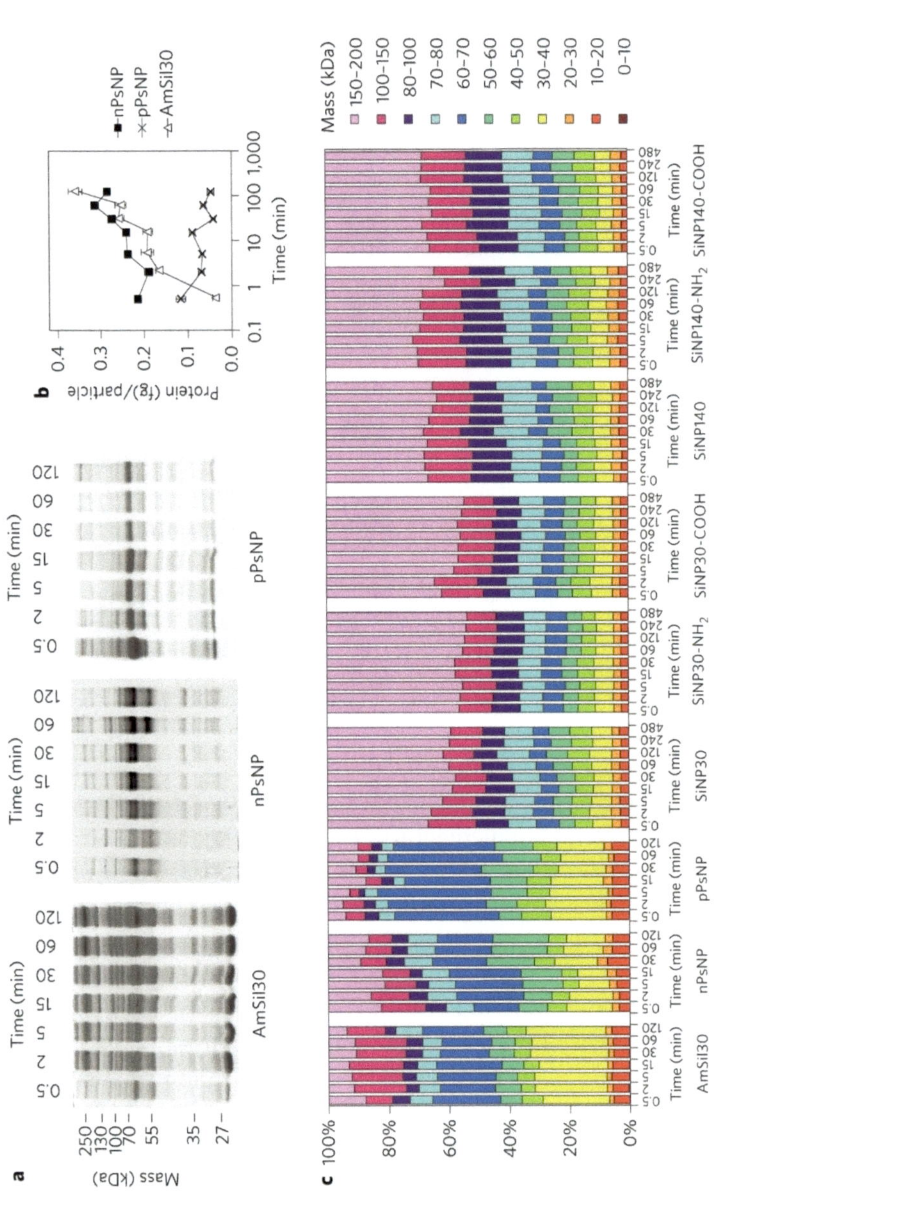
a
Mass (kDa)
250
130
100
70
55
35
27
Time (min)
0.5
2
5
15
30
60
120
AmSil30
nPsNP
pPsNP
b
Protein (fg)/particle
0.4
0.3
0.2
0.1
0.0
0.1
1
10
100
1,000
Time (min)
nPsNP
pPsNP
AmSil30
c
100%
80%
60%
40%
20%
0%
Mass (kDa)
150–200
100–150
80–100
70–80
60–70
50–60
40–50
30–40
20–30
10–20
0–10
Time (min)
AmSil30
nPsNP
pPsNP
SiNP30
SiNP30-NH2
SiNP30-COOH
SiNP140
SiNP140-NH2
SiNP140-COOH

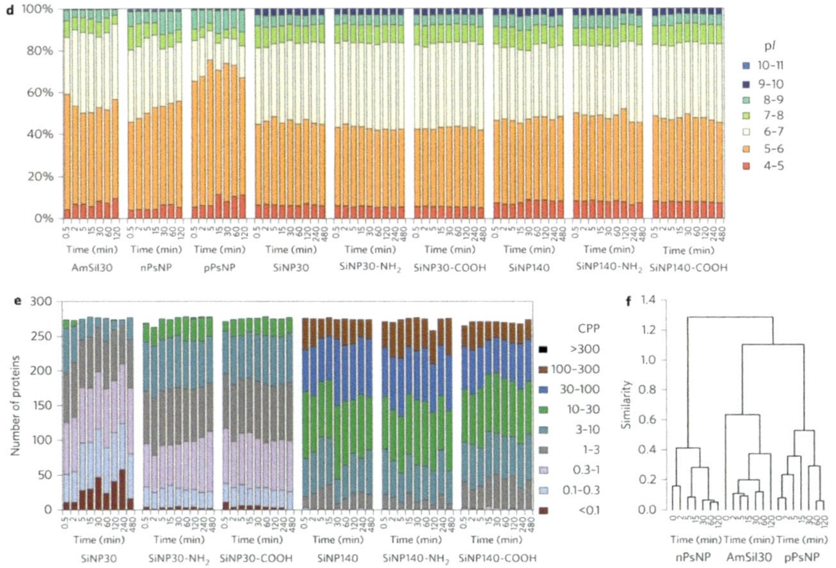

Figure 9.1 Protein coronas and their composition are established rapidly. (a) SDS–PAGE was used to visualize nanoparticle-bound plasma proteins. Molecular mass and time points are indicated. (b) Corona quantification (protein (fg) per particle) at the indicated time points. Continuously increased protein binding was observed for AmSil30 and also, slightly, for nPsNPs, whereas pPsNPs showed decreased binding over time. Values are mean ± s.d. from two independent experiments. (c,d) Classification of corona proteins identified on nine different nanoparticles by LC-MS according to their calculated molecular mass (c) or isoelectric point (d). Relative percentages are shown. (e) Plasma exposure time modulates protein abundance (averaged molecule number per nanoparticle) on the indicated silica nanoparticles. Compared with SiNP30, proteins bound to the larger SiNP140 in higher copy numbers. Relative numbers of proteins present at the indicated copy numbers per indicated nanoparticle are shown. (f) Dendrogram illustrating sample similarities between protein binding profiles, which shows that they are correctly kinetically classified and that significant changes in the corona composition occurred at early rather than at late exposure time periods. Detailed values are available in Supplementary Table S7. CPP, copy number per particle.

tively charged silanol and silanolate residues on the silica surface at physiological pH 7.3 (Table 9.1).

Further, our data revealed that the similarity between kinetic protein-binding profiles of the negatively charged silica (AmSil30) and the pPsNPs was higher than between the profiles of both polystyrene particle types (Fig. 9.1f). Other studies reported that the surface properties of different PsNPs were more relevant for protein absorption than particle size [7, 16].

Collectively, our analyses demonstrate that none of the nanoparticles' physicochemical properties alone was able exclusively to control formation, composition and evolution of the protein corona. As protein exposure time was identified as a critical factor, most probably a multiparameter classifier will be required to model and predict nanoparticle–protein interaction profiles in general.

Corona composition changes only quantitatively. Previous studies proposed a rather dynamic protein corona that changes in composition over time because of continuous protein association and dissociation events controlled predominantly by the Vroman effect [7, 16, 19]. Interestingly, bioinformatic analysis of our comprehensive data sets now revealed that particle-specific protein fingerprints were already determined by the earliest exposure time point (0.5 min), and did not significantly change qualitatively even on prolonged plasma exposure (480 min, Supplementary Tables S3 and S4). The few proteins that were below the detection limit at certain time points belonged exclusively to the group of low-abundance proteins. Whether a secondary 'soft corona' layer, which may not be recovered by our analytical approach, also exists remains to be proved [7].

However, as already evident from the heat maps, which illustrate the kinetic corona protein abundance, we found that the amounts of bound proteins changed significantly over time, that is, quantitatively (Supplementary Figs. S2 and S5–S7). As predicted from the Vroman effect, we observed protein groups that displayed increased or reduced binding over time (Fig. 9.2a,c (protein groups I and II) and Supplementary Fig. S7), which implies that these proteins energetically displace or are displaced by other proteins [28]. Our analysis uncovered novel binding kinetics for biological relevant protein groups for all nanoparticles investigated, which cannot be explained solely by the Vroman effect. Classification of protein-binding modalities unexpectedly identified proteins characterized by low abundance at the beginning of plasma exposure and at later time points, but displaying peak abundance at intermediate time points (Fig. 9.2b,c (protein group III) and Supplementary Fig. S8). Other proteins, however, showed exactly the opposite behaviour, being highly abundant at early and late time points, but not at intermediate time points (Fig. 9.2c (protein group IV) and Supplementary Fig. S8). Previous kinetic studies did not employ quantitative LC-MS-based proteomics, and thus these complex binding kinetics went unnoticed until now [7, 14, 16, 29]. Importantly, representative candidates for all binding modalities were verified independently by immunoblot (Fig. 9.2), which underlines the reliability of our analysis. Comparison of the protein-binding groups using Venn diagrams showed no nanoparticle-independent overlap, indicative of particle- specific binding characteristics (Supplementary Figs. S2, S9 and S10). Although the effects of crowding proteins are certainly important [7, 30], the binding patterns observed in a complex physiologically relevant environment cannot be explained by current mathematical corona-evolution models obtained in simplified experimental systems [7, 29–32]. As reviewed by Vogler [33], our knowledge on the regulation of protein binding to nanosized objects in complex environments is, indeed, still limited.

Rapidly formed coronas modulate pathobiological effects. The rapid formation and complexity of the coronas was unexpected. As, according to our data, an early time-dependent impact on nanoparticle-induced biological effects is expected (Fig. 9.1f and Supplementary Fig. S4), we next employed human-cell models of the blood system.

First, the pathobiological relevance and kinetic impact of the corona formation was demonstrated by using freshly isolated human thrombocytes. Thrombocytes were activated rapidly on exposure to pristine silica nanoparticles, that is particles not covered with a protein corona. Although the final extent of thrombocyte activation, which is reflected by their aggregation, was slightly less effective when compared to that of the natural stimulus collagen

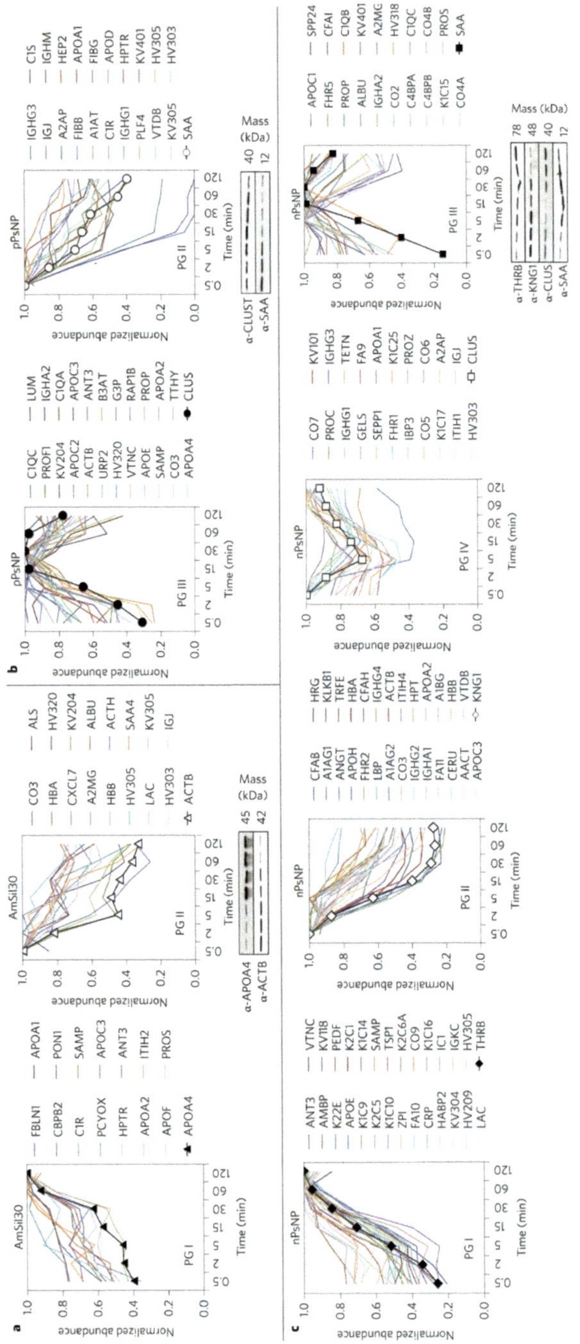
a
AmSil30
PG I
PG II
b
pPsNP
PG III
c
nPsNP
PG IV
Normalized abundance
Time (min)
Mass (kDa)
α-APOA4
α-ACTB
α-CLUST
α-SAA
α-THRB
α-KNG1
α-CLUS

Figure 9.2 Correlation analysis reveals distinct kinetic protein-binding modalities during corona evolution. a–c, Time-smoothed normalized protein abundance profiles of AmSil30 (a), pPsNPs (b) and nPsNPs (c) nanoparticle coronas were classified into four groups by correlation analysis. Relative values normalized to the maximum amount (set to 1) across all time points for each protein are shown. Protein groups PG I and PG II displayed increasing or reduced binding over time, respectively. 'Peak' proteins (PG III) are characterized by low abundance at the beginning of plasma exposure and at later time points, but displaying peak abundance at intermediate time points. Other proteins (PG IV) show exactly the opposite behaviour, being highly abundant at early and late time points, but not at intermediate time points. Profiles were verified independently for representative candidates (marked by symbols) by immunoblot analysis. Molecular mass is indicated. APOA4, apolipoprotein A-IV; ACTB, cytoplasmic actin-1; CLUS, clusterin; SAA, serum amyloid A-4; THRB, prothrombin; KNG, kininogen-1 (Supplementary Table S3 lists values and full names for individual proteins).

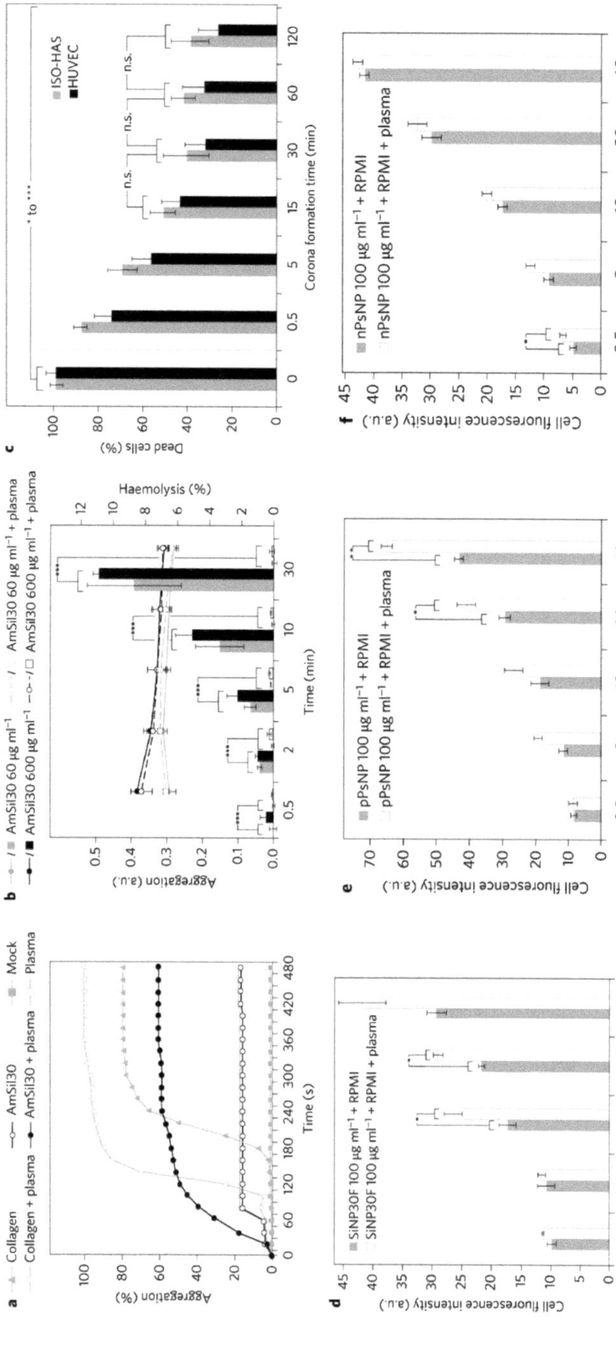
a
Collagen
Collagen + plasma
AmSil30
AmSil30 + plasma
Mock
Plasma
Aggregation (%)
Time (s)
b
AmSil30 60 μg ml⁻¹
AmSil30 60 μg ml⁻¹ + plasma
AmSil30 600 μg ml⁻¹
AmSil30 600 μg ml⁻¹ + plasma
Aggregation (a.u.)
Haemolysis (%)
Time (min)
c
ISO-HAS
HUVEC
Dead cells (%)
Corona formation time (min)
n.s.
d
SiNP30F 100 μg ml⁻¹ + RPMI
SiNP30F 100 μg ml⁻¹ + RPMI + plasma
Cell fluorescence intensity (a.u.)
Time (min)
e
pPsNP 100 μg ml⁻¹ + RPMI
pPsNP 100 μg ml⁻¹ + RPMI + plasma
Cell fluorescence intensity (a.u.)
Time (min)
f
nPsNP 100 μg ml⁻¹ + RPMI
nPsNP 100 μg ml⁻¹ + RPMI + plasma
Cell fluorescence intensity (a.u.)
Time (min)

Figure 9.3 Rapid corona formation kinetically impacts nanopathophysiology. (a) Thrombocytes were exposed to AmSil30 (600 μg ml^{-1}) in buffer or in plasma. Aggregation was quantitated by aggregometry. Collagen (0.7 μg ml^{-1}) alone or together with plasma demonstrated thrombocyte vitality. Buffer or plasma showed no effects. Results from a representative experiment are shown. Whereas thrombocytes were activated rapidly on exposure to pristine nanoparticles, the formation of the plasma protein corona rapidly prevented thrombocyte aggregation. (b) Erythrocytes were exposed to the indicated concentrations of AmSil30 in buffer (filled columns and circles) or in plasma (open columns and circles). Erythrocyte aggregation (lines) or haemolysis (columns) was quantified by spectrophotometric assays at the indicated time points (see Methods for details). For negative controls treatment was with PBS or plasma alone. Values are mean ± s.d. from three independent experiments that involved two independent blood donors. Haemolysis, but not erythrocyte aggregation, was affected by rapid plasma protein corona formation. (c) Silica nanoparticles (AmSil30, 600 μg ml^{-1}) were precoated with human plasma by exposure for the indicated time periods, washed and resuspended in protein-free medium (time 0 indicates uncoated nanoparticles). Primary human endothelial cells (HUVEC) or a microvascular endothelial cell line (ISO-HAS1) were exposed to the nanoparticles for 4 h and further incubated in serum-containing medium for 24 h. Vitality was determined by measuring ATP content. Cells treated with uncoated nanoparticles (time 0) were set at 100%. Columns show the mean+s.d. from three independent experiments. Time-dependent corona formation reduced nanoparticle-induced toxicity most significantly at early plasma coating times. All time points differ significantly from the control (0 min); significance ranges from $P < 0.05$ to $P < 0.005$. There is also a significant difference in the percentage of dead cells between the early-exposure time points (0 versus 0.5 min, 0.5 min versus 5 min, 5 min versus 15 min). In contrast, the later-exposure time points (15 min versus 30 min, 30 min versus 60 min, 60 min versus 120 min) show no significant difference (n.s.) in the percentage of dead cells. (d–f) Automated high-throughput microscopy to quantify time-dependent cellular uptake of fluorescent silica (SiNP30F, 100 μg ml^{-1}) or PsNPs (n/pPsNPs, 100 μg ml^{-1}). ISO-HAS1 endothelial cells were exposed to the nanoparticles suspended in RPMI (filled columns) or RPMI that contained 10% human plasma (open columns) for the indicated time points. Total cellular fluorescence is displayed as arbitrary units (a.u.). Assays were performed in triplicate. Columns show the mean ± s.d. from three independent experiments. Corona formation enhanced early uptake of silica (d) and positively charged pPsNPs (e). Negatively charged nPsNPs were affected less (f). $^{*}P < 0.05$; $^{**}P < 0.01$; $^{***}P < 0.005$; n.s., not significant.

(Fig. 9.3a), activation by the nanoparticles occurred even more rapidly. In contrast, nanoparticles that had been exposed to human plasma for 0.5 min did not activate thrombocytes because of the presence of the plasma protein corona. Prolonged exposure and corona formation times did not affect aggregation further. As a control, addition of collagen demonstrated the viability and reactivity of thrombocytes even in the presence of plasma.

Erythrocytes are present in the blood at high concentrations and are one of the first cell types to come into contact with nanoparticles that enter the blood system [17]. In contrast to thrombocytes (Fig. 9.3a) [34], assembly of the plasma protein corona on silica nanoparticles had no effect on erythrocyte aggregation (Fig. 9.3b and Supplementary Fig. S11). The rapid corona formation, however, efficiently prevented haemolysis, which only occurred when human erythrocytes were exposed to pristine nanoparticles (Fig. 9.3b).

Next, we analysed the time-dependent impact of corona formation on nanoparticle-induced toxicity. Silica nanoparticles (AmSil30) were precoated with human plasma by exposure for the indicated time period, separated from plasma components (Supplementary Fig. S2), washed and resuspended in protein-free medium. When primary human endothelial cells or a microvascular endothelial cell line (ISO-HAS1) were exposed to the corona-coated nanoparticles, we observed a reduction of cell-death induction, which was dependent on corona formation time and was most significant for the early corona (Fig. 9.3c). Coronas formed by prolonged exposure to plasma (>30 min) did not counteract toxicity further, which underlines the biological relevance of early exposure time points.

Nanoparticle uptake is an important determinant for nanopathology [7, 22, 35]. Automated high-throughput and confocal laser scanning microscopy showed that corona formation increased attachment and cellular uptake for fluorescent silica as well as for the pPsNPs in endothelial cells, particularly at early exposure time points (Fig. 9.3d,e and Supplementary Fig. S12). In contrast, uptake of nPsNPs was affected less by corona formation (Fig. 9.3f). Hence, covering nanoparticles with a 'physiological coating' can, indeed, promote their interaction with the cellular uptake machinery, whereas the surface charge of the bare nanoparticles studied was less important [27]. Notably, compared to previous studies that investigated cow serum or human plasma proteins on prolonged exposure in epithelial cells [7, 19, 22], significant differences were observed, which underlines the necessity to dissect molecular mechanisms responsible for early as well as late exposure effects.

As there is a current debate as to whether corona formation affects the dispersion state of nanoparticles and subsequent nanobiological effects, we analysed the nanoparticle dispersity on exposure to plasma. These measurements indicate that the studied nanoparticles were mostly monodisperse at early exposure time points (<10 min, Table 9.1), albeit aggregates started to form on prolonged plasma exposure (>30 min, Supplementary Table S1). However, the size polydispersity of the aggregates was narrow, which rather indicates monodispersity of the formed morphologies. The extent of aggregation was dependent on the physicochemical properties of the respective nanoparticles, and similar observations are reported for most nanoparticles studied so far [7]. Although we conclude from our data that the early biological effects observed were triggered by single nanoparticles or small nanoparticle oligomers, we cannot exclude completely the impact of larger nanoparticle aggregates, particularly at later exposure time points. In general, nanoparticle-induced biological effects are most probably caused by a mixture of single nanoparticles and nanoparticle aggregates in physiological environments.

Collectively, depending on the individual biological characteristics of the respective cell type studied, clearly the rapid formation of the plasma protein corona is capable of kinetically modulating distinct pathophysiological effects. Mechanistically, these may be mediated by the modulation of cellular signalling events by bioactive corona components.

Although nanoparticle-induced pathological effects, including cell death, coagulation or thrombosis, have been described for the blood system [7, 17], neither distinct nanoparticle characteristics nor corona proteins causally involved have been identified so far. In contrast to the current view, which assumes that the protein corona consists of only a few tens of proteins, the corona complexity uncovered here precludes a rapid experimental testing of all potential bioactive corona components.

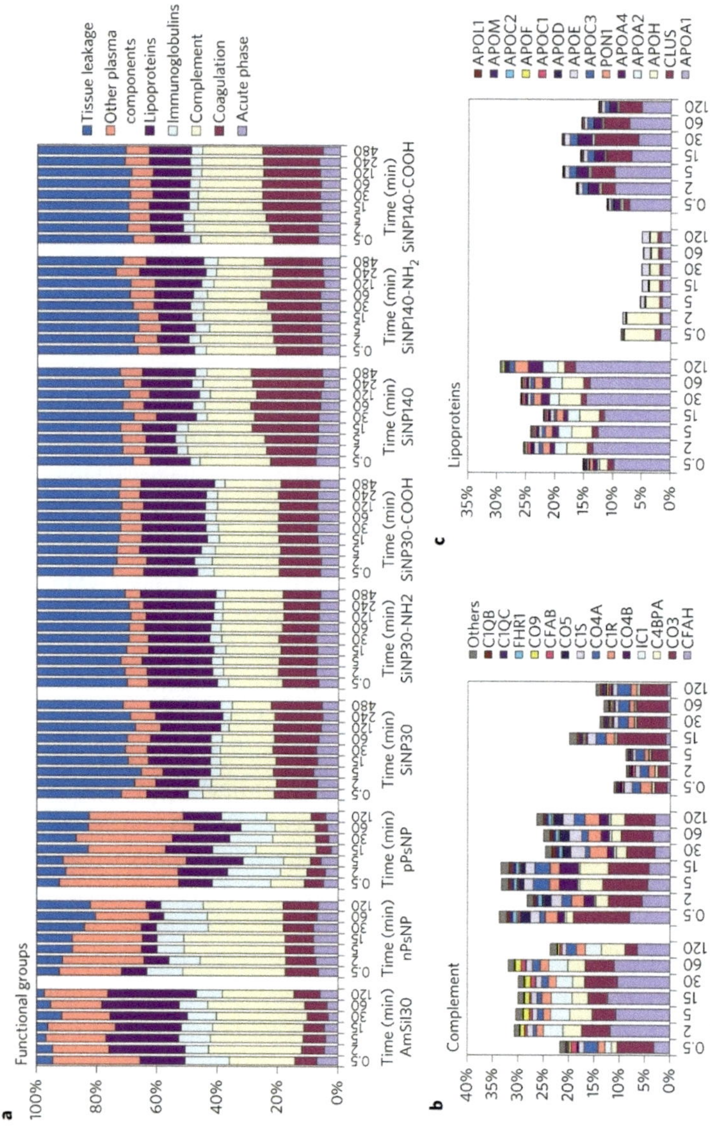
a
Functional groups
100%
80%
60%
40%
20%
0%
Time (min) AmSil30
Time (min) nPsNP
Time (min) pPsNP
Time (min) SiNP30
Time (min) SiNP30-NH2
Time (min) SiNP30-COOH
Time (min) SiNP140
Time (min) SiNP140-NH$_2$
Time (min) SiNP140-COOH
Tissue leakage
Other plasma components
Lipoproteins
Immunoglobulins
Complement
Coagulation
Acute phase
b
Complement
40%
35%
30%
25%
20%
15%
10%
5%
0%
Time (min) AmSil30
Time (min) nPsNP
Time (min) pPsNP
Others
C1QB
C1QC
FHR1
CO9
CFAB
CO5
C1S
CO4A
C1R
CO4B
IC1
C4BPA
CO3
CFAH
c
Lipoproteins
35%
30%
25%
20%
15%
10%
5%
0%
Time (min) AmSil30
Time (min) nPsNP
Time (min) pPsNP
APOL1
APOM
APOC2
APOF
APOC1
APOD
APOE
APOC3
PON1
APOA4
APOA2
APOH
CLUS
APOA1

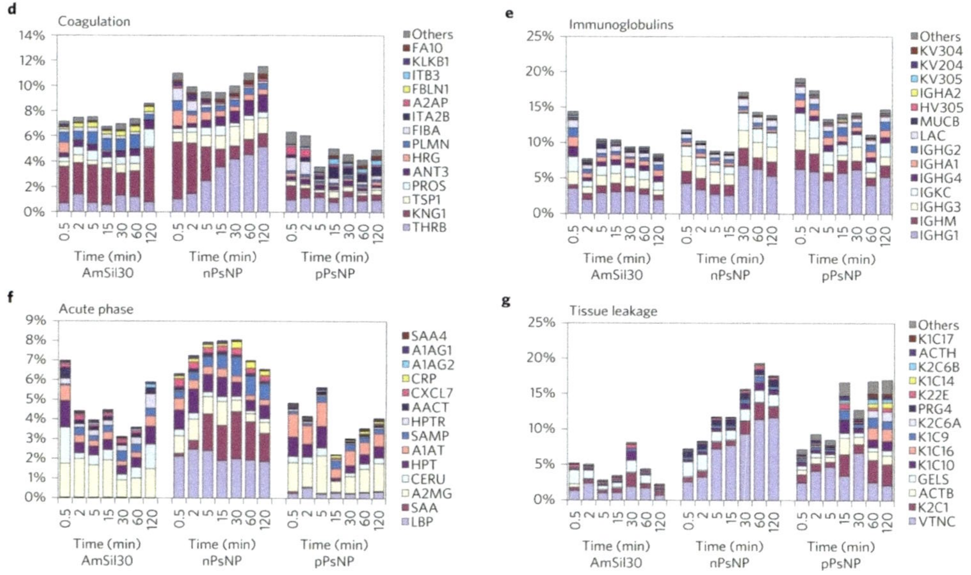

Figure 9.4 Bioinformatic classification of corona components. (a) Proteins identified in the respective nanoparticles' coronas by quantitative LC-MS at the indicated time points were grouped according to biological processes of the blood system. The relative percentages of total proteins are shown. In addition to physicochemical nanoparticle characteristics, incubation time significantly affected the corona composition for all functional groups and particle types investigated. (b–g) The concentration of proteins involved in complement activation (b), coagulation (d), acute phase (f) and tissue leakage (g) were particularly strongly affected by plasma exposure time. Lipoproteins bound preferentially to silica nanoparticles (c), whereas enhanced binding of tissue leakage proteins was observed for n/pPsNPs (g). Although immunoglobulins are present in high amounts in the plasma, these proteins displayed only moderate binding to all nanoparticles (e). Detailed values for all individual proteins are available in Supplementary Tables S3 and S4.

However, to facilitate their continuative rational identification, corona proteins were classified bioinformatically according to their physiological molecular functions (Fig. 9.4a). Not only nanoparticle characteristics, but also plasma exposure time clearly affected the abundance of factors associated with processes such as complement activation, cell death, immune response [36], coagulation and acute-phase response (Fig. 9.4b–g and Supplementary Figs. S13 and S14). Our quantitative analysis also allowed calculations of protein copy number/nanoparticle (Fig. 9.1e, Supplementary Fig. S15 and Tables S6 and S7), and thereby facilitated rational selection of candidate proteins for subsequent studies [1, 8, 11, 37]. Assuming a random distribution, low-abundance proteins (<1 molecule/nanoparticle) are less likely to trigger nanoparticle-induced biological responses. Most identified proteins were, however, present at >1 molecule/nanoparticle, which underlines their relevance. Supplementary Table S8 lists the most-abundant coagulation-relevant corona proteins for each nanoparticle that contain not only factors that promote, but also those that counteract coagulation cascades [17]. Whereas kininogen-1, which plays a central role for the function of the kallikrein–kinin–kininogen system [38], remains predominantly adsorbed on silica nanoparticles, on polystyrene particles it was replaced rapidly by prothrombin or integrin α-IIb.

A functionally diverse group of plasma proteins highly enriched in the coronas of silica nanoparticles are the apolipoproteins, involved in lipid and cholesterol transport (Fig. 9.4c and Supplementary Figs. S13 and S14). Particularly, the amounts of apolipoprotein E and A-I were enriched significantly in the coronas at 2 min of exposure to plasma compared to 0.5 min (Supplementary Tables S3 and S4), and both apolipoproteins are reported to promote nanoparticle uptake into endothelial cells [39–41]. To provide additional evidence for the biological relevance of the kinetic corona modulation, silica nanoparticles were corona coated by exposure to plasma for 0.5 min or 2 min, washed and resuspended in protein-free medium. When primary human endothelial cells were exposed to the precoated nanoparticles, increased cellular uptake was observed for the 2 min corona-coated nanoparticles. Immunoblot analysis verified a higher concentration of apolipoprotein E in the corona, which suggests that corona apolipoproteins are, indeed, biologically active (Supplementary Fig. S16).

Conclusions

We demonstrate here that in a physiological relevant environment, highly complex protein coronas are established rapidly, kinetically change only quantitatively and modulate early pathobiological effects at the nanoparticle–biological interface, which has important implications for nanomedicine, nanotoxicology and nanoecology. In contrast to the current view [7], pristine nanoparticles may interact with cellular components of the blood system only for a short time prior to being coated with a complex protein corona (Supplementary Fig. S17); this cannot be explained by the current simple core–shell models [7, 29–32]. As evolution of the biomolecule corona can, indeed, modulate nanoparticle-induced biological processes, this 'natural functionalization' with bioactive proteins may be exploited by a rational design of nanomaterials to optimize biotechnological applications and reduce potential nanohazards. Quantitative proteomic data are needed as a resource for model building, to predict and simulate the kinetics of protein-binding to nanosized objects in physiological systems.

Methods

Reagents were purchased from Sigma Aldrich and Nycomed.

Nanoparticles. PsNPs and different silica nanoparticles were synthesized or purchased (silica nanoparticles, Kisker-Products; Am-Sil30, Nyacol Nano Technologies). Their average size, polydispersity index (PDI) and zeta potential were characterized by TEM, DLS and zeta potential measurements as reported [15, 21] (details in the Supplementary Information).

Plasma protein binding. Human plasma samples were obtained from the citrated blood of 15 healthy individuals (according to institutional bioethics approval), pooled, stored in liquid nitrogen and used for all experiments comprising proteomics as well as cell studies [15]. Nanoparticles in buffer-A [15] were incubated with a tenfold excess of human plasma for different time points, loaded onto a sucrose cushion (0.7 M in buffer-A) and centrifuged through the cushion to separate nanoparticle–protein complexes from plasma (20 min at 15,300*g*, 4°C; Supplementary Fig. S2). Pellets were washed three times with buffer-A, and proteins eluted by adding polyacrylamide gel electrophoresis (PAGE)/LC-MS buffer as described previously [15].

Bioinformatics and statistical analysis. Hierarchical cluster analysis was performed based on absolute label-free protein quantification results provided by ISOQUANT [15]. Protein and sample dendrograms were constructed by agglomerative average-linkage clustering using one-degree Pearson correlation of protein $\log_2$ ratios as the distance metric. For both clustering approaches, Pearson correlation was determined based on pairwise complete observations. Heat maps of protein $\log_2$ ratios were plotted using the heatmap.2 function from *R*-package gplots, based on previously calculated protein–sample dendrograms. For experiments stating *P* values, a paired Student's *t*-test was performed as described previously [15], assuming significance at $P < 0.05$.

Thrombocyte preparation and aggregometry. Preparation of thrombocytes and aggregometry measurements were performed as described previously [42]. Briefly, blood was collected and centrifuged (100*g*, 10 min) to obtain plasma-containing thrombocytes (PCTs). PCTs were adjusted with Tyrode's buffer (137 mM NaCl, 12 mM $NaHCO_3$, 2.6 mM KCl, 5.5 mM glucose, 1 g l^{-1} BSA, pH 7.4) to a concentration of 2×10^8 ml^{-1}. For preparation of plasma-free thrombocytes (PFTs), Tyrode's buffer (pH 6.5) was added to PCTs, thrombocytes were collected by centrifugation (1,200*g*, 10 min), washed and resuspended in Tyrode's buffer (pH 7.4) at a final concentration of 2×10^8 ml^{-1}. Measurements were performed on an APACT 4S PLUS Aggregometer (Greiner) in a total volume of 200 μl that contained 20 μl AmSil30 nanoparticles at final concentrations of 600 μg ml^{-1} or buffer, directly added to the platelets (3.6×10^7) prior to measurements. For PFTs, 1 mM $CaCl_2$ was added before measurements. As agonist, collagen (0.7 μg ml^{-1}; Nycomed) was used as the positive control. Buffer/plasma without nanoparticles served as the negative control.

Immunoblotting. Immunoblotting was carried out as described previously [15]. Antibodies were α-actin (Sigma Aldrich), α-apoA4 (Cell Signaling), α-apoE, α-clusterin, α-gelsolin, α-prothrombin, goat- α-rabbit immunoglobulin G (IgG) Ab and rabbit- α-mouse IgG Ab conjugated with horseradish peroxidase (Santa Cruz), α-kininogen-1 and α-serum amyloid A (Abcam).

Cultured cell experiments. The human microvascular endothelial cell line (ISOHAS1) and human umbilical vein endothelial cells (Lonza) were isolated and maintained as described previously [43] (details in the Supplementary Information). Cell vitality was assessed by the AlamarBlue assay as described previously [44]. Alternatively, the impact of nanoparticles exposure on cell vitality was assessed by using the LIVE/DEAD viability/cytotoxicity kit (Invitrogen) and the Target Activation V4 assay (Thermo Fisher) as outlined previously [45].

Confocal laser scanning microscopy. This was employed to determine the cellular localization of nanoparticles as described [44]. Images were acquired using Leica LAS AF software on a Leica SP5 II system (Leica). Further experimental details are provided in the Supplementary Information.

Automated high-content microscopy. This was employed to quantify nanoparticles uptake using the automated ArrayScanVTI imaging platform (Thermo Fisher) as described previously [45]. Briefly, ISO-HAS1 cells were seeded into black-walled 96-well thin-bottom μClear plates (Greiner)and further cultivated for 24 h. Cells were washed with PBS and either protein-free RPMI1640 medium or RPMI1640 containing 10% human plasma was added, and cells were exposed to 100 μg ml^{-1} fluorescent nanoparticles. Subsequently, cells were washed with PBS, fixed for 15 min with 4% paraformaldehyde and cell nuclei stained with Hoechst 33342. To quantify the amount of cell-associated nanoparticles, images were analysed using Target Activation V4 assay [45]. For every cell, a binary image mask was created from the Hoechst 33342 staining signal to define the region of interest, marking the nucleus. In the second channel (red/green fluorescence), this circular mask was dilated (four pixels) to cover the whole cell, and the intensity of the red fluorescence signal was quantified within this mask. Scans were performed with settings to give subsaturating fluorescence intensity. A minimum of 1,000 cells well^{-1} was recorded. PBS-treated cells served as a negative control to correct for background fluorescence.

Haemocompatibility assays. The impact of nanoparticles on erythrocyte aggregation and haemolysis was analysed as described previously [46]. Experimental details are provided in the Supplementary Information.

Protein separation and identification. Gel electrophoresis and mass spectrometry including data processing were performed as described previously [15, 47]. Further experimental details are provided in the Supplementary Information.

Supplementary Information

Available at: https://www.nature.com/articles/nnano.2013.181#Sec17

Acknowledgements

This work was supported by Grants DFG-SPP1313 and DFG-SFB490/Z3 and by BMBFNanoKon/NanoMed/MRCyte, Zeiss-ChemBioMed, University Mainz Forschungszentrum Immunologie, Research Center for Immunology and Stiftung Rheinland-Pfalz (NANOSCH, NanoScreen). We thank R. Spohrer for preparation of the manuscript sample, S. Schneider for technical assistance and R. Zellner for discussions.

Author Contributions

R.H.S., S.T. and D.D. conceived and designed the study. S.T., D.D., J.K., V.F., R.H., F.S., K.K. and S.K.K. performed the experiments. A.M.,

F.S., D.F., K.L. and M.M. synthesized and characterized nanoparticles. S.T., J.K., D.F., C.R., K.L., H.S., M.M., S.K.K. and R.H.S. analysed data. R.H.S., S.K.K. and S.T. wrote the manuscript and Supplementary Information.

Competing Financial Interests

The authors declare no competing financial interests.

References

1. Riehemann, K. et al. Nanomedicine—challenge and perspectives. *Angew. Chem. Int. Ed.* **48**, 872–897 (2009).
2. Chauhan, V. P. et al. Normalization of tumour blood vessels improves the delivery of nanomedicines in a size-dependent manner. *Nature Nanotech.* **7**, 383–388 (2012).
3. von Maltzahn, G. et al. Nanoparticles that communicate in vivo to amplify tumour targeting. *Nature Mater.* **10**, 545–552 (2012).
4. Minchin, R. Nanomedicine: sizing up targets with nanoparticles. *Nature Nanotech.* **3**, 12–13 (2008).
5. Teli, M. K., Mutalik, S., and Rajanikant, G. K. Nanotechnology and nanomedicine: going small means aiming big. *Curr. Pharm. Des.* **16**, 1882–1892 (2010).
6. Smita, S. et al. Nanoparticles in the environment: assessment using the causal diagram approach. *Environ. Health* **11** (suppl. 1), S13 (2012).
7. Monopoli, M. P., Aberg, C., Salvati, A., and Dawson, K. A. Biomolecular coronas provide the biological identity of nanosized materials. *Nature Nanotech.* **7**, 779–786 (2012).
8. Nel, A. E. et al. Understanding biophysicochemical interactions at the nano–bio interface. *Nature Mater.* **8**, 543–557 (2009).
9. Nystrom, A. M., and Fadeel, B. Safety assessment of nanomaterials: implications for nanomedicine. *J. Control. Release* **161**, 403–408 (2012).
10. Thomas, C. R. et al. Nanomaterials in the environment: from materials to high-throughput screening to organisms. *ACS Nano* **5**, 13–20 (2011).
11. Oberdörster, G. Safety assessment for nanotechnology and nanomedicine: concepts of nanotoxicology. *J. Intern. Med.* **267**, 89–105 (2011).

12. Xia, X. R., Monteiro-Riviere, N. A., and Riviere, J. E. An index for characterization of nanomaterials in biological systems. *Nature Nanotech.* **5**, 671–675 (2010).
13. Monopoli, M. P., Bombelli, F. B., and Dawson, K. A. Nanobiotechnology: nanoparticle coronas take shape. *Nature Nanotech.* **6**, 11–12 (2011).
14. Monopoli, M. P. et al. Physical-chemical aspects of protein corona: relevance to in vitro and in vivo biological impacts of nanoparticles. *J. Am. Chem. Soc.* **133**, 2525–2534 (2011).
15. Tenzer, S. et al. Nanoparticle size is a critical physicochemical determinant of the human blood plasma corona: a comprehensive quantitative proteomic analysis. *ACS Nano* **5**, 7155–7167 (2011).
16. Zhang, H. et al. Quantitative proteomics analysis of adsorbed plasma proteins classifies nanoparticles with different surface properties and size. *Proteomics* **11**, 4569–4577 (2011).
17. Dobrovolskaia, M. A., Germolec, D. R., and Weaver, J. L. Evaluation of nanoparticle immunotoxicity. *Nature Nanotech.* **4**, 411–414 (2009).
18. Salvati, A. et al. Transferrin-functionalized nanoparticles lose their targeting capabilities when a biomolecule corona adsorbs on the surface. *Nature Nanotech.* **8**, 137–143 (2013).
19. Casals, E., Pfaller, T., Duschl, A., Oostingh, G. J., and Puntes, V. Time evolution of the nanoparticle protein corona. *ACS Nano* **4**, 3623–3632 (2010).
20. Jaskiewicz, K. et al. Probing bioinspired transport of nanoparticles into polymersomes. *Angew. Chem. Int. Ed.* **51**, 4613–4617 (2012).
21. Lunov, O. et al. Differential uptake of functionalized polystyrene nanoparticles by human macrophages and a monocytic cell line. *ACS Nano* **5**, 1657–1669 (2011).
22. Lesniak, A. et al. Effects of the presence or absence of a protein corona on silica nanoparticle uptake and impact on cells. *ACS Nano* **6**, 5845–5857 (2012).
23. Van Hoecke, K. et al. Ecotoxicity and uptake of polymer coated gold nanoparticles. *Nanotoxicology* **7**, 37–47 (2013).
24. Cedervall, T. et al. Understanding the nanoparticle–protein corona using methods to quantify exchange rates and affinities of proteins for nanoparticles. *Proc. Natl Acad. Sci. USA* **104**, 2050–2055 (2007).
25. Walczyk, D., Bombelli, F. B., Monopoli, M. P., Lynch, I., and Dawson, K. A. What the cell 'sees' in bionanoscience. *J. Am. Chem. Soc.* **132**, 5761–5768 (2010).

26. Lundqvist, M. et al. Nanoparticle size and surface properties determine the protein corona with possible implications for biological impacts. *Proc. Natl Acad. Sci. USA* **105**, 14265–14270 (2008).
27. Huhn, D. et al. Polymer-coated nanoparticles interacting with proteins and cells: focusing on the sign of the net charge. *ACS Nano* **7**, 3253–3263 (2013).
28. Vroman, L. Effect of absorbed proteins on the wettability of hydrophilic and hydrophobic solids. *Nature* **196**, 476–477 (1962).
29. Lundqvist, M. et al. The evolution of the protein corona around nanoparticles: a test study. *ACS Nano* **5**, 7503–7509 (2011).
30. Gebauer, J. S. et al. Impact of the nanoparticle–protein corona on colloidal stability and protein structure. *Langmuir* **28**, 9673–9679 (2012).
31. Dell'Orco, D., Lundqvist, M., Oslakovic, C., Cedervall, T., and Linse, S. Modeling the time evolution of the nanoparticle–protein corona in a body fluid. *PLoS One* **5**, e10949 (2010).
32. Calzolai, L., Franchini, F., Gilliland, D., and Rossi, F. Protein–nanoparticle interaction: identification of the ubiquitin–gold nanoparticle interaction site. *Nano Lett.* **10**, 3101–3105 (2010).
33. Vogler, E. A. Protein adsorption in three dimensions. *Biomaterials* **33**, 1201–1237 (2012).
34. Hamad, I. et al. Distinct polymer architecture mediates switching of complement activation pathways at the nanosphere–serum interface: implications for stealth nanoparticle engineering. *ACS Nano* **4**, 6629–6638 (2010).
35. Lunov, O. et al. Lysosomal degradation of the carboxydextran shell of coated superparamagnetic iron oxide nanoparticles and the fate of professional phagocytes. *Biomaterials* **31**, 9015–9022 (2010).
36. Sim, R. B., and Wallis, R. Surface properties: immune attack on nanoparticles. *Nature Nanotech.* **6**, 80–81 (2011).
37. Rivera Gil, P., Oberdörster, G., Elder, A., Puntes, V., and Parak, W. J. Correlating physico-chemical with toxicological properties of nanoparticles: the present and the future. *ACS Nano* **4**, 5527–5531 (2011).
38. Schousboe, I., and Nystrom, B. High molecular weight kininogen binds to laminin—characterization and kinetic analysis. *FEBS J.* **276**, 5228–5238 (2009).
39. Zensi, A. et al. Human serum albumin nanoparticles modified with apolipoprotein A-I cross the blood–brain barrier and enter the rodent brain. *J. Drug Target.* **18**, 842–848 (2010).

40. Zensi, A. et al. Albumin nanoparticles targeted with Apo E enter the CNS by transcytosis and are delivered to neurones. *J. Control. Release* **137**, 78–86 (2009).
41. Aggarwal, P., Hall, J. B., McLeland, C. B., Dobrovolskaia, M. A., and McNeil, S. E. Nanoparticle interaction with plasma proteins as it relates to particle biodistribution, biocompatibility and therapeutic efficacy. *Adv. Drug Deliv. Rev.* **61**, 428–437 (2009).
42. Reinhardt, C. et al. Tissue factor and PAR1 promote microbiota-induced intestinal vascular remodelling. *Nature* **483**, 627–631 (2011).
43. Engels, K. et al. NO signaling confers cytoprotectivity through the surviving network in ovarian carcinomas. *Cancer Res.* **68**, 5159–5166 (2008).
44. Knauer, S. K. et al. Functional characterization of novel mutations affecting survivin (BIRC5)-mediated therapy resistance in head and neck cancer patients. *Hum. Mutat.* **34**, 395–404 (2012).
45. Bier, C. et al. Allosteric inhibition of Taspase1's pathobiological activity by enforced dimerization in vivo. *FASEB J.* **26**, 3421–3429 (2012).
46. Bauer, M. et al. Poly(2-ethyl-2-oxazoline) as alternative for the stealth polymer poly(ethylene glycol): comparison of in vitro cytotoxicity and hemocompatibility. *Macromol. Biosci.* **12**, 986–998 (2012).
47. Tenzer, S. et al. Proteome-wide characterization of the RNA-binding protein RALY-interactome using the in vivo-biotinylation-pulldown-quant (iBioPQ) approach. *J. Proteome Res.* **12**, 2869–2884 (2013).

Chapter 10

Normalization of Tumour Blood Vessels Improves the Delivery of Nanomedicines in a Size-Dependent Manner

Vikash P. Chauhan,[a,b,*] Triantafyllos Stylianopoulos,[a,c,*] John D. Martin,[a,d] Zoran Popović,[e] Ou Chen,[e] Walid S. Kamoun,[a] Moungi G. Bawendi,[e] Dai Fukumura,[a] and Rakesh K. Jain[a]

[a] *Edwin L. Steele Laboratory, Department of Radiation Oncology, Massachusetts General Hospital and Harvard Medical School, Boston, Massachusetts 02114, USA*
[b] *Harvard School of Engineering and Applied Sciences, Harvard University, Cambridge, Massachusetts 02138, USA*
[c] *Department of Mechanical and Manufacturing Engineering, University of Cyprus, Nicosia 1678, Cyprus*
[d] *Department of Chemical Engineering, Massachusetts Institute of Technology, Cambridge, Massachusetts 02139, USA*
[e] *Department of Chemistry, Massachusetts Institute of Technology, Cambridge, Massachusetts 02139, USA*
jain@steele.mgh.harvard.edu

The blood vessels of cancerous tumours are leaky [1–3] and poorly organized [4–7]. This can increase the interstitial fluid pressure inside tumours and reduce blood supply to them, which impairs drug delivery [8, 9]. Anti-angiogenic therapies—which

*These authors contributed equally to this work.
Reproduced with permission from *Nature Nanotechnology*, **7**, 383–388, 2012.

Nano-Enabled Medical Applications
Edited by Lajos P. Balogh

ISBN 978-981-4800-32-7 (Hardcover), 978-0-429-39903-9 (eBook)
www.jennystanford.com

'normalize' the abnormal blood vessels in tumours by making them less leaky—have been shown to improve the delivery and effectiveness of chemotherapeutics with low molecular weights [10], but it remains unclear whether normalizing tumour vessels can improve the delivery of nanomedicines. Here, we show that repairing the abnormal vessels in mammary tumours, by blocking vascular endothelial growth factor receptor-2, improves the delivery of smaller nanoparticles (diameter, 12 nm) while hindering the delivery of larger nanoparticles (diameter, 125 nm). Using a mathematical model, we show that reducing the sizes of pores in the walls of vessels through normalization decreases the interstitial fluid pressure in tumours, thus allowing small nanoparticles to enter them more rapidly. However, increased steric and hydrodynamic hindrances, also associated with smaller pores, make it more difficult for large nanoparticles to enter tumours. Our results further suggest that smaller (~12 nm) nanomedicines are ideal for cancer therapy due to their superior tumour penetration.

Elevated tumour interstitial fluid pressure (IFP) hinders drug delivery by abolishing the fluid pressure gradients that lead to rapid convective (flow-driven) penetration into tumours [11]. This limits drug penetration across vessel walls into tumours (transvascular) and through tumour tissue (interstitial) to slow diffusion [8]. Antiangiogenic therapies can repair tumour vessel abnormalities such as the large heterogeneous pores that facilitate leakiness by inducing vessel maturation [12, 13]. This 'vascular normalization' reduces the IFP to induce convective penetration of molecules up to the size of immunoglobulin G (IgG, ~11 nm) (Supplementary Tables 10.S1 and 10.S2) [12–14]. With this normalization process, anti-angiogenic therapies seem to benefit patients with colorectal [15] and brain tumours [16, 17], potentially through improved drug delivery, reduced chemoresistance and immune reprogramming [10]. Whether normalizing vessels can improve the delivery of nanomedicines (ranging in size from 10 to 125 nm) is not known. These slow-diffusing large therapeutics are providing new hope for cancer treatment [18, 19] and would greatly benefit from convective delivery. Unfortunately, the increased hydrodynamic and steric hindrance resulting from the smaller vessel pores caused by normalization may compromise the advantage gained from enhanced convection.

To determine how vascular normalization affects nanomedicine delivery, we studied whether the vascular endothelial growth factor (VEGF) receptor-2 blocking antibody DC101 modulates nanoparticle penetration rates in orthotopic mammary tumours in vivo. We used intravital multiphoton microscopy and as probes used a system of quantum dot-based nanoparticles with tunable size and fluorescence emission wavelength but identical surface chemistry [20]. With these tools we measured the effects of DC101 on the real-time delivery of particles ranging from 12 to 125 nm in diameter (the range for approved nanomedicines). Using the resulting data, we applied a novel physiologically based mathematical model for drug delivery to tumours to determine how anti-angiogenics affect pore size distributions. We also used this model to study how pore size distributions can be therapeutically modulated to optimize the delivery of different sizes of nanomedicines.

We found that a 5 mg kg^{-1} dose of DC101 transiently decreases vessel diameter in orthotopic E0771 tumours, consistent with the structural vascular normalization 'window' measured in mice [12, 13] and patients [16] (Supplementary Fig. 10.S1). We imaged nanoparticle delivery with or without DC101 in orthotopic 4T1 and E0771 mammary tumours (Fig. 10.1a, Supplementary Fig. 10.S2). We quantified nanoparticle penetration rates as transvascular mass flux per unit vascular surface area and transvascular concentration difference, often termed the effective permeability [21]. Vascular normalization with 10 mg kg^{-1} DC101 led to a 3.1-fold enhancement of transvascular flux in 4T1 tumours for the smallest (12 nm) nanoparticles, with no improvement in penetration for the larger (60 nm and 125 nm) nanoparticles (Fig. 10.1b). Similarly, 5 mg kg^{-1} DC101 led to 2.7-fold improvement in transvascular flux in E0771 tumours for the 12 nm particles, with no enhancement for the larger nanoparticles (Fig. 10.1c). Some individual control tumours exhibited zero or near-zero penetration rates for the larger nanoparticles, with DC101 therapy apparently shutting down large nanoparticle delivery in several tumours (Fig. 10.1b,c). Consistent with the structural normalization 'window', we characterized a

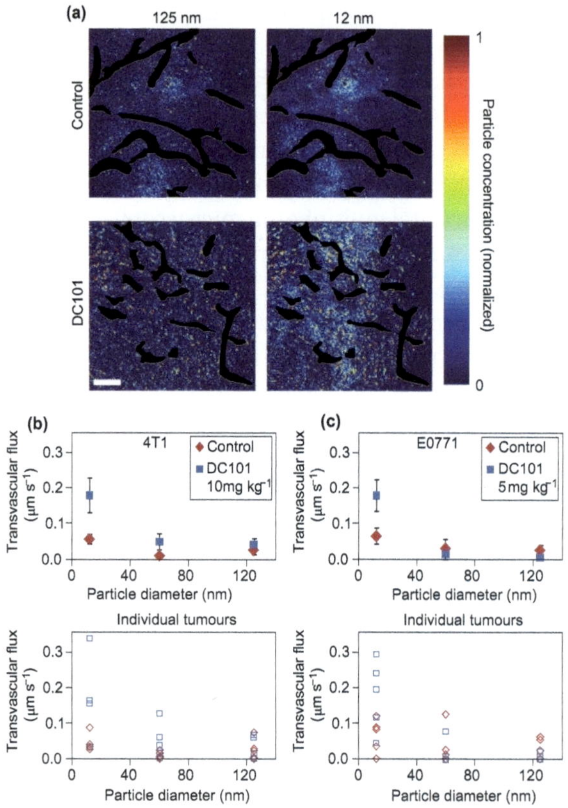

Figure 10.1 Effects of vascular normalization on nanoparticle delivery in tumours. (a) Nanoparticle penetration versus particle size in orthotopic 4T1 mammary tumours in response to normalizing therapy with DC101. Nanoparticle concentrations (denoted by pseudocolour) are relative to initial intravascular levels, with vessels shown in black. Normalization improves 12 nm particle penetration while not affecting 125 nm penetration. Scale bar, 100 μm. (b,c) Penetration rates (transvascular flux) for nanoparticles in orthotopic 4T1 (b) and E0771 (c) mammary tumours in mice treated with 10 mg kg^{-1} or 5 mg kg^{-1} DC101, respectively. Filled symbols (top) denote averages by mouse, and open symbols (bottom) are individual tumours. Normalization improves the transvascular flux of 12 nm particles on day 2 by a factor of 3.1 in 4T1 ($P = 0.042$, Student's t-test) and 2.7 in E0771 ($P = 0.049$, Student's t-test), while not improving delivery for larger nanoparticles. Normalization also reduces the flux of larger nanoparticles to zero in several individual tumours. Animal number, $n = 5$ for all groups.

functional normalization window for nanoparticle delivery. During treatment of mice bearing E0771 tumours with 5 mg kg^{-1} DC101 every three days, the transvascular flux of 12 nm particles was enhanced on days 2 and 5 but returned to baseline levels by day 8 (Fig. 10.2). Importantly, a 10 mg kg^{-1} dose seemed to hinder nanoparticle delivery in E0771 tumours (Supplementary Fig. 10.S3), suggesting a need for judicious dosing based on vascular sensitivity in each tumour and host.

To study how changes in vascular pore size distribution can bring about this complex size-dependent improvement in nanoparticle penetration rates, we developed a mathematical model of drug delivery to tumours (for details see Supplementary Information). The tumour vasculature was represented by a two-dimensional percolation network with one inlet and one outlet, which has been shown to resemble the vascular structure and function of tumours (Fig. 10.3a) [6, 22]. This involved a series of interconnected nodes representing vessel segments. Each node was assigned a pore size, assuming a unimodal pore size distribution throughout the tumour vasculature, based on previous studies [1, 23]. We assumed axial Poiseuille-type blood flow [24, 25]. Drug exchange with the interstitial space followed Starling's approximation for both diffusive and convective mass flux [24]. Interstitial drug transport also occurred by diffusion and convection, with the interstitial fluid flow driving convection calculated using Darcy's law. We used pore theory for the transport of spherical particles through cylindrical pores [26, 27] to calculate the hindrances to diffusion and convection for each pore size [24]. We first solved the steady-state fluid problem, requiring the net fluid accumulation at each node to be zero, and determined the microvascular pressure (MVP) and IFP (Supplementary Figs. 10.S4–S6). Subsequently, we solved the transient drug delivery problem and calculated the transvascular flux versus particle size as in the experiment. Model parameters were based on previous studies (Supplementary Tables 10.S3 and 10.S4).

Applying the model, we studied how changes in the mean and heterogeneity (standard deviation) of the vascular pore size distribution affect fluid pressure profiles and drug delivery in tumours. Consistent with previous studies, large heterogeneous

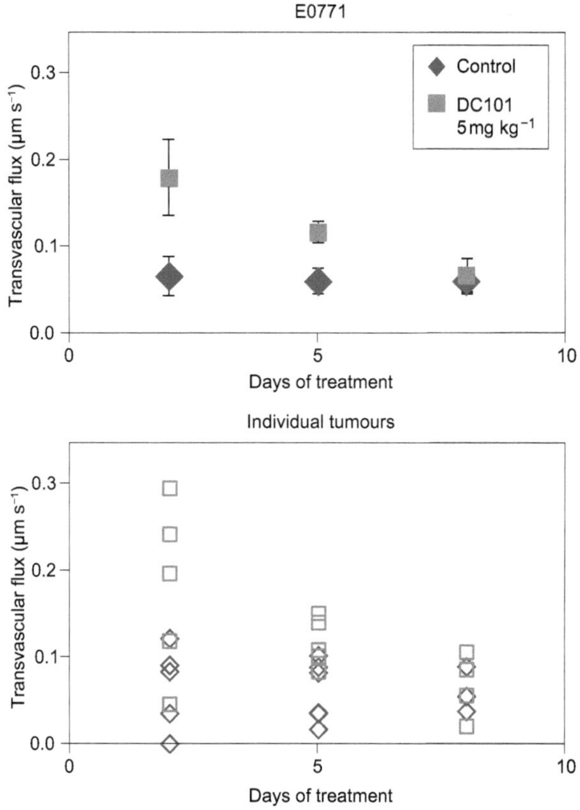

Figure 10.2 Functional vascular normalization window for nanomedicine delivery. Penetration rates (transvascular flux) for 12 nm nanoparticles in orthotopic E0771 mammary tumours. Measurements over an eight-day course of treatment with either 5 mg kg^{-1} DC101 or non-specific rat IgG every three days starting on day 0. Filled symbols (top) denote averages by mouse, and open symbols (bottom) are individual tumours. Treatment with DC101 enhances nanoparticle transvascular flux on days 2 ($P = 0.049$, Student's t-test) and 5 ($P = 0.017$, Student's t-test), with no difference in the treatment groups by day 8. Animal number, $n = 4$–5 for all groups.

pores result in elevated IFP in the centre of the tumour with no transvascular pressure difference, ΔP = MVP – IFP (Fig. 10.3b). Smaller, more homogeneous pores produce low IFP with a non-zero ΔP (the result of vascular normalization) that induces convection (Fig. 10.3b). We simulated the penetration rates for therapeutics ranging from 1 to 250 nm in size while varying the mean pore sizes from 40 to 1,000 nm (diameter) [1] for homogeneous, moderate or heterogeneous pores corresponding to 20 nm, 60 nm or 100 nm pore size standard deviations, respectively (area fractions held constant; Fig. 10.3c). Generally, the smaller therapeutics (1–12 nm) showed the most rapid tumour penetration, with the largest therapeutics (125–250 nm) not leaving the vasculature to any appreciable extent (Fig. 10.3d). Importantly, convection is dominant at small mean pore sizes. Increasing the mean pore size past a certain point (>140 nm) hinders nanoscale therapeutic delivery due to the rising IFP, leading to limited convection and leaving diffusion as the dominant mechanism (Fig. 10.4a). This effect combines with increasing hydrodynamic and steric hindrance to transport as pore sizes approach therapeutic particle sizes (Fig. 10.4b). Therefore, each size of therapeutic has its own ideal mean pore size for maximal delivery to tumours. This ideal mean pore size becomes larger with therapeutic size, although increasing pore size heterogeneity broadens these maxima towards smaller mean pore sizes for therapeutics larger than 12 nm.

To investigate the potential translational implications of these findings, we studied whether vascular normalization enhances the effectiveness of anticancer nanomedicines. We compared two clinically used nanomedicines with widely varying sizes: Doxil, which has a diameter of ~100 nm, and Abraxane, which has a size of ~10 nm following dilution in plasma (Supplementary Fig. 10.S7). Taking into account the five-day normalization window we characterized for nanoparticles, we treated mice bearing orthotopic E0771 mammary tumours with DC101 (5 mg kg^{-1}) or non-specific IgG on days 0 and 3, and with Doxil (2 mg kg^{-1}) or Abraxane (10 mg kg^{-1}) on days 1–5 (Fig. 10.5a). We quantified the effect of each combination therapy based on the time to reach double the initial volume (Fig. 10.5b). Although both Doxil and Abraxane monotherapy induced a similar growth delay, vascular normalization with DC101 enhanced the effectiveness of only the ~10 nm Abraxane, and did not affect that of the ~100 nm Doxil.

Combining our experimental and simulated results, several conclusions can be drawn. The experiments suggest that vascular normalization with anti-angiogenic therapies will only enhance

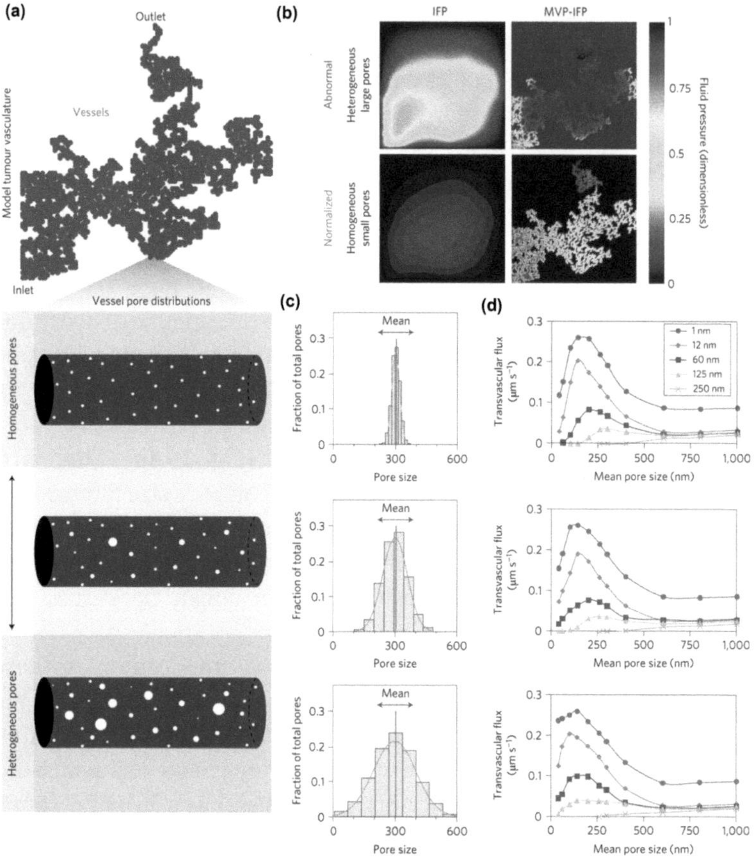

Figure 10.3 Mathematical model predictions of how changes in vascular pore size distribution affect delivery for different sizes of drugs. (a) Model tumour vasculature, formed as a percolation network, with a schematic of vessel pore structure. (b) Effect of pore size distribution on fluid pressure (dimensionless by division by the inlet pressure). Large heterogeneous pores produce an elevated IFP that approaches the MVP, resulting in a near-zero transvascular pressure gradient (MVP–IFP) for central tumour vessels. Small homogeneous pores result in a near-zero IFP and a high transvascular pressure gradient that can drive convective drug delivery. (c) Mean pore size (diameter) and pore size standard deviation are varied to predict how pore size changes affect drug delivery. Three standard deviations, at 20 nm, 60 nm or 100 nm, are selected to represent homogeneous, moderate and heterogeneous pores, respectively. (d) Simulations of transvascular flux versus mean pore size and pore size standard deviation for drugs from 1 to 250 nm in diameter.

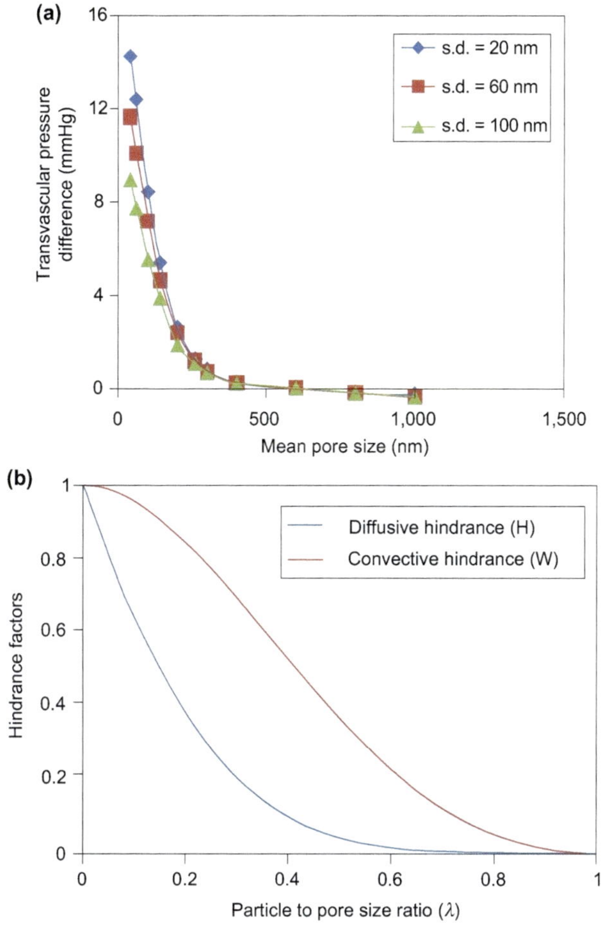

Figure 10.4 Dependence of transvascular pressure gradient and transport hindrance on pore size. (a) Transvascular pressure difference as a function of vessel wall pore size. The plot presents the model predictions for three different standard deviations (s.d. = 20, 60 and 100 nm). The standard deviation of the distribution affects the transvascular pressure difference only for small pore sizes. For pore size distributions with a mean >400 nm the pressure difference, and thus the fluid flux across the vessel wall, is practically zero. (b) Hindrance factors for transport through pores versus particle-to-pore size ratio. The diffusive (H) and convective (W) hindrance factors, which represent hydrodynamic and steric transport hindrance through pores, depend strongly on the particle to pore size ratio (λ). A hindrance factor of 1 indicates no hindrance and a factor of zero denotes no transport whatsoever. Both diffusion and convection are increasingly hindered as particle size approaches pore size.

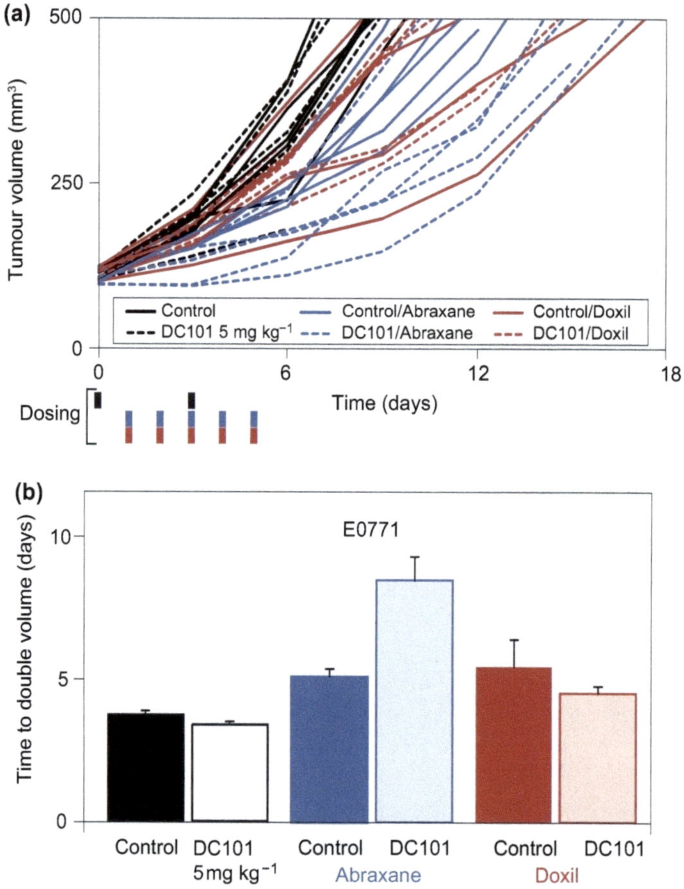

Figure 10.5 Improvement of cytotoxic nanomedicine effectiveness by vascular normalization. (a) Volumes of orthotopic E0771 mammary tumours in response to treatment with DC101 or non-specific rat IgG control (5 mg kg^{-1} on days 0 and 3) in combination with either the ~10 nm nanomedicine Abraxane (10 mg kg^{-1} on days 1–5) or the ~100 nm nanomedicine Doxil (2 mg kg^{-1} on days 1–5). (b) Quantification of tumour growth rates, based on the time to reach double the initial volume. Abraxane ($P = 0.040$, Mann-Whitney U-test) and Doxil ($P = 0.040$, Mann-Whitney U-test) monotherapy both induce growth delays versus the control treatment. Normalization with DC101 enhances the effectiveness of the ~10 nm Abraxane ($P = 0.040$, Mann-Whitney U-test), but does not affect that of the ~100 nm Doxil. Animal number, $n = 4$–5 for all groups.

delivery and effectiveness for relatively small therapeutics, including small-molecule chemotherapeutics, biologics and small nanoparticles. Together with the model, these data support the general, although experimentally unproven, concept that vascular normalization reduces vessel pore sizes (Supplementary Fig. 10.S8). This reduces the IFP, leading to a non-zero ΔP throughout the tumours (an effect seen experimentally [12, 15]), which our model shows can restore convective drug penetration. Unfortunately, smaller and more homogeneous pores can also hinder transport for large therapeutics, resulting in diminished or unimproved tumour penetration for larger nanoparticles in both our experiments and simulations. The simulations also predict that remodelling vessels to increase pore sizes would enhance delivery for only the largest ($>$125 nm) therapeutics, which are size-excluded by small pores. Perhaps most importantly, they also show that tumour penetration rates decrease with increasing nanoparticle size—an effect most pronounced for tumours with smaller vessel pores, as with normalization.

Our findings emphasize the importance of size in nanomedicine design by demonstrating that 12 nm particles penetrate tumours better than larger particles. Physical principles dictate that both diffusive and convective penetration—transvascular and interstitial—are faster for smaller particles [8, 18, 26, 27]. Importantly, most normal organs have a non-sinusoid continuous epithelium that may be either fenestrated or non-fenestrated, with pore cutoff sizes of up to 6–12 nm (ref. 28), suggesting that 12 nm particles are the smallest that can take advantage of the enhanced permeability and retention (EPR) effect that leads to favourable toxicity profiles for nanomedicine [19]. Indeed, the smallest probe demonstrating selective delivery (a plurality of the injected dose reaching the tumour) through passive EPR is the $\sim$11 nm IgG [29]. Although vascular-targeted tumour-penetrating ligands can enhance nanoparticle penetration [30] and tumour cell-targeting can improve uptake and retention [19], targeting ligands cannot fully overcome tumour penetration barriers made worse by large size [8]. Considering the superior mass flux into tumours and long circulation times for small nanoparticles, together with the large number of patients receiving normalizing anti-angiogenics [5], small size may represent an important new design constraint for anticancer nanomedicine.

Methods

Tumour models. Orthotopic mammary tumour models were prepared by implanting a small piece (1 mm^3) of viable tumour tissue from a source tumour animal into severe combined immunodeficient (SCID) mice bearing mammary fat pad chambers [21]. The tumours were allowed to grow to 3 mm in diameter. All animal procedures were carried out following the Public Health Service Policy on Humane Care of Laboratory Animals and approved by the Institutional Animal Care and Use Committee of Massachusetts General Hospital.

Treatment. Mice were treated with 5 or 10 mg kg^{-1} of DC101 (ImClone Systems), using non-specific rat IgG for control treatments, as a 5 mg ml^{-1} solution by intraperitoneal injection on days 0, 3 and 6. Imaging studies were carried out on days 2, 5 and 8.

Nanoparticle synthesis. Nanoparticles were prepared and characterized as described previously [20].

In vivo imaging. A mixture of nanoparticles with diameters of 12 nm (emission, 476 nm), 60 nm (emission, 540 nm) and 125 nm (emission, 625 nm) was prepared for intravenous injection. Concentrations were adjusted by means of in vitro calibration to result in roughly equal photoluminescence intensities for all three nanoparticle samples under 800 nm multiphoton excitation. Following 200 μl retroorbital injections of these concentrations, multiphoton imaging was carried out as described previously [21] on a custom-built multiphoton laser-scanning microscope using a confocal laser-scanning microscope body (Olympus 300; Optical Analysis) and a broadband femtosecond laser source (High Performance MaiTai, Spectra- Physics). Image slices were taken at ~60 mW at the sample surface with depths ranging from 0 to 201 μm, with steps of 2.76 μm and 2.76 × 2.76 μm pixels. Mosaic images were taken in raster pattern using a motorized stage (H101,

Prior Scientific) and customized automation software (LabView, National Instruments). Imaging studies were performed with a × 20 magnification, 0.95 NA water immersion objective (Olympus XLUMPlanFl, 1-UB965, Optical Analysis).

Image analysis. Images were analysed using custom analysis software developed in Matlab (The Mathworks) as described previously [21]. The analysis approach involved three-dimensional vessel tracing to create vessel metrics and a three-dimensional map of voxel intensity versus distance to the nearest vessel over time. Images were also corrected for sample movement over time with three-dimensional image registration. The normalized transvascular flux was calculated using

$$\frac{J_t}{S_v(C_v - C)} = P_{\text{eff}} = \lim_{t\to 0} \frac{\partial \int_{r=R}^{\infty} C(r) r \partial r}{\partial t (C_V - C) R},$$

where J_t is the transvascular flux, S_v the vessel surface area, C_v the concentration of the probe in the vessel, C the concentration of the probe immediately extravascular, P_{eff} the effective permeability [21], t time after the initial image, r the distance from the vessel central axis, and R the vessel radius at that point along the vessel. Fluorescence intensities were used as these concentrations. The calculation was made as an average over the entire imaged volume for each tumour.

Model equations. Details of the model and corresponding equations are provided in the Supplementary Information.

Tumour growth studies. E0771 mammary tumours were orthotopically implanted in female SCID mice. The mice were split into treatment groups, time-matched for time after implantation and size-matched for tumour volume at this time (110–111 mm^3). The mice were treated at this initial size with 5 mg kg^{-1} DC101 or non-specific IgG on days 0 and 3 by intraperitoneal injection. The mice were simultaneously treated with either 2 mg kg^{-1} Doxil (diameter, ~100 nm) or 10 mg kg^{-1} Abraxane (diameter, ~10 nm) on days 1–5 by retro-orbital injection. These relative doses are similar to the relative doses for these two nanomedicines in patients. The primary tumours were then measured every three days, beginning on day

0, using calipers. Tumour growth was quantified using the time for each to reach double its initial volume.

Statistical analysis. The data are presented as means with standard errors. Groups were compared using a Student's *t*-test, except for tumour growth data, for which we anticipated a non-normal distribution and used an (exact) Mann–Whitney U-test. In pairwise comparisons of groups in tumour growth experiments, *P*-values were adjusted using Holm's method.

Acknowledgements

The authors thank J. Kahn and S. Roberge for technical assistance, J. Baish for assistance with the mathematical model and M. Ancukiewicz for assistance with statistical analysis. The authors acknowledge ImClone Systems for generously providing DC101, the National Institutes of Health (P01-CA080124, R01-CA126642, R01-CA115767, R01-CA096915, R01- CA085140, R01-CA098706, T32-CA073479), a DoD Breast Cancer Research Innovator award (W81XWH-10-1-0016) and an FP7 Marie-Curie IRG grant (PIRG08-GA-2010-276894).

Author Contributions

V.P.C. and R.K.J. conceived and designed the experiments. T.S. and R.K.J. designed and developed the mathematical model and its simulations. V.P.C., J.D.M. and O.C. performed the experiments. T.S. carried out the mathematical model simulations. V.P.C., T.S., J.D.M. and W.S.K. analysed the data. Z.P., O.C., W.S.K., M.G.B. and D.F. contributed materials/analysis tools. V.P.C., T.S. and R.K.J. co-wrote the paper. All authors discussed the results and commented on the manuscript.

Additional Information

R.K.J. receives research support from Dyax, MedImmune and Roche, is a consultant for Dyax and Noxxon, is on the Scientific Advisory

Board for Enlight and SynDevRx, is on the Board of Trustees for H&Q Capital Management and is a co-founder of Xtuit. The other authors declare no competing financial interests.

References

1. Hobbs, S. K. et al. Regulation of transport pathways in tumor vessels: role of tumor type and microenvironment. *Proc. Natl Acad. Sci. USA* **95**, 4607–4612 (1998).
2. Nagy, J. A., Dvorak, A. M., and Dvorak, H. F. VEGF-A and the induction of pathological angiogenesis. *Annu. Rev. Pathol.* **2**, 251–275 (2007).
3. Yuan, F. et al. Microvascular permeability and interstitial penetration of sterically stabilized (stealth) liposomes in a human tumor xenograft. *Cancer Res.* **54**, 3352–3356 (1994).
4. Baish, J. W. et al. Scaling rules for diffusive drug delivery in tumor and normal tissues. *Proc. Natl Acad. Sci. USA* **108**, 1799–1803 (2011).
5. Carmeliet, P., and Jain, R. K. Molecular mechanisms and clinical applications of angiogenesis. *Nature* **473**, 298–307 (2011).
6. Gazit, Y., Berk, D. A., Leunig, M., Baxter, L. T., and Jain, R. K. Scale-invariant behavior and vascular network formation in normal and tumor tissue. *Phys. Rev. Lett.* **75**, 2428–2431 (1995).
7. Nagy, J. A., Chang, S. H., Dvorak, A. M., and Dvorak, H. F. Why are tumour blood vessels abnormal and why is it important to know? *Br. J. Cancer* **100**, 865–869 (2009).
8. Chauhan, V. P., Stylianopoulos, T., Boucher, Y., and Jain, R. K. Delivery of molecular and nanoscale medicine to tumors: transport barriers and strategies. *Annu. Rev. Chem. Biomol. Eng.* **2**, 281–298 (2011).
9. Jain, R. K. Normalization of tumor vasculature: an emerging concept in antiangiogenic therapy. *Science* **307**, 58–62 (2005).
10. Goel, S. et al. Normalization of the vasculature for treatment of cancer and other diseases. *Physiol. Rev.* **91**, 1071–1121 (2011).
11. Jain, R. K., and Baxter, L. T. Mechanisms of heterogeneous distribution of monoclonal antibodies and other macromolecules in tumors: significance of elevated interstitial pressure. *Cancer Res.* **48**, 7022–7032 (1988).
12. Tong, R. T. et al. Vascular normalization by vascular endothelial growth factor receptor 2 blockade induces a pressure gradient across the

vasculature and improves drug penetration in tumors. *Cancer Res.* **64**, 3731–3736 (2004).

13. Winkler, F. et al. Kinetics of vascular normalization by VEGFR2 blockade governs brain tumor response to radiation: role of oxygenation, angiopoietin-1, and matrix metalloproteinases. *Cancer Cell* **6**, 553–563 (2004).
14. Nakahara, T., Norberg, S. M., Shalinsky, D. R., Hu-Lowe, D. D., and McDonald, D. M. Effect of inhibition of vascular endothelial growth factor signaling on distribution of extravasated antibodies in tumors. *Cancer Res.* **66**, 1434–1445 (2006).
15. Willett, C. G. et al. Direct evidence that the VEGF-specific antibody bevacizumab has antivascular effects in human rectal cancer. *Nature Med.* **10**, 145–147 (2004).
16. Batchelor, T. T. et al. AZD2171, a pan-VEGF receptor tyrosine kinase inhibitor, normalizes tumor vasculature and alleviates edema in glioblastoma patients. *Cancer Cell* **11**, 83–95 (2007).
17. Sorensen, A. G. et al. Increased survival of glioblastoma patients who respond to anti-angiogenic therapy with elevated blood perfusion. *Cancer Res.* **72**, 402–407 (2012).
18. Jain, R. K., and Stylianopoulos, T. Delivering nanomedicine to solid tumors. *Nature Rev. Clin. Oncol.* **7**, 653–664 (2010).
19. Peer, D. et al. Nanocarriers as an emerging platform for cancer therapy. *Nature Nanotech.* **2**, 751–760 (2007).
20. Popovic, Z. et al. A nanoparticle size series for in vivo fluorescence imaging. *Angew. Chem. Int. Ed.* **49**, 8649–8652 (2010).
21. Chauhan, V. P. et al. Fluorescent nanorods and nanospheres for real-time in vivo probing of nanoparticle shape-dependent tumor penetration. *Angew. Chem. Int. Ed.* **50**, 11417–11420 (2011).
22. Gazit, Y. et al. Fractal characteristics of tumor vascular architecture during tumor growth and regression. *Microcirculation* **4**, 395–402 (1997).
23. Hashizume, H. et al. Openings between defective endothelial cells explain tumor vessel leakiness. *Am. J. Pathol.* **156**, 1363–1380 (2000).
24. Baish, J. W., Netti, P. A., and Jain, R. K. Transmural coupling of fluid flow in microcirculatory network and interstitium in tumors. *Microvasc. Res.* **53**, 128–141 (1997).
25. Pozrikidis, C., and Farrow, D. A. A model of fluid flow in solid tumors. *Ann. Biomed. Eng.* **31**, 181–194 (2003).

26. Bungay, P. M., and Brenner, H. The motion of a closely-fitting sphere in a fluid-filled tube. *Int. J. Multiphase Flow* **1**, 25–56 (1973).
27. Deen, W. M. Hindered transport of large molecules in liquid-filled pores. *AIChE J.* **33**, 1409–1425 (1987).
28. Sarin, H. Physiologic upper limits of pore size of different blood capillary types and another perspective on the dual pore theory of microvascular permeability. *J. Angiogenes. Res.* **2**, 14 (2010).
29. Matsumura, Y., and Maeda, H. A new concept for macromolecular therapeutics in cancer chemotherapy: mechanism of tumoritropic accumulation of proteins and the antitumor agent Smancs. *Cancer Res.* **46**, 6387–6392 (1986).
30. Ruoslahti, E., Bhatia, S. N., and Sailor, M. J. Targeting of drugs and nanoparticles to tumors. *J. Cell Biol.* **188**, 759–768 (2010).

Supplementary Information: Chapter 10

10.S1 Mathematical Model

Equations and assumptions: We represent the tumour vasculature as a percolation network at the percolation threshold, with one inlet and one outlet as shown in Fig. 10.3a in the main text. These network structures have been found to represent tumour vasculature in animal models well [1–2], though real angiography images could be used to represent additional heterogeneity present in unusual vascular networks. We assume that blood flow (Q_{vascular}) is axial and follows Poiseuille's law [3–4], $Q_{\text{vascular}} = \frac{\pi d^4}{128\mu} \nabla p_v$, where d is the vessel diameter, p_v is the vascular pressure, and μ is the blood viscosity. Fluid exchange across the vessel wall ($Q_{\text{transmural}}$) follows Starling's law with no osmotic pressure difference [3, 5], $Q_{\text{transmural}} = L_{\text{p}} S(p_{\text{v}} - p_{\text{i}})$, where L_p is the hydraulic conductivity of the vessel wall, S is the surface area of the vessel, and p_i is the interstitial fluid pressure. The effects of decreased interstitial oncotic pressure, which occurs with vascular normalization [6] and can reduce convection, were not considered due to their relatively small effects compared to fluid pressure changes based on previous studies [7]. Interstitial fluid transport (Q_{tissue}) follows Darcy's law, $Q_{\text{tissue}} = -KA_c \nabla p_i$, where K is the hydraulic conductivity of the interstitial space and A_c is the tissue cross-sectional area. The tissue cross-sectional area can be related to the vascular density, Sv, and the diameter of the vessel, d, by $A_c = \frac{\pi d}{S_V}$ [3]. The inlet microvascular pressure is assumed to be invariable, as has been measured with DC101 therapy [6].

Inside vessels, transport is assumed to be dominated by convection and is governed by $\frac{dc_v}{dt} = Q_{\text{vascular}} c_v$, where c_v is the intravascular concentration of the nanoparticle. In the interstitial space, transport can be both diffusive and convective such that $\frac{dc_i}{dt} = D\nabla^2 c_i - v\nabla c_i$, where c_i is the concentration of the nanoparticle in the interstitial space, D is the diffusion coefficient, and v is the interstitial fluid velocity given by Darcy's law. Neither binding of nanoparticles to cancer cells or extracellular matrix components, nor uptake by cells, is considered, as the effects of these processes on transvascular flux are assumed negligible on short time scales. The hindrance to diffusion from transport around and through cells is taken into account through the use of SFA-FRAP diffusion data, which measures the hindrance from both cells and matrix components due to its length scale [8]. The mass flux, φ, at the interface between vascular and interstitial space is given by Starling's approximation as, $\varphi = L_p$ (p_v - *pi1*- σ*CvPe*–*CiePe* - 1) with *Pe* = *Lp1* – σ *pv* -*piP*, where *Pe* is the Péclet number across the vessel wall, and P is the vascular permeability of the nanoparticle through the pores of the wall. We assume the pores of the vessel walls to be cylindrical and have a unimodal distribution of sizes. We calculate the hydraulic conductivity L_p, vascular permeability P, and reflection coefficient σ, by the size and the surface area of the pores of the vessel walls and the size of the spherical nanoparticles as [9] $L_p = \frac{\gamma r_o^2}{8\mu L}$, $P \frac{\gamma H D_o}{L}$, $\sigma = 1 - W$, where γ is the fraction of wall surface area occupied by pores, r_o is the pore radius, L is the thickness of the vessel wall, D_o is the diffusion coefficient of the particle in free solution at 37°C given by the Stokes–Einstein relationship, and the diffusive, H, and convective, W, hindrance factors are given in Bungay and Brenner [10]. These hindrance factors depend on the relative size of the particles to the pores (Fig. 10.4b).

Solution strategy: For the solution strategy we employ a two-dimensional percolation network with one inlet and one outlet to represent the structure of the tumour vasculature [1–2] (Fig. 10.3a), and follow a procedure similar to that employed by Baish et al. [3]. We discretize the intravascular and interstitial space with computational nodes as shown in Supplementary Fig. 10.S4. For the vascular space we assign a different pore size to each of the computational nodes taken by a unimodal distribution with given mean and standard deviation. Therefore, each of these nodes has its own values of L_p, P, and σ.

We first solve the coupled steady state fluid problem to calculate the vascular and interstitial fluid pressures, assuming that the fluid flow entering each computational node equals the fluid flow exiting the node, i.e., $\sum_i Q_i = 0$ for each node i. For boundary conditions, we specify the vascular pressure at the inlet and outlet of the network. The boundaries of the interstitial space, are assumed to be surrounded by lymphatic vessels and specify the pressure there to be zero.

Subsequently, we solve the transient nanoparticle transport problem with a finite difference method using central differencing for diffusion, upwind differencing for convection and an explicit Euler time integration scheme. We model a bolus injection of the nanoparticles according to the experimental procedure. The dimensionless concentration of the particles at the inlet is one and it is decaying exponentially based on our experimental measurements of the blood-half-life of the particles. The concentration at the outlet and at the boundary of the interstitial space is set to zero.

Calculation of effective vascular permeability: To use the model in order to calculate the effective vascular permeability, P_{eff}, we follow the same methodology as is being done in the experiment. We define a region at the centre of the interstitial space surrounding the blood vessels and record the average concentration of nanoparticles through time. The effective permeability should satisfy the equation:

$$\frac{dC}{dt} = P_{\text{eff}} S_{\text{v}} (C_{\text{v}} - C)$$

where C is the average concentration in the interstitial space, S_{v} the vascular density, and C_{v} the concentration of nanoparticles in the blood given by $C_{\text{v}} = C_{\text{vo}} e^{-t/K_{\text{d}}}$. C_{vo} is the initial concentration, t the time, and K_{d} is the blood half-life. Combining the two equations we get:

$$\frac{C}{C_{\mathrm{vo}}} = \frac{P_{\mathrm{eff}} S_{\mathrm{v}} K_{\mathrm{d}}}{1 - P_{\mathrm{eff}} S_{\mathrm{v}} K_{\mathrm{d}}} \left(e^{-P_{\mathrm{eff}} S_{\mathrm{v}} t} - e^{-t/K_{\mathrm{d}}}\right).$$

We fit the above equation to the model calculations of the average concentrations through time and derive the value of the effective vascular permeability.

Values of model parameters: The values of the model parameters are summarized in Supplementary Table 10.S3. The blood half-lives of these nanoparticles were measured in a previous study [11], while the interstitial diffusion coefficient of the particles was obtained by published experimental measurements of macromolecules with similar hydrodynamic diameter as the diameter of the nanoparticles employed in the current study [12] (Supplementary Table 10.S4).

Fluid pressure calculations: Typical fluid pressure calculations for vessels with small and large pores are shown in the following Supplementary Figs. 10.S5 and 10.S6. For small pores (Supplementary Fig. 10.S5), the IFP is low and a transvascular pressure difference is established across the vessel wall. For large pores (Supplementary Fig. 10.S6), the IFP at the centre of the tumour increases and becomes comparable to the microvascular pressure. As a result, the transvascular pressure difference across the vessel wall is no longer maintained. Moreover, according to our simulations it is likely for large pores the IFP to locally exceed the microvascular pressure.

Supplementary Tables

Table 10.S1 Studies reporting the impact of anti-VEGF pathway normalization strategies on the delivery of therapeutic compounds/systemically administered molecules into tumours

Systemically administered molecule	Anti-angiogenic therapy	Tumour model(s)	Effect on delivery
Conventional cytotoxics			
Irinotecan	A4.6.1	Colon carcinoma	↑ [14]
Topotecan, etoposide	Bevacizumab	Neuroblastoma	↑ [15]
Temozolomide	Sunitinib	Glioma	↑ [16–17]*
Antibodies			
Non-specific IgG, anti-E-cadherin Ab	Axitinib	Lung carcinoma, pancreatic tumour	↑ (per vessel) [18]
Other molecules			
BSA	DC101	Breast carcinoma, colon carcinoma	↑ [19]
FDG	Bevacizumab	Rectal carcinoma	↑ (per vessel) [20–21]**

*Increased delivery of temozolomide noted with sunitinib 20 mg/kg, but not at 60 mg/kg.
**Study performed in human subjects.
Abbreviations: BSA, bovine serum albumin; FDG, fluorodeoxyglucose.
Adapted from Goel et al. [13].

Table 10.S2 Studies reporting anti-VEGF and downstream pathway inhibition-induced improvements in tumour oxygenation

Anti-angiogenic therapy	Tumour model	Effect on oxygenation	Time window of improved oxygenation
Antibody therapy			
Bevacizumab	Melanoma, breast carcinoma, ovarian carcinoma	↑	2–4 days after start of therapy [22]
Bevacizumab	GBM	↑	Up to 5 days [23]
DC101	GBM	↑	2–8 days after start of therapy [24]
Anti-PlGF Ab	Pancreatic carcinoma	No change	N/A [25]
TKI therapy			
Sunitinib	Squamous carcinoma	↑	O_2 measured 4 days after start of therapy [26]
Semaxanib	Melanoma	↑	O_2 measured 3 days after start of therapy [27]
PI-103 (PI3K inhibitor)	Fibrosarcoma, squamous carcinoma	↑	O_2 measured 10 days after start of therapy 28
Other therapies			
FTIs (Ras inhibitors)	Prostate carcinoma, bladder carcinoma, glioma, fibrosarcoma, squamous carcinoma	↑	O_2 increased up to 7–10 days [28–30]
Nelfinavir (AKT inhibitor)	Fibrosarcoma, squamous carcinoma	↑	O_2 measured 10 days after start of therapy [28]
PLX4720 (BRAF inhibitor)	Melanoma, colorectal carcinoma	↑	O_2 measured 14 days after start of therapy [31]

Genetic models			
$VEGF^{-/-}$ (myeloid cells)	Lung carcinoma	↑	N/A [32]
$nNOS^{-/-}$ (tumour cells)	Glioblastoma	↑	N/A [33]
$\alpha_v\beta_3/\alpha_v\beta_5$ integrin-FAK-Rho knockdown (tumour cells)	Glioblastoma	↑	N/A [34]

Abbreviations: FAK, Focal Adhesion Kinase; GBM, glioblastoma multiforme; nNOS, neuronal Nitric Oxide synthase; PI3K, Phosphoinostide-3-kinase; PlGF, Placental Growth Factor; TKI, tyrosine kinase inhibitor; VEGF, Vascular Endothelial Growth Factor.
Adapted from Goel et al. [13].

Table 10.S3 Physiological values of the model parameters

Model parameters	Value	Reference
Size of the domain	1 cm	[7]
Blood viscosity	3×10^{-5} (mmHg-sec)	[3]
Vascular pressure at the inlet	25 mmHg	[35]
Vascular pressure at the outlet	5 mmHg	[35]
Vascular density	200 cm^{-1}	[36]
Vessel wall thickness	5 μm	[37]
Vessel diameter	15 μm	[38]
Interstitial Space Conductivity	8×10^{-7} cm^2/mmHg-sec	[39]

Table 10.S4 Diffusion coefficients [12] and blood half-life data [11] used in the mathematical model

Particle size	1 nm	12 nm	60 nm	125 nm	250 nm
Diffusion coefficient (cm^2/s)	2×10^{-6}	2×10^{-7}	5×10^{-8}	6×10^{-9}	1×10^{-9}
Blood Half-life (min)	1550*	1480 ± 241	995 ± 97	582 ± 48	500*

*Extrapolation from experimental data.

Supplementary Figures

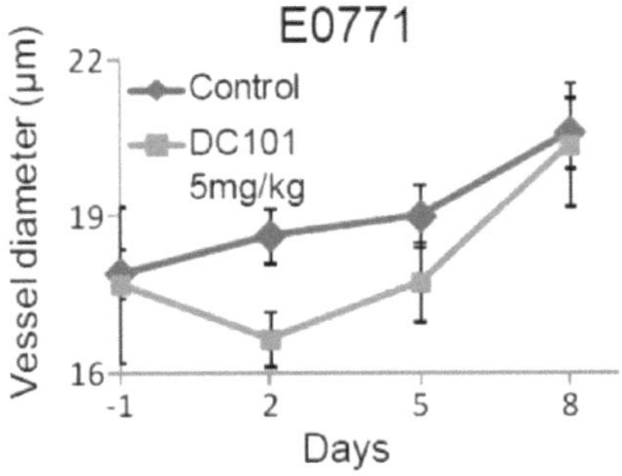

Figure 10.S1 Structural vascular normalization "window" based on tumour vessel diameter. We imaged blood vessel networks in SCID mice bearing orthotopic E0771 mammary tumours in mammary fat pad windows, and we treated these mice with either 5 mg/kg DC101 therapy or non-specific rat IgG (control) every three days starting on day 0. Treatment with DC101 resulted in a reduction in mean vessel diameter on day 2 (p = 0.03, Student's t-test), with no difference by day 8, indicating the presence of a normalization window. Animal number $n = 4$–6 for all groups.

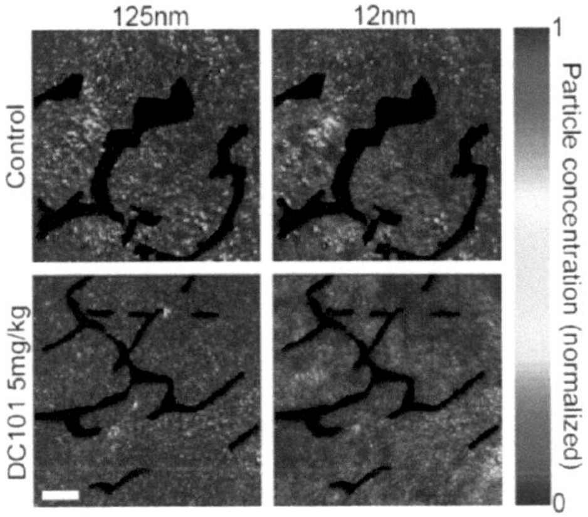

Figure 10.S2 Nanoparticle penetration versus particle size in orthotopic E0771 mammary tumours in response to normalizing therapy with DC101. Intensities are relative to initial intravascular levels, with vessels in black. Normalization on day 2 after 5 mg/kg DC101 improves 12 nm particle penetration while not affecting 125 nm penetration. Scale −100 μm.

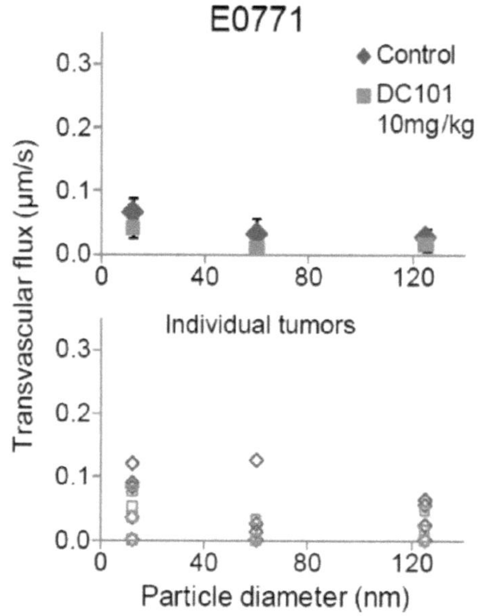

Figure 10.S3 Effect of a higher DC101 dose on nanoparticle delivery. A 10 mg/kg dose of DC101 does not improve transvascular flux for any size of nanoparticle in orthotopic E0771 mammary tumours. Animal number $n = 4$–5 for all groups.

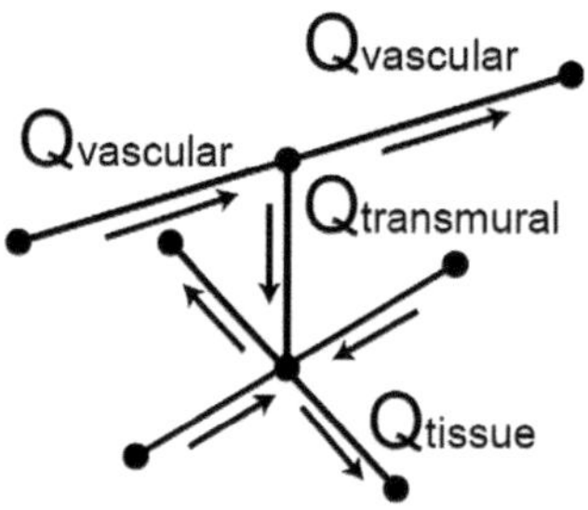

Figure 10.S4 Local connectivity of the network. The arrows show the directions of possible fluid flow.

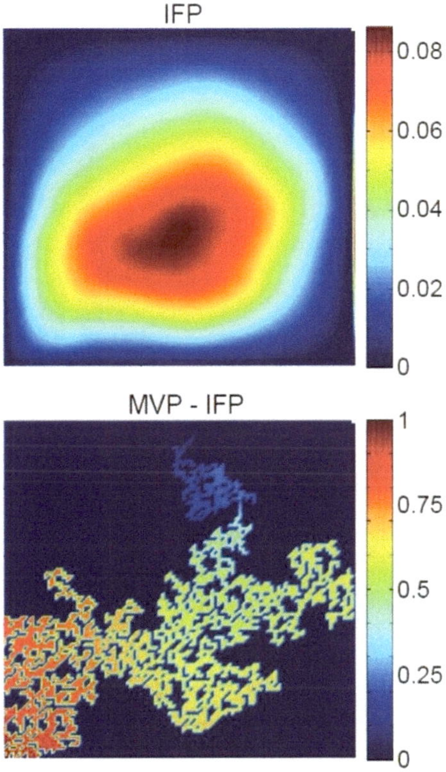

Figure 10.S5 Predicted fluid pressure for small pores. Dimensionless IFP (top) and transvascular pressure difference (bottom) for a unimodal pore size distribution with mean: 40 nm and standard deviation: 20 nm. The pressure values are dimensionless by division with the vascular pressure at the inlet.

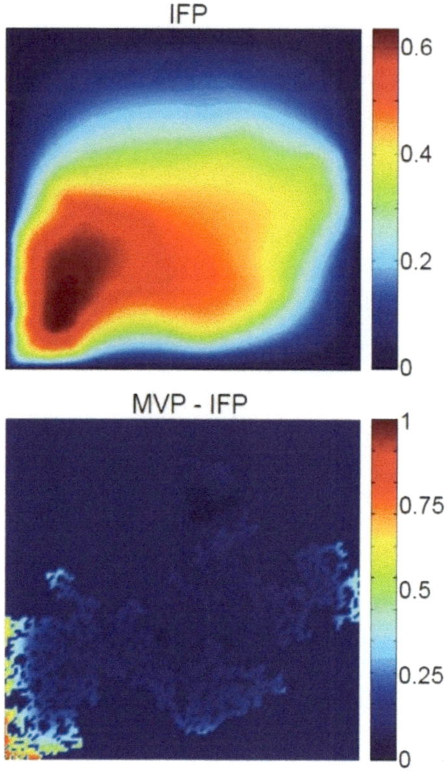

Figure 10.S6 Predicted fluid pressure for large pores. Dimensionless IFP (top) and transvascular pressure difference (bottom) for a unimodal pore size distribution with mean: 400 nm and standard deviation: 200 nm. The pressure values are dimensionless by division with the vascular pressure at the inlet.

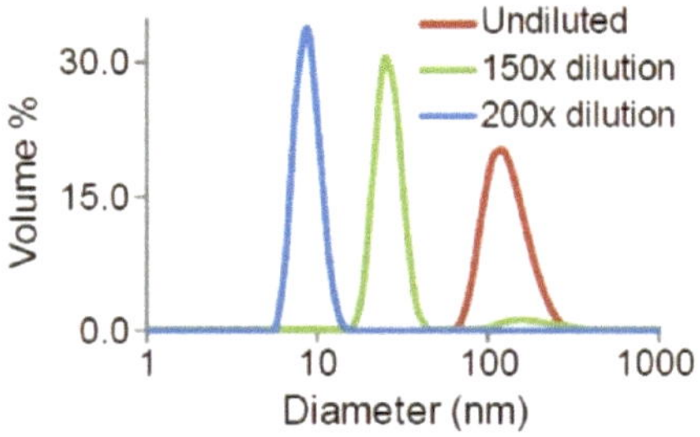

Figure 10.S7 Size of Abraxane nanomedicine upon dilution. We measured the hydrodynamic diameter of Abraxane upon dilution in PBS using dynamic light scattering. We found that Abraxane had an initial size of 130.7 nm before dilution, but immediately decreased in size to 8.8 nm upon dilution. We also characterized an intermediate size of 24.4 nm. Furthermore, we found that Abraxane similarly breaks down in bovine serum albumin solution – a simulated plasma – indicating that it reaches a ~10 nm size in vivo.

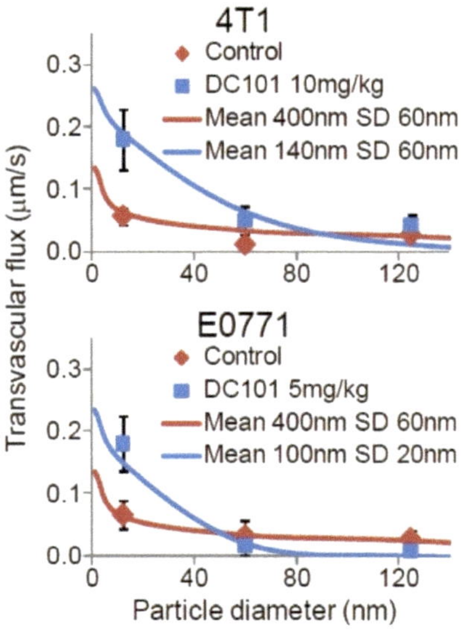

Figure 10.S8 Fits of the model predictions to the experimental data. The lines represent the predictions, while the points show the experimental data. For 4T1 tumours, 10 mg/kg DC101 seems to reduce the mean pore size from 400 nm to 140 nm while maintaining the pore size standard deviation at 60 nm. For E0771 tumours, 5 mg/kg DC101 appears to reduce the mean pore size from 400 nm to 100 nm and the pore size standard deviation from 60 nm to 20 nm. These data suggest that normalizing therapy may reduce both mean pore sizes and pore size heterogeneity to achieve reductions in IFP.

Supplementary References

1. Baish, J. W., and Jain, R. K. Fractals and cancer. *Cancer Res.* **60**, 3683–3688 (2000).
2. Gazit, Y. et al. Fractal characteristics of tumor vascular architecture during tumor growth and regression. *Microcirculation* **4**, 395–402 (1997).
3. Baish, J. W., Netti, P. A., and Jain, R. K. Transmural coupling of fluid flow in microcirculatory network and interstitium in tumors. *Microvasc. Res.* **53**, 128–141 (1997).
4. Pozrikidis, C., and Farrow, D. A. A model of fluid flow in solid tumors. *Ann. Biomed. Eng.* **31**, 181–194 (2003).
5. Baxter, L. T., and Jain, R. K. Transport of fluid and macromolecules in tumors. I. Role of interstitial pressure and convection. *Microvasc. Res.* **37**, 77–104 (1989).
6. Tong, R. T. et al. Vascular normalization by vascular endothelial growth factor receptor 2 blockade induces a pressure gradient across the vasculature and improves drug penetration in tumors. *Cancer Res.* **64**, 3731–3736 (2004).
7. Jain, R. K., Tong, R. T., and Munn, L. L. Effect of vascular normalization by antiangiogenic therapy on interstitial hypertension, peritumor edema, and lymphatic metastasis: insights from a mathematical model. *Cancer Res.* **67**, 2729–2735 (2007).
8. Chauhan, V. P. et al. Multiscale measurements distinguish cellular and interstitial hindrances to diffusion in vivo. *Biophys. J.* **97**, 330–336 (2009).
9. Deen, W. M. Hindered transport of large molecules in liquid-filled pores. *AIChE J.* **33**, 1409–1425 (1987).
10. Bungay, P. M., and Brenner, H. The motion of a closely fitting sphere in a fluid-filled tube. *Int. J. Multiphase Flow* **1**, 25–56 (1973).
11. Popovic, Z. et al. A nanoparticle size series for in vivo fluorescence imaging. *Angew. Chem. Int. Ed.* **49**, 8649–8652 (2010).
12. Pluen, A. et al. Role of tumor-host interactions in interstitial diffusion of macromolecules: cranial vs. subcutaneous tumors. *Proc. Natl. Acad. Sci. USA* **98**, 4628–4633 (2001).
13. Goel, S. et al. Normalization of the vasculature for treatment of cancer and other diseases. *Physiol. Rev.* **91**, 1071–1121 (2011).

14. Wildiers, H. et al. Effect of antivascular endothelial growth factor treatment on the intratumoral uptake of CPT-11. *Br. J. Cancer* **88**, 1979–1986 (2003).
15. Dickson, P. V. et al. Bevacizumab-induced transient remodeling of the vasculature in neuroblastoma xenografts results in improved delivery and efficacy of systemically administered chemotherapy. *Clin. Cancer Res.* **13**, 3942–3950 (2007).
16. Zhou, Q., and Gallo, J. M. Differential effect of sunitinib on the distribution of temozolomide in an orthotopic glioma model. *Neuro-Oncology* **11**, 301–310 (2009).
17. Zhou, Q., Guo, P., and Gallo, J. M. Impact of angiogenesis inhibition by sunitinib on tumor distribution of temozolomide. *Clin. Cancer Res.* **14**, 1540–1549 (2008).
18. Nakahara, T., Norberg, S. M., Shalinsky, D. R., Hu-Lowe, D. D., and McDonald, D. M. Effect of inhibition of vascular endothelial growth factor signaling on distribution of extravasated antibodies in tumors. *Cancer Res.* **66**, 1434–1445 (2006).
19. Tong, R. T. et al. Vascular normalization by vascular endothelial growth factor receptor 2 blockade induces a pressure gradient across the vasculature and improves drug penetration in tumors. *Cancer Res.* **64**, 3731–3736 (2004).
20. Willett, C. G. et al. Efficacy, safety, and biomarkers of neoadjuvant bevacizumab, radiation therapy, and fluorouracil in rectal cancer: a multidisciplinary phase II study. *J. Clin. Oncol.* **27**, 3020–3026 (2009).
21. Willett, C. G. et al. Direct evidence that the VEGF-specific antibody bevacizumab has antivascular effects in human rectal cancer. *Nature Med.* **10**, 145–147 (2004).
22. Dings, R. P. et al. Scheduling of radiation with angiogenesis inhibitors anginex and Avastin improves therapeutic outcome via vessel normalization. *Clin. Cancer Res.* **13**, 3395–3402 (2007).
23. McGee, M. C. et al. Improved intratumoral oxygenation through vascular normalization increases glioma sensitivity to ionizing radiation. *Int. J. Radiat. Oncol. Biol. Phys.* **76**, 1537–1545 (2010).
24. Winkler, F. et al. Kinetics of vascular normalization by VEGFR2 blockade governs brain tumor response to radiation: role of oxygenation, angiopoietin-1, and matrix metalloproteinases. *Cancer Cell* **6**, 553–563 (2004).

25. Fischer, C. et al. Anti-PlGF inhibits growth of VEGF(R)-inhibitor-resistant tumors without affecting healthy vessels. *Cell* **131**, 463–475 (2007).
26. Batra, S., Matsumoto, S., Hyodo, F., Mitchell, J., and Krishna, M. C. MRI assessment of angiogenesis inhibitor sunitinib's influence on tumor oxygenation to identify an optimal chemoradiotherapeutic window. *Int. J. Radiat. Oncol. Biol. Phys.* **75**, S43 (2009).
27. Eichhorn, M. E. et al. Contrast enhanced MRI and intravital fluorescence microscopy indicate improved tumor microcirculation in highly vascularized melanomas upon short-term anti-VEGFR treatment. *Cancer Biol. Ther.* **7**, 1006–1013 (2008).
28. Qayum, N. et al. Tumor vascular changes mediated by inhibition of oncogenic signaling. *Cancer Res.* **69**, 6347–6354 (2009).
29. Delmas, C. et al. The farnesyltransferase inhibitor R115777 reduces hypoxia and matrix metalloproteinase 2 expression in human glioma xenograft. *Clin. Cancer Res.* **9**, 6062–6068 (2003).
30. Cohen-Jonathan, E. et al. The farnesyltransferase inhibitor L744,832 reduces hypoxia in tumors expressing activated H-ras. *Cancer Res.* **61**, 2289–2293 (2001).
31. Bottos, A. et al. Targeting oncogenic serine/threonine-protein kinase BRAF in cancer cells inhibits angiogenesis and abrogates hypoxia. *Proc. Natl. Acad. Sci. USA* **109**, E353–E359 (2012).
32. Stockmann, C. et al. Deletion of vascular endothelial growth factor in myeloid cells accelerates tumorigenesis. *Nature* **456**, 814–818 (2008).
33. Kashiwagi, S. et al. Perivascular nitric oxide gradients normalize tumor vasculature. *Nature Med.* **14**, 255–257 (2008).
34. Skuli, N. et al. Alphavbeta3/alphavbeta5 integrins-FAK-RhoB: a novel pathway for hypoxia regulation in glioblastoma. *Cancer Res.* **69**, 3308–3316 (2009).
35. Boucher, Y., and Jain, R. K. Microvascular pressure is the principal driving force for interstitial hypertension in solid tumors: implications for vascular collapse. *Cancer Res.* **52**, 5110–5114 (1992).
36. Yuan, F. et al. Microvascular permeability and interstitial penetration of sterically stabilized (stealth) liposomes in a human tumor xenograft. *Cancer Res.* **54**, 3352–3356 (1994).
37. Jain, R. K. Transport of molecules across tumor vasculature. *Cancer Metastasis Rev.* **6**, 559–593 (1987).

38. Yuan, F. et al. Vascular permeability and microcirculation of gliomas and mammary carcinomas transplanted in rat and mouse cranial windows. *Cancer Res.* **54**, 4564–4568 (1994).

39. Netti, P. A., Berk, D. A., Swartz, M. A., Grodzinsky, A. J., and Jain, R. K. Role of extracellular matrix assembly in interstitial transport in solid tumors. *Cancer Res.* **60**, 2497–2503 (2000).

Chapter 11

The Properties and Applications of Nanodiamonds

Vadym Mochalin,[a] Olga Shenderova,[b] Dean Ho,[c] and Yury Gogotsi[d]

[a] *Missouri University of Science & Technology, 232 Schrenk Hall, 400 W. 11th St, Rolla, Missouri 65409, USA*
[b] *Adámas Nanotechnologies, Inc., 8100 Brownleigh Drive, Raleigh, North Carolina 27617, USA*
[c] *SINAPSE, Departments of Biomedical Engineering and Pharmacology, Biomedical Institute for Global Health Research and Technology, National University of Singapore, Singapore*
[d] *Department of Materials Science and Engineering and A. J. Drexel Nanotechnology Institute, Drexel University, Philadelphia, Pennsylvania 19104, USA*
gogotsi@drexel.edu

Nanodiamonds have excellent mechanical and optical properties, high surface areas and tunable surface structures. They are also non-toxic, which makes them well suited to biomedical applications. Here we review the synthesis, structure, properties, surface chemistry and phase transformations of individual nanodiamonds and clusters of nanodiamonds. In particular we discuss the rational control of the mechanical, chemical, electronic and optical properties of nanodiamonds through surface doping, interior doping and

Reproduced with permission from *Nature Nanotechnology*, **7**, 11–23, 2012.

Nano-Enabled Medical Applications
Edited by Lajos P. Balogh

ISBN 978-981-4800-32-7 (Hardcover), 978-0-429-39903-9 (eBook)
www.jennystanford.com

the introduction of functional groups. These little gems have a wide range of potential applications in tribology, drug delivery, bioimaging and tissue engineering, and also as protein mimics and a filler material for nanocomposites.

Nanoscale diamond particles were first produced by detonation in the USSR in the 1960s [1], but they remained essentially unknown to the rest of the world until the end of the 1980s [2]. Then, beginning in the late 1990s, a number of important breakthroughs led to wider interest in these particles, which are now known as nanodiamonds. First, colloidal suspensions of individual diamond particles with diameters of 4–5 nm ('single-digit' nanodiamonds) became available [3]. Second, researchers started to use fluorescent nanodiamonds as a non-toxic alternative to semiconductor quantum dots for biomedical imaging [4, 5]. Third, nanoscale magnetic sensors [6] based on nanodiamonds were developed. Fourth, the chemical reactivity of the surface of nanodiamonds allowed a variety of wet [7, 8] and gas [9] chemistry techniques to be employed to tailor the properties of nanodiamonds for use in composites [10–14] and also for other applications, such as attaching drugs and biomolecules [15–17]. Fifth, new environmentally benign purification techniques were developed, and these allowed high-purity nanodiamond powders with controlled surface chemistry to be produced in large volumes at a low cost [18, 19]. Finally, nanodiamond was found to be less toxic than other carbon nanoparticles [20–22] and, as a result, is currently being considered for applications in biomedical imaging, drug delivery and other areas of medicine.

Today there is a baffling array of nanodiamonds available for research. They have been synthesized by the detonation technique (Fig. 11.1), laser ablation [23], high-energy ball milling of high-pressure high-temperature (HPHT) diamond microcrystals [24], plasma-assisted chemical vapour deposition (CVD) [25], autoclave synthesis from supercritical fluids [26], chlorination of carbides [27], ion irradiation of graphite [28], electron irradiation of carbon 'onions' [29] and ultrasound cavitation [30], with the first three of these methods being used commercially. Astronomical observations suggest that nanodiamonds are present in the protoplanetary disks of certain types of stars [31, 32], although the origins of these cosmic sources are still under investigation. Meanwhile, we need to be able to produce nanodiamonds in large quantities on the Earth for research and for industrial applications. Here we review the production, properties and applications of nanodiamonds with sizes between 2 and 10 nm, larger than the higher diamondoids [33] and smaller than the diamond particles used in abrasives. Occasionally, however, we will refer to measurements on larger particles [34–37] that provide insights relevant to sub-10 nm nanodiamonds.

11.1 Formation, Thermodynamics and Phase Transformations

It is known that graphite is the most stable form of carbon at ambient temperatures and pressures, and that diamond is metastable. Although the energy difference between the two phases is only 0.02 eV per atom, they are separated by a high energy barrier (~0.4 eV per atom), so high temperatures and pressures and/or catalysts are needed to interconvert graphite and diamond. At the nanoscale, however, the carbon phase diagram must also include cluster size as a third parameter (alongside pressure and temperature) because the Gibbs free energy depends on the surface energy, and this leads to changes in the phase diagram [38, 39] (Fig. 11.1b).

The stability of various phases of nanoscale carbon has been the focus of numerous theoretical and computational studies. Atomistic models have shown that, for sizes below 3–6 nm, tetrahedral hydrocarbons are more stable than polyaromatics [40]. Later it was discovered that the morphology has an important role in the stability of nanodiamonds by influencing surface reconstruction and the formation of sp^2 carbon [41]. Whereas the bare (non-functionalized) surfaces of cubic crystals exhibit structures similar to bulk diamond, the surfaces of octahedral, cuboctahedral and spherical clusters show a transition from sp^3 carbon to sp^2 carbon. Preferential exfoliation of the (111) surfaces begins for clusters in the subnanometre range and promotes the cluster transition

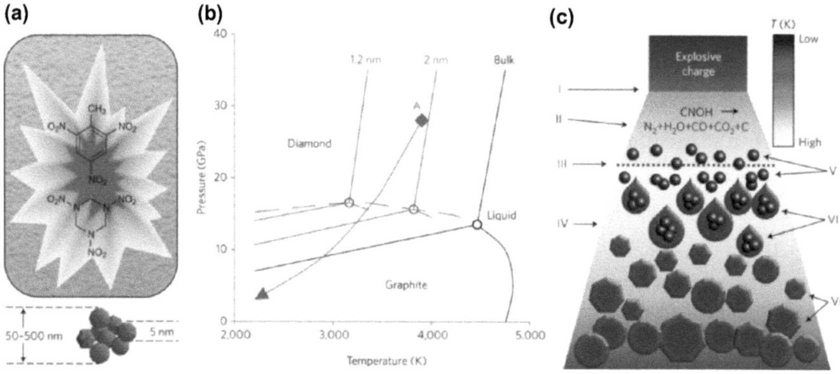

Figure 11.1 Detonation synthesis of nanodiamonds. (a) To synthesize nanodiamonds, explosives with a negative oxygen balance (for example a mix of 60 wt% TNT ($C_6H_2(NO_2)_3CH_3$) and 40 wt% hexogen ($C_3H_6N_6O_6$)) are detonated in a closed metallic chamber in an atmosphere of N_2, CO_2 and liquid or solid H_2O. After detonation, diamond-containing soot is collected from the bottom and the walls of the chamber. (b) Phase diagram showing that the most stable phase of carbon is graphite at low pressures, and diamond at high pressures, with both phases melting when at temperatures above 4,500 K (with the precise melting temperature for each phase depending on the pressure). The phase diagrams for nanoscale carbon are similar, but the liquid phase is found at lower temperatures [38, 39]. During detonation, the pressure and temperature rise instantaneously, reaching the Jouguet point (point A), which falls within the region of liquid carbon clusters of 1–2 nm in size for many explosives. As the temperature and pressure decrease along the isentrope (red line), carbon atoms condense into nanoclusters, which further coalesce into larger liquid droplets and crystallize [39]. When the pressure drops below the diamond–graphite equilibrium line, the growth of diamond is replaced by the formation of graphite. (c) Schematic of the detonation wave propagation showing (I) the front of the shock wave caused by the explosion; (II) the zone of chemical reaction in which the explosive molecules decompose; (III) the Chapman–Jouguet plane (where *P* and *T* correspond to point A in Fig. 11.1b, indicating the conditions when reaction and energy release are essentially complete); (IV) the expanding detonation products; (V) the formation of carbon nanoclusters; (VI) the coagulation into liquid nanodroplets; and (VII) the crystallization, growth and agglomeration of nanodiamonds [39].

to endofullerene in the case of small clusters (tens of atoms) or onion-like shells with diamond cores ('buckydiamond') for larger clusters (hundreds of atoms) [41]. Large clusters (~1–3.3 nm) of various shapes with less than 76% surface graphitization are likely to be stable in the diamond structure, possessing a thin (either single or double layer) graphitic shell [42].

The surface of sp^3 clusters must be either stabilized through termination with functional groups or reconstructed into sp^2 carbon. Therefore, in addition to size and shape, the stability of carbon nanoparticles also depends on their surface terminations. So far, only the role of hydrogen terminations has been studied in depth. Hydrogen-terminated nanodiamonds with sizes under 1.5 nm (known as diamondoids) are stable molecules that occur naturally in oil [33]. According to first-principles simulations [43], as the size of diamond reaches 3 nm or more, buckydiamonds become energetically preferred over hydrogenated nanodiamonds. First results of modelling nanodiamonds terminated with oxygen-containing [44] and nitrogen-containing functional groups [45] showed complex behaviour, in which different groups favoured different nanodiamond facets depending on the temperature and the environment of the particle, as well as the size and morphology of the nanodiamonds.

When building a model of a nanodiamond, it is necessary to account for the presence of surface functional groups and their variability, sp^2 carbon, and the shape of the particle. Well-purified nanodiamond grains can have almost perfect crystalline structure with negligible fractions of non-diamond carbon (Fig. 11.2). Observations by transmission electron microscopy [18, 19, 46] show that nanodiamond particles are polyhedra consisting of a diamond core built up of sp^3 carbon, which may be partially coated by a graphitic shell or amorphous carbon with dangling bonds terminated by functional groups (Figs. 11.2 and 11.3). Additional characteristic features of the nanodiamond not included in this model are nitrogen impurities (up to 2–3 wt%), which can form complexes in the core of nanodiamond particles [47], and the presence of twins and grain boundaries in the crystallites (Fig. 11.2d). The latter can be responsible for the broadening of the X-ray diffraction peaks that was earlier attributed to disordered sp^3 carbon.

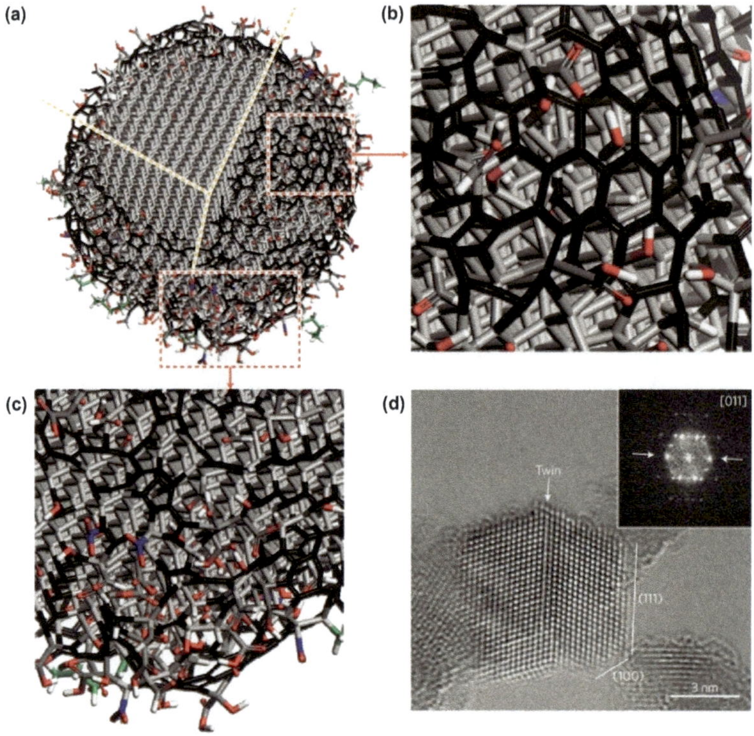

Figure 11.2 Structure of a single nanodiamond particle. (a) Schematic model illustrating the structure of a single ~5 nm nanodiamond after oxidative purification. The diamond core is covered by a layer of surface functional groups, which stabilize the particle by terminating the dangling bonds. The surface can also be stabilized by the conversion of sp^3 carbon to sp^2 carbon. A section of the particle has been cut along the amber dashed lines and removed to illustrate the inner diamond structure of the particle. (b,c) Close-up views of two regions of the nanodiamond shown in (a). The sp^2 carbon (shown in black) forms chains and graphitic patches (b). The majority of surface atoms are terminated with oxygen-containing groups ((c); oxygen atoms are shown in red, nitrogen in blue). Some hydrocarbon chains (green, lower left of (a,c)) and hydrogen terminations (hydrogen atoms are shown in white) are also seen. (d) Each nanodiamond is made up of a highly ordered diamond core. Some nanodiamonds are faceted, such as the one shown in this transmission electron micrograph, whereas most have a rounded shape, as shown in (a). The inset is a fast Fourier transform of the micrograph, which confirms that this nanodiamond has a highly ordered diamond core. Panel (d), reproduced with permission from [19], © 2011 ACS.

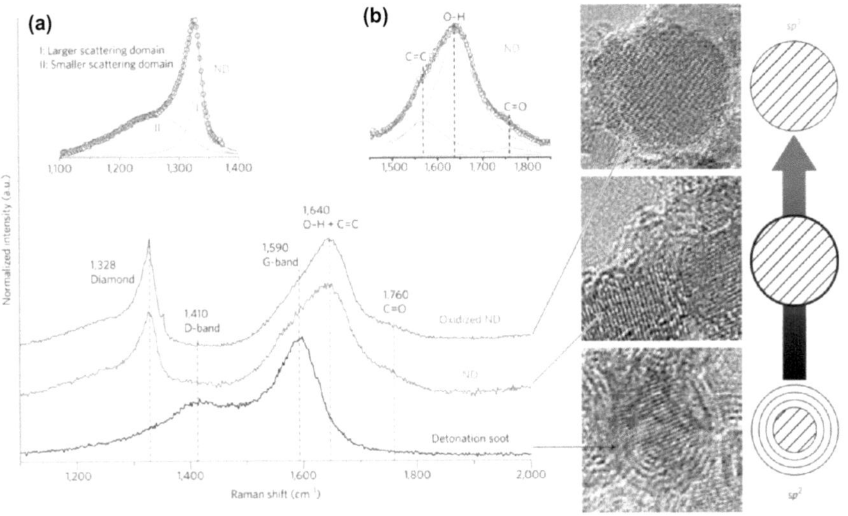

Figure 11.3 Raman spectroscopy and structure of nanodiamond. Electron micrographs showing detonation soot (bottom), purified nanodiamond (middle) and oxidized nanodiamond (top). The diamond cores in detonation soot seem to be completely covered by graphitic shells, and this is confirmed by the Raman spectrum (black line), which is dominated by the G-band of graphitic carbon at 1,590 cm^{-1} and has no diamond peak. Purified nanodiamonds are partially covered by a thin layer of graphite, so a diamond peak can be seen at 1,328 cm^{-1} in the Raman spectrum (blue line). This thin layer of graphite is completely removed by oxidation in air, so the Raman spectrum of oxidized nanodiamonds has an even stronger diamond peak (red line). The diamond peak in the Raman spectrum of purified and air oxidized nanodiamond (inset **a**) is a combination of peaks originating from larger (I) and smaller (II) coherence scattering domains. The phonon confinement model [84] gives a good fit (blue line) to experimental data (open circles). The broad feature at 1,500–1,800 cm^{-1} in the spectrum of air oxidized nanodiamond (inset **b**) originates from surface functional groups and adsorbed molecules, with some contribution from sp^2 carbon atoms. The Raman spectra were recorded following excitation by an ultraviolet laser (325 nm).

A single model cannot describe all kinds of nanodiamonds. Different models should be used depending on the size, shape and surface chemistry—parameters controlled by manufacturing and purification methods. The ‘universal’ model of nanodiamond shown in Fig. 11.2 captures only the most common and important features.

11.2 Synthesis and Purification

Nanodiamonds can be produced from molecules of explosives, which provide both a source of carbon and energy for the conversion (Fig. 11.1a) [39, 48, 49]. This is an environmentally friendly and economically viable method for disposing of old munitions, such as Composition B, although other explosives may be used too. The detonation takes place in a closed chamber filled with an inert gas or water (ice) coolant, called 'dry' or 'wet' synthesis, respectively. The resultant product—detonation soot—is a mixture of diamond particles 4–5 nm in diameter with other carbon allotropes and impurities. Detonation soot contains up to 75 wt% of diamond [48, 49]. The carbon yield is 4–10% of the weight of the explosive, depending on cooling media [48, 49].

Danilenko proposed the mechanism of nanodiamond formation during detonation [39, 48]. Pressures and temperatures at the Jouguet point (point A in Fig. 11.1b) are not high enough to produce liquid bulk carbon, but they are high enough to produce liquid carbon at the nanoscale (Fig. 11.1b). The region of liquid carbon is shifted to lower temperatures for nanocarbon, and the region of nanodiamond stability is slightly shifted to higher pressures (Fig. 11.1b). Thus, it is suggested that nanodiamond is formed by homogeneous nucleation in the volume of the supersaturated carbon vapour by condensation and crystallization of liquid carbon (Fig. 11.1c). Other techniques that use explosives (such as the use of shock waves to produce nanodiamonds from graphite) yield nanodiamonds with crystallite sizes over 10 nm.

In addition to the diamond phase, the detonation soot contains both graphitic carbon (25–85 wt%) and incombustible impurities (metals and oxides, 1–8 wt%) [18, 49]. It must be purified for most applications. The metal impurities originate from the igniter used to initiate detonation (typically azides of lead, silver or copper) and the steel walls of the detonation chamber (iron and other metals). The impurities can be inside the nanodiamond aggregates or attached to their outer surface, and the nanodiamond aggregates should be disintegrated in order to remove the trapped impurities.

On an industrial scale, detonation soot is purified using liquid oxidants (such as HNO_3, a mixture of H_2SO_4 and HNO_3, $K_2Cr_2O_7$ in

H_2SO_4, KOH/KNO_3, Na_2O_2, HNO_3/H_2O_2 under pressure, or $HClO_4$) to remove non-diamond carbon [48, 49]. To remove non-carbon impurities to <0.5 wt% level, the purified product is subjected to HCl and other treatments. Liquid-phase purification is both hazardous and expensive, contributing up to 40% to the product cost.

Alternatively, non-diamond carbon can be oxidized in a more environmentally friendly manner by air or ozone-enriched air at elevated temperatures [18, 19]. Oxidation in air is robust, cost-effective and the most environmentally benign purification technique, capable of increasing the diamond content from ~25 wt% (detonation soot) to >95 wt% (Fig. 11.3). At the same time, oxidation removes various functional groups present on the nanodiamond surface and produces oxygen-containing surface species, mainly anhydrides and carboxylic acids [18], converting various grades of nanodiamond powder into a material with a high content of diamond and a unified surface chemistry. Ozone-purified nanodiamonds [19] have a small aggregate size in aqueous dispersions (~160–180 nm) and a substantially higher content of faceted 3–5 nm particles compared with the acid-purified nanodiamond. The ozone-purified hydrosols also have a very low pH (1.6–2 for 10 wt% hydrosol) and an electrokinetic potential (ζ-potential) (−50 mV for polydispersed sample and down to −100 mV for the 20–30 nm fraction) that is constant over a pH range of 2–12, apparently owing to highly acidic surface groups. Thus, gas oxidation is the most promising method for nanodiamond purification. Surface reduction in hydrogen atmosphere has also been attempted as a purification technique, but non-diamond carbon was not completely removed by this method [50].

11.3 De-Aggregation and Surface Modification

Nanodiamonds have diameters of 4–5 nm, but they tend to aggregate and typical commercial suspensions of nanodiamonds contain larger aggregates (which can withstand ultrasonic treatment). The nanodiamond powders used in many of the early experiments were also prone to aggregation, and the results of these experiments are therefore more relevant to aggregates than to single nanodiamonds

[51]. Although these aggregated samples may be useful in chromatography [52] or drug delivery [53], in many applications de-aggregation into individual primary particles is often needed to benefit fully from the advantages of nanodiamond.

De-aggregation of nanodiamond in suspensions by milling with ceramic microbeads (ZrO_2 or SiO_2) or by microbead-assisted ultrasonic disintegration was developed by Osawa and co-workers, yielding colloidal solutions of individual nanodiamonds 4–5 nm in diameter [3]. The disadvantages of microbead milling are contamination with bead material and generation of graphitic layers on the nanodiamond surface [54]. Amorphous carbon and metal contaminants confined within nanodiamond aggregates and released during bead milling also need to be removed. Attempts to purify bead-milled nanodiamond with liquid oxidizers lead to re-aggregation of the primary particles [54]. The most recent studies [55] suggest that sufficient purification and oxidation in air allow for subsequent isolation of a stable hydrosol of particles 4–5 nm in diameter by centrifugation.

De-aggregation from micrometre-sized aggregates down to stable nanodiamond particles of diameters 5–20 nm has recently been achieved by dry milling [56], using low-cost, abundant and cheap milling media (such as water-soluble salts and sugars) that do not introduce contaminants (Fig. 11.4a,b). Several other methods for de-aggregation have been proposed. Nanodiamond reduction in borane accompanied by ultrasonic treatment resulted in significantly smaller aggregates [57]. Surface graphitization with subsequent functionalization allows dispersion into $\sim$20 nm aggregates [58]. High-temperature hydrogen treatment resulted in stable single-particle aqueous nanodiamond colloids from which a fraction of 2–4 nm nanodiamonds was isolated by centrifugation at $>$10,000 rpm [50].

One concern that needs to be addressed is the possible re-aggregation of the nanodiamonds after they have been subjected to further surface functionalization. Drying nanodiamonds (for storage) can also promote re-aggregation due to capillary forces pulling the individual nanoparticles together. Attractive van der Waals forces also lead to re-aggregation, which makes it more difficult to functionalize the nanodiamonds. Ultrasound-assisted treatment of nanodiamonds in NaCl solution [59] can prevent re-aggregation after drying, possibly because Na^+ ions become attached to the nanodiamond surface. This may also explain why re-aggregation does not occur when nanodiamonds are produced through NaCl-assisted milling [56].

Centrifugation separates the nanodiamond particles into fractions by weight and size [60, 61]. It is a contamination-free approach (unlike, for example, bead milling), and nanodiamonds of different sizes (known as fractions) can be isolated for different applications. For example, only nanodiamonds with aggregate sizes of greater than 100 nm can form photonic structures that diffract light [62] (Fig. 11.4c), but much smaller particles are required for drug delivery. If air- or ozone-purified nanodiamonds (which have a higher content of small aggregates than acid-purified nanodiamonds) are used as starting materials, the production of pure samples of small nanodiamonds by centrifugation becomes economically feasible: for example, 5 wt% aqueous suspensions of the 25 nm nanodiamond fraction have been produced this way [61]. Ultracentrifugation can extract single-digit nanodiamond particles [63], but the yield is low. Therefore, the search continues for simple, non-contaminating and scalable methods for dispersing nanodiamond into single particles.

A distinct feature of nanodiamonds compared with carbon nanotubes and other graphitic nanoparticles is that many different functional groups can be attached to their surface (Figs. 11.2 and 11.3), allowing quite sophisticated surface functionalizations without compromising the useful properties of the diamond core [64]. It is also essential to understand, however, how these groups will interact with their surroundings, and to reduce any detrimental effects (such as aggregation [7, 65]).

Although the various functional groups present on commercial nanodiamond powders can be used for covalent functionalization, it is more convenient to start with carboxylated nanodiamond produced by air or ozone purification, and then take advantage of the rich chemistry of COOH groups (Fig. 11.5). Treatment in a hydrogen microwave CVD plasma at temperatures above 700°C completely reduces COOH groups and completely removes oxygen from the surface to produce hydrogenated nanodiamond [66].

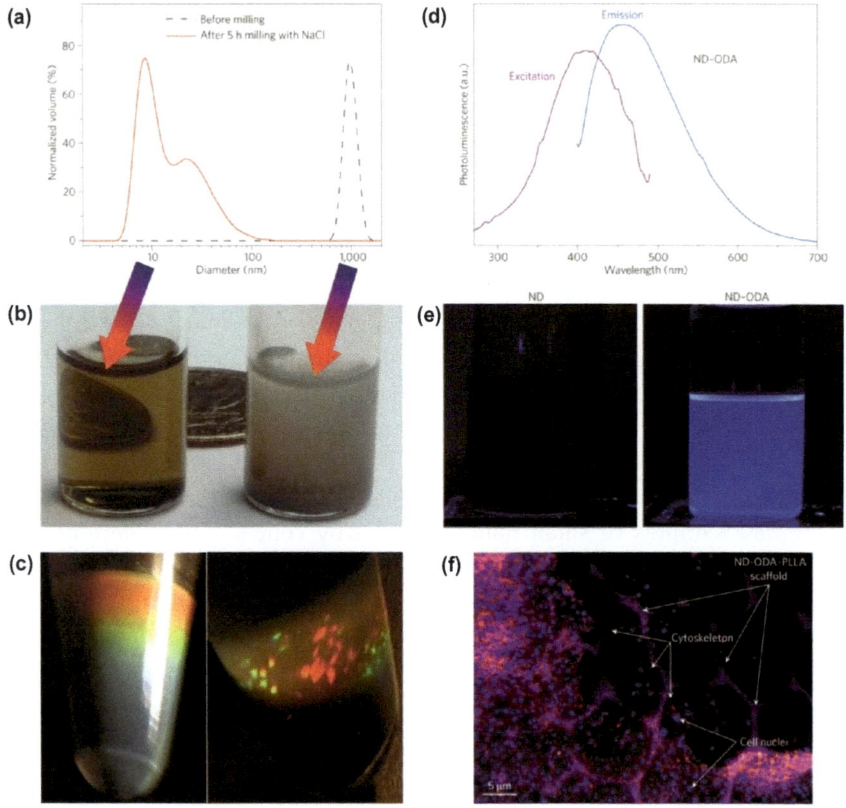

Figure 11.4 Optical properties of nanodiamonds. (a,b) De-aggregation by salt-assisted dry milling reduces the size of diamond particles from ~1 μm to less than 10 nm (a), and makes suspensions of the particles both darker and more transparent (b).The changes in colour are not related to the presence of graphitic carbon [56]. (c) Photonic structures formed by centrifugation of suspensions of nanodiamonds in deionized water. (d,e) Covalently attaching ODA to nanodiamonds changes their optical properties. ND–ODA absorbs and re-emits light over a wide range of wavelengths, as can be seen in these excitation (purple) and emission (blue) spectra (d). Moreover, and in contrast to non-functionalized nanodiamond, ND–ODA is strongly blue fluorescent when illuminated with ultraviolet light (e). (f) ND–ODA can be used for bio-imaging, as illustrated by this confocal micrograph of the fluorescent scaffold made of ND–ODA–PLLA with 7F2 osteoblasts grown on it (see main text for details). Panel (c) reproduced with permission from [62], © 2008 IOP.

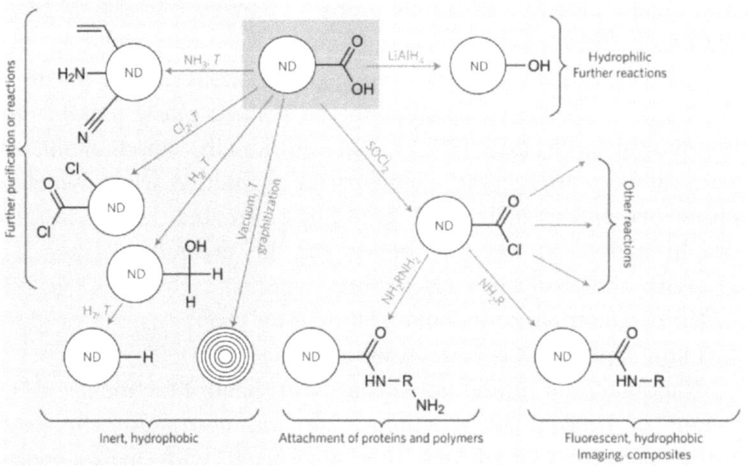

Figure 11.5 Surface modification. Precise control over surface chemistry requires a sample of purified nanodiamond with only one kind of functional group attached to its surface. Nanodiamond terminated with carboxylic groups (ND–COOH; green region) is a common starting material (and is made by air oxidation or ozone treatment of nanodiamond, followed by treatment in aqueous HCl to hydrolyse anhydrides and remove metal impurities). The surface of ND–COOH can be modified by high-temperature gas treatments (red) or ambient-temperature wet chemistry techniques (blue). Heating in NH_3, for example, can result in the formation of a variety of different surface groups including NH_2, C–O–H, C≡N and groups containing C=N [9, 48]. Heating in Cl_2 produces acylchlorides, and F_2 treatment forms C–F groups (not shown) [67, 137, 138]. Treatment in H_2 completely reduces C=O to C–O–H and forms additional C–H groups. Hydroxyl (OH) groups may be removed at higher temperatures or with longer hydrogenation times, or by treatment in hydrogen plasma [66]. Annealing in N_2, Ar or vacuum completely removes the functional groups and converts the nanodiamonds into graphitic carbon nano-onions [139, 140]. A wide range of surface groups and functionalized nanodiamonds can also be produced using wet chemistry treatments.

Compared with gas treatment (at 400–850°C), wet chemistry requires milder conditions and provides a better selectivity through a large number of functional group conversions known in organic chemistry (Fig. 11.5). Reactive C–F and C–Cl surface species created by halogen annealing and photochemical chlorination have also been used in numerous wet chemical reactions [67, 68].

Nanodiamond with O–H terminations was involved in esterification with acyl-chlorides yielding nanodiamond terminated by long alkyl chains [69], and in silanization/de-aggregation [70]. Other reactions were also used [7, 65], including those particularly suitable for biomedical applications [57, 71, 72]. Additionally, nanodiamond can be modified through the chemistry of graphitic carbon which is either present intrinsically (Fig. 11.2b) or created through surface graphitization. Strong C–C bonds can be created between the graphitic shell and the surface group, whereas techniques that rely on the chemistry of nanodiamond functional groups usually produce C–X bonds (where X is N, O, S and so on).

Nanodiamond graphitic shells have been functionalized by means of Diels–Alder reactions [73] and diazonium chemistry [58]. Diazonium chemistry was also used with hydrogenated nanodiamond [74] to form C–C bonds between the attached moiety and diamond core, and with hydroxylated nanodiamond [70] to form C–O–C bonds. Thus, nanodiamond provides numerous options for surface functionalization, but the outcome strongly depends on the purity and uniformity of the surface chemistry of the starting material. As already mentioned, functionalization also affects the stability of diamond surfaces [44, 45]. One challenge in the field is to develop techniques for quantitative analysis of various surface groups of nanodiamond. So far, most of the data are qualitative.

11.4 Vibrational Spectroscopy of Nanodiamond

Raman spectroscopy and Fourier transform infrared (FTIR) spectroscopy provide valuable insights into phase composition and surface terminations of nanodiamonds. FTIR can detect functional groups and adsorbed molecules on the surface [18, 50, 75–78], and it can also detect changes in the surface chemistry of functionalized nanodiamond [5, 8, 14, 79]. Nitrogen defects in nanodiamond manifest themselves as two broad bands in the region 1,100–2,500 cm^{-1}, which overlaps with the peaks of surface functional groups. The most characteristic features of nanodiamonds after oxidative purification include: O–H stretch (3,200–3,600 cm^{-1}) and bend (1,630–1,640 cm^{-1}) with bands originating from both adsorbed species and from O–H groups covalently attached to the nanodiamond surface; C=O stretch at 1,700–1,800 cm^{-1} where C=O can be part of ketone, aldehyde, carboxylic acid, ester, anhydride, cyclic ketone, lactone or lactam; C–H stretch at 2,850–3,000 cm^{-1}, with bands originating from asymmetric and symmetric C–H stretch vibrations in CH_2 and CH_3 groups.

Many nanodiamonds also have a very broad absorption feature at 1,000–1,500 cm^{-1} (which is known as the 'fingerprint region') that is a combination of many overlapping peaks. These include: O–H deformational and C–O–C stretch vibrations, epoxy C–O stretch, C–C stretch, amide C–N stretch and C–N–H deformational vibrations; peaks related to nitrogen defects; and vibrations of NO_2, SO_2OH and other groups.

Raman spectroscopy, on the other hand, provides information on the structure, composition and homogeneity of the material, and also information on the surface groups. But its interpretation for nanodiamond is not straightforward [80]. For example, the peak at ~1,640 cm^{-1} in the spectrum of detonation nanodiamond was only recently associated with O–H vibrations [75]. Raman spectra depend on structure, purity, sp^3/sp^2 ratio, crystal size and surface chemistry (Fig. 11.3). The spectra of nanodiamond powders with a high sp^2 carbon content are dominated by the D and G bands of graphitic carbon and the diamond signal is weak or absent. With increasing sp^3 content, the intensity of the diamond peak increases, while the D band weakens. Because of the small Raman scattering cross-section of diamond and the shielding effect of graphitic and amorphous carbon around the diamond core, ultraviolet lasers with excitation energy close to the bandgap of diamond (5.5 eV) are needed to amplify the Raman signal of nanodiamond and suppress the D band of graphitic carbon that may overlap with a weak diamond peak. The nanodiamond peak is broadened (full width at half maximum, FWHM, >30 cm^{-1}) and down-shifted (~1,326 cm^{-1}) with respect to bulk diamond, with a shoulder at ~1,250 cm^{-1} [18, 81–83] that originates from smaller nanodiamond particles or smaller coherent scattering domains separated by defects in larger nanodiamond particles. These factors were accounted for by an improved phonon confinement model [84] (Fig. 11.3a).

First-principles techniques have been applied to compute the size-dependent evolution of the diamond Raman spectrum in the <3 nm range [85, 86], and we hope that it will be possible to compute the properties of larger nanoparticles within a few years. Until then, the phonon confinement model can be used to determine size from Raman data more reliably than is possible with transmission electron microscopy.

Another prominent feature in the Raman spectra of nanodiamond is a broad asymmetric peak between 1,500 and 1,800 cm^{-1}, which is a superposition of graphitic carbon band at 1,590 cm^{-1}, an O–H bending peak at 1,640 cm^{-1} and a shoulder at 1,740–1,760 cm^{-1} due to a C=O stretch coming from surface functional groups [75] (Fig. 11.3b). It is not yet clear, however, if the O–H contribution originates from nanodiamond functional groups or adsorbed water [87]. The interference of O–H bending with G-band of graphitic carbon should be taken into account when using water cooling of nanodiamond samples during Raman spectra acquisition.

11.5 Properties of Nanodiamond

Diamond is an outstanding material in many respects, and nanodiamond inherits most of the superior properties of bulk diamond and delivers them at the nanoscale. These properties include superior hardness and Young's modulus, biocompatibility, optical properties and fluorescence, high thermal conductivity and electrical resistivity, chemical stability, and the resistance of nanodiamond to harsh environments. Fluorescence and biocompatibility are discussed in this section, and the other useful properties of nanodiamond are covered in the section on applications.

Fluorescence. The presence of nitrogen-vacancy (NV) centres—a nitrogen atom next to a vacancy—in nanodiamond leads to useful fluorescence properties. NV centres can be created by irradiating nanodiamond with high-energy particles (electrons, protons, helium ions), followed by vacuum annealing at 600–800°C [4, 88]. Irradiation forms vacancies in the diamond structure [89], and during annealing these vacancies migrate and get trapped by the nitrogen atoms that are always present in diamond. Two types of NV centres are formed—neutral (NV^0) and negatively charged (NV^-)—and these have different emission spectra. The NV^- centre is of particular interest because it has an $S = 1$ spin ground state that can be spin-polarized by optical pumping and manipulated using electron paramagnetic resonance. Moreover, it has a long spin coherence time. Fluorescent NV centres in isotopically clean diamond are of particular interest for quantum computing [90], and NV centres in nanodiamond are also being investigated for applications in high-resolution magnetic sensing [6, 91, 92], fluorescence resonance energy transfer [93] and biomedical imaging [4].

The development of bright photoluminescent nanodiamonds has been mostly done on materials produced by high-pressure, high-temperature synthesis [4, 24] (containing up to hundreds of parts per million of native substitutional nitrogen), which are then crushed down to nanosized particles, including single-digit nanodiamonds [24]. Although the concentration of the fluorescent NV defects produced by electron irradiation does not, to a first approximation, depend on the size of the nanocrystals [94], the fraction of NV^- defects produced by the same technique decreases with decreasing size of the nanodiamond [88], presumably because of electron traps located at the surface.

Up to now, nanodiamonds synthesized from explosives were not among the preferred candidates for imaging applications based on NV centres. Recently, intermittent luminescence originating from NV centres has been observed in individual pristine 5 nm detonation nanodiamonds produced from a TNT and hexogen precursor [92]. Stable luminescence has also recently been detected from NV centres trapped in larger ($>$20 nm) nanodiamonds produced from TNT and hexogen [47], and graphite and hexogen [95] precursors. These results suggest that various factors (such as the amount of nitrogen in the precursors and the cooling conditions) have a role in the nitrogen incorporation into the nanodiamond core and in possible in situ formation of NV centres [92, 95]. Size-related proximity of surface defects and the presence of numerous internal defects might contribute to the low fluorescence intensity of nanodiamond produced from explosives. This is supported by the demonstration of a very high content (up to 1%) of NV defects in detonation nanodiamond after sintering at temperature $T = 800°C$ and pressure $P = 6$ GPa [96], which enlarges the nanodiamond sizes.

Fluorescent particles can also be produced by linking [21, 97] or adsorbing [98] various fluorophores onto nanodiamond. Fluorophore-conjugated nanodiamond can travel through cellular compartments of varying pH without degradation of the surface-conjugated fluorophore or alteration of cell viability over extended periods of time [99]. Bright blue fluorescent nanodiamond has recently been produced by covalent linking of octadecylamine to carboxylic groups on nanodiamond surface [5] (Fig. 11.4d,e).

Fluorescent nanodiamonds combine the advantages of semi-conductor quantum dots (small size, high photostability, bright multicolour fluorescence) with biocompatibility, non-toxicity and rich surface chemistry, which means that they have the potential to revolutionize in vivo imaging applications [4, 100–101].

11.5.1 Biocompatibility and Fate in the Body

Diamond and glassy carbon are known to be non-toxic, but we cannot assume that carbon nanoparticles are also non-toxic. Owing to the different purification procedures used by different manufacturers, and the multiple options for surface modification, the toxicity of nanodiamonds is of legitimate concern [20]. In vitro and in vivo studies have been conducted to examine characteristics as diverse as cell viability, gene programme activity, and in vivo mechanistic and physiological behaviour [11, 20–22, 102–104]. Nanodiamonds instilled within the trachea were reported to be of low pulmonary toxicity, with the amount of nanodiamond in the alveolar region decreasing with time, and macrophages burdened with nanodiamonds were observed in the bronchia for 28 days after exposure [102]. Intravenously administered nanodiamond complexes at high dosages did not change serum indicators of liver and systemic toxicity [104].

To evaluate the fate of nanodiamonds and their impact on stress response activity and worm reproduction, fluorescent nanodiamond aggregates with average hydrodynamic size of ~120 nm were fed and microinjected in the translucent *Caenorhabditis elegans* worm, and then tracked for several days [103]. Bare nanodiamonds typically remained in the worm lumen, whereas nanodiamonds coated with dextran or bovine serum albumin (BSA) were absorbed into the intestinal cells. Nanodiamonds that were microinjected into worm gonads were transferred into the larvae and offspring, but this had no impact on the reproductive capabilities or survival of the worms. Further experiments involving DAF-16:GFP (DAF-16 is a gene group that regulates the stress and immune response of cells; GFP is green fluorescent protein) confirmed that fluorescent nanodiamond is not toxic and does not induce stress in the worm model, thus providing support for its use in in vivo imaging. But given the number of surface modifications that are possible, it is important to be certain that the functionalized nanodiamonds intended for biomedical applications remain safe. Therefore, we have recently compared the cytotoxicity and osteoblast proliferation and gene expression effects of carboxylated nanodiamond, octadecylamine modified nanodiamond (ND–ODA), and composites of poly(L-lactic acid) with ND–ODA ([11]; Fig. 11.4f). Although no harmful effects were found in any of these materials, toxicity and biocompatibility testing of new nanodiamond-based materials should continue.

11.6 Current and Future Applications

Nanodiamond additives have been used for electrolytic and electroless metal plating for many years [49]. More recently, they have been used in other applications such as the chemical vapour deposition of diamond films [105], in magnetic resonance imaging [106], chromatography [52, 107] and, proteomics and mass spectrometry [108]. Carbon 'onions' (which are produced by the graphitization of nanodiamond) have shown potential for applications in energy storage [109], composites [110, 111] and catalysis [112] (nanodiamond itself has also demonstrated catalytic activity [113]); and boron-doped nanodiamonds (which are conducting) can be used in electroanalysis, electrochemical double-layer capacitors and batteries. Furthermore, undoped non-conducting nanodiamond demonstrates redox activity in electrochemical systems [114]. In this section we focus on applications in five areas: tribology and lubrication; nanocomposites; drug delivery; protein mimics; and tissue scaffolds and surgical implants. The last three of these applications are made possible by the ability of nanodiamonds to self assemble [115] and the fact that a wide range of small molecules, proteins, antibodies, therapeutics and nucleic acids can bind to the surface of nanodiamonds. Other potential biomedical applications [21] not covered below include the use of nanodiamonds as supports for solid-phase peptide synthesis and as sorbents for detoxification and separation. The use of fluorescent nanodiamonds for biomedical imaging is discussed above (see also Fig. 11.4f).

11.6.1 Tribology and Lubrication

The addition of diamond-containing detonation soot to lubricants [49] decreases fuel consumption by ~5% and makes engines last longer. It was assumed that this happened because the graphite in the soot lubricated while the nanodiamonds reduced friction by polishing away asperities on sliding surfaces. However, purified nanodiamond itself provides enhanced tribological performance when dispersed alone or with polytetrafluoroethylene (PTFE) or metal nanoparticles in greases or oils [116]. Initially it was assumed that the nanodiamonds acted as 'ball bearings', but this has not been confirmed as universal in more recent studies [117], which suggests that different lubrication mechanisms could be at work in different systems. For example, embedding nanodiamond from a lubricant into a carbon steel surface may explain reduced friction and wear, whereas the wear mechanism for an aluminium alloy is dominated by the viscosity of the nanodiamond suspension [117].

The versatile surface chemistry of nanodiamond means that it can be tailored so that it disperses in a variety of different systems, including oil and water [118]. Carbon onions can also act as an efficient lubricant [119], probably owing to the microscopic ball-bearing action. Overall, lubrication is more complex than it seems at first, but it is reasonable to assume that both nanodiamonds and carbon onions embedded into metal surfaces separate the sliding surfaces and prevent wear caused by metal–metal adhesion.

11.6.2 Nanocomposites

The superior mechanical and thermal properties of diamond, combined with the rich surface chemistry of nanoscale diamond particles, make nanodiamond an excellent filler material for composites. Moreover, the biocompatibility and chemical stability of the diamond core make these composites very well suited for biomedical applications. Substantial improvements have been reported in the mechanical strength [10, 11, 14, 120–122], wear resistance [123], adhesion [124], electromagnetic shielding [110] and thermal conductivity [111, 120] of polymers on the addition of nanodiamond. However, degradation in properties occurs when non-purified or aggregated nanodiamonds are used, emphasizing the need to use a well-dispersed and properly functionalized material.

The maximum improvement in mechanical properties was recently demonstrated [120] when a nanodiamond porous scaffold was infiltrated by a polymer. Transparent poly(vinyl alcohol) nanocomposites with improved mechanical properties have been produced by adding small amounts of nanodiamond [125]. The interactions at the interface between the diamond nanoparticles and the matrix, and also the dispersion of the nanodiamonds in the matrix, can be controlled through surface chemistry (Fig. 11.6). For example, hydrophobic ND–ODA is highly dispersible in hydrophobic polymers such as poly(L-lactic acid) (PLLA), resulting in almost an order of a magnitude increase in hardness, together with reduced creep and three times higher Young's modulus at 10 wt% of ND–ODA ([11]; Fig. 11.6b). Further improvement in composite properties can be achieved when nanodiamond is modified in such a way that it forms a strong covalent interface with ceramic, metal or polymer matrix [12, 14, 126] (Fig. 11.6a,c). Nanocomposites are a rapidly growing area in nanodiamond research.

11.6.3 Drug Delivery

The properties needed for a drug delivery platform include biocompatibility, the ability to carry a broad range of therapeutics (Fig. 11.7), dispersability in water and scalability. The potential for

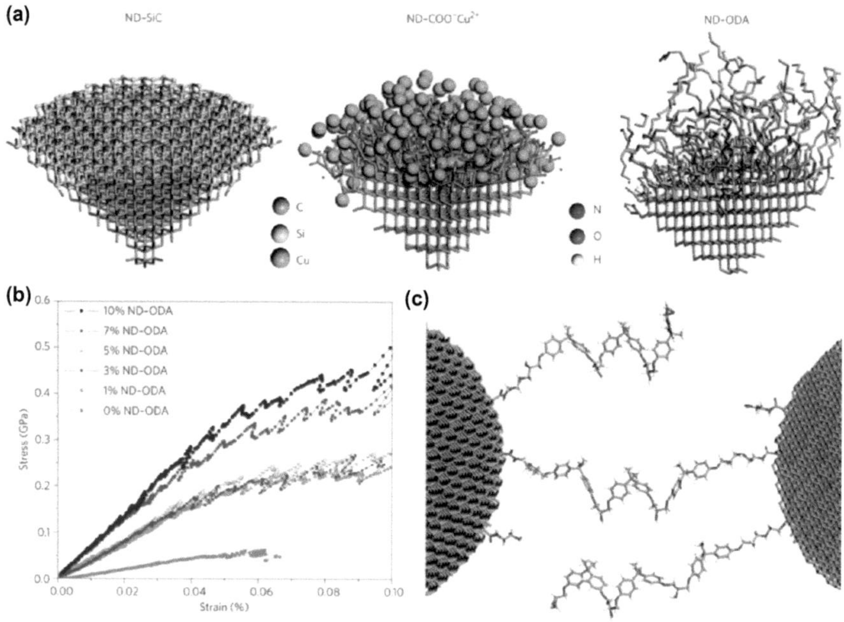

Figure 11.6 Advanced atomic-level composite design with nanodiamond. (a) Three examples of the interfaces between nanodiamond and different matrices. Nanodiamond can bind to SiC through C–Si bonds between the surface of the nanodiamond and the Si atoms in the SiC to produce ND–SiC (left). Carboxylic groups present on the nanodiamond surface can form salts by ion exchange reactions with different metal ions, such as Cu^{2+} (middle; [141]). Metal ions can be later reduced, forming an atomically thin metal layer around the particle. These metallized particles can be used as a means to disperse nanodiamonds in metals that do not wet carbon, and also to produce wear-resistant ND–Cu sliding contacts. Nanodiamonds with surface carboxylic groups can be functionalized through covalent attachment of ODA by amide bond formation (right). (b) Stress–strain curves for six ND-ODA–PLLA composites that contain different amounts of ND–ODA [11]. Young's modulus of a given composite is proportional to the slope of its stress–strain curve. (c) Aminated nanodiamond, produced through covalent attachment of ethylenediamine to carboxylic groups on the surface of the nanodiamond, can replace traditional epoxy curing agents (amines) in reaction with epoxy resin. This results in the covalent incorporation of the nanodiamond into the epoxy polymer network at a molecular level, improving the mechanical properties of the polymer matrix [14].

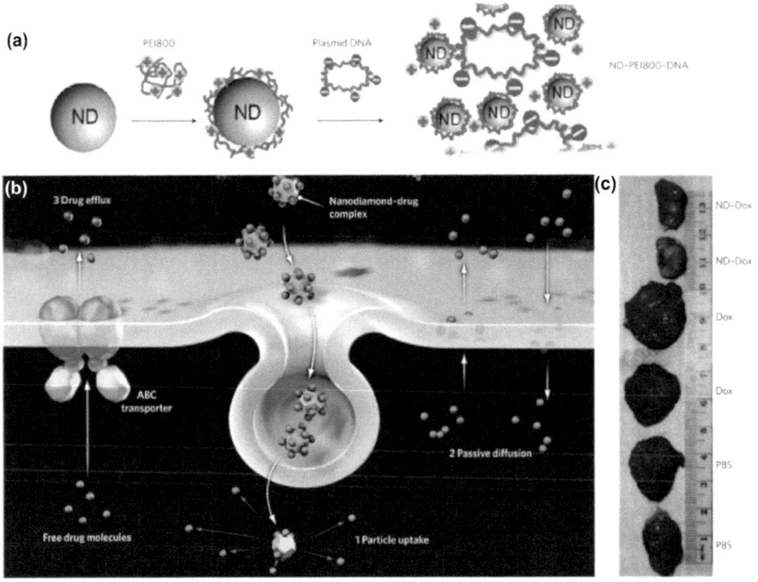

Figure 11.7 Nanodiamonds and drug delivery. (a) DNA can be electrostatically attached to nanodiamonds by first covering negatively charged carboxylated nanodiamonds with positively charged PEI800 molecules. A similar electrostatic binding strategy has been used to attach siRNA and doxorubicin (Dox) to nanodiamond [104]. (b) Schematic representation of a proposed mechanism for ND–Dox complexes interacting with a cell. (1) Endocytosis of the ND–drug complexes. (2) Diffusion of free drug molecules across the cell membrane. (3) ABC transporter proteins efflux free drug molecules out of the cell, whereas ND–drug complexes are able to remain inside the cell and deliver a steady, lethal dose of the drug to the tumour. (c) Photographs of breast-cancer tumours after treatment with ND–Dox (top), Dox (middle) and a control (PBS; bottom). Two representative tumours are shown in each case. The large size of the tumours excised after long-term treatment with Dox or PBS illustrates a reduced ability of Dox to inhibit tumour growth owing to the extreme resistance of the 4T1 breast cancer to chemotherapy. In contrast, treatment with ND–Dox clearly reduces the size of the tumours. Figure reproduced with permission from: (a) [127], © 2009 ACS; (b) [142], © 2011 AAAS; (c) [104], © 2011 AAAS.

targeted therapy—possibly in combination with imaging—is also important. Nanodiamonds are capable of meeting most, if not all, of these requirements [15, 17, 53, 127, 128].

Following initial studies [53, 129] demonstrating nanodiamond-mediated delivery of doxorubicin (a drug used to treat a range of cancers), the efficacy and safety of nanodiamonds has recently been validated in mice [104]. Nanodiamond–doxorubicin complexes (ND–Dox) were used to treat drug-resistant breast cancer (4T1) and liver cancer (LT2-M) models. The nanodiamond reduced the capacity of the tumours to expel the doxorubicin (Fig. 11.7b), and the circulation half-time of the ND–Dox complexes was found to be 10 times that of unmodified doxorubicin [104]. Other advantages of ND–Dox complexes include the absence of myelosuppression (which is high when free Dox molecules are used), the absence of mortality when high doses are delivered (high doses of free Dox generally kill the mice in these experiments) and significant reductions in the sizes of the tumours [104] (Fig. 11.7c). In addition to delivering small molecules, nanodiamonds coated with polyethylenimine 800 (PEI800) were studied for delivery of nucleic acids. These studies have shown a 70-fold enhancement of GFP plasmid transfection efficacy while maintaining the less toxic properties of PEI800 [127]. And ND–PEI800 delivery of siRNA to silence GFP expression resulted in increased efficacy over lipofectamine (a widely used delivery platform) in physiological conditions [128]. A range of other cargos have been delivered including covalently attached drugs [130, 131], proteins [15, 16], small molecules under acidic conditions (which are commonly observed in tumours) [16] and siRNA for specific cancers [17]. As yet, nanodiamonds have been mostly studied as potential injectable therapeutic agents for generalized drug delivery, but it has also been shown that films of parylene–nanodiamond composites can be used for the localized sustained release of drugs over periods ranging from two days to one month [132, 133].

11.6.4 Protein Mimics

The small size, stable core, rich surface chemistry, ability to self-assemble [115] and low cytotoxicity of nanodiamonds have led to suggestions that they could be used to mimic globular proteins [134]. As examples of mimicking the transport function of proteins, in addition to delivering drugs into cells, functionalized nanodiamonds could be used to transport other molecules, including genetic material, across cellular membranes, including the blood–brain barrier. Other protein functions could also be mimicked by nanodiamond particles. For instance, histones are highly alkaline proteins that are involved in DNA folding and unfolding: these processes involve the DNA strand spooling around a bio-molecular complex called a nucleosome, which has histones at its core. Although nanodiamonds are smaller than histones ($\sim$5 nm compared with $\sim$10 nm), their size can be increased by selective air oxidation [135]. Moreover, their surfaces can be made alkaline by covalent attachment of amino groups [14] (Fig. 11.5) and then wrapped with DNA (similar to Fig. 11.7a) to make artificial nucleosomes. The catalytic properties of nanodiamond (recently demonstrated in the synthesis of styrene in mild conditions [112]) imply that the enzymatic functions of some proteins could also be mimicked by functionalized nanodiamond.

11.6.5 Tissue Scaffolds and Surgical Implants

Tissue engineering and regenerative medicine are areas of significant interest, particularly given their potential for restoring damaged tissue. Nanodiamond monolayers have been shown to act as a platform for neuronal growth similar to protein-coated materials [136]. The superior mechanical properties of nanodiamond—in combination with tunable surface chemistry, ability to deliver drugs and biologically active molecules, and biocompatibility—are beneficial for reinforcement of biodegradable polymers to create multifunctional tissue engineering scaffolds. One nanodiamond–polymer composite being explored for biomedical applications is ND–ODA–PLLA. Although PLLA is biocompatible and bioresorbable, it is not mechanically robust for load bearing implants: however, the addition of well-dispersed ND–ODA leads to hardness and Young's modulus values that are close to those of human cortical bone [11] (Fig. 11.6b). Moreover, murine 7F2 osteoblast morphology and proliferation were unaffected when cultured with the ND–ODA–PLLA (Fig. 11.4f). Thus, these composites possess clinically relevant properties while being non-toxic and highly scalable to produce. The significant improvement in characteristics produced by nanodiamond incorporation in this material represent an outcome that may be used in a broad range of other biomaterials.

11.7 Outlook

Many challenges remain, such as producing particles with sizes smaller than 4 nm, preventing (re-)aggregation, large-volume manufacturing of colloidal solutions of single-digit particles, and better control of surface chemistry. There is also a long list of fundamental questions. What is the shape of as-produced nanodiamond? (Most researchers study nanodiamonds after oxidative purification, which may change their shape.) What is the content of twinned crystals in nanodiamond and how do they affect the reported grain size values? What causes the dark brown colour of single-digit nanodiamond aqueous colloids? What causes the high ζ-potential of hydrogen-treated, ZrO_2-milled or ozone-purified nanodiamonds in water? What causes positive ζ-potential of hydrogen-treated and partially graphitized nanodiamonds in water? How do hydrogen-treated nanodiamonds, which seem to be hydrophobic, disperse in water? What is the mechanism of increased wear resistance and lower friction coefficients when nanodiamond colloids are used? How can nanodiamonds dispersed in water or ethylene glycol replace oil? How does surface chemistry influence the brightness of the fluorescence from NV centres? What is the mechanism responsible for the fluorescence of ND–ODA and similar modified nanodiamonds? What causes nanodiamonds to agglomerate? Are the agglomerates strong and permanent, or weakly bonded and dynamic? What are the differences in structure and agglomeration strength between detonation nanodiamonds produced by wet or dry methods?

The wide range of potential applications for nanodiamonds will continue to drive research in this field forward. Better understanding of their structure and surface chemistry will lead to greater control over their properties, and also help to increase manufacturing volumes, possibly to levels that will surpass those of fullerenes and other carbon nanomaterials. The search for new ways to make nanodiamonds will also continue, and any increase in supply will almost certainly lead to new applications.

Acknowledgements

We thank our students and post-docs who helped to collect data and to write and revise the paper, and V. Danilenko for useful discussions. V.N.M. and Y.G. acknowledge support from the National Science Foundation (CMMI-0927963, nanodiamond–polymer composites) and from FIRST (Fluid Interface Reactions, Structures and Transport), an Energy Frontier Research Center funded by the US Department of Energy Office of Science, Office of Basic Energy Sciences (nanodiamond chemistry, graphitization and carbon nano-onions). O.S. was supported in part by the Space and Naval Warfare Systems Centers (N66001-04-1-8933) and the Army Research Laboratory (W911NF-04-2-0023). D.H. was supported by the National Science Foundation (CMMI-0846323, CMMI- 0856492, DMI-0327077, DMR-1105060), the National Center for Learning and Teaching, the V Foundation for Cancer Research Scholars Award, the Wallace H. Coulter Foundation Translational Research Award, National Cancer Institute (U54CA151880 and 1R01CA159178-01) and the EU Framework Programme (FP7-KBBE-2009-3).

References

1. Danilenko, V. V. On the history of the discovery of nanodiamond synthesis. *Phys. Solid State* **46**, 595–599 (2004); this paper, by the inventor of detonation nanodiamond, describes the history of the discovery of nanodiamonds.
2. Greiner, N. R., Phillips, D. S., Johnson, J. D., and Volk, F. Diamonds in detonation soot. *Nature* **333**, 440–442 (1988).
3. Ozawa, M., et al. Preparation and behavior of brownish, clear nanodiamond colloids. *Adv. Mater.* **19**, 1201–1206 (2007).
4. Chang, Y. R., et al. Mass production and dynamic imaging of fluorescent nanodiamonds. *Nature Nanotech.* **3**, 284–288 (2008); luminescent

nanodiamonds can be mass-produced by irradiating synthetic diamond nanocrystallites with He ions; nanodiamonds produced with this method have been used in a commercial product.

5. Mochalin, V. N., and Gogotsi, Y. Wet chemistry route to hydrophobic blue fluorescent nanodiamond. *J. Am. Chem. Soc.* **131**, 4594–4595 (2009).
6. Maze, J. R., et al. Nanoscale magnetic sensing with an individual electronic spin in diamond. *Nature* **455**, 644–647 (2008).
7. Krueger, A. Diamond nanoparticles: Jewels for chemistry and physics. *Adv. Mater.* **20**, 2445–2449 (2008).
8. Zheng, W. W., et al. Organic functionalization of ultradispersed nanodiamond: Synthesis and applications. *J. Mater. Chem.* **19**, 8432–8441 (2009).
9. Spitsyn, B. V., et al. Inroad to modification of detonation nanodiamond. *Diamond Relat. Mater.* **15**, 296–299 (2006).
10. Behler, K. D., et al. Nanodiamond–polymer composite fibers and coatings. *ACS Nano* **3**, 363–369 (2009).
11. Zhang, Q., et al. Fluorescent PLLA-nanodiamond composites for bone tissue engineering. *Biomaterials* **32**, 87–94 (2011); this paper reports that a nanodiamond-poly(L-lactic acid) composite can be used for bone tissue engineering.
12. Wang, D. H., Tan, L. S., Huang, H. J., Dai, L. M., and Osawa, E. In-situ nanocomposite synthesis: Arylcarbonylation and grafting of primary diamond nanoparticles with a poly(ether–ketone) in polyphosphoric acid. *Macromolecules* **42**, 114–124 (2009).
13. Cheng, J. L., He, J. P., Li, C. X., and Yang, Y. L. Facile approach to functionalize nanodiamond particles with V-shaped polymer brushes. *Chem. Mater.* **20**, 4224–4230 (2008).
14. Mochalin, V. N., et al. Covalent incorporation of aminated nanodiamond into an epoxy polymer network. *ACS Nano* **5**, 7494–7502 (2011); demonstration of covalent incorporation of aminated nanodiamond into epoxy, resulting in a composite with improved mechanical properties.
15. Shimkunas, R. A., et al. Nanodiamond–insulin complexes as pH-dependent protein delivery vehicles. *Biomaterials* **30**, 5720–5728 (2009).
16. Purtov, K. V., Petunin, A. I., Burov, A. E., Puzyr, A. P., and Bondar, V. S. Nanodiamonds as carriers for address delivery of biologically active substances. *Nanoscale Res. Lett.* **5**, 631–636 (2010).

17. Alhaddad, A., et al. Nanodiamond as a vector for siRNA delivery to Ewing sarcoma cells. *Small* **7**, 3087–3095 (2011).
18. Osswald, S., Yushin, G., Mochalin, V., Kucheyev, S. O., and Gogotsi, Y. Control of sp^2/sp^3 carbon ratio and surface chemistry of nanodiamond powders by selective oxidation in air. *J. Am. Chem. Soc.* **128**, 11635–11642 (2006); nanodiamonds can be purified by the selective oxidation of non-diamond carbon in air; this method has been used to produce high-purity nanodiamond powders on an industrial scale.
19. Shenderova, O., et al. Surface chemistry and properties of ozone-purified detonation nanodiamonds. *J. Phys. Chem. C* **115**, 9827–9837 (2011); comprehensive description of the distinctive properties of ozone-purified detonation nanodiamond.
20. Schrand, A. M., et al. in *Safety of Nanoparticles. From Manufacturing to Medical Applications. Nanostructure Science and Technology* (ed. Webster, T. J.) 159–187 (Springer, 2009).
21. Schrand, A. M., Hens, S. A. C., and Shenderova, O. A. Nanodiamond particles: Properties and perspectives for bioapplications. *Crit. Rev. Solid State Mater. Sci.* **34**, 18–74 (2009).
22. Schrand, A. M., et al. Are diamond nanoparticles cytotoxic? *J. Phys. Chem. B* **111**, 2–7 (2007).
23. Yang, G. W., Wang, J. B., and Liu, Q. X. Preparation of nano-crystalline diamonds using pulsed laser induced reactive quenching. *J. Phys. Condens. Mat.* **10**, 7923–7927 (1998).
24. Boudou, J. P., et al. High yield fabrication of fluorescent nanodiamonds. *Nanotechnology* **20**, 235602 (2009).
25. Frenklach, M., et al. Induced nucleation of diamond powder. *Appl. Phys. Lett.* **59**, 546–548 (1991).
26. Gogotsi, Y. G., et al. Structure of carbon produced by hydrothermal treatment of β-SiC powder. *J. Mater. Chem.* **6**, 595–604 (1996).
27. Welz, S., Gogotsi, Y., and McNallan, M. J. Nucleation, growth, and graphitization of diamond nanocrystals during chlorination of carbides. *J. Appl. Phys.* **93**, 4207–4214 (2003).
28. Daulton, T. L., Kirk, M. A., Lewis, R. S., and Rehn, L. E. Production of nanodiamonds by high-energy ion irradiation of graphite at room temperature. *Nucl. Instrum. Meth. B* **175**, 12–20 (2001).
29. Banhart, F., and Ajayan, P. M. Carbon onions as nanoscopic pressure cells for diamond formation. *Nature* **382**, 433–435 (1996).

30. Galimov, É., et al. Experimental corroboration of the synthesis of diamond in the cavitation process. *Dokl. Phys.* **49**, 150–153 (2004).
31. Guillois, O., Ledoux, G., and Reynaud, C. Diamond infrared emission bands in circumstellar media. *Astrophys. J.* **521**, L133–L136 (1999).
32. Goto, M., et al. Spatially resolved 3 μm spectroscopy of Elias 1: Origin of diamonds in protoplanetary disks. *Astrophys. J.* **693**, 610–616 (2009).
33. Dahl, J. E., Liu, S. G., and Carlson, R. M. K. Isolation and structure of higher diamondoids, nanometer-sized diamond molecules. *Science* **299**, 96–99 (2003).
34. Vaijayanthimala, V., and Chang, H. C. Functionalized fluorescent nanodiamonds for biomedical applications. *Nanomedicine* **4**, 47–55 (2009).
35. Xing, Y., and Dai, L. Nanodiamonds for nanomedicine. *Nanomedicine* **4**, 207–218 (2009).
36. Barnard, A. S. Diamond standard in diagnostics: Nanodiamond biolabels make their mark. *Analyst* **134**, 1751–1764 (2009).
37. Hui, Y. Y., Cheng, C-L., and Chang, H-C. Nanodiamonds for optical bioimaging. *J. Phys. D* **43**, 374021 (2010).
38. Viecelli, J. A., Bastea, S., Glosli, J. N., and Ree, F. H. Phase transformations of nanometer size carbon particles in shocked hydrocarbons and explosives. *J. Chem. Phys.* **115**, 2730–2736 (2001).
39. Danilenko, V. V. in *Synthesis, Properties and Applications of Ultra-nanocrystalline Diamond* (Proceedings of NATO Advanced Research Workshop) (eds Gruen, D. Shenderova O., and Vul', A.) 181–198 (Springer, 2005).
40. Badziag, P., Verwoerd, W. S., Ellis, W. P., and Greiner, N. R. Nanometre-sized diamonds are more stable than graphite. *Nature* **343**, 244–245 (1990).
41. Barnard, A. S., Russo, S. P., and Snook, I. K. Structural relaxation and relative stability of nanodiamond morphologies. *Diamond Relat. Mater.* **12**, 1867–1872 (2003).
42. Barnard, A. S., and Sternberg, M. Crystallinity and surface electrostatics of diamond nanocrystals. *J. Mater. Chem.* **17**, 4811–4819 (2007).
43. Raty, J. Y., and Galli, G. Ultradispersity of diamond at the nanoscale. *Nature Mater.* **2**, 792–795 (2003).
44. Lai, L., and Barnard, A. S. Modeling the thermostability of surface functionalisation by oxygen, hydroxyl, and water on nanodiamonds. *Nanoscale* **3**, 2566–2575 (2011).

45. Lai, L., and Barnard, A. S. Stability of nanodiamond surfaces exposed to N, NH, and NH_2. *J. Phys. Chem. C* **115**, 6218–6228, (2011).
46. Aleksenskiy, A., Baidakova, M., Osipov, V., and Vul', A. in *Nanodiamonds. Applications in Biology and Nanoscale Medicine* (ed. Ho, D.) 55–79 (Springer, 2010).
47. Vlasov, I. I., et al. Nitrogen and luminescent nitrogen-vacancy defects in detonation nanodiamond. *Small* **6**, 687–694 (2010).
48. Shenderova, O. A., and Gruen, D. M. *Ultrananocrystalline Diamond: Synthesis, Properties, and Applications* (William Andrew, 2006); first book in English on Russian research on detonation nanodiamonds.
49. Dolmatov, V. Y. Detonation synthesis ultradispersed diamonds: Properties and applications. *Usp. Khim.* **70**, 687–708 (2001).
50. Williams, O. A., et al. Size-dependent reactivity of diamond nanoparticles. *ACS Nano* **4**, 4824–4830 (2010).
51. Ho, D. *Nanodiamonds Applications in Biology and Nanoscale Medicine.* (Springer, 2010).
52. Fedyanina, O. N., and Nesterenko, P. N. Regularities of chromatographic retention of phenols on microdispersed sintered detonation nanodiamond in aqueous-organic solvents. *Russ. J. Phys. Ch. A* **84**, 476–480 (2010).
53. Huang, H., Pierstorff, E., Osawa, E., and Ho, D. Active nanodiamond hydrogels for chemotherapeutic delivery. *Nano Lett.* **7**, 3305–3314 (2007).
54. Osawa, E. Recent progress and perspectives in single-digit nanodiamond. *Diamond Relat. Mater.* **16**, 2018–2022 (2007); review article on the deagglomeration of detonation nanodiamond into 4–5 nm nanoparticles.
55. Aleksenskiy, A. E., Eydelman, E. D., and Vul, A. Y. Deagglomeration of detonation nanodiamonds. *Nanosci. Nanotechnol. Lett.* **3**, 68–74 (2011).
56. Pentecost, A., Gour, S., Mochalin, V., Knoke, I., and Gogotsi, Y. Deaggregation of nanodiamond powders using salt- and sugar-assisted milling. *ACS Appl. Mater. Interfaces* **2**, 3289–3294 (2010).
57. Krueger, A., Stegk, J., Liang, Y. J., Lu, L., and Jarre, G. Biotinylated nanodiamond: Simple and efficient functionalization of detonation diamond. *Langmuir* **24**, 4200–4204 (2008).
58. Liang, Y. J., et al. Deagglomeration and surface modification of thermally annealed nanoscale diamond. *J. Colloid Interface Sci.* **354**, 23–30 (2011).

59. Bondar, V. S., and Puzyr, A. P. Nanodiamonds for biological investigations. *Phys. Solid State* **46**, 716–719 (2004).
60. Shenderova, O., et al. Modification of detonation nanodiamonds by heat treatment in air. *Diamond Relat. Mater.* **15**, 1799–1803 (2006).
61. Larionova, I., et al. Properties of individual fractions of detonation nanodiamond. *Diamond Relat. Mater.* **15**, 1804–1808 (2006).
62. Grichko, V., Tyler, T., Grishko, V. I., and Shenderova, O. Nanodiamond particles forming photonic structures. *Nanotechnology* **19**, 225201 (2008).
63. Morita, Y., et al. A facile and scalable process for size-controllable separation of nanodiamond particles as small as 4 nm. *Small* **4**, 2154–2157 (2008).
64. Meinhardt, T., Lang, D., Dill, H., and Krueger, A. Pushing the functionality of diamond nanoparticles to new horizons: Orthogonally functionalized nanodiamond using click chemistry. *Adv. Funct. Mater.* **21**, 494–500 (2011).
65. Krueger, A. The structure and reactivity of nanoscale diamond. *J. Mater. Chem.* **18**, 1485–1492 (2008).
66. Arnault, J. C. Surface chemical modifications and surface reactivity of nanodiamonds hydrogenated by CVD plasma. *Phys. Chem. Chem. Phys.* **13**, 11481–11487 (2011).
67. Liu, Y., Gu, Z. N., Margrave, J. L., and Khabashesku, V. N. Functionalization of nanoscale diamond powder: Fluoro-, alkyl-, amino-, and amino acid- nanodiamond derivatives. *Chem. Mater.* **16**, 3924–3930 (2004).
68. Lisichkin, G., Korol'kov, V., Tarasevich, B., Kulakova, I., and Karpukhin, A. Photochemical chlorination of nanodiamond and interaction of its modified surface with C-nucleophiles. *Russ. Chem. Bull.* **55**, 2212–2219 (2006).
69. Krueger, A., and Boedeker, T. Deagglomeration and functionalisation of detonation nanodiamond with long alkyl chains. *Diamond Relat. Mater.* **17**, 1367–1370 (2008).
70. Liang, Y. J., Ozawa, M., and Krueger, A. A general procedure to functionalize agglomerating nanoparticles demonstrated on nanodiamond. *ACS Nano* **3**, 2288–2296 (2009).
71. Krueger, A. New carbon materials: Biological applications of functionalized nanodiamond materials. *Chem. Eur. J.* **14**, 1382–1390 (2008).
72. Kruger, A., Liang, Y. J., Jarre, G., and Stegk, J. Surface functionalisation of detonation diamond suitable for biological applications. *J. Mater. Chem.* **16**, 2322–2328 (2006).

73. Jarre, G., Liang, Y. J., Betz, P., Lang, D., and Krueger, A. Playing the surface game- Diels-Alder reactions on diamond nanoparticles. *Chem. Commun.* **47**, 544–546 (2011).
74. Yeap, W. S., Chen, S. M., and Loh, K. P. Detonation nanodiamond: An organic platform for the Suzuki coupling of organic molecules. *Langmuir* **25**, 185–191 (2009).
75. Mochalin, V., Osswald, S., and Gogotsi, Y. Contribution of functional groups to the Raman spectrum of nanodiamond powders. *Chem. Mater.* **21**, 273–279 (2009).
76. Kulakova, I. Surface chemistry of nanodiamonds. *Phys. Solid State* **46**, 636–643 (2004).
77. Jiang, T. L., Xu, K., and Ji, S. F. FTIR studies on the spectral changes of the surface functional groups of ultradispersed diamond powder synthesized by explosive detonation after treatment in hydrogen, nitrogen, methane and air at different temperatures. *J. Chem. Soc. Faraday Trans.* **92**, 3401–3406 (1996).
78. Ji, S. F., Jiang, T. L., Xu, K., and Li, S. B. FTIR study of the adsorption of water on ultradispersed diamond powder surface. *Appl. Surf. Sci.* **133**, 231–238 (1998).
79. Korolkov, V. V., Kulakova, I. I., Tarasevich, B. N., and Lisichkin, G. V. Dual reaction capacity of hydrogenated nanodiamond. *Diamond Relat. Mater.* **16**, 2129–2132 (2007).
80. Ferrari, A. C., and Robertson, J. Raman spectroscopy of amorphous, nanostructured, diamond-like carbon, and nanodiamond. *Phil. Trans. R. Soc. A* **362**, 2477–2512, (2004).
81. Mykhaylyk, O. O., Solonin, Y. M., Batchelder, D. N., and Brydson, R. Transformation of nanodiamond into carbon onions: A comparative study by high-resolution transmission electron microscopy, electron energy-loss spectroscopy, X-ray diffraction, small-angle X-ray scattering, and ultraviolet Raman spectroscopy. *J. Appl. Phys.* **97**, 074302 (2005).
82. Obraztsova, E. D., et al. Raman and photoluminescence investigations of nanograined diamond films. *Nanostruct. Mater.* **6**, 827–830 (1995).
83. Yushin, G. N., Osswald, S., Padalko, V. I., Bogatyreva, G. P., and Gogotsi, Y. Effect of sintering on structure of nanodiamond. *Diamond Relat. Mater.* **14**, 1721–1729 (2005).
84. Osswald, S., Mochalin, V. N., Havel, M., Yushin, G., and Gogotsi, Y. Phonon confinement effects in the Raman spectrum of nanodiamond. *Phys. Rev. B* **80**, 075419 (2009).

85. Li, W. F., Irle, S., and Witek, H. A. Convergence in the evolution of nanodiamond Raman spectra with particle size: A theoretical investigation. *ACS Nano* **4**, 4475–4486 (2010).
86. Filik, J., et al. Raman spectroscopy of nanocrystalline diamond: An ab initio approach. *Phys. Rev. B* **74**, 035423 (2006).
87. Korobov, M. V., Avramenko, N. V., Bogachev, A. G., Rozhkova, N. N., and Osawa, E. Nanophase of water in nano-diamond gel. *J. Phys. Chem. C* **111**, 7330–7334 (2007).
88. Rondin, L., et al. Surface-induced charge state conversion of nitrogen-vacancy defects in nanodiamonds. *Phys. Rev. B* **82**, 115449 (2010).
89. Slepetz, B., Laszlo, I., Gogotsi, Y., Hyde-Volpe, D., and Kertesz, M. Characterization of large vacancy clusters in diamond from a generational algorithm using tight binding density functional theory. *Phys. Chem. Chem. Phys.* **12**, 14017–14022 (2010).
90. Neumann, P., et al. Single-shot readout of a single nuclear spin. *Science* **329**, 542–544 (2010).
91. Balasubramanian, G., et al. Nanoscale imaging magnetometry with diamond spins under ambient conditions. *Nature* **455**, 648–651 (2008).
92. Bradac, C., et al. Observation and control of blinking nitrogen-vacancy centres in discrete nanodiamonds. *Nature Nanotech.* **5**, 345–349 (2010).
93. Tisler, J., et al. Highly efficient FRET from single NV center in nanodiamonds to single organic molecule. *ACS Nano* **5**, 7893–7898 (2011).
94. Tisler, J., et al. Fluorescence and spin properties of defects in single digit nanodiamonds. *ACS Nano* **3**, 1959–1965 (2009).
95. Shenderova, O., et al. Nitrogen control in nanodiamond produced by detonation shock-wave-assisted synthesis. *J. Phys. Chem. C* **115**, 14014–14024 (2011).
96. Baranov, P. G., et al. Enormously high concentrations of fluorescent nitrogen- vacancy centers fabricated by sintering of detonation nanodiamonds. *Small* **7**, 1533–1537 (2011).
97. Hens, S. C., et al. Nanodiamond bioconjugate probes and their collection by electrophoresis. *Diamond Relat. Mater.* **17**, 1858–1866 (2008).
98. Huang, L. C. L., and Chang, H. C. Adsorption and immobilization of cytochrome C on nanodiamonds. *Langmuir* **20**, 5879–5884 (2004).

99. Schrand, A. M., Lin, J. B., Hens, S. C., and Hussain, S. M. Temporal and mechanistic tracking of cellular uptake dynamics with novel surface fluorophore-bound nanodiamonds. *Nanoscale* **3**, 435–445 (2011).
100. Faklaris, O., et al. Photoluminescent diamond nanoparticles for cell labeling: Study of the uptake mechanism in mammalian cells. *ACS Nano* **3**, 3955–3962 (2009).
101. McGuinness, L. P., et al. Quantum measurement and orientation tracking of fluorescent nanodiamonds inside living cells. *Nature Nanotech.* **6**, 358–363 (2011).
102. Yuan, Y., et al. Pulmonary toxicity and translocation of nanodiamonds in mice. *Diamond Relat. Mater.* **19**, 291–299 (2010).
103. Mohan, N., Chen, C. S., Hsieh, H. H., Wu, Y. C., and Chang, H. C. In vivo imaging and toxicity assessments of fluorescent nanodiamonds in *Caenorhabditis elegans*. *Nano Lett.* **10**, 3692–3699 (2010); demonstration of the biocompatibility of nanodiamond in worm models.
104. Chow, E. K., et al. Nanodiamond therapeutic delivery agents mediate enhanced chemoresistant tumor treatment. *Sci. Transl. Med.* **3**, 73ra21 (2011); demonstration that nanodiamonds can enable significant improvements to drug delivery efficacy and safety in multiple tumour models in vivo.
105. Shenderova, O., Hens, S., and McGuire, G. Seeding slurries based on detonation nanodiamond in DMSO. *Diamond Relat. Mater.* **19**, 260–267 (2010).
106. Manus, L. M., et al. Gd(III)-nanodiamond conjugates for MRI contrast enhancement. *Nano Lett.* **10**, 484–489 (2010).
107. Saini, G., et al. Core-shell diamond as a support for solid-phase extraction and high-performance liquid chromatography. *Anal. Chem.* **82**, 4448–4456 (2010).
108. Wu, C. C., Han, C. C., and Chang, H. C. Applications of surface-functionalized diamond nanoparticles for mass-spectrometry-based proteomics. *J. Chin. Chem. Soc.* **57**, 583–594 (2010).
109. Pech, D., et al. Ultrahigh-power micrometre-sized supercapacitors based on onion-like carbon. *Nature Nanotech.* **5**, 651–654 (2010).
110. Shenderova, O., et al. Detonation nanodiamond and onion-like carbon: Applications in composites. *Phys. Status Solidi A* **205**, 2245–2251 (2008).
111. Shenderova, O., et al. Nanodiamond and onion-like carbon polymer nanocomposites. *Diamond Relat. Mater.* **16**, 1213–1217 (2007).

112. Su, D. S., et al. Oxidative dehydrogenation of ethylbenzene to styrene over ultra-dispersed diamond and onion-like carbon. *Carbon* **45**, 2145–2151 (2007).
113. Zhang, J. A., et al. Surface chemistry and catalytic reactivity of a nanodiamond in the steam-free dehydrogenation of ethylbenzene. *Angew. Chem. Int. Ed.* **49**, 8640–8644 (2010).
114. Holt, K. B. Diamond at the nanoscale: Applications of diamond nanoparticles from cellular biomarkers to quantum computing. *Phil. Trans. Roy. Soc. A* **365**, 2845–2861 (2007).
115. Huang, H., Dai, L., Wang, D. H., Tan, L-S., and Osawa, E. Large-scale self-assembly of dispersed nanodiamonds. *J. Mater. Chem.* **18**, 1347–1352 (2008).
116. Ivanov, M. G., Pavlyshko, S. V., Ivanov, D. M., Petrov, I., and Shenderova, O. Synergistic compositions of colloidal nanodiamond as lubricant-additive. *J Vac. Sci. Technol. B* **28**, 869–877 (2010).
117. Chou, C. C., and Lee, S. H. Tribological behavior of nanodiamond-dispersed lubricants on carbon steels and aluminum alloy. *Wear* **269**, 757–762 (2010).
118. Kato, T., Lin, W. M., and Osawa, E. Lubrication property of single-digit-nanodiamond in an aqueous colloid. *J. Jpn. Soc. Tribol.* **54**, 122–129 (2009).
119. Matsumoto, N., Joly-Pottuz, L., Kinoshita, H., and Ohmae, N. Application of onion-like carbon to micro and nanotribology. *Diamond Relat. Mater.* **16**, 1227–1230 (2007).
120. Neitzel, I., Mochalin, V., Knoke, I., Palmese, G. R., and Gogotsi, Y. Mechanical properties of epoxy composites with high contents of nanodiamond. *Compos. Sci. Technol.* **71**, 710–716 (2011).
121. Maitra, U., Prasad, K. E., Ramamurty, U., and Rao, C. N. R. Mechanical properties of nanodiamond-reinforced polymer-matrix composites. *Solid State Commun.* **149**, 1693–1697 (2009).
122. Zhang, Q., Naito, K., Tanaka, Y., and Kagawa, Y. Grafting polyimides from nanodiamonds. *Macromolecules* **41**, 536–538 (2008).
123. Lee, J. Y., Lim, D. P., and Lim, D. S. Tribological behavior of PTFE nanocomposite films reinforced with carbon nanoparticles. *Composites B* **38**, 810–816 (2007).
124. Stravato, A., Knight, R., Mochalin, V., and Picardi, S. C. HVOF-sprayed nylon-11 + nanodiamond composite coatings: Production and characterization. *J. Therm. Spray Technol.* **17**, 812–817 (2008).

125. Morimune, S., Kotera, M., Nishino, T., Goto, K., and Hata, K. Poly(vinyl alcohol) nanocomposites with nanodiamond. *Macromolecules* **44**, 4415–4421 (2011).
126. Li, L., Davidson, J. L., and Lukehart, C. M. Surface functionalization of nanodiamond particles via atom transfer radical polymerization. *Carbon* **44**, 2308–2315 (2006).
127. Zhang, X. Q., et al. Polymer-functionalized nanodiamond platforms as vehicles for gene delivery. *ACS Nano* **3**, 2609–2616 (2009).
128. Chen, M., et al. Nanodiamond vectors functionalized with polyethylenimine for siRNA delivery. *J. Phys. Chem. Lett.* **1**, 3167–3171 (2010).
129. Li, X., et al. TAT-conjugated nanodiamond for the enhanced delivery of doxorubicin. *J. Mater. Chem.* **21**, 7966–7973 (2011).
130. Liu, K. K., et al. Covalent linkage of nanodiamond-paclitaxel for drug delivery and cancer therapy. *Nanotechnology* **21**, 315106 (2010).
131. Zhang, X-Q., et al. Multimodal nanodiamond drug delivery carriers for selective targeting, imaging, and enhanced chemotherapeutic efficacy. *Adv. Mater.* **23**, 4770–4775 (2011).
132. Huang, H. J., Pierstorff, E., Osawa, E., and Ho, D. Protein-mediated assembly of nanodiamond hydrogels into a biocompatible and biofunctional multilayer nanofilm. *ACS Nano* **2**, 203–212 (2008).
133. Lam, R., et al. Nanodiamond-embedded microfilm devices for localized chemotherapeutic elution. *ACS Nano* **2**, 2095–2102 (2008).
134. Kotov, N. A. Inorganic nanoparticles as protein mimics. *Science* **330**, 188–189 (2010).
135. Osswald, S., Havel, M., Mochalin, V., Yushin, G., and Gogotsi, Y. Increase of nanodiamond crystal size by selective oxidation. *Diamond Relat. Mater.* **17**, 1122–1126 (2008).
136. Thalhammer, A., Edgington, R. J., Cingolani, L. A., Schoepfer, R., and Jackman, R. B. The use of nanodiamond monolayer coatings to promote the formation of functional neuronal networks. *Biomaterials* **31**, 2097–2104 (2010).
137. Liu, Y., Khabashesku, V. N., and Halas, N. J. Fluorinated nanodiamond as a wet chemistry precursor for diamond coatings covalently bonded to glass surface. *J. Am. Chem. Soc.* **127**, 3712–3713 (2005).
138. Lisichkin, G. V., Kulakova, I. I., Gerasimov, Y. A., Karpukhin, A. V., and Yalkovlev, R. Y. Halogenation of detonation-synthesised nanodiamond surfaces. *Mendeleev Commun.* **19**, 309–310 (2009).

139. Kuznetsov, V. L., Chuvilin, A. L., Butenko, Y. V., Malkov, I. Y., and Titov, V. M. Onion-like carbon from ultra-disperse diamond. *Chem. Phys. Lett.* **222**, 343–348 (1994).
140. Portet, C., Yushin, G., and Gogotsi, Y. Electrochemical performance of carbon onions, nanodiamonds, carbon black and multiwalled nanotubes in electrical double layer capacitors. *Carbon* **45**, 2511–2518 (2007).
141. Panich, A. M., et al. Proton magnetic resonance study of diamond nanoparticles decorated by transition metal ions. *J. Phys. D* **44**, 125303 (2011).
142. Merkel, T. J., and DeSimone, J. M. Dodging drug-resistant cancer with diamonds. *Sci. Transl. Med.* **3**, 73ps8 (2011).

Chapter 12

Nanotechnological Strategies for Engineering Complex Tissues

Tal Dvir,[a,b] Brian P. Timko,[b,c] Daniel S. Kohane,[b] and Robert Langer[a]

[a] *Department of Chemical Engineering, Massachusetts Institute of Technology, Cambridge, Massachusetts 02139, USA*
[b] *Laboratory for Biomaterials and Drug Delivery, Department of Anesthesiology, Division of Critical Care Medicine, Children's Hospital Boston, Harvard Medical School, 300 Longwood Avenue, Boston, Massachusetts 02115, USA*
[c] *Harvard-Massachusetts Institute of Technology, Division of Health Sciences and Technology, Cambridge, Massachusetts 02139, USA*
rlanger@mit.edu

Tissue engineering aims at developing functional substitutes for damaged tissues and organs. Before transplantation, cells are generally seeded on biomaterial scaffolds that recapitulate the extra-cellular matrix and provide cells with information that is important for tissue development. Here we review the nanocomposite nature of the extracellular matrix, describe the design considerations for different tissues and discuss the impact of nanostructures on the properties of scaffolds and their uses in monitoring the behaviour of engineered tissues. We also examine the different nanodevices used to trigger certain processes for tissue development, and offer

---

Reproduced with permission from *Nature Nanotechnology*, **6**, 13–22, 2011.

Nano-Enabled Medical Applications
Edited by Lajos P. Balogh

ISBN 978-981-4800-32-7 (Hardcover), 978-0-429-39903-9 (eBook)
www.jennystanford.com

our view on the principal challenges and prospects of applying nanotechnology in tissue engineering.

Organ transplantation is limited by the number of available donors and high process cost, leaving thousands of people each year on the transplant waiting lists in the United States alone. Many die before an organ donor becomes available. Tissue engineering has evolved as an interdisciplinary technology combining principles from the life, material and engineering sciences with the goal of developing functional substitutes for these damaged tissues and organs [1]. Rather than simply introducing cells into a diseased area to repopulate a defect and/or restore function, in tissue engineering the cells are often seeded in or onto biomaterials before transplantation. These materials serve as temporary scaffolds and promote the reorganization of the cells to form a functional tissue [1] (Fig. 12.1).

Until recently, it was believed that the macroporous features of scaffolds used in tissue engineering mimicked the dimension scale of the extracellular matrix (ECM), and that the matrix itself (natural or artificial) only served as a support for the cells; morphogenesis was controlled passively by defining tissue boundaries. Emphasis was placed on critical engineering and material issues, such as improving mass transfer into the core of the cell constructs and designing biocompatible and biodegradable scaffolds with mechanical properties suitable for engineering various tissues [2].

As the field evolved, attention focused on the biology of the scaffolds (reviewed by Place and colleagues [3], and by Lutolf and Hubbell [4]) and how they affect various cell types. Tissue engineers had recognized that some of the widely used scaffolds do not fairly recapitulate the cell microenvironment and that the ECM is a dynamic and hierarchically organized nanocomposite that regulates essential cellular functions such as morphogenesis, differentiation, proliferation, adhesion and migration [5]. As a consequence, researchers developed and used existing nanotechnological tools for tissue engineering to design advanced nanocomposite scaffolds that can better mimic the ECM and eventually assemble more complex and larger functional tissues.

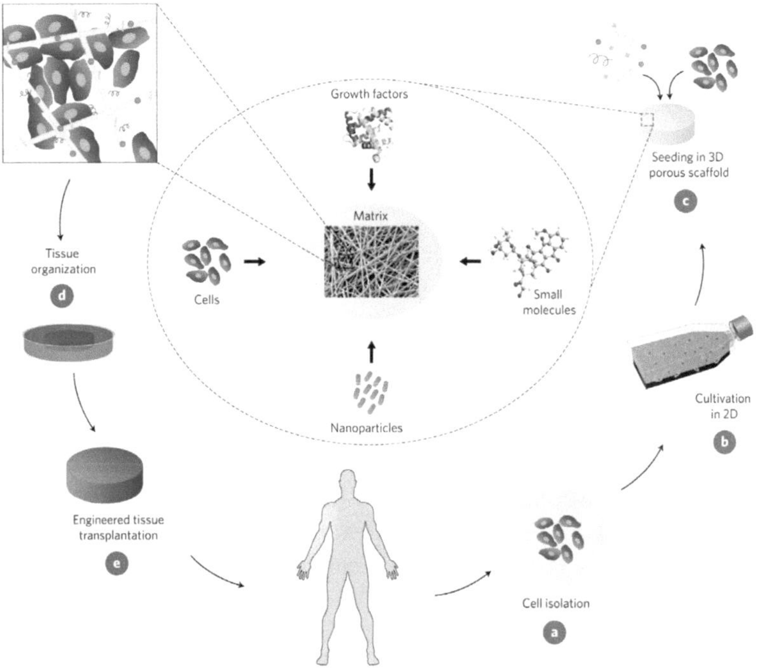

Figure 12.1 An example of a tissue engineering concept that involves seeding cells within porous biomaterial scaffolds. (a), Cells are isolated from the patient and may be cultivated (b) in vitro on two-dimensional surfaces for efficient expansion. (c) Next, the cells are seeded in porous scaffolds together with growth factors, small molecules, and micro- and/or nanoparticles. The scaffolds serve as a mechanical support and a shape-determining material, and their porous nature provides high mass transfer and waste removal. (d) The cell constructs are further cultivated in bioreactors to provide optimal conditions for organization into a functioning tissue. (e) Once a functioning tissue has been successfully engineered, the construct is transplanted on the defect to restore function.

To explain the synthesis of nanoscience with tissue engineering, we will begin by describing the nanocomposite nature of the ECM and discuss how recreating its nanostructure could enhance functional tissue organization, noting the differences in properties and design criteria between various engineered tissues. We will review the impact of nanostructures on matrix properties and their use in monitoring the behaviour of engineered tissues. Finally, we will discuss the principal challenges and prospects in the application of nanotechnology to the field.

12.1 Recreating the Extracellular Microenvironment

Engineering functional tissue requires effective organization of cells into tissue with morphological and physiological features resembling those in vivo. This task is difficult because the signalling factors that drive tissue assembly have not yet been fully identified. Morphogenesis in the three-dimensional (3D) scaffold should occur in a similar way to natural organ development. The cells reorganize owing to interaction with the ECM on the basis of topography [6], mechanical properties (such as matrix stiffness, elasticity and viscosity [7–9]), or concentration gradients of immobilized growth factors [10] or ECM molecules [11]. Recently, Ott and co-workers [12] reported a study emphasizing the importance of the ECM structure in guiding the seeded cells and promoting morphogenesis. Rat hearts were decellularized by perfusion of detergents to preserve the underlying ECM and then reseeded with cardiac and endothelial cells [12]. The cells migrated and self-organized in their natural location in the matrix and by day 8, under physiological load and electrical stimulation, the constructs were able to generate pump function [12]. Similar studies have shown successful engineering of liver, bone, arteries and lung [13–16]. These reports stress the significance of the underlying ECM in promoting a unique microenvironment that fosters tissue organization.

The ECM, which provides cells with a wealth of information, is composed of an intricate interweaving of protein fibres such as fibrillar collagens and elastins, ranging from 10 to several hundreds of nanometres. The mesh is covered with nanoscale adhesive proteins such as laminin and fibronectin that provide specific binding sites for cell adhesion (interacting with integrins, cadherins and so forth) and have been shown to regulate important cell behaviours such as growth, shape, migration and differentiation [17]. Polysaccharides such as hyaluronic acid and heparan sulphate fill the interstitial space between the fibres and act as a compression buffer against the stress placed on the ECM or serve as a growth factor depot [18] (Fig. 12.2).

The significance of the overall fibrillar and porous nanoscale topography of the ECM in promoting essential cellular processes has led tissue engineers to replace macroporous scaffolds with biomimetic materials with nanoscale features. For example, to engineer the ECM fibrous mesh, researchers fabricated scaffolds composed of fibres with appropriate diameters that formed a highly interconnected, porous architecture allowing high mass transfer to and waste removal from the developing tissue [19]. In the past decade, there has been tremendous interest in exploring the potency of biomimetic fibres in tissue engineering, and several techniques originally developed for different purposes have been used for this application.

Electrospinning [20, 21] and molecular self-assembly [22] are common nanofabrication techniques to create 3D scaffolds composed of interwoven fibres. Electrospinning is a simple method that uses an electric field to control the deposition of polymer fibres on a target substrate. The fibres are continuous, with high porosity and high spatial interconnectivity. Moreover, the direction of the fibres can be selectively controlled by spinning speed and polymer concentration to fit the needs of various engineered tissues. The fibre bioactivity can also be designed to include delivery systems that control the release of cues for tissue assembly such as growth factors, peptides, enzymes, drugs, DNA and RNA [23–27]. A recent example is the method developed for fabricating biocompatible polymeric electrospun fibres that contained nanoparticles able to control the release of two proteins in parallel (Fig. 12.3a) [28]. These scaffolds, which enable sequential release of multiple growth factors or cytokines, may have important applications in tissue engineering, such as promoting vasculogenesis or influencing the fate of stem cells [29, 30].

Although considered as a 3D substrate for cell cultivation, electrospun fibres usually promote the assembly of thin tissues on their surface and do not allow proper infiltration of the cells to the core of the matrix. To address this challenge, researchers have combined micro- and nanotechnologies. Hybrid scaffolds that incorporated

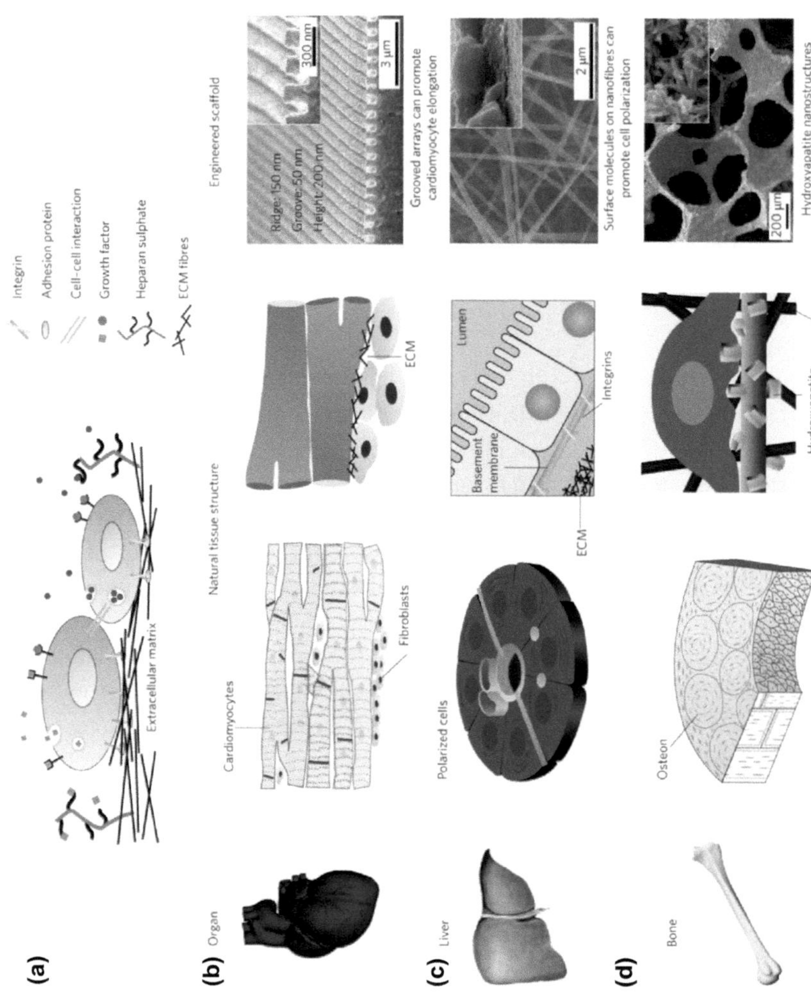
(a)
Extracellular matrix
Integrin
Adhesion protein
Cell–cell interaction
Growth factor
Heparan sulphate
ECM fibres
(b)
Organ
Natural tissue structure
Engineered scaffold
Cardiomyocytes
Fibroblasts
ECM
Ridge: 150 nm
Groove: 50 nm
Height: 200 nm
300 nm
3 μm
Grooved arrays can promote cardiomyocyte elongation
(c)
Liver
Polarized cells
Lumen
Basement membrane
Integrins
ECM
2 μm
Surface molecules on nanofibres can promote cell polarization
(d)
Bone
Osteon
Hydroxyapatite crystals
ECM
200 μm
Hydroxyapatite nanostructures can enhance osteogenesis

Figure 12.2 The information provided to cells by the extracellular matrix (ECM). (a) ECM fibres provide cells with topographical features that trigger morphogenesis. Adhesion proteins such as fibronectin and laminin located on the fibres interact with the cells through their transmembrane integrin receptors to initiate intracellular signalling cascades, which affect most aspects of cell behaviour. Polysaccharides such as hyaluronic acid and heparan sulphate act as a compression buffer against the stress, or serve as a growth factor depot. (b–d) Illustrations of the heart, liver and bone at the level of organ (left) and tissue and cell/matrix interaction (centre), followed by scanning electron micrographs of engineered scaffolds (right). The ECMs of various tissues have different composition and spatial organization of molecules to maintain specific tissue morphologies. For example (b), the ECM of muscle tissues, such as the heart, forces the heart cells (cardiomyocytes) to couple mechanically to each other and to form elongated and aligned cell bundles that create an anisotropic syncytium. Nanogrooved surfaces (SEM image) are suitable matrices for cardiac tissue engineering because they force cardiomyocytes to align. (c) Cells composing epithelial tissues are polarized and contact three types of surfaces for efficient mass transfer: the ECM, other cells and a lumen. Nanofibres modified with surface molecules can promote cell adhesion and tissue polarity (SEM images). (d) Bone is a nanocomposite material consisting primarily of a collagen-rich organic matrix and inorganic hydroxyapatite nanocrystallites, which serve as a chelating agent for mineralization of osteoblasts. The scaffold structure (SEM image), stiffness and hydroxyapatite nanopatterning on the surface (inset) can enhance osteoblast spreading and bone tissue formation. Figures reproduced with permission from: (b) ref. 56, © 2010 NAS; (c) ref. 59, © 2009 Elsevier; (d) ref. 65, © 2010 Elsevier.

both electrospun fibres and 3D microprinting were fabricated to present 3D macroporous structures with interconnected pores for cell culture that also mimicked the ECM [31]. This new scaffold was able to improve cell entrapment and proliferation significantly. In addition, it increased the amount of ECM proteins produced, promoted better cell differentiation and, overall, enhanced tissue regeneration [31].

Despite the simplicity of electrospinning in creating fibrous scaffolds, one substantial disadvantage is that the diameters of the fibres are usually at the upper limits of the 50–500 nm range seen in natural ECM. To emulate natural ECM better, both structurally and functionally, and to promote cell–matrix interaction at the molecular level, 3D scaffolds were created by molecular self-assembly of fibres. This technique involves the spontaneous organization of individual components into an ordered and stable structure with pre-programmed non-covalent bonds (Fig. 12.3b) [32]. The most commonly investigated self-assembled nanofibre matrix for tissue engineering applications is the peptide amphiphile, a chemical compound possessing both hydrophobic and hydrophilic properties [33]. The peptides that assemble to the 3D scaffolds can have a fibre diameter as small as 10 nm and the scaffold pore size can range between 5 and 200 nm, significantly smaller than those produced by electrospinning [22]. An important advantage of this approach is the ability to include functional motif sequences (such as short peptides) that promote adherence, differentiation and maturation [34–36]. This class of scaffolds can provide both mechanical support and instructive cues to the developing tissue. For example, neural progenitor cells were encapsulated within a 3D network of nanofibers [37], designed to present to cells the neurite-promoting laminin epitope IKVAV. Cultivation within this scaffold induced very rapid differentiation of cells into neurons, while discouraging the differentiation to other cell lineages [37], emphasizing the importance and simplicity of controlling cell fate by pre-designing the fibre composition. A challenge in this field is to engineer fibres effectively to provide adhesion and organization motifs that the cells can interact with, and provide multiple cues for intracellular signalling that lead to differentiation and control of gene expression.

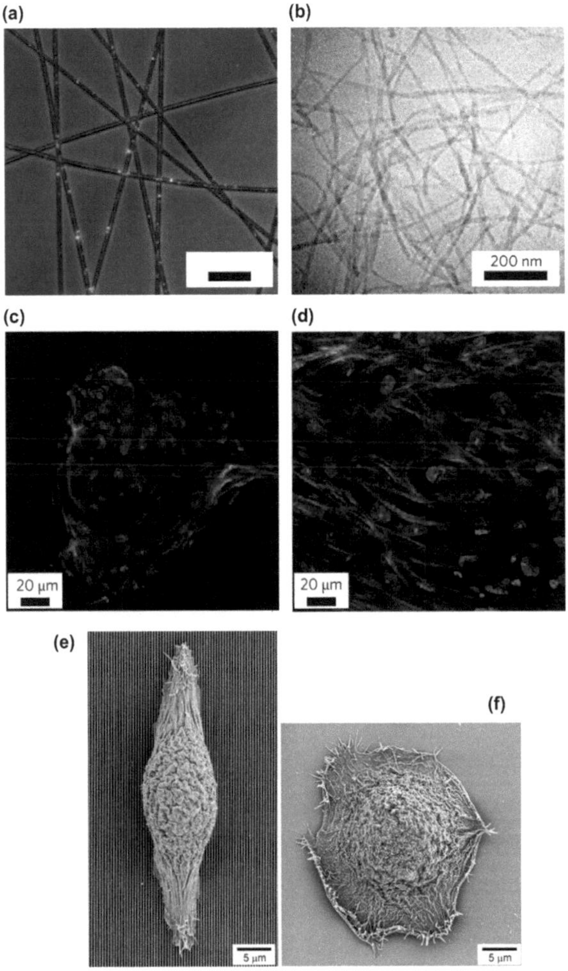

Figure 12.3 Recreating ECM components using nanoscale tools. (a) ECM nanofibres produced by electrospinning polymeric fibres contain nanoparticles that release epidermal growth factor (green) and bovine serum albumin (red) in parallel. Scale bar, 20 μm. (b) Self-assembled peptide amphiphile nanofibres. (c,d) Alginate scaffolds containing short motifs of ECM adhesion proteins such as RGD encouraged mesenchymal stem cells to spread and attach to the matrix (c), whereas on unmodified scaffolds (d) only cell–cell interactions were seen (collagen fibres, green; nuclei, red). (e) Epithelial cells respond to nanopatterning by alignment and elongation along the grating axis. (f) On smooth substrates, cells are mostly rounded. Figures reproduced with permission from: (a) ref. 28, © 2009 Wiley; (b) ref. 32, © 2009 AAAS; (c, d), ref. 42, © 2009 Elsevier; (e, f), ref. 48, © 2003 Company of Biologists.

Other non-fibrous parts of the ECM also have an impact on cell behaviour, and their incorporation into scaffolds is essential. For example, cells interact with the ECM via nanoscale transmembrane integrin receptors which tether the cell cytoskeleton to adhesion molecules, such as fibronectin and laminin, located on the fibres [38]. These focal adhesions initiate the intracellular signalling cascade, which affects most aspects of cell behaviour [7, 39]. Therefore, immobilizing short motifs of the ECM adhesion proteins such as RGD and YIGSR to 3D matrix surfaces enhanced functionality in terms of cell spreading and differentiation (Fig. 12.3c,d) [40–42].

In an attempt to increase the local concentration of growth factors and cytokines in the engineered tissue microenvironment, researchers recapitulated their electrostatic interactions with heparan sulphate proteoglycans [43–45], which act as a reservoir for growth factors [17]. For example, an alginate hydrogel was modified to mimic the heparin/heparan sulphate binding groups by sulphating the uronic acids in the saccharide backbone [45]. The researchers demonstrated strong binding of 10 known heparin binding proteins to the matrix and studied their controlled release [29]. The scaffolds were then used to promote pre-vascularization of engineered cardiac patches after releasing angiogenic and pro-survival factors, and after transplantation on the infarcted heart, promoted its regeneration [46].

In vivo, in addition to the biochemical properties of the ECM, cells encounter and interact with many topographical features ranging from folded protein to banded collagen through a phenomenon known as contact guidance. A mimetic approach to study this effect and to create the structure and length scale (5 nm to micrometre scale) of the native ECM is by nanopatterning of different geometries such as nanogroove, nanopost and nanopit arrays [47]. Although the patterning is only applied on 2D surfaces, this technique reveals aspects of cell behaviour and reactions to various substrate topographies that might be relevant to future 3D scaffolds. The topography of substrates influences a variety of cellular processes, including changes in shape, differentiation and adhesion [47]. For example, epithelial cells elongate and align along patterns of grooves and ridges with feature dimensions as small as 70 nm, whereas on smooth substrates, cells are mostly rounded (Fig. 12.3e,f) [48]. A few years ago, the control of human mesenchymal stem cell (MSC) differentiation using nanoscale symmetry and disorder was demonstrated. Highly ordered topographies produced low to negligible cellular adhesion and osteoblastic differentiation, whereas nanoscale topographic disorder stimulated MSCs to produce bone mineral [49]. The precise mechanism responsible for the morphological response to different nanotopographies is still debated, although a prevalent view is that it arises from the generation of anisotropic stresses. Recent comprehensive reviews of nanoscale topographies and their effect on cells are provided by Bettinger et al. [47] and Kotov et al. [50].

12.2 Scaffold Design Considerations

The ECMs of various tissues in the body differ in the composition and spatial organization of the collagens, elastins, proteoglycans and adhesion molecules, to maintain specific tissue morphologies and organ specific shape and function, and to supply specific instructive cues. Therefore, the design considerations for scaffolds should vary according to the desired engineered tissue. For example, the biochemical, electrical and mechanical functions of the heart are uniquely tangled with biological nanostructures [51]. The heart's 3D ECM network is composed of an intricate, micro- and nanoscale interweaving pattern of fibrillar collagen and elastin bundles that form a dense, elastic network with proteoglycans and with adhesive and non-adhesive molecules. In this defined mesh, the cardiomyocytes are forced to couple mechanically to each other, to form elongated and aligned cell bundles that interact with each other or with neighbouring capillaries and nerves (Fig. 12.2b). This multi-bundled elongated structure is essential for creating an anisotropic syncytium that provides the muscle with unique electrical and mechanical properties, allowing blood pumping to provide the body with oxygen.

Post-isolation cells lose their ultrastructural elongated morphology and their interaction with their surroundings, and they adopt a random distribution on the flat surface of the scaffold, which compromises many of their physiological properties. In recent years, various approaches to promoting elongation and alignment of engineered cardiac tissue have been used, including applying mechanical stretching [52], interstitial fluid flow [53], electrical stimulation [54] or microcontact printing [55], but these have had limited success in engineering a completely aligned, thick tissue. In attempt to promote aligned structures, a nanotopographically controlled model of cardiac muscle was developed that mimics the structural and functional properties of native myocardial tissue and specifically the ECM architecture. Cardiac cells were cultured on polyethylene glycol hydrogel grooved arrays with ridge (groove) widths ranging from 150 (50) to 800 (800) nm. These cells self-assembled in the direction of the grated surfaces, became elongated and formed a highly anisotropic cell array that could produce the aligned contraction essential for heart function [56]. Although this technique promotes a proper cardiac-cell organization, the challenge of patterning such arrays within 3D scaffolds and engineering thick, aligned tissues still remains.

Most of the tissues in the body do not require an aligned mesh of ECM. In epithelial tissues such as the liver, pancreas or kidney, each cell is polarized and contacts three types of surfaces: a basal surface, which contacts the ECM, a lateral surface, which contacts other cells, and an apical surface, which faces the lumen (Fig. 12.2c). Cells that do not contact the ECM or ones lacking apical surfaces will undergo apoptosis or generate a lumen at a region of contact with other cells [57]. Engineering this precise structure is essential for the function of the tissue. In the body, epithelial tissues are involved in secretion, absorption and transcellular transport, and therefore reorganizing the cells in this complex structure, which includes layers of epithelial cells facing blood vessels, will result in functioning tissues allowing efficient diffusion to and from the adjacent lumens. To induce the interaction of the basal surface of the liver with ECM fibres, Bettinger and co-workers sought to mimic the nanotopography of native collagen films [58]. Because collagen fibrils can extend to tens of micrometres in length and have a diameter between 260 and 410 nm, poly(ester amide) substrates were nanofabricated to produce pillar geometries in this size scale. Cultivation of primary hepatocytes on these substrates enhanced cell attachment and spreading, and most importantly, maintained metabolic function [58]. To prepare this engineered tissue for in vivo applications, it is also important to pattern special geometries between the polarized tissues that induce endothelial cell assembly to lumens. A different approach to providing hepatocytes with a microenvironment that more closely resembles the native one is by culturing them on nanofibrous galactosylated chitosan scaffolds. Nanoscale surface receptors on hepatocytes can interact with the embedded galactose ligands and lead to the formation of flat aggregates, which show a higher level of liver-specific functions [59].

Among the most investigated tissues in the field of tissue engineering is bone. It contains a unique nanocomposite material mainly made of inorganic hydroxyapatite nanocrystallites and a collagen- rich organic matrix [60]. Hydroxyapatite serves as a chelating agent for mineralization of osteoblasts in bone regeneration while the collagen provides mechanical support, promoting adhesion and proliferation. The crystals, 50 nm long, 25 nm wide and 2–5 nm thick, reside in the grooved regions formed by the 3D organization of the fibrils (Fig. 12.2d) [61]. In attempts to orchestrate a microenvironment that will promote the assembly of bone tissue, several groups have investigated the effect of scaffolds based on a combination of natural bone ECM and hydroxyapatite, reporting superior osteoblast adhesion, growth and stimulation for mineralization, compared with matrices without hydroxyapatite [62–64]. In the past year, scaffolds containing various shapes and sizes of hydroxyapatite particles have been fabricated and their bioactivity investigated. The researchers found that nanocomposite coated scaffold of needle-shaped hydroxyapatite particles showed the strongest osteoblast differentiation profile compared with rod and spherical shaped particles [65]. Such observations emphasize that in addition to the dimension scale of the particles, their specific topography is of great importance for the interaction with cells.

Table 12.1 Mimicking ECM components to create cell microenvironment

Natural component	Method	Material used	Role	Refs
Fibres: collagens, elastins	Electrospinning	Carbon, PLA, PGA, PLGA, polycaprolactone, polydioxanone, chitosan	3D cell scaffolding, promote adhesion	19–21, 30, 31
	Self-assembly	Peptide amphiphile	3D cell scaffolding, promote adhesion, differentiation	32–34, 36, 37
	Nanopatterning	PEG, poly(ester amide)	Induce morphogenesis	56, 58
Adhesion proteins: laminin, fibronectin	Conjugation of adhesion motifs to the matrix backbone	RGD and IKVAV amino acid sequences	Promote cell adhesion to matrices	41, 42, 59
Proteoglycans: heparan sulphate	Sulphating matrix backbone	Alginate-sulphate, perlecan, sulphated chitosan	Control the release and presentation of growth factors and cytokines	29, 43–45
Minerals: hydroxyapatite		Hydroxyapatite	Induce mineralization	62–65

12.3 Nanocomposites to Compensate for Scaffold Limitations

In addition to creating ECM-like structures (Table 12.1), nanomaterials can be useful for other purposes. The rationale behind incorporating nanostructures is to compensate for other scaffold limitations such as weak mechanical properties, lack of electrical conductivity, the absence of adhesive and microenvironment-defining moieties, and the inability of cells to self-assemble to 3D tissues (Table 12.2). For example, carbon nanotubes show viscoelastic behaviour similar to that observed in soft-tissue membranes [66], so they have been used to increase the Young's modulus and tensile strength of the hybrid biomaterial [67]. Carbon nanotubes were also shown to support the cultivation of neurons. In recent years, several studies have shown that conjugation of these nanotubes to different substrates can affect cell behaviour and promote attachment, growth, differentiation and long-term survival of neurons [68–70].

Table 12.2 Nanomaterials to compensate for matrix limitations

Matrix limitation	Nanomaterial used	Effect	Refs
Weak mechanical properties	Carbon nanotubes	Increased Young's modulus and tensile strength	67
Electrical resistance	Carbon nanotubes	Increased conductivity	74
Weak cell-matrix interaction	Nanotitanate wires	Cell adhesion to the matrix	78
Lack of biochemical cues	Synthetic and natural nanospheres	Controlled release of biomolecules	79
	Gold nanowires	Control over localization of biomolecules	80
Inability of cells to assemble to 3D structures	Magnetic nanoparticles	Engineering of tubular structures	84
	Phage, magnetic iron oxide and gold nanoparticles	Control over geometry of cell mass	85

One of the main obstacles in neural tissue engineering for the regeneration of a nerve tissue such as the spinal cord may be the loss of anisotropic conduction within the cell-seeded construct owing to lack of tissue consistency or to the non-conductive nature of the biomaterial. One approach to addressing this problem is to incorporate conducting nanostructures into the cell culture. Neurons that grow on a conductive nanotube meshwork display more efficient signal transmission [70–72]. In a recent study, Cellot and co-workers provided new mechanistic insight into how nanotubes target the integrative properties of neurons, showing that nanotubes can improve the responsiveness of neurons by forming tight contacts with the cell membranes that might favour electrical shortcuts between different compartments of the neuron [73]. Such neuronal/nanotube network hybrids may allow one to predict or engineer the interactions between nanomaterials and neurons, and guide the design of smart biomaterials for the engineering of electrically propagating tissues.

Carbon nanotubes have also been used to create 3D conductive structures. Gui and colleagues assembled nanotubes with a diameter of 44 nm into an interconnected 3D framework with ~99% porosity [74]. The porous structure allowed direct polymer infiltration without disturbing the nanotube interconnection, resulting in a sponge composite with high electrical conductivity [74]. The potential of these hybrid scaffolds to support the cultivation of neurons or cardiomyocytes should be further explored.

Nanostructures can also be used to increase the viability and adhesiveness of cells to pre-formed microporous scaffolds [75–77]. Bioactive nanotitanate wires and belts were naturally grown directly inside 3D macrostructures to obtain a hierarchical architecture similar to that of human bone [78]. Large amounts of nanowires and nanobelts, with an average width of 20 to 1,300 nm, covered the pores of the scaffold, growing almost perpendicular to the pore walls. The resulting nanostructures increased the overall hydrophilicity of the porous scaffold, promoting more efficient cell seeding and adhesion [78].

The process of tissue growth and development requires a constant supply of instructive cues. Another promising application for nanoscale structures in tissue engineering is the incorporation of controlled release systems into scaffolds. Controlled delivery of biomolecules, such as growth factors and cytokines in vitro or in vivo, is crucial in the support and enhancement of tissue morphogenesis, viability and functionality. Advances in nanotechnology provided the basis for fabrication of nanoparticulate delivery systems with large ratios of surface area to volume, rendering them very effective within the scaffold microenvironment. Examples of nanoparticles for controlled release of biomolecules include synthetic polymeric nanospheres, nanotubes, nanowires, liposomes and dendrimers (for a comprehensive review see ref. 79).

Recently, Fan and co-workers reported that gold nanowires can be functionalized with a cytokine and manipulated to specific locations using electric fields [80]. The nanowires were able to deliver and release their payload to a pre-specific cell with subcellular resolution and activate desirable signalling pathways. Such control over localization of biomolecules in specific zones, if applied in 3D constructs, can assist in precise engineering of the cellular microenvironment by delivering factors to specific cell types in co-cultures. For example, this technique can be used to engineer the complex microenvironment of the stem cell niche, which is composed of a variety of materials and cell types, originating from stem cells. Factors inducing self-renewal or differentiation of the cells can be manipulated to affect specific cells in different regions of the construct.

In recent years, nanoparticles have also been used to manipulate biomaterials and cells in order to create desired tissue structures. For instance, Alsberg and co-workers used magnetic forces to fabricate fibrin gels with defined nanoarchitecture. Thrombin-coated magnetic beads were manipulated to create an initial 2D hexagonal array that served as nucleation sites for the growth of fibril fibres that assembled into an ordered scaffold. Endothelial cells seeded on these scaffolds adhered to the fibres and their actin filaments aligned according to the scaffold fibres [81].

Nanomagnetic actuation has also been used to culture and detach monolayers of cells. Cells were conjugated to magnetic nanoparticles and seeded on culture plates with a magnet placed underneath. The cells formed a multilayered tissue that could be detached by removing the magnet, or formed into a tubular structure by rolling a cylindrical magnet [82–84]. Souza and co-workers reported a new approach to control the assembly of 3D tissues that could provide an alternative to biodegradable scaffolds and protein matrices [85]. The technology is based on the cellular uptake and subsequent magnetic levitation of a hydrogel composed of phage, magnetic iron oxide and gold nanoparticles. By spatially controlling the magnetic field while cells divide and grow, the researchers were able to manipulate the geometry of the cell mass and engineer 3D structures. This approach can be used to engineer complex tissues composed of several cell types, organized in specialized niches by transferring cells to their specific locations.

12.4 Nanodevices for Tissue Engineering

After successfully engineering a tissue, nanodevices can be useful for triggering desired processes and for following tissue development, where functionality is dependent on the specific material chosen (Table 12.3). Moreover, the interaction of nanomaterials with existing microtechnologies may improve their function. For example, electrodes designed to stimulate neural tissue can be improved by coating them with nanostructures such as carbon nanotubes or other inorganic crystals [86]. Individual carbon nanotubes have been shown to form spontaneous junctions with lipid bilayers that are tight compared with the size of a single protein [87]. Such tight interfaces can improve the quality of neural stimulation and recording by reducing the impedance between the device and cell membrane [50]. Tungsten or stainless steel electrodes coated with carbon nanotubes were shown to be more effective at stimulating and recording from brain tissue than their uncoated counterparts. This is mostly attributed to impedance considerations as well as the higher surface area for charge transfer and favourable stability in vivo [86]. Other types of nanoparticle coatings enable more exquisite functionality such as light sensitivity. For instance, photovoltaic mercury telluride particles assembled into layer-by-layer electrodes were able to stimulate cultured neurons when exposed to visible light [88].

Table 12.3 Inorganic nanoparticles in cellular interfaces and detection

Material	Forms	Applications	Physical attributes
Silicon nanowires	Intrinsic or doped composition, straight or kinked morphologies	Electronic detection	High sensitivity to local charges when operated as transistors
Carbon nanotubes	Semiconducting or metallic	Surface coatings for cell interfaces, electronic detection, optical detection	Exhibit sensitivity to local charges when used as transistors Are fluorescent in near-infrared range
Metallic nanostructures	Spheres, shells or rods. Can be composed of metals such as Au or Ag, or can be synthesized as heterostructures	Optical detection	Exhibit surface plasmon resonance and surface enhanced Raman scattering (SERS) that is dependent on ligand coupling
Optically active semiconductors	Direct-bandgap materials such as HgTe	Surface coatings for optical cellular stimulation	Enable conversion of light into electric field

In addition to stimulation, nanostructures can be used to record the electronic signals that are transmitted through cells such as neurons and cardiomyocytes. One way to record these signals is by lithographically defining nanostructures as field-effect transistors, which are sensitive to local electric field changes. In particular, silicon nanowire transistors are useful for measuring extracellular signals because they exhibit particularly exquisite field-effect sensitivity compared with conventional, planar devices; they are just tens of nanometres in diameter and can therefore interface with cells and tissue at a subcellular level; and they show nanotopographic features that encourage tight interfaces with biological systems, a prerequisite for recording signals with high signal-to-noise ratio [89]. In a recent study, arrays of nanowire devices were interfaced with patterned, cultured neurons and used to measure extracellular signal fluctuations locally at individual axons, dendrites and soma [90]. Nanowire transistors were also interfaced with cardiomyocytes that were either cultured directly on top of nanowire arrays [91] or cultured separately on polymer films that were then brought into direct contact with nanowire arrays [92, 93]. In each case, nanowire devices recorded signals that were substantially larger than those attainable with planar, microscale transistors, demonstrating the added benefit of being orders of magnitude smaller than a cell. Nanowire transistors are also useful for recording signal transmission across the surface of cardiac tissue [94] or throughout brain slices [95].

The synthesis and fabrication techniques associated with nanowire electronics are versatile and could be tailored to specific applications. Nanowire devices have been fabricated on a wide range of substrates, including silicon oxide and flexible plastic [89]. Inorganic or organic transistors have also been demonstrated on resorbable substrates such as silk fibroin [96] or poly(lactic-co-glycolic) acid [97]. The geometry and composition of nanowires themselves has also been elucidated, encompassing complex core–shell wires and even kinked structures with complex chemical profiles [98]. These kinked nanowires together with flexible substrate materials have been used to fabricate 3D, freestanding nanowire transistor probes that were used to penetrate the membranes of living cells and measure intra-cellular signals (Fig. 12.4a) [99]. These nanowire structures may be useful in future for creating 3D array of sensors inside macroporous scaffolds.

Aside from forming electronic interfaces with cells, nanostructured devices can be used to monitor, in real time, the concentration of biomolecules or other chemical species of interest present within the scaffold. Silicon nanowire transistors, for example, have been used to selectively detect proteins at femtomolar-level concentrations or viruses at the single-particle level. In these systems, selectivity was achieved by antibodies that were covalently coupled to the nanowire surface, enabling analyte-specific detection even in complex media, such as blood serum [89]. By detecting small amounts of biomolecules secreted by the engineered tissue, this technique can be used to report on the condition of the

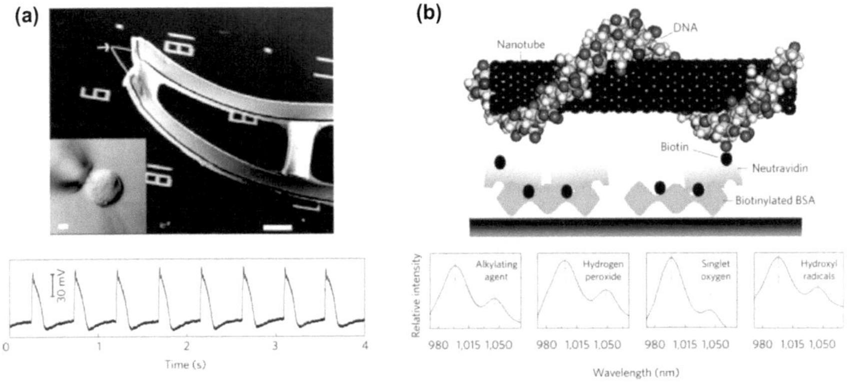

Figure 12.4 Nanodevices in tissue engineering. (a) Three-dimensional, free-standing nanowire transistor probe for electrical recording. The probe is composed of a kinked nanowire (yellow arrow) and a flexible substrate material. The device is used to penetrate the membrane of living cells (inset) and measure intracellular signals (lower panel). Scale bars, 5 μm. (b) Biosensors based on carbon nanotubes are used for the detection of genotoxic analytes, including chemotherapeutic drugs and reactive oxygen species. Upper figure shows a schematic of a sensor made from DNA a single-walled carbon nanotube, attached to a glass surface. Lower figures reveal the spectral changes arising from the interaction of the nanotube sensor with (from left to right): a chemotherapeutic agent, hydrogen peroxide, singlet oxygen and hydroxyl radicals (blue curve, before addition of analytes; green curve, after addition of analytes). Figures reproduced with permission from: (a) ref. 99, © 2010 AAAS; (b) ref. 101, © 2010 NPG.

tissue before transplantation. Other types of nanoparticles can be probed with light to yield optical information that is correlated to the chemical environment. Metallic nanostructures, for example, plasmonically couple to incident light and can be engineered to produce strong surface-enhanced Raman scattering (SERS)-based signals. In a recent example, pairs of gold–silver core–shell nanoparticles were tethered with DNA to form 'nano- dumbbells' with SERS signals emerging only from coupled particles. Hence, nano-dumbbells behave as single-molecule DNA sensors and could also be engineered for other bioassays [100].

Carbon nanotubes have been used as multimodal sensors for chemotherapeutic drugs or reactive oxygen species, where adsorption of these species on the surface of the nanotubes uniquely altered their photoluminescence spectra (Fig. 12.4b) [101]. They have also been used as sensors inside hydrogels, where their emission spectra shifted with the hydration state or cross-linker density in the hydrogel. Gels that were cross-linked with apo-glucose oxidase, a glucose-sensitive cross-linker, were used to reversibly detect glucose, both in vitro and when subdermally implanted in a rat model [102]. In the future such sensors could be used as part of a control system incorporated within insulin-secreting patches for the treatment of diabetes.

12.5 Challenges and Prospects

Nanotechnologies clearly have had an impact on tissue engineering and still have great potential to advance therapeutic methods based on tissue engineering. The synthesis of new nanostructures and their incorporation into existing macro- and microtechnologies have led to improvements in the ability to provide a true biomimetic microenvironment to the developing tissue. But challenges still need to be addressed. It is believed that engineered nanomaterials such as carbon nanotubes, nanowires and other inorganic materials will be increasingly used in tissue engineering [103, 104]. In recent years, several reports have indicated possible negative effects of carbon nanotubes on cells and their potential to provoke oxidative stress, inflammation, genetic damage and long-term pathological effects. Therefore the biocompatibility and biodegradation of inorganic nanomaterials need to be thoroughly investigated before they can be safely applied in clinical trials. (For a comprehensive review on the toxicology of nanomaterials, see Kunzmann et al. [105].) Another challenge that must be addressed is the complexity of creating 3D porous scaffolds with nanotopographies. The interaction of cells with various nanoscale topographies has proved to be important for creating monolayers of functioning tissues [47, 56, 58], but 'scaled-up' versions of these nanopatterning technologies remain to be achieved. Moreover, it is crucial to discover the key factors promoting the assemblies of different tissues and to create specific microenvironments. Developing nanotechnological tools for controlling and guiding cells to desired locations in 3D matrices will be useful for engineering complex multicellular constructs such as epithelial and vascularized tissues.

We believe that future strategies could involve incorporating intelligent nanoscale biosensors inside scaffolds to follow the development of engineered tissues after transplantation, and according to biochemical composition or tissue behaviour, trigger the release of compensating cues. For example, when oxygen levels are low, the system would release angiogenic factors for fast new blood vessel formation. We also believe that incorporating nanomaterials, such as nanotubes or wires on the outer surfaces of 3D scaffolds, could serve to reduce inflammatory responses to transplanted engineered tissues by modulating macrophage adhesion and viability. For example, zinc oxide nanorods (50 nm in diameter and around 500 nm long) sputtered on polyethylene terephthalate discs inhibited the formation of the typical foreign body capsule occurring after subcutaneous transplantation in mice [106].

Finally, we envisage the use of smart nanoparticulate systems that will recruit stem cells to desired sites in the body and instruct the formation of tissues in vivo, on demand, by triggering the release of chemical attractants. Smart controllable nanorobots could potentially circulate inside the body, find diseased tissues and repair them by destroying defective cells and molecules, or encourage cells to regain function [107]. These robots could be powered by biologically inspired nanomotors based on conversion of chemical energy (usually stored as adenosine triphosphate) into mechanical energy. We foresee nanowire-based brain-machine interfaces that could assist paralysed patients by re-routing movement-related signals around injured parts of the nervous system.

Acknowledgements

This work was supported by the National Institutes of Health Grants DE13023 and DE016516. T.D. acknowledges a Postdoctoral Fellowship from the American Heart Association. We thank S. Cohen, S. McAllister and B. Tian for their comments.

Additional Information

R.L. has a financial interest in Pervasis and Fibrocell Science, Inc.

References

1. Langer, R., and Vacanti, J. P. Tissue engineering. *Science* **260**, 920–926 (1993).
2. Freed, L. E. et al. Advanced tools for tissue engineering: Scaffolds, bioreactors, and signaling. *Tissue Eng.* **12**, 3285–3305 (2006).
3. Place, E. S., Evans, N. D., and Stevens, M. M. Complexity in biomaterials for tissue engineering. *Nature Mater.* **8**, 457–470 (2009).
4. Lutolf, M. P., and Hubbell, J. A. Synthetic biomaterials as instructive extracellular microenvironments' for morphogenesis in tissue engineering. *Nature Biotech.* **23**, 47–55 (2005).
5. Tsang, K. Y., Cheung, M. C., Chan, D., and Cheah, K. S. The developmental roles of the extracellular matrix: beyond structure to regulation. *Cell Tissue Res.* **339**, 93–110 (2010).
6. Bauer, A. L., Jackson, T. L., and Jiang, Y. Topography of extracellular matrix mediates vascular morphogenesis and migration speeds in angiogenesis. *PLoS Comp. Biol.* **5**, e1000445 (2009).
7. Levental, K. R. et al. Matrix crosslinking forces tumor progression by enhancing integrin signaling. *Cell* **139**, 891–906 (2009).
8. Evans, N. D. et al. Substrate stiffness affects early differentiation events in embryonic stem cells. *Eur. Cell Mater.* **18**, 1–13; discussion 13–14 (2009).
9. Discher, D. E., Janmey, P., and Wang, Y. L. Tissue cells feel and respond to the stiffness of their substrate. *Science* **310**, 1139–1143 (2005).
10. Cohen, E. D. et al. Wnt signaling regulates smooth muscle precursor development in the mouse lung via a tenascin C/PDGFR pathway. *J. Clin. Invest.* **119**, 2538–2549 (2009).
11. Rozario, T., and DeSimone, D. W. The extracellular matrix in development and morphogenesis: A dynamic view. *Dev. Biol.* **341**, 126–140 (2010).
12. Ott, H. C. et al. Perfusion-decellularized matrix: Using nature's platform to engineer a bioartificial heart. *Nature Med.* **14**, 213–221 (2008).

13. Uygun, B. E. et al. Organ reengineering through development of a transplantable recellularized liver graft using decellularized liver matrix. *Nature Med.* **16**, 814–820 (2010).
14. Grayson, W. L. et al. Engineering anatomically shaped human bone grafts. *Proc. Natl Acad. Sci. USA* **107**, 3299–3304 (2010).
15. Gui, L., Muto, A., Chan, S. A., Breuer, C. K., and Niklason, L. E. Development of decellularized human umbilical arteries as small-diameter vascular grafts. *Tissue Eng. A* **15**, 2665–2676 (2009).
16. Petersen, T. H. et al. Tissue-engineered lungs for in vivo implantation. *Science* **329**, 538–541.
17. Hynes, R. O. The extracellular matrix: Not just pretty fibrils. *Science* **326**, 1216–1219 (2009).
18. Sasisekharan, R., Shriver, Z., Venkataraman, G., and Narayanasami, U. Roles of heparan-sulphate glycosaminoglycans in cancer. *Nature Rev. Cancer* **2**, 521–528 (2002).
19. Lee, S. et al. Preparation of macroporous carbon nanofibres with macroscopic openings in the surfaces and their applications. *Nanotechnology* **20**, 445702 (2009).
20. Ayres, C. E., Jha, B. S., Sell, S. A., Bowlin, G. L., and Simpson, D. G. Nanotechnology in the design of soft tissue scaffolds: innovations in structure and function. *Wiley Interdiscip. Rev. Nanomed. Nanobiotechnol.* **2**, 20–34 (2010).
21. Barnes, C. P., Sell, S. A., Boland, E. D., Simpson, D. G., and Bowlin, G. L. Nanofiber technology: Designing the next generation of tissue engineering scaffolds. *Adv. Drug Deliv. Rev.* **59**, 1413–1433 (2007).
22. Zhang, S. Fabrication of novel biomaterials through molecular self-assembly. *Nature Biotechnol.* **21**, 1171–1178 (2003).
23. Ma, Z. W., Kotaki, M., Inai, R., and Ramakrishna, S. Potential of nanofiber matrix as tissue-engineering scaffolds. *Tissue Eng.* **11**, 101–109 (2005).
24. Zeng, J. et al. Poly(vinyl alcohol) nanofibres by electrospinning as a protein delivery system and the retardation of enzyme release by additional polymer coatings. *Biomacromolecules* **6**, 1484–1488 (2005).
25. Sun, Z. C., Zussman, E., Yarin, A. L., Wendorff, J. H., and Greiner, A. Compound core–shell polymer nanofibres by co-electrospinning. *Adv. Mater.* **15**, 1929–1932 (2003).
26. Ionescu, L. C., Lee, G. C., Sennett, B. J., Burdick, J. A., and Mauck, R. L. An anisotropic nanofiber/microsphere composite with controlled release

of biomolecules for fibrous tissue engineering. *Biomaterials* **31**, 4113–4120 (2010).

27. Cao, H. Q., Jiang, X., Chai, C., and Chew, S. Y. RNA interference by nanofiber-based siRNA delivery system. *J. Cont. Release* **144**, 203–212 (2010).
28. Dong, B., Smith, M. E., and Wnek, G. E. Encapsulation of multiple biological compounds within a single electrospun fiber. *Small* **5**, 1508–1512 (2009).
29. Freeman, I., and Cohen, S. The influence of the sequential delivery of angiogenic factors from affinity-binding alginate scaffolds on vascularization. *Biomaterials* **30**, 2122–2131 (2009).
30. Yilgor, P., Tuzlakoglu, K., Reis, R. L., Hasirci, N., and Hasirci, V. Incorporation of a sequential BMP-2/BMP-7 delivery system into chitosan-based scaffolds for bone tissue engineering. *Biomaterials* **30**, 3551–3559 (2009).
31. Moroni, L., Schotel, R., Hamann, D., de Wijn, J. R., and van Blitterswijk, C. A. 3D fiber-deposited electrospun integrated scaffolds enhance cartilage tissue formation. *Adv. Funct. Mater.* **18**, 53–60 (2008).
32. Hartgerink, J. D., Beniash, E., and Stupp, S. I. Self-assembly and mineralization of peptide-amphiphile nanofibres. *Science* **294**, 1684–1688 (2001).
33. Cui, H., Webber, M. J., and Stupp, S. I. Self-assembly of peptide amphiphiles: From molecules to nanostructures to biomaterials. *Biopolymers* **94**, 1–18 (2010).
34. Gelain, F., Bottai, D., Vescovi, A., and Zhang, S. Designer self-assembling peptide nanofibrescaffolds for adult mouse neural stem cell 3-dimensional cultures. *PLoS One* **1**, e119 (2006).
35. Xu, J. et al. Endothelial cells anchoring by functionalized yeast polypeptide. *J. Biomed. Mater. Res. A* **87**, 819–824 (2008).
36. Rexeisen, E. L. et al. Self-assembly of fibronectin mimetic peptide-amphiphile nanofibres. *Langmuir* **26**, 1953–1959 (2009).
37. Silva, G. A. et al. Selective differentiation of neural progenitor cells by high-epitope density nanofibers. *Science* **303**, 1352–1355 (2004).
38. Heino, J., and Kapyla, J. Cellular receptors of extracellular matrix molecules. *Curr. Pharm. Des.* **15**, 1309–1317 (2009).
39. Stevens, M. M., and George, J. H. Exploring and engineering the cell surface interface. *Science* **310**, 1135–1138 (2005).

40. Schofer, M. D. et al. Effect of direct RGD incorporation in PLLA nanofibres on growth and osteogenic differentiation of human mesenchymal stem cells. *J. Mater. Sci. Mater. Med.* **20**, 1535–1540 (2009).
41. Jeon, O., Powell, C., Ahmed, S. M., and Alsberg, E. Biodegradable, photocrosslinked alginate hydrogels with independently tailorable physical properties and cell adhesivity. *Tissue Eng. A* **16**(9): 2915–2925 (2010).
42. Re'em, T., Tsur-Gang, O., and Cohen, S. The effect of immobilized RGD peptide in macroporous alginate scaffolds on TGFbeta1-induced chondrogenesis of human mesenchymal stem cells. *Biomaterials* **31**, 6746–6755 (2010).
43. Casper, C. L., Yang, W., Farach-Carson, M. C., and Rabolt, J. F. Coating electrospun collagen and gelatin fibres with perlecan domain I for increased growth factor binding. *Biomacromolecules* **8**, 1116–1123 (2007).
44. Zhou, H. et al. Enhanced bioactivity of bone morphogenetic protein-2 with low dose of 2-N, 6-O-sulfated chitosan in vitro and in vivo. *Biomaterials* **30**, 1715–1724 (2009).
45. Freeman, I., Kedem, A., and Cohen, S. The effect of sulfation of alginate hydrogels on the specific binding and controlled release of heparin-binding proteins. *Biomaterials* **29**, 3260–3268 (2008).
46. Dvir, T. et al. Prevascularization of cardiac patch on the omentum improves its therapeutic outcome. *Proc. Natl Acad. Sci. USA* **106**, 14990–14995 (2009).
47. Bettinger, C. J., Langer, R., and Borenstein, J. T. Engineering substrate topography at the micro- and nanoscale to control cell function. *Angew. Chem. Int. Ed.* **48**, 5406–5415 (2009).
48. Teixeira, A. I., Abrams, G. A., Bertics, P. J., Murphy, C. J., and Nealey, P. F. Epithelial contact guidance on well-defined micro- and nanostructured substrates. *J. Cell. Sci.* **116**, 1881–1892 (2003); this work is among the first to document that the nanotopographic features encountered in the native basement membrane can profoundly affect cell behaviour.
49. Dalby, M. J. et al. The control of human mesenchymal cell differentiation using nanoscale symmetry and disorder. *Nature Mater.* **6**, 997–1003 (2007).
50. Kotov, N. A. et al. Nanomaterials for neural interfaces. *Adv. Mater.* **21**, 3970–4004 (2009).

51. Parker, K. K., and Ingber, D. E. Extracellular matrix, mechanotransduction and structural hierarchies in heart tissue engineering. *Phil. Trans. R. Soc. Lond. B* **362**, 1267–1279 (2007).
52. Zimmermann, W. H. et al. Tissue engineering of a differentiated cardiac muscle construct. *Circ. Res.* **90**, 223–230 (2002).
53. Dvir, T., Levy, O., Shachar, M., Granot, Y., and Cohen, S. Activation of the ERK1/2 cascade via pulsatile interstitial fluid flow promotes cardiac tissue assembly. *Tissue Eng.* **13**, 2185–2193 (2007).
54. Radisic, M. et al. Functional assembly of engineered myocardium by electrical stimulation of cardiac myocytes cultured on scaffolds. *Proc. Natl Acad. Sci. USA* **101**, 18129–18134 (2004).
55. McDevitt, T. C., Woodhouse, K. A., Hauschka, S. D., Murry, C. E., and Stayton, P. S. Spatially organized layers of cardiomyocytes on biodegradable polyurethane films for myocardial repair. *J. Biomed. Mater. Res. A* **66**, 586–595 (2003).
56. Kim, D. H. et al. Nanoscale cues regulate the structure and function of macroscopic cardiac tissue constructs. *Proc. Natl Acad. Sci. USA* **107**, 565–570 (2010).
57. Bryant, D. M., and Mostov, K. E. From cells to organs: building polarized tissue. *Nature Rev. Mol. Cell Biol.* **9**, 887–901 (2008).
58. Bettinger, C. J., Kulig, K. M., Vacanti, J. P., Langer, R., and Borenstein, J. T. Nanofabricated collagen-inspired synthetic elastomers for primary rat hepatocyte culture. *Tissue Eng. A* **15**, 1321–1329 (2009).
59. Feng, Z. Q. et al. The effect of nanofibrous galactosylated chitosan scaffolds on the formation of rat primary hepatocyte aggregates and the maintenance of liver function. *Biomaterials* **30**, 2753–2763 (2009).
60. Chan, C. K. et al. Biomimetic nanocomposites for bone graft applications. *Nanomedicine (Lond.)* **1**, 177–188 (2006).
61. Sachlos, E., Gotora, D., and Czernuszka, J. T. Collagen scaffolds reinforced with biomimetic composite nano-sized carbonate-substituted hydroxyapatite crystals and shaped by rapid prototyping to contain internal microchannels. *Tissue Eng.* **12**, 2479–2487 (2006).
62. Zhang, Y. et al. Enhanced biomineralization in osteoblasts on a novel electrospun biocomposite nanofibrous substrate of hydroxyapatite/collagen/ chitosan. *Tissue Eng. A* **16**(6): 1949–1960 (2010).
63. Bhattacharyya, S. et al. Biodegradable polyphosphazene-nanohydroxyapatite composite nanofibres: scaffolds for bone tissue engineering. *J. Biomed. Nanotechnol.* **5**, 69–75 (2009).

64. Zhang, L. et al. Biologically inspired rosette nanotubes and nanocrystalline hydroxyapatite hydrogel nanocomposites as improved bone substitutes. *Nanotechnology* **20**, 175101 (2009).

65. Roohani-Esfahani, S. I., Nouri-Khorasani, S., Lu, Z., Appleyard, R., and Zreiqat, H. The influence hydroxyapatite nanoparticle shape and size on the properties of biphasic calcium phosphate scaffolds coated with hydroxyapatite–PCL composites. *Biomaterials* **31**, 5498–5509 (2010).

66. Suhr, J. et al. Fatigue resistance of aligned carbon nanotube arrays under cyclic compression. *Nature Nanotech.* **2**, 417–421 (2007).

67. Wang, S. F., Shen, L., Zhang, W. D., and Tong, Y. J. Preparation and mechanical properties of chitosan/carbon nanotubes composites. *Biomacromolecules* **6**, 3067–3072 (2005).

68. Mattson, M. P., Haddon, R. C., and Rao, A. M. Molecular functionalization of carbon nanotubes and use as substrates for neuronal growth. *J. Mol. Neuro.* **14**, 175–182 (2000).

69. Hu, H., Ni, Y. C., Montana, V., Haddon, R. C., and Parpura, V. Chemically functionalized carbon nanotubes as substrates for neuronal growth. *Nano Lett.* **4**, 507–511 (2004).

70. Lovat, V. et al. Carbon nanotube substrates boost neuronal electrical signaling. *Nano Lett.* **5**, 1107–1110 (2005).

71. Massobrio, G., Massobrio, P., and Martinoia, S. Modeling the neuron-carbon nanotube–ISFET junction to investigate the electrophysiological neuronal activity. *Nano Lett.* **8**, 4433–4440 (2008).

72. Mazzatenta, A. et al. Interfacing neurons with carbon nanotubes: Electrical signal transfer and synaptic stimulation in cultured brain circuits. *J. Neurosci.* **27**, 6931–6936 (2007).

73. Cellot, G. et al. Carbon nanotubes might improve neuronal performance by favouring electrical shortcuts. *Nature Nanotech.* **4**, 126–133 (2009); the authors suggest a mechanism for how carbon nanotubes affect the collective electrical activity of neuronal networks in vitro.

74. Gui, X. et al. Soft, highly conductive nanotube sponges and composites with controlled compressibility. *ACS Nano* **4**, 2320–2326 (2010).

75. Bianco, A. et al. Microstructure and cytocompatibility of electrospun nanocomposites based on poly(epsilon-caprolactone) and carbon nanostructures. *Int. J. Artif. Organs* **33**, 271–282 (2010).

76. Heo, S. J. et al. In vitro and animal study of novel nano-hydroxyapatite/poly(epsilon-caprolactone) composite scaffolds fabricated by layer manufacturing process. *Tissue Eng. A* **15**, 977–989 (2009).

77. Pek, Y. S., Gao, S., Arshad, M. S., Leck, K. J., and Ying, J. Y. Porous collagen– apatite nanocomposite foams as bone regeneration scaffolds. *Biomaterials* **29**, 4300–4305 (2008).
78. Wu, S. L. et al. A biomimetic hierarchical scaffold: natural growth of nanotitanates on three-dimensional microporous Ti-based metals. *Nano Lett.* **8**, 3803–3808 (2008).
79. Zhang, S. F., and Uludag, H. Nanoparticulate systems for growth factor delivery. *Pharm. Res.* **26**, 1561–1580 (2009).
80. Fan, D. et al. Subcellular-resolution delivery of a cytokine through precisely manipulated nanowires. *Nature Nanotech.* **5**, 545–551 (2010).
81. Alsberg, E., Feinstein, E., Joy, M. P., Prentiss, M., and Ingber, D. E. Magnetically- guided self-assembly of fibrin matrices with ordered nano-scale structure for tissue engineering. *Tissue Eng.* **12**, 3247–3256 (2006).
82. Ito, A., Ino, K., Kobayashi, T., and Honda, H. The effect of RGD peptide-conjugated magnetite cationic liposomes on cell growth and cell sheet harvesting. *Biomaterials* **26**, 6185–6193 (2005).
83. Ito, A. et al. Novel methodology for fabrication of tissue-engineered tubular constructs using magnetite nanoparticles and magnetic force. *Tissue Eng.* **11**, 1553–1561 (2005).
84. Pislaru, S. V. et al. Magnetic forces enable rapid endothelialization of synthetic vascular grafts. *Circulation* **114**, I314–I318 (2006).
85. Souza, G. R. et al. Three-dimensional tissue culture based on magnetic cell levitation. *Nature Nanotech.* **5**, 291–296 (2010). The authors show a culture system in which the geometry and cell mass can be manipulated by spatially controlling the magnetic field.
86. Keefer, E. W., Botterman, B. R., Romero, M. I., Rossi, A. F., and Gross, G. W. Carbon nanotube coating improves neuronal recordings. *Nature Nanotech.* **3**, 434–439 (2008).
87. Zhou, X. J., Moran-Mirabal, J. M., Craighead, H. G., and McEuen, P. L. Supported lipid bilayer/carbon nanotube hybrids. *Nature Nanotech.* **2**, 185–190 (2007).
88. Pappas, T. C. et al. Nanoscale engineering of a cellular interface with semiconductor nanoparticle films for photoelectric stimulation of neurons. *Nano Lett.* **7**, 513–519 (2007).
89. Timko, B. P., Cohen-Karni, T., Qing, Q., Tian, B. Z., and Lieber, C. M. Design and implementation of functional nanoelectronic interfaces

with biomolecules, cells, and tissue using nanowire device arrays. *IEEE Trans. Nanotechnol.* **9**, 269–280 (2010).

90. Patolsky, F. et al. Detection, stimulation, and inhibition of neuronal signals with high-density nanowire transistor arrays. *Science* **313**, 1100–1104 (2006).
91. Eschermann, J. F. et al. Action potentials of HL-1 cells recorded with silicon nanowire transistors. *Appl. Phys. Lett.* **95**, 083703 (2009).
92. Cohen-Karni, T., Timko, B. P., Weiss, L. E., and Lieber, C. M. Flexible electrical recording from cells using nanowire transistor arrays. *Proc. Natl Acad. Sci. USA* **106**, 7309–7313 (2009).
93. Cohen-Karni, T., Qing, Q., Li, Q., Fang, Y., and Lieber, C. M. Graphene and nanowire transistors for cellular interfaces and electrical recording. *Nano Lett.* **10**, 1098–1102 (2010).
94. Timko, B. P. et al. Electrical recording from hearts with flexible nanowire device arrays. *Nano Lett.* **9**, 914–918 (2009).
95. Qing, Q. et al. Nanowire transistor arrays for mapping neural circuits in acute brain slices. *Proc. Natl Acad. Sci. USA* **107**, 1882–1887 (2010).
96. Kim, D-H. et al. Dissolvable films of silk fibroin for ultrathin conformal bio-integrated electronics. *Nature Mater.* **9**, 511–517 (2010).
97. Bettinger, C. J., and Bao, Z. A. Organic thin-film transistors fabricated on resorbable biomaterial substrates. *Adv. Mater.* **22**, 651–655 (2010).
98. Tian, B. Z., Xie, P., Kempa, T. J., Bell, D. C., and Lieber, C. M. Single-crystalline kinked semiconductor nanowire superstructures. *Nature Nanotech.* **4**, 824–829 (2009).
99. Tian, B. et al. Three-dimensional, flexible nanoscale field-effect transistors as localized bioprobes. *Science* **329**, 830–834 (2010); this article describes a nanoscale field-effect transistor cell probe that is capable of observing and monitoring the intracellular signals of living electrogenic cells.
100. Lim, D. K., Jeon, K. S., Kim, H. M., Nam, J. M., and Suh, Y. D. Nanogap-engineerable Raman-active nanodumbbells for single-molecule detection. *Nature Mater.* **9**, 60–67 (2010).
101. Heller, D. A. et al. Multimodal optical sensing and analyte specificity using single-walled carbon nanotubes. *Nature Nanotech.* **4**, 114–120 (2009).
102. Barone, P. W. et al. Modulation of single-walled carbon nanotube photoluminescence by hydrogel swelling. *ACS Nano* **3**, 3869–3877 (2009).

103. Farokhzad, O. C., and Langer, R. Impact of nanotechnology on drug delivery. *ACS Nano* **3**, 16–20 (2009).
104. Riehemann, K. et al. Nanomedicine—challenge and perspectives. *Angew. Chem. Int. Ed.* **48**, 872–897 (2009).
105. Kunzmann, A. et al. Toxicology of engineered nanomaterials: Focus on biocompatibility, biodistribution and biodegradation. *Biochim. Biophys. Acta* **1810**, 361–373 (2011); this article describes the interaction between engineered nanomaterials and biological systems and may help in designing safer and more compatible nanomaterials for future applications in medicine.
106. Zaveri, T. D. et al. Contributions of surface topography and cytotoxicity to the macrophage response to zinc oxide nanorods. *Biomaterials* **31**, 2999–3007 (2010).
107. Mallouk, T. E., and Sen, A. Powering nanorobots. *Sci. Am.* **300**, 72–77 (2009).

Chapter 13

Promises, Facts and Challenges for Carbon Nanotubes in Imaging and Therapeutics

Kostas Kostarelos,[a] Alberto Bianco,[b] and Maurizio Prato[c]

[a] *Nanomedicine Laboratory, Centre for Drug Delivery Research, The School of Pharmacy, University of London, London WC1N 1AX, UK*
[b] *CNRS, Immunologie et Chimie Thérapeutiques, Institut de Biologie Moléculaire et Cellulaire UPR9021, 67084 Strasbourg, France*
[c] *Dipartimento di Scienze Farmaceutiche, Università di Trieste, 34127 Trieste, Italy*
kostas.kostarelos@manchester.ac.uk, a.bianco@ibmc.u-strasbg.fr, prato@units.it

The use of carbon nanotubes in medicine is now at the crossroads between a proof-of-principle concept and an established preclinical candidate for a variety of therapeutic and diagnostic applications. Progress towards clinical trials will depend on the outcomes of efficacy and toxicology studies, which will provide the necessary risk-to-benefit assessments for carbon-nanotube-based materials. Here we focus on carbon nanotubes that have been studied in preclinical animal models, and draw attention to the promises, facts and challenges of these materials as they transition from research to the clinical phase. We address common

---

Reproduced with permission from *Nature Nanotechnology*, **4**, 627–633, 2009.

Nano-Enabled Medical Applications
Edited by Lajos P. Balogh

ISBN 978-981-4800-32-7 (Hardcover), 978-0-429-39903-9 (eBook)
www.jennystanford.com

questions regarding the use of carbon nanotubes in disease imaging and therapy, and highlight the opportunities and challenges ahead.

One of the key advantages of carbon nanotubes in biomedical applications is that they can be easily internalized by cells, and therefore can act as delivery vehicles for a variety of molecules relevant to therapy and diagnosis. Moreover, their unique electrical, thermal and spectroscopic properties in a biological context offer further advances in the detection, monitoring and therapy of diseases. The therapeutic and diagnostic applications of nanotubes are intensively explored by many academic and industrial laboratories around the world, in parallel with their toxicological profile and any potential adverse pathogenic reactions from exposure. The risk–benefit balance for these materials—to be reached by combining both areas of investigation—will ultimately determine their clinical fate.

As-produced carbon nanotubes are insoluble in most organic or aqueous solvents, therefore for any type of biological application the nanotube surface needs to be modified. For example, chemically functionalized carbon nanotubes have been shown to act as unique non-viral delivery systems for the transfer of nucleic acids. They have been explored for targeted delivery of small organic molecules (for example, antibiotics and anticancer agents), the development of peptide-based synthetic vaccines, as a platform for antibody targeting, and for the transportation of proteins and sugar mimetics [1]. However, carbon nanotubes are still at the very early stages of their clinical development, and their efficacy and limitations must be carefully determined and addressed. Concerns about the toxic effects of these materials are under intensive debate [2]. It is imperative to determine the impact of nanotube exposure on biological components at the cellular, tissue and overall physiological level.

A further challenge is related to the lack of an accepted protocol to determine the degree of purity of the carbon nanotube material used [3]. Standard chromatographic techniques—such as thin-layer chromatography, high-pressure liquid chromatography and gel permeation chromatography—have achieved limited success, with no general or reproducible outcomes. Such technical limitations will need to be addressed, and standard procedures for the production and purification of nanotubes or functionalized nanotube materials should be developed to enable a move into large-scale multi-centre clinical trials.

As these challenges are confronted, it is necessary to show unequivocally the therapeutic efficacy of these nanotubes over established alternatives if industrial biotechnology and pharmaceutical investments are to continue and grow. Demonstrating the benefits of nanotubes in medicine is also needed to avoid common unrealistic expectations that may prove to be counter-productive to the development of the field overall. In this article, we examine the state-of-the-art in the use of carbon nanotubes for imaging and therapeutics, focusing only on studies that use in vivo models, and highlight the opportunities and challenges as these materials move closer to the clinical setting.

13.1 Carbon Nanotubes Used in Imaging and Therapy

Carbon nanotubes can be single-walled [4] or multiwalled [5] and are now produced in substantial quantities for a variety of commercial applications [6]. They have diameters in the nanometre range and their length can reach several microns. In terms of their use in biomedical applications, the initial hurdle has been the insolubility of carbon nanotubes in most solvents, and particularly in biologically compatible buffers and conditions. Different strategies have been developed to make carbon nanotubes compatible with the biological milieu [7]. The two main methodologies are based on the non-covalent coating of nanotubes with amphiphilic molecules (for example, lipids and polymers), and the covalent functionalization of the nanotube surface by grafting various chemical groups directly onto the backbone.

Figure 13.1 illustrates the types of carbon nanotubes that have been explored in biomedical studies using in vivo preclinical models. The three categories shown have different structural and surface characteristics that critically influence their biological performance.

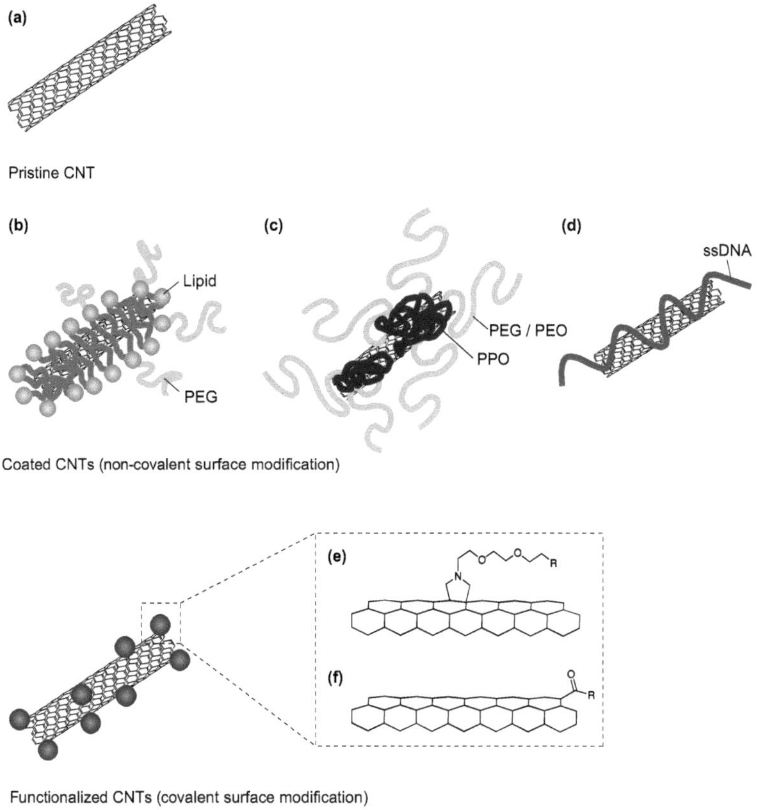

Figure 13.1 Types of carbon nanotube studied in vivo for imaging and therapy. (a) Pristine carbon nanotubes (CNTs) are those without any surface modification. (b) Lipid-coated nanotubes (primarily single-walled nanotubes) with or without PEGylated lipids and other versions of further modified lipid molecules. (c) Copolymer or surfactant-coated nanotubes (primarily single-walled nanotubes). PEO is polyethylene oxide; PPO is polypropylene oxide. (d) Single-stranded DNA (ssDNA)-coated nanotubes (primarily single-walled nanotubes). (e, f) Chemically functionalized nanotubes (both single-walled and multiwalled nanotubes) by 1,3 dipolar cycloaddition (e) and by acid oxidation (f).

Pristine nanotubes (Fig. 13.1a) are the prototype materials produced and are the most difficult to handle biologically because they are hardly dispersed in aqueous solutions and have a strong tendency to interact hydrophobically and aggregate. Interestingly, pristine nanotubes—mostly poor-quality aqueous dispersions—have been the predominant type used in most toxicology studies [2]. Their dispersion is dramatically improved by coating the nanotube surface with different amphiphilic macromolecules such as lipid- polyethylene glycol (PEG) conjugates (Fig. 13.1b) [8], copolymers and surfactants (for example, Kentera and Pluronic F127) (Fig. 13.1c) [9, 10] and single-stranded DNA (Fig. 13.1d) [11]. Covalently functionalized nanotubes that have been used in biomedical applications are made of pristine materials that have undergone surface modification by using either cycloaddition reactions to attach ammonium groups (Fig. 13.1e) or strong acid treatment to generate carboxylic acid groups (Fig. 13.1f) [7]. Both types of chemical functionalization strategies remarkably improve the water dispersibility of the nanotubes, and at the same time offer a flexible platform for further derivatization.

Of the many strategies developed to modify the surface of nanotubes used in medicine, it is clear that the degree of aggregation and the individualization of nanotube materials in the biological milieu (blood, intraperitoneal, interstitial fluids, and so on) have an important role in their pharmacological performance.

13.2 Lessons Learnt from Preclinical in vivo Studies

All in vivo studies using carbon nanotubes reported so far have used one of the types described above (Fig. 13.1), and the majority of preclinical models have focused on oncology—making cancer the primary disease candidate for future clinical trials. Carbon nanotubes offer many potential advantages over other types of nanoparticles used or developed for cancer therapy. For example, covalently functionalized carbon nanotubes are able to evade the endosomal compartment and translocate directly into the cytoplasm of different types of cells [12]. Furthermore, their unique physical properties permit efficient electromagnetic stimulation and highly sensitive detection using various imaging modalities. Their large surface area and internal volume also allows drugs and a variety of small molecules, such as contrast agents, to be loaded onto the nanotube. Carbon nanotubes have been used to halt tumour growth in the context of various therapeutic modalities including chemotherapy, hyperthermia and gene silencing. Also, in vivo tumour targeting by systemic administration has been described using both coated and covalently functionalized nanotubes. Despite the advances made, comparative in vivo studies against other 'benchmark' nanoparticles with a proven clinical record (such as liposomes) are lacking, with only a single such study recently reported [13]. Table 13.1 describes all the in vivo studies published so far according to the type of carbon nanotube used.

The strategy of coating nanotubes with lipopolymer molecules has been pioneered by the Dai group [14], and more recently explored as a delivery system for cancer therapy. According to this strategy, nanotubes are surface-coated with the lipopolymers to achieve aqueous dispersibility. The distal end of the PEG chain is generally used to link all other molecules, such as targeting moieties (for example, the RGD sequence) [8, 15], radionuclides (^{64}Cu) [8] and drugs (Paclitaxel) [16]. Tumour accumulation has been reported for nanotubes coated with a phospholipid–PEG–RGD peptide conjugate, and imaged by using the photoacoustic and Raman signatures of the carbon nanotubes. In the only therapeutic study using lipopolymer-coated nanotubes, the water-insoluble drug Paclitaxel was conjugated at the tip of the PEG chain and reportedly led to relative tumour volume suppression in a murine cancer model [16], even in the absence of a targeting ligand. However, direct comparisons with approved therapeutics based on alternative drug delivery systems (for example, Abraxane and Doxil) are needed to establish improved efficacy and to illustrate the advantages offered by carbon nanotubes.

Another strategy using coated carbon nanotubes to treat cancer has been based on the capacity of nanotubes to convert electromagnetic radiation into heat. Carbon-nanotube-mediated hyperthermic treatment of tumour tissue by radiofrequency activation after intratumoural administration of pristine tubes coated with Kentera (a polymer based on polyphenylene ethynylene) has been described [10]. Even though different radiofrequency pulse sequences seemed to ablate cancer cells when the nanotubes were injected directly into the tumour, no quantitative efficacy data has yet been reported with this methodology. More in vivo studies using nanotubes to achieve established solid tumour ablation by hyperthermia are expected in the near future by different groups.

Chemical functionalization of carbon nanotubes offers the advantage that the functional groups, further modified with a therapeutic, targeting or imaging molecule, are stably attached on the nanotube backbone, and therefore avoid the risk of macromolecule desorption or exchange with serum proteins and other blood components following administration. As can be seen from Table 13.1, a wider range of therapeutic models have been explored using functionalized nanotubes, compared with coated ones.

One type of functionalized carbon nanotube explored in biomedical applications is based on the covalent surface modification methodologies developed by our laboratories. These nanotubes have now been studied for various applications, including imaging using various radionuclides (^{111}In, ^{86}Y) [17–19], vaccination using immune-activating peptides (FMDV-derived peptide, AMA-1 peptide) [20, 21], and cancer therapy using monoclonal antibody (Rituximab) [22] and small interfering RNA (siRNA) [13] therapeutics (using two different human-tumour models of lymphoma and lung cancer, respectively). The lung cancer study by Podesta and colleagues [13] indicated that carbon nanotubes may be more effective in prolonging the survival of tumour-bearing animals, compared with cationic liposomes, when administered locally (intratumorally)—presumably owing to their more facile translocation into the tumour cell cytoplasm. Another type of functionalized nanotube that has been explored in vivo is based on the chemical modification of carboxylic acid groups at the nanotube tips and sidewalls introduced after strong acid treatment. These nanotubes have also been studied using tracing radionuclides (^{125}I, ^{14}C) [23, 24], as well as for cancer treatment using vaccination strategies (tumour-lysate proteins) [25], siRNA gene-silencing (TERT gene) [26] and chemotherapy (cisplatin) [27]. All of these studies have reported tumour growth delay with no adverse reactions.

Overall, the in vivo studies performed so far indicate that carbon nanotubes could potentially have a significant contribution to efficacious treatment of diseases. However, even in the case of

Table 13.1 Preclinical in vivo studies using carbon nanotubes

Type of carbon nanotube*	Biomedical application	Therapeutic/imaging agent	Preclinical model	Refs
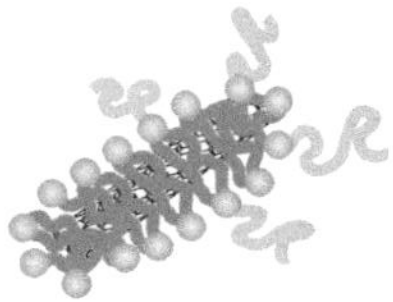	Cancer imaging (systemic)	Radionuclide (^{64}Cu)	Solid-tumour model	[8]
	Cancer treatment (systemic)	Paclitaxel	Solid-tumour model	[16]
	Cancer imaging (systemic)	Intrinsic Raman and ultrasound signal	Solid-tumour model	[15]
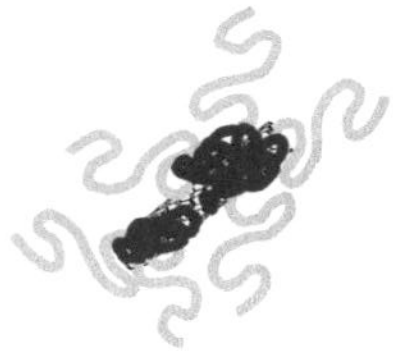	Cancer imaging (localized)	Hyperthermia by radiofrequency activation	Solid-tumour model	[10]
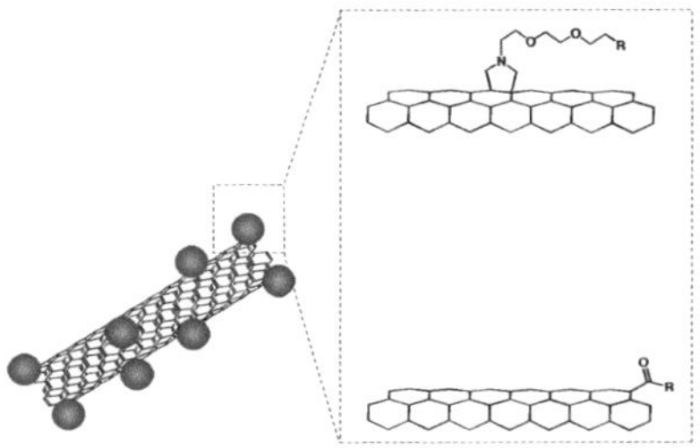	Vaccination	FMDV-derived peptides	Normal mice	[21]
	Vaccination	(AMA-1) peptide	Rodent malaria	[20]
	Imaging (systemic)	Radionuclide (^{111}In and ^{86}Y)	Normal mice	[17–19]
	Cancer imaging (systemic)	Monoclonal antibody (Rituximab)	Bone-marrow-tumour model	[22]
	Cancer treatment (localized)	Gene silencing (siRNA)	Solid-tumour model	[13]
	Cancer imaging (systemic)	Radionuclide (^{125}I and ^{14}C)	Normal mice	[23, 24]
	Cancer treatment (localized)	Gene silencing (siRNA)	Solid-tumour model	[26]
	Tumour vaccination	Tumour-lysate proteins	Solid-tumour model	[25]
	Cancer treatment (systemic)	Cisplatin	Solid-tumour model	[27]

*See Fig. 13.1 for explanation of carbon nanotube types.

cancer therapy where most in vivo studies have been attempted, it is extremely difficult to correlate and compare the data generated. Different tumour models, administration routes, nanotube doses, ways to describe tumour-volume data and therapeutic modalities (chemotherapy, hyperthermia, gene silencing, immunotherapy) have been used. Moreover, the lack of comparisons with well-established alternatives (for example, liposomes, polymers) in the design of most studies prevents a clear determination of the advantages and limitations of carbon nanotubes compared with existing technologies.

13.3 Current Status of Carbon Nanotube Toxicity in Biomedicine

Generally, very limited toxicity data exist on the types of carbon nanotubes described above that have been studied for medical applications. The majority of toxicological studies have focused on pristine non-functionalized nanotubes that are dispersed in different buffers with a huge variation in the quality of the dispersions. These studies mostly consider nanotube exposure in the context of occupational and environmental health, and are mainly designed by administration of nanotubes through the pulmonary route (for example, intratracheal, intranasal) [28–31]. A few of these studies have indicated serious risks associated with carbon nanotube exposure, however, the material, dosing and administration routes used are not directly relevant to those explored for medical applications. There is evidence based on some toxicological studies that prolonged accumulation of long (>10 μm), rigid, pristine nanotubes in tissues may be associated with health risks such as carcinogenesis and should be avoided [32, 33]. These studies assessed the risk for mesothelioma formation following exposure (through intraperitoneal injection) and, even though they have their limitations, they have generated serious discussion regarding the overall toxicological profile of uncoated and coated nanotubes when these consist of long pristine materials.

Table 13.2 Toxicity studies of carbon nanotubes developed for medical imaging and therapy

Type of carbon nanotube*	Route and method of detection	Organ accumulation and excretion route	Model and adverse effects	Refs
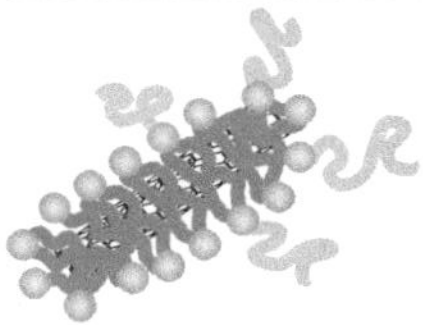	Intravenous • Raman spectroscopy • Optical microscopy (Hematoxylin–eosin histology)	Liver accumulation Slow faecal excretion	Normal animals (mouse) No adverse effects	[36, 37]
	Intravenous • Near–infrared fluorescence • Mass spectrometry (^{13}C)	Liver accumulation No excretion studied	Normal animals (rabbit and mouse) No adverse effects	[34, 35, 38]
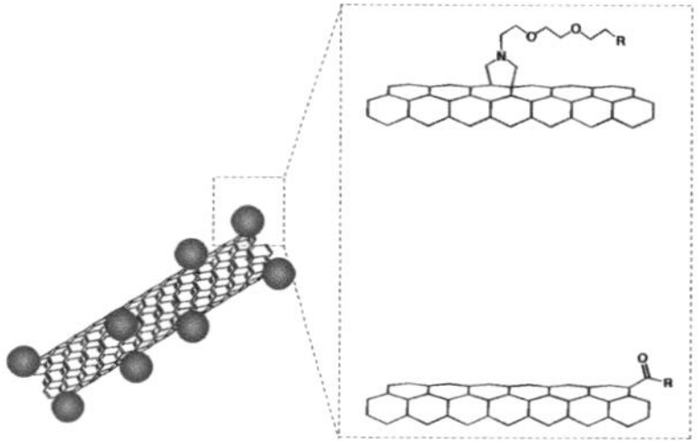	Intravenous • Optical microscopy (Hematoxylin–eosin histology) • Electron microscopy (urine, kidney sections)	Limited organ (liver and lung) accumulation Urinary excretion	Normal animals (mouse and rat) No adverse effects	[19, 40, 41]
	Intravenous • Electron microscopy (liver sections)	Liver and lung accumulation Slow faecal and urinary excretion	Normal animals (mouse) No adverse effects	[23]

*See Fig. 13.1 for explanation of carbon nanotube types.

In Table 13.2, the in vivo studies that address some of the toxicity considerations of the nanotubes explored for therapy and diagnostics are described. These reports aim mainly at elucidating the tissue distribution, organ accumulation, excretion route and any physiological abnormalities caused from intravenous administration of carbon nanotubes, using a variety of detection methodologies. An overall conclusion from these studies is the absence of acute or other adverse reactions between one week and three months following nanotube administration. However, none of these studies were designed with a toxicology model or specific mechanism under consideration. This is needed to determine the overall toxicity profile of carbon nanotubes—particularly in comparison with known toxins and other nanoparticle types—and to elucidate structure–function relationships that will help engineer effective and safe nanotube materials.

Among the studies examining the lipopolymer- or surfactant-coated nanotubes, there is evidence that (1) some degree of desorption of the molecules coating the nanotube backbone occurs in vivo on administration [34]; (2) the liver is the predominant organ where coated nanotubes tended to accumulate [34–37]; and (3) new methodologies based on the Raman signature of carbon nanotubes can help determine the tissue accumulation of coated nanotubes [37, 38]. Prolonged (months-long) accumulation of coated nanotubes in the liver has been described [36, 38], with slow excretion mainly through the bile and the faecal pathway [37]. Although these studies confirm that carbon nanotubes are not biodegradable and are not metabolized by the liver, as is the case with small molecules or other delivery systems such as liposomes, the overall toxicological implications from such in vivo behaviour remains unresolved.

Shortening of nanotubes (to <1 μm) from chemical treatment may be responsible for the improved toxicological profile of functionalized nanotubes that have been used in in vivo studies [39]. The excretion of a considerable fraction of intravenously administered functionalized nanotubes has been shown independently by two groups [18, 19] to be taking place by means of rapid translocation through the glomerular filter [40]. This effect is thought to depend mainly on how well the nanotubes are individually separated in the administered functionalized nanotube dispersion; the higher the number of individually separated nanotubes, the higher their urinary excretion and the lower their accumulation in tissues. However, in view of medical applications based on systemic administration, the rapid excretion of functionalized nanotubes poses the challenge of maintaining the injected material within the blood circulation for longer periods of time to allow targeting and interaction with diseased tissue. The type and degree of chemical functionalization on the nanotube surface is considered critically important in determining tissue accumulation and excretion [41], particularly for functionalization chemistries that may not lead to shortened length and dispersions of individual nanotubes. Overall, there is good agreement between studies that urinary excretion of functionalized nanotubes and low organ accumulation is thought to be due to shortened and better dispersed nanotubes [18, 19, 24].

Adverse reactions following intravenous administration of the types of nanotubes studied in vivo for medical applications have not been reported so far. There is agreement among studies using coated nanotubes that indicate mainly liver accumulation whereas, studies using functionalized nanotubes show significant urinary excretion that is strongly dependent on the fraction of individualized nanotubes in the injected dispersion. The route of administration is also a determining factor for tissue accumulation and possible toxic responses. More studies exploring different routes are therefore needed. Moreover, toxicological studies using the types of nanotubes explored for therapy and diagnostics should be done in collaboration with toxicologists, using toxicology models to determine whether the mechanisms of activation that are described for pristine nanotubes can be alleviated by coating or covalent functionalization.

13.4 Opportunities and Challenges

As clearly illustrated in the sections above, intense efforts and interest have been invested by a growing number of laboratories to explore carbon nanotubes in medicine. This is based on the range of advantages that carbon nanotubes may offer over alternative systems in therapeutic or diagnostic applications. However, it is important to recognize that translating carbon nanotubes from an interesting nanomaterial to an effective pharmaceutical product is still at the nascent stages of development. At the same time, blinded 'fear' of the risks associated with any new nanomaterial should not impede investment and investigation.

Common questions regarding the use of carbon nanotubes in medicine are addressed in Box 13.1. The answers to these questions are based on peer-reviewed published data. Interestingly, each of these questions raise a set of challenges and opportunities associated with the use of nanotubes in the diagnosis and treatment of diseases.

The use of carbon nanotubes in medicine, as first suggested a few years ago, has already resulted in a number of studies using various in vivo models. Many more studies from a wider spectrum of laboratories and applications are expected in the immediate future as more disease-oriented groups will be motivated to study the utility of carbon nanotubes in their models. Moreover, there are other biomedical applications using carbon nanotube materials that are also being explored—mainly in the engineering of devices such as electrodes (for example, neurological tissue stimulation) [42], scaffolds for tissue regeneration (for example, orthopaedic and dental implants) [43, 44] and ex vivo biosensors [45]—which were deemed outside the scope of this article. Such applications, particularly those that will not require interaction with living tissue (such as biosensors) may indeed be developed sooner than others. However, irrespective of application and type of carbon nanotubes used, their clinical development will be determined by solid evidence on the advantages offered compared with established technologies.

In this chapter, we have attempted to highlight the progress made so far, focusing only on nanotubes that have been explored for medical imaging and therapeutics using preclinical models. It is now apparent that the use of carbon nanotubes in therapy and diagnosis of diseases will be determined by the systematic determination of the benefit they offer against the risk they pose. The biomedical and toxicological investigations of these materials will have to take place simultaneously to transform carbon nanotubes into a clinical reality.

Box 13.1 Carbon nanotubes in medicine: questions and answers

Q: Are carbon nanotubes really useful in medicine?
A: Proof-of-principle studies using in vitro and in vivo models indicate that carbon nanotubes may help treat various diseases (cancer, AIDS, malaria, metabolic diseases), but only one study [13] so far has reported a therapeutic outcome (prolonged survival) in a preclinical human-tumour model.

Challenges:

- Nanotubes may not treat disease more effectively than established technologies.
- The risk-to-benefit ratio offered by nanotube-based therapeutics and diagnostics may weigh towards the risk.

Opportunities:

- The possible contributions of nanotubes in medicine are almost unlimited and wide-ranging, from advanced delivery systems, electrodes and biosensors to probes for diagnostics and treatment- monitoring devices.

Q: Can carbon nanotubes help cure cancer?
A: It is too early to determine because only early-stage preclinical studies are available and at present there are no clinical studies underway.

Challenges:

- Unrealistic claims and expectations of tumour-targeted carbon nanotubes could cause a backlash.
- Carbon nanotubes may offer a poor risk-to-benefit ratio in oncology settings.

Opportunities:

- Cancer is a challenging disease. Patients and clinicians will be keen to explore new therapeutic concepts such as carbon-nanotube-based treatments, as long as these offer benefits.

Q: Can carbon nanotubes act as 'nanorobots' in the blood stream?

A: Injectable nanorobots have not yet been developed, and active navigation of nanoparticles in the blood stream has not been achieved. Therefore, nanotubes can neither act as nanorobots nor be navigated in the blood stream.

Challenges:

- The association between the fear of 'self-replicating nanorobots' and carbon nanotubes.
- Nanotubes as components of nanorobots and other nanomachines that may accumulate and intoxicate the body.

Opportunities:

- Carbon nanotubes can act as components of nanofabricated machinery and offer tremendous capabilities—for example, in wireless communication and monitoring between the patient and the clinician.

Q: Are carbon nanotubes biocompatible and what does that mean?

A: The term 'biocompatibility' is commonly ill-defined. In most cases, it implies the ability to interact with the biological milieu without adverse reactions. Chemically functionalized nanotubes have been shown by many groups to be more biocompatible (no immune or acute inflammatory responses) [21] than pristine nanotubes.

Challenges:

- Some types of carbon nanotubes or their impurities may accumulate in the body, leading to deposits that may cause unwanted side effects in the long-term.

Opportunities:

- New carbon nanotube materials and strategies to make them biocompatible are actively pursued.

Q: Are carbon nanotubes toxic?
A: Toxicity depends strongly on the type of nanotube, the dose, the route of administration and the tissue that is most affected. Pristine nanotubes have been shown to activate various mechanisms associated with toxicity, however these effects are shown to be remarkably reduced when properly functionalized with chemical groups. So far, no in vivo study using the types of nanotubes developed for medical purposes has reported adverse effects, however, studies with established toxicology models are much needed.

Challenges:

- The structural similarity and association between carbon nanotubes and the carcinogenic asbestos fibres.
- Absolute statements such as 'carbon nanotubes are toxic' can be very damaging.

Opportunities:

- Systematic toxicological studies of carbon nanotubes to make them the 'standard' fibrilar nanomaterial.
- Need to determine the extent of toxicological risks from using nanotubes, their doses, types and route of administration

Q: Should regulatory authorities restrict the use of carbon nanotubes?
A: No regulatory authority around the world has recommended or enforced a ruling on restricted use of carbon nanotubes. For medical applications, the very strict regulatory and authorization framework for new drugs is considered adequate.

Challenges:

- Restrictions posed on the use of carbon nanotubes based on premature evidence of adverse effects and unsubstantiated 'nanofear'.

Opportunities:

- Clinical development of a carbon-nanotube-based therapeutic or diagnostic will act as a proof-of-principle nanomedical product.

Acknowledgements

This work was supported by the University of Trieste, INSTM, Italian Ministry of Education MUR (cofin Prot. 2006034372 and Firb RBIN04HC3S), Regione Friuli Venezia-Giulia, CNRS and the Agence Nationale de la Recherche (grant ANR-05-JCJC-0031-01). Partial support from the European Union FP7 ANTICARB (HEALTH-2007-201587) research programme is also acknowledged.

References

1. Prato, M., Kostarelos, K., and Bianco, A. Functionalized carbon nanotubes in drug design and discovery. *Acc. Chem. Res.* **41**, 60–68 (2008).
2. Boczkowski, J., and Lanone, S. Potential uses of carbon nanotubes in the medical field: how worried should patients be? *Nanomedicine* **2**, 407–410 (2007).
3. Itkis, M. E. et al. Comparison of analytical techniques for purity evaluation of single-walled carbon nanotubes. *J. Am. Chem. Soc.* **127**, 3439–3448 (2005).
4. Iijima, S., and Ichihashi, T. Single-shell carbon nanotubes of 1-nm diameter. *Nature* **363**, 603–605 (1993).
5. Iijima, S. Helical microtubules of graphitic carbon. *Nature* **354**, 56–58 (1991).
6. Jorio, A., Dresselhaus, G., and Dresselhaus, M. S. *Carbon Nanotubes: Advanced Topics in the Synthesis, Structure, Properties and Applications* (Springer-Verlag, 2008).
7. Tasis, D., Tagmatarchis, N., Bianco, A., and Prato, M. Chemistry of carbon nanotubes. *Chem. Rev.* **106**, 1105–1136 (2006).
8. Liu, Z. et al. In vivo biodistribution and highly efficient tumour targeting of carbon nanotubes in mice. *Nature Nanotech.* **2**, 47–52 (2007).
9. Ali-Boucetta, H. et al. Multiwalled carbon nanotube-doxorubicin supramolecular complexes for cancer therapeutics. *Chem. Commun.* **12**, 459–461 (2008).
10. Gannon, C. J. et al. Carbon nanotube-enhanced thermal destruction of cancer cells in a noninvasive radiofrequency field. *Cancer* **110**, 2654–2665 (2007).

11. Zheng, M. et al. DNA-assisted dispersion and separation of carbon nanotubes. *Nature Mater.* **2**, 338–342 (2003).
12. Kostarelos, K. et al. Cellular uptake of functionalized carbon nanotubes is independent of functional group and cell type. *Nature Nanotech.* **2**, 108–113 (2007).
13. Podesta, J. P. et al. Antitumor activity and prolonged survival by carbon nanotube-mediated therapeutic siRNA silencing in a human lung xenograft model. *Small* **5**, 1176–1185 (2009).
14. Liu, Z., Tabakman, S., Welsher, K., and Dai, H. Carbon nanotubes in biology and medicine: In vitro and in vivo detection, imaging and drug delivery. *Nano Res.* **2**, 85–120 (2009).
15. De la Zerda, A. et al. Carbon nanotubes as photoacoustic molecular imaging agents in living mice. *Nature Nanotech.* **3**, 557–562 (2008).
16. Liu, Z. et al. Drug delivery with carbon nanotubes for in vivo cancer treatment. *Cancer Res.* **68**, 6652–6660 (2008).
17. Lacerda, L. et al. Dynamic imaging of functionalized multi-walled carbon nanotube systemic circulation and urinary excretion. *Adv. Mater.* **20**, 225–230 (2008).
18. McDevitt, M. R. et al. PET imaging of soluble yttrium-86-labeled carbon nanotubes in mice. *PLoS ONE* **2**, e907 (2007).
19. Singh, R. et al. Tissue biodistribution and blood clearance rates of intravenously administered carbon nanotube radiotracers. *Proc. Natl Acad. Sci. USA* **103**, 3357–3362 (2006).
20. Yandar, N. et al. Immunological profile of a Plasmodium vivax AMA-1 N-terminus peptide-carbon nanotube conjugate in an infected Plasmodium berghei mouse model. *Vaccine* **26**, 5864–5873 (2008).
21. Pantarotto, D. et al. Immunization with peptide-functionalized carbon nanotubes enhances virus-specific neutralizing antibody responses. *Chem. Biol.* **10**, 961–966 (2003).
22. McDevitt, M. R. et al. Tumor targeting with antibody-functionalized, radiolabeled carbon nanotubes. *J. Nucl. Med.* **48**, 1180–1189 (2007).
23. Deng, X. et al. Translocation and fate of multi-walled carbon nanotubes in vivo. *Carbon* **45**, 1419–1424 (2007).
24. Wang, H. et al. Biodistribution of carbon single-wall carbon nanotubes in mice. *J. Nanosci. Nanotechnol.* **4**, 1019–1024 (2004).
25. Meng, J. et al. Carbon nanotubes conjugated to tumor lysate protein enhance the efficacy of an antitumor immunotherapy. *Small* **4**, 1364–1370 (2008).

26. Zhang, Z. et al. Delivery of telomerase reverse transcriptase small interfering RNA in complex with positively charged single-walled carbon nanotubes suppresses tumor growth. *Clin. Cancer Res.* **12**, 4933–4939 (2006).
27. Bhirde, A. A. et al. Targeted killing of cancer cells in vivo and in vitro with EGF-directed carbon nanotube-based drug delivery. *ACS Nano* **3**, 307–316 (2009).
28. Ryman-Rasmussen, J. P. et al. Inhaled multiwalled carbon nanotubes potentiate airway fibrosis in murine allergic asthma. *Am. J. Resp. Cell Mol.* **40**, 349–358 (2009).
29. Shvedova, A. A. et al. Inhalation vs. aspiration of single-walled carbon nanotubes in C57BL/6 mice: inflammation, fibrosis, oxidative stress, and mutagenesis. *Am. J. Physiol. Lung C.* **295**, 552–565 (2008).
30. Chou, C. C. et al. Single-walled carbon nanotubes can induce pulmonary injury in mouse model. *Nano Lett.* **8**, 437–445 (2008).
31. Mitchell, L. A. et al. Pulmonary and systemic immune response to inhaled multiwalled carbon nanotubes. *Toxicol. Sci.* **100**, 203–214 (2007).
32. Takagi, A. et al. Induction of mesothelioma in p53± mouse by intraperitoneal application of multi-wall carbon nanotube. *J. Toxicol. Sci.* **33**, 105–116 (2008).
33. Poland, C. A. et al. Carbon nanotubes introduced into the abdominal cavity of mice show asbestos-like pathogenicity in a pilot study. *Nature Nanotech.* **3**, 423–428 (2008).
34. Cherukuri, P. et al. Mammalian pharmacokinetics of carbon nanotubes using intrinsic near-infrared fluorescence. *Proc. Natl Acad. Sci. USA* **103**, 18882–18886 (2006).
35. Yang, S. T. et al. Biodistribution of pristine single-walled carbon nanotubes in vivo. *J. Phys. Chem.* **111**, 17761–17764 (2007).
36. Schipper, M. L. et al. A pilot toxicology study of single-walled carbon nanotubes in a small sample of mice. *Nature Nanotech.* **3**, 216–221 (2008).
37. Liu, Z. et al. Circulation and long-term fate of functionalized, biocompatible single-walled carbon nanotubes in mice probed by Raman spectroscopy. *Proc. Natl Acad. Sci. USA* **105**, 1410–1415 (2008).
38. Yang, S. T. et al. Long-term accumulation and low toxicity of single-walled carbon nanotubes in intravenously exposed mice. *Toxicol. Lett.* **181**, 182–189 (2008).

39. Kostarelos, K. The long and short of carbon nanotube toxicity. *Nature Biotechnol.* **26**, 774–776 (2008).
40. Lacerda, L. et al. Carbon-nanotube shape and individualization critical for renal excretion. *Small* **4**, 1130–1132 (2008).
41. Lacerda, L. et al. Tissue histology and physiology following intravenous administration of different types of functionalized multiwalled carbon nanotubes. *Nanomedicine* **3**, 149–161 (2008).
42. Keefer, E. W. et al. Carbon nanotube coating improves neuronal recordings. *Nature Nanotech.* **3**, 434–439 (2008).
43. Sitharaman, B. et al. In vivo biocompatibility of ultra-short single-walled carbon nanotube/biodegradable polymer nanocomposites for bone tissue engineering. *Bone* **43**, 362–370 (2008).
44. Khang, D., Park, G. E., and Webster, T. J. Enhanced chondrocyte densities on carbon nanotube composites: the combined role of nanosurface roughness and electrical stimulation. *J. Biomed. Mater. Res. A* **86**, 253–260 (2008).
45. Heller, D. A. et al. Multimodal optical sensing and analyte specificity using single-walled carbon nanotubes. *Nature Nanotech.* **4**, 114–120 (2009).

Chapter 14

In vivo Biodistribution and Highly Efficient Tumour Targeting of Carbon Nanotubes in Mice

Zhuang Liu,[a,*] Weibo Cai,[b,*] Lina He,[b] Nozomi Nakayama,[a] Kai Chen,[b] Xiaoming Sun,[a] Xiaoyuan Chen,[b] and Hongjie Dai[a]

[a] *Department of Chemistry, Stanford University, Stanford, California 94305, USA*
[b] *The Molecular Imaging Program at Stanford, Department of Radiology and Bio-X Program, Stanford University School of Medicine, Stanford, California 94305, USA*
hdai@stanford.edu, shawchen@stanford.edu

Single-walled carbon nanotubes (SWNTs) exhibit unique size, shape and physical properties [1–3] that make them promising candidates for biological applications. Here, we investigate the biodistribution of radio-labelled SWNTs in mice by in vivo positron emission tomography (PET), ex vivo biodistribution and Raman spectroscopy. It is found that SWNTs that are functionalized with phospholipids bearing polyethylene-glycol (PEG) are surprisingly stable in vivo. The effect of PEG chain length on the biodistribution and circulation of the SWNTs is studied. Effectively PEGylated SWNTs exhibit relatively long blood circulation times and low uptake by the

*These authors contributed equally to this work.
Reproduced with permission from *Nature Nanotechnology*, **2**, 47–52, 2007.

Nano-Enabled Medical Applications
Edited by Lajos P. Balogh

ISBN 978-981-4800-32-7 (Hardcover), 978-0-429-39903-9 (eBook)
www.jennystanford.com

reticuloendothelial system (RES). Efficient targeting of integrin positive tumour in mice is achieved with SWNTs coated with PEG chains linked to an arginine–glycine–aspartic acid (RGD) peptide. A high tumour accumulation is attributed to the multivalent effect of the SWNTs. The Raman signatures of SWNTs are used to directly probe the presence of nanotubes in mice tissues and confirm the radio-label-based results.

An emerging field in nanotechnology is the exploration of interesting structural, mechanical, electrical and optical properties of SWNTs [1–3] for biological applications including biosensors [4], molecular transporters for drug delivery [5–12] and potential new therapies [11]. Water-solubilized carbon nanotubes are found to traverse the cell membrane via endocytosis to deliver molecular cargoes including proteins [13] and nucleic acids such as plasmids [9] and short interfering RNA [10]. The intrinsic physical properties of SWNTs can also be exploited. The high optical absorbance of SWNTs in the near-infrared regime causes heating under laser irradiation, which is useful for destroying cancer cells that are selectively internalized with nanotubes [11]. Various groups have found that well water-solubilized nanotubes with high hydrophilicity are non-toxic, even at high concentrations [5–12, 14, 16].

The fate and biological effects of carbon nanotubes in animals are critical to potential applications in vivo. There are only two reports on the biodistribution of chemically functionalized nanotubes intravenously injected into animals [17, 18]. In these, Wang et al. [17] and Singh et al. [18] reported that two differently functionalized SWNTs behaved like small molecules in mice and freely cleared through the urine with little uptake by the liver or other organs of the RES. It therefore appears, based on the information to date, that SWNTs defy the general trend of high RES uptake observed for nanomaterials in general [19, 20]. The validity and generality of these results deserve careful investigation. Also, much effort is needed to achieve targeted accumulation of SWNTs in vivo, a goal central to potential therapeutics [19, 20], which has not yet been achieved with nanotubes.

In this work, we investigated the biodistribution and tumour targeting ability of SWNTs in mice, using Hipco nanotubes nonco-

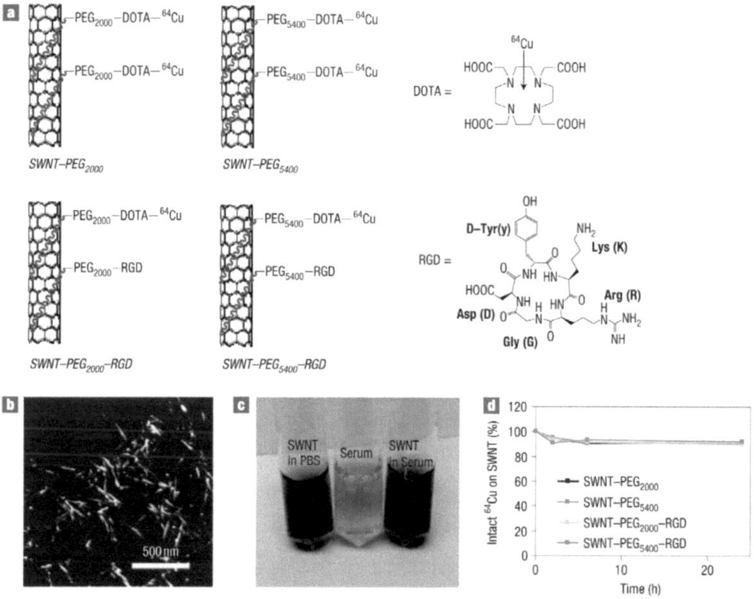

Figure 14.1 Water-soluble carbon nanotubes functionalized with PEG, radio labels and RGD peptide. (a) Schematic drawings of non-covalently functionalized SWNT–PEG_{2000}, SWNT–PEG_{5400}, SWNT–PEG_{2000}–RGD, SWNT–PEG_{5400}–RGD with DOTA–^{64}Cu. The hydrophobic carbon chains (blue segments) of the phospholipids strongly bind to the sidewalls of the SWNTs, and the PEG chains render water solubility to the SWNTs. The DOTA molecules on the SWNTs are used to chelate ^{64}Cu for radio labelling. (b) An atomic force microscope AFM image of SWNT–PEG_{5400} deposited on a silicon substrate. (c) A photograph of stable SWNT–PEG_{2000} suspensions in PBS and full fetal bovine serum. (d) Serum stability test showing that ^{64}Cu remains intact on carbon nanotubes over 24 h incubation in full mouse serum. The slight reduction during the early two time points was due to the removal of residual free ^{64}Cu radio labels in the nanotube solution by filtration.

valently functionalized with phospholipid–PEG (PL–PEG), the PEG chains being one of two different lengths (molecular weight of PEG chains = 2,000 and 5,400, respectively) (Fig. 14.1a). The diameters and lengths of the PL–PEG-functionalized SWNTs were 1–5 nm and 100–300 nm, respectively (Fig. 14.1b) and were well solubilized and suspended in buffers and whole serum without any aggregation (Fig. 14.1c; also see Supplementary Information, Fig. 14.S1).

Macrocyclic chelating agent DOTA (1,4,7,10-tetraazacyclododecane-1,4,7,10-tetraacetic acid) was attached to the termini of the PEG chains and used to conjugate positron emitting radionuclide ^{64}Cu (radio decay half-life $t_{1/2} = 12.7$ h) (Fig. 14.1a). Owing to the importance of integrin $\alpha_v\beta_3$ to tumour angiogenesis and metastasis [21, 22], we conjugated SWNTs to both ^{64}Cu and c(RGDyK) (Fig. 14.1a), a potent integrin $\alpha_v\beta_3$ antagonist, aimed at in vivo targeting of integrin $\alpha_v\beta_3$-positive tumours in mice via specific RGD–integrin $\alpha_v\beta_3$ binding [23, 24]. We measured the numbers of DOTA (by isotope dilution assay [25]) and RGD conjugated to SWNTs (see Supplementary Information, Table 14.S1), carried out a receptor binding assay, and observed the multivalent binding of SWNTs to cell-surface receptors via multiple RGD on each tube (see Supplementary Information, Fig. 14.S4). Importantly, the ^{64}Cu radio labels remained intact on SWNTs after incubation in full mouse serum over 24 h (Fig. 14.1d). No detachment from nanotubes was observed owing to strong binding of the phospholipids onto SWNTs [10, 11], even when heated to high temperatures, as demonstrated by the lack of nanotube aggregation at 70°C over more than 1 week (see Supplementary Information, Fig. 14.S1).

Mice bearing subcutaneous integrin $\alpha_v\beta_3$-positive U87MG tumours were intravenously injected with SWNT–PEG_{2000} and SWNT–PEG_{5400}, followed by microPET scans at multiple time points (up to 24 h). Both SWNT conjugates exhibited prominent uptake in the liver (Fig. 14.2a, b), with lower uptake for SWNT–PEG_{5400} than SWNT–PEG_{2000} (Fig. 14.2b). Blood sampling revealed that SWNT–PEG_{5400} exhibited a much longer blood circulation time ($t_{1/2} \approx 2$ h) than SWNT–PEG_{2000} ($t_{1/2} \approx 0.5$ h) (Fig. 14.2c). Biodistribution in various organs (Fig. 14.3a) after killing the mice at 24 h post-injection (p.i.) revealed prominent SWNT uptake in the liver and spleen (consistent with PET data, Fig. 14.3c) and low uptake in the tumour, muscle, bone, skin and other organs.

The fact that much of the radioactivity remained in the mice at 24 h p.i. (Fig. 14.3d) suggested relatively slow excretion of SWNTs, which differed from the previous finding where SWNTs acted as small molecules, with little RES uptake and free excretion from mice [17, 18]. Note that the dimensions of our SWNTs (diameter

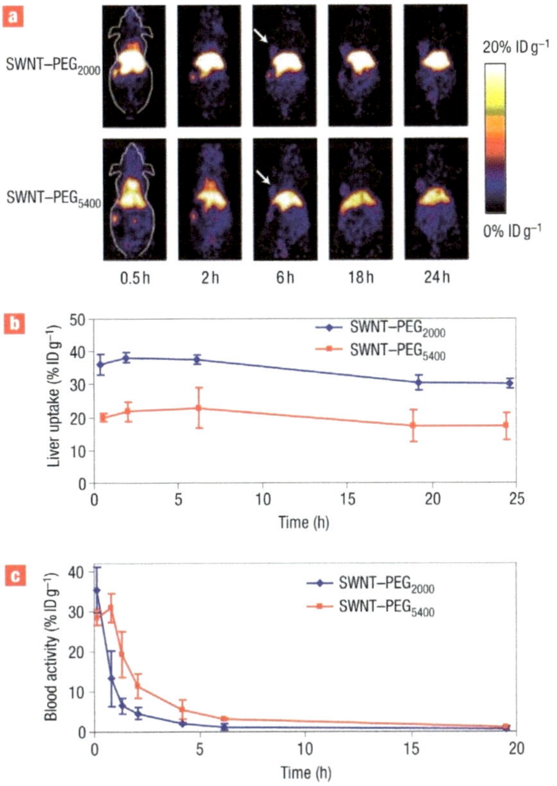

Figure 14.2 Functionalization-dependent biodistribution and blood circulation of intravenously injected SWNTs in mice bearing the U87MG human glioblastoma tumour. (a) MicroPET images of two mice at various time points post tail-vein injection of ^{64}Cu-labelled SWNT-PEG$_{2000}$ and SWNT–PEG$_{5400}$, respectively. The arrows point to the tumours. (b) Liver uptake curves over time as measured by PET for the two SWNT conjugates. (c) Blood activity curves for the two conjugates. All data points represent three animals per group (four mice per group for c).

~1–5 nm, length ~100–300 nm) were in fact smaller than the SWNT bundles (diameter ~10–40 nm, length ~1 mm or larger) used in previous reports [17, 18]. It is known that RES uptake of nanomaterials is size-dependent and should be higher for larger sizes [20]. Our biodistribution data were accurate, because there

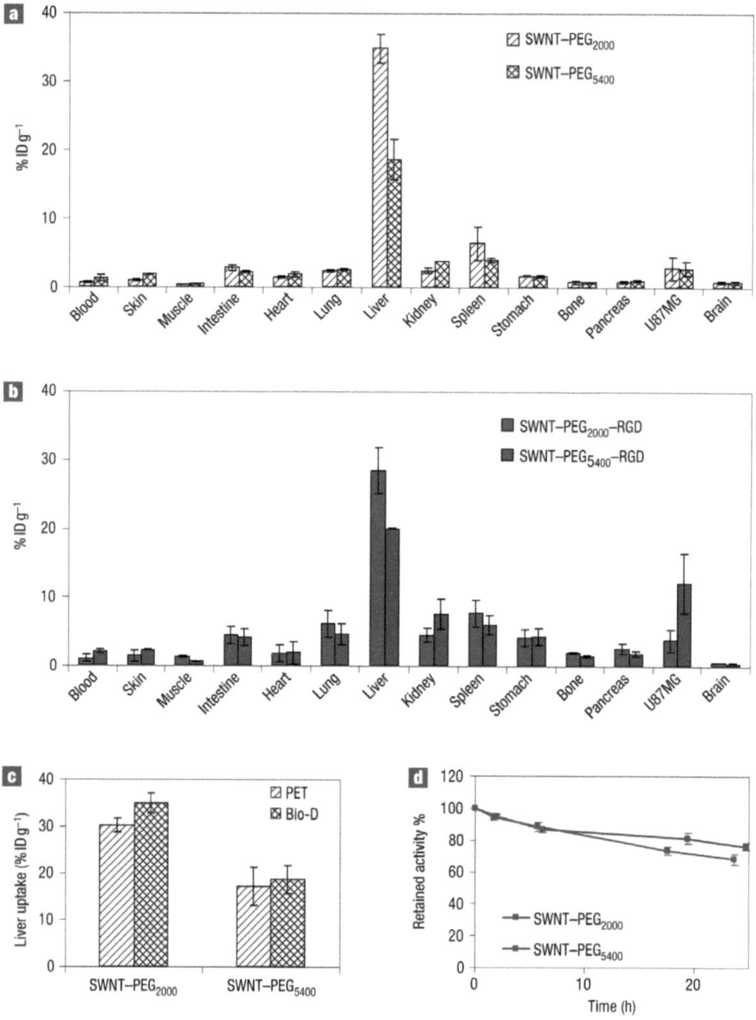

Figure 14.3 Biodistribution and retained activity of ^{64}Cu-labelled SWNTs in mice. (a) Biodistribution of SWNT–PEG$_{2000}$ and SWNT–PEG$_{5400}$ in various organs at 24 h p.i. (b) Biodistribution of SWNT–PEG$_{2000}$–RGD and SWNT–PEG$_{5400}$–RGD at 24 h p.i. (c) Comparison of liver uptake of SWNTs measured by microPET and biodistribution at 24 h p.i. (d) Total ^{64}Cu activity left in the mice at different time points p.i. of the two different SWNT conjugates. The radio-decay of ^{64}Cu with a half-life of 12.7 h was corrected in the data. All data shown are based on three mice per group.

were few free unconjugated ^{64}Cu radio labels in our SWNT solutions and the ^{64}Cu-conjugated SWNTs were serum stable (Fig. 14.1c,d). Otherwise, it is known that small molecules (such as free ^{64}Cu ions and PEG–^{64}Cu) would indeed be rapidly excreted from mice via the urine over several hours [26].

The biodistribution of our SWNTs is similar to other nanomaterials (sizes ~10–100 nm) with a tendency to undergo RES uptake, including in the liver [19, 20]. RES uptake occurs via osponization, that is, antibody binding to nanomaterials in the plasma for recognition by phagocytes in the RES. We found that PEGylation by PL–PEG_{5400} imparted to SWNTs high hydrophilicity and resistance to protein nonspecific binding (NSB) (PL–PEG_{2000} functionalization was insufficient to prevent protein NSB to SWNTs; see Supplementary Information, Fig. 14.S3), consistent with the reduced RES uptake and longer blood circulation of SWNT–PEG_{5400} [19, 20].

Next, we show the efficient targeting of integrin $\alpha_v\beta_3$-positive U87MG tumours via RGD-functionalization of SWNT–PEG_{5400} and specific RGD–integrin $\alpha_v\beta_3$ recognition (Fig. 14.4, and see Fig. 14.3b for biodistribution of SWNT–PEG_{5400}–RGD). Although SWNT–PEG_{2000}–RGD showed only a slight increase in tumour uptake compared to that without RGD (Fig. 14.4a, first column and Fig. 14.4b), SWNT–PEG_{5400}–RGD exhibited a high tumour uptake of ~10–15% injected dose (ID) g^{-1} (Fig. 14.4a, second column), a significant increase from ~3–4% ID g^{-1} for SWNT–PEG_{5400} free of RGD (Fig. 14.4c). Tumour uptake of SWNTs was rapid, reaching a plateau at about 6 h p.i., and then levelled off in the next 20 h. The tumour uptake levels derived from PET images were consistent with what was measured by biodistribution (Fig. 14.3b), and the tumour/muscle uptake ratio was calculated to be >15 for SWNT–PEG_{5400}–RGD. The high tumour accumulation was a result of the long blood circulation time, the specific high tumour-binding affinity of RGD-functionalized SWNT–PEG_{5400}, and the multivalency effect (that is, multiple RGDs along a SWNT binding to multiple integrin $\alpha_v\beta_3$ simultaneously; see Supplementary Information, Fig. 14.S4). Note that without nanotubes, PEG–RGD–^{64}Cu species were known to be rapidly cleared from mice through the renal route in just a few hours, and U87MG tumour accumulation peaked at around

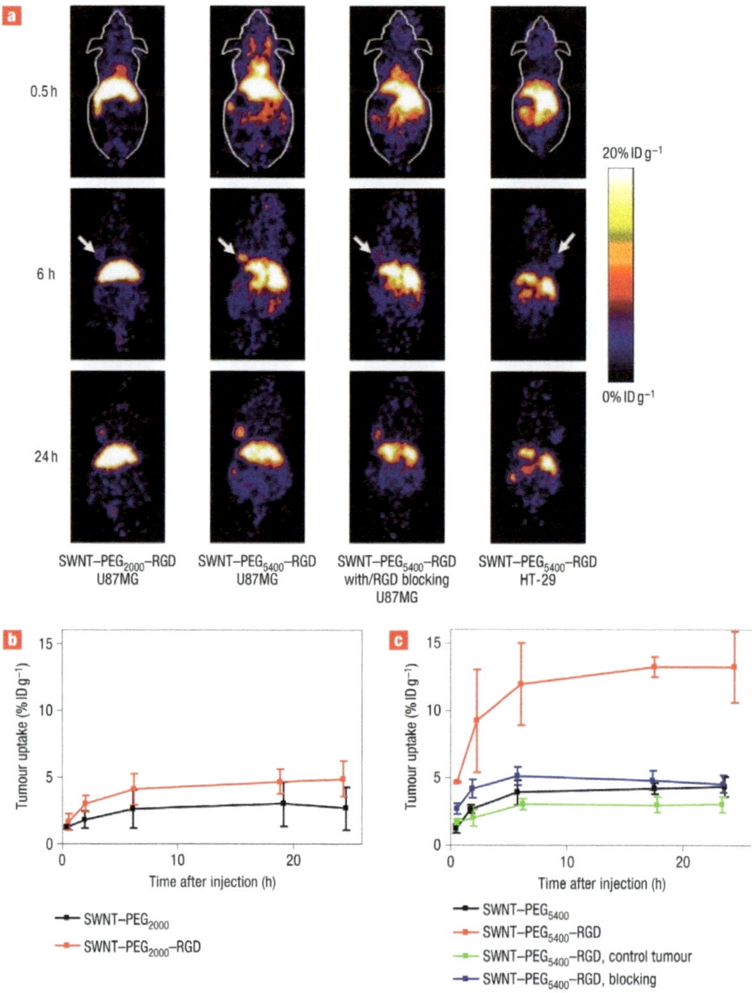

Figure 14.4 Targeting of integrin $\alpha_v\beta_3$-positive U87MG tumour in mice by RGD-functionalized SWNTs. (a) MicroPET images of mice. The arrows point to the tumours. High tumour uptake (~15% ID g^{-1}) of SWNT–PEG$_{5400}$–RGD is observed in the U87MG tumour (second column), in contrast to the low tumour uptake (first column) of SWNT–PEG$_{2000}$–RGD. The third column is a control experiment showing blocking of SWNT–PEG$_{5400}$–RGD tumour uptake by co-injection of free c(RGDyK). The fourth column is a control experiment showing low uptake of SWNT–PEG$_{5400}$–RGD in an integrin $\alpha_v\beta_3$-negative HT-29 tumour. (b,c) U87MG tumour uptake curves for mice injected with SWNT–PEG$_{2000}$ (b) and SWNT–PEG$_{5400}$ (c), with and without RGD. All data shown represent three mice per group.

~3% ID g^{-1} 30 min after injection, and decreased afterwards [26]. In control experiments, we injected a blocking dose of c(RGDyK) (15 mg kg^{-1}) into U87MG tumour-bearing mice along with SWNT–PEG_{5400}–RGD, and observed that the uptake of nanotubes in the tumour was significantly blocked (Fig. 14.4a, third column). Mice bearing integrin $\alpha_v\beta_3$-negative HT-29 tumours (see Supplementary Information, Fig. 14.S5) were also injected with SWNT–PEG_{5400}–RGD, and little tumour uptake of nanotubes was observed (~3% ID g^{-1}, Fig. 14.4a, fourth column).

To take advantage of the intrinsic optical properties of SWNTs, we used Raman spectroscopy to directly detect SWNTs in the various tissues of a mouse. Non-covalently functionalized SWNTs exhibit strong resonance Raman bands, including the G band at ~1,580 cm^{-1}, characteristic of graphitic carbon (Fig. 14.5a) (refs 1 and 2). A U87MG tumour-bearing mouse injected with a high dose of SWNT–PEG_{5400}–RGD (0.5 mg kg^{-1}) was killed after PET imaging (Fig. 14.5b) at 8 h p.i. Raman spectroscopy clearly revealed the existence of SWNTs in the liver and tumour samples with high G-band Raman intensities, a slight signal in the kidney sample, and no SWNT Raman signal in muscle (Fig. 14.5c). The amounts of SWNT in each of the tissue samples, derived from Raman intensities calibrated against SWNT solutions with known concentrations, were in reasonable agreement with PET data based on radioactivity of nanotubes (Fig. 14.5d). Thus, our Raman data provide direct proof of tumour uptake of SWNTs. The co-localization of ^{64}Cu radio labels and SWNTs in various mouse tissues also suggests a proof of the in vivo stability of our noncovalent functionalization of SWNTs. Such functionalization retains the intrinsic properties of SWNTs (for example, without the Raman intensity degradation seen in covalently modified tubes) for biological and medical applications such as Raman probing [27] and fluorescence imaging [8] in vitro, ex vivo and potentially in vivo.

We observed no obvious toxicity or negative health effects (such as weight loss and fatigue), with many mice injected with SWNT–PEG at dosages up to 2 mg kg^{-1} over monitoring periods of up to several months. However, the long-term fate of SWNT–PEG injected into mice requires further investigation. Based on the retained

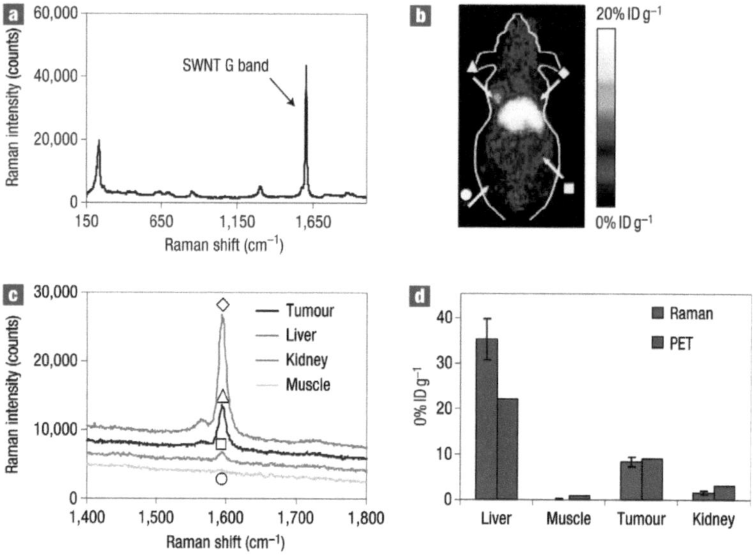

Figure 14.5 Detecting nanotubes in mice tissues using characteristic Raman signatures of SWNTs. (a) A Raman spectrum of a solution of SWNT–PEG_{5400}. The G band of the SWNTs is indicated by the arrow. The peak at the lower Raman shift near 230 cm^{-1} corresponds to the radial breathing modes of the SWNTs. (b) A two-dimensional projection of the microPET image of a U87MG tumour-bearing mouse 8 h p.i. of a high dose of SWNT–PEG_{5400}–RGD solution. The arrows point to the tumour and several organs used for Raman measurements (symbols correspond with those in c). (c) Raman spectra in the G-band region of SWNTs recorded on lyophilized tumour and tissue powder samples suspended in surfactant solutions. The data provides direct evidence of the existence of SWNTs in the tumour. (d) Comparison of the biodistribution data obtained by PET imaging and ex vivo Raman measurements. The error bars in the Raman data were obtained with several measurements performed over different parts of the tissues.

radioactivity in mice, we observed relatively slow excretion of SWNTs. Longer-term monitoring is currently ongoing to investigate SWNT clearance from mice.

Our current work establishes strong non-covalently functionalized SWNTs for in vivo applications. Surveying the literature, we found that the nanotube tumour accumulation of ~13% ID g^{-1} over long periods (>24 h) is among the best achieved using nanomaterials. The unique one-dimensional shape and flexible structure of SWNTs enables a polyvalency effect (see Supplementary Information, Fig. 14.S4) and enhances tumour binding affinity. The one-dimensional shape may also facilitate SWNTs leaking out from blood microvessels to reach cancer cells in the tumour through vascular and interstitial barriers [28]. The exact distribution of our SWNTs within the tumour requires further investigation. SWNTs of various lengths will be ideal vehicles for elucidating size and shape effects on nanomaterial distribution within tumours. These discoveries will be useful for the application of the structural and physical properties of SWNTs for therapeutic approaches.

14.1 Methods

14.1.1 Preparation of PEGylated SWNTs

Raw Hipco SWNTs were sonicated in an aqueous solution of PL–PEG_{2000}–NH_2, that is, DSPE–PEG_{2000}–Amine (1,2-distearoyl-*sn*-glycero-3-phosphoethanolamine-N-(amino (polyethyleneglycol) 2000)) (Avanti Polar Lipids) or PL–PEG_{5400}–NH_2, DSPE–PEG_{5400}–Amine for 1 h, centrifuged at 24,000 *g* for 6 h to obtain short, PL–PEG-functionalized SWNTs in supernatant. Filtration through 100-kDa filters (Millipore) removed excess phospholipids. PL–PEG_{5400}–NH_2 was synthesized by linking NHS–PEG_{3400}–Boc (Nektar) with PL–PEG_{2000}–NH_2 in aqueous solution followed by deprotection of Boc. The optical absorbance of the SWNTs was measured to determine their concentration (see Supplementary Information) [11].

14.1.2 DOTA and RGD Conjugation to SWNTs

DOTA–SNHS (Sulfo-NHS) was synthesized by reacting DOTA with *N*-hydroxysulphonosuccinimide in the presence of 1-ethyl-3-(3-(dimethylamino)-propyl) carbodiimide (EDC) at 1:1:1 molar ratio [29]. PL–PEG_{2000}–NH_2 and PL–PEG_{5400}–NH_2 functionalized SWNTs were mixed with DOTA–SNHS at pH 7.4 and incubated for 4 h. The SWNT solution was purified by filtration to remove unconjugated DOTA.

For conjugation of both RGD and DOTA, DOTA–SNHS and sulpho-SMCC (sulphosuccinimidyl 4-N-maleimidomethyl cyclohexane-1-carboxylate) were mixed at 1:5 molar ratios and incubated in SWNT–PEG–NH_2 solutions at pH 7.4 for 2 h. Upon removal of excess reagents, the SWNTs were reacted overnight with 0.2 mM of thiolated RGD [24] in the presence of 10 mM Tris(2-carboxyethyl) phosphine hydrochloride (TCEP) at pH 7.4, yielding SWNT–PEG_{2000}–RGD and SWNT–PEG_{5400}–RGD with both RGD and DOTA on the SWNTs.

14.1.3 ^{64}Cu-Labelling of SWNTs and Serum Stability Test

$^{64}CuCl_2$ (supplied by the University of Wisconsin-Madison) was diluted in 300 μl of 0.1 M sodium acetate buffer (NaOAc, pH 6.5). Two mCi of ^{64}Cu was added to 100 μl of each of the SWNT conjugates (~150 nM) buffered by 300 μl of 0.1 M NaOAc solution (pH 6.5). The reaction mixture was incubated for 1 h at 40°C after which excess unconjugated ^{64}Cu was removed. The final radio-labelling yield was 60–80%.

The serum stability of radio labels on SWNTs was investigated by incubating ^{64}Cu-labelled SWNTs in full mouse serum at 37°C for up to 24 h. Portions of SWNT–serum suspension were collected at different time points and filtered through 300-kDa cutoff filters. The filtrates were collected and the radioactivity was measured. The percentages of retained (intact) ^{64}Cu on the SWNTs were calculated by using the relation (total activity–activity in filtrate)/total activity.

14.1.4 Cell Lines and the Animal Model

U87MG human glioblastoma and HT-29 human colorectal cancer cell lines (from American Type Culture Collection, ATCC) were cultured under standard conditions. The U87MG and HT-29 tumour models were generated by subcutaneous injection of 5×10^6 cells in 50 μl PBS into the front left and front right legs of the mice, respectively. The mice were used for the study when the tumour volume reached 200–300 mm^3.

14.1.5 Micropet Imaging

PET imaging was carried out on a microPET R4 rodent model scanner (Concorde Microsystems) [30]. About 1 mg of SWNTs (~30 nM) in 150–200 μl PBS with 200–300 μCi of ^{64}Cu were injected into each mouse at the tail vein, except for one mouse, which was injected with ~10 mg of SWNTs for the Raman study. The tissue/organ uptake (% ID g^{-1}) based on PET imaging was calculated as described previously [25, 30]. Note that all the PET data and biodistribution data have been decay corrected by the half-life of ^{64}Cu. Three mice in each group were used for the data presented in this work.

14.1.6 Biodistribution and Blood Circulation Studies

Nude mice bearing U87MG tumours were injected with ~200–300 μCi of ^{64}Cu-labelled SWNTs. The mice were killed at 24 h p.i. The major organs were collected and wet weighed. The radioactivity in the tissues was measured using a γ-counter (Packard). For each mouse, the radioactivity of the tissue samples was calibrated against a known aliquot of the injectate, and normalized to a body weight of 25 g. Data presented were from three animals per group.

Approximately 3–5 μl of blood was taken from the tail vein of the mice injected with 200–300 μCi of ^{64}Cu-labelled SWNTs at different time points p.i., and the radioactivity was measured by a γ-counter. The data shown were based on four mice in each group. The values of blood circulation half-life were obtained by first-order exponential decay fits.

14.1.7 Ex vivo Raman Spectroscopic Measurements of Mouse Tissues

A U87MG tumour-bearing mouse injected with a high dose of SWNT-PEG_{5400}-RGD (0.05 mg ml^{-1}, 300 nM in 200 μl PBS) was killed at 8 h p.i. The tumour, liver, kidney and a portion of the muscle were lyophilized and sonicated in 1% SDS and 1% Triton X-100 solution to form homogenous suspensions. Raman spectra of the suspensions were acquired with a Renishaw micro-Raman instrument. SWNT suspensions in lysis buffer with known concentrations were measured to obtain a standard calibration curve (see Supplementary Information, Fig. 14.S6), against which the SWNT concentrations in the tissue samples were calculated and compared to the microPET data. More experimental details can be found in the Supplementary Information.

Acknowledgements

This work was supported in part by a Ludwig Translational Research Grant at Stanford University and NIH-NCI CCNE-TR at Stanford (H.D.), National Institute of Biomedical Imaging and Bioengineering (NIBIB) (R21 EB001785), National Cancer Institute (NCI) (R21 CA102123, P50 CA114747, U54 CA119367, R24 CA93862), Department of Defense (DOD) (W81XWH-04-1-0697, W81XWH-06-1-0665, W81XWH-06-1-0042, DAMD17-03-1-0143) and a Benedict Cassen Postdoctoral Fellowship from the Education and Research Foundation of the Society of Nuclear Medicine (to W.C.).

Author Contributions

H.D., X.C., Z.L. and W.C. conceived and designed the experiments. Z.L., W.C., X.C., L.H., N.N., K.C. and X.S. performed the experiments. H.D., Z.L., W.C. and X.C. co-wrote the manuscript. All authors discussed the results and commented on the manuscript.

Competing Financial Interests

The authors declare that they have no competing financial interests.

References

1. Dresselhaus, M., and Dai, H. (eds) MRS 2004 *Carbon Nanotube Special Issue* (2004).
2. Dresselhaus, M. S., Dresselhaus, G., and Avouris, P. (eds) *Carbon Nanotubes* (Springer, Berlin, 2001).
3. Dai, H. Carbon nanotubes: opportunities and challenges. *Surf. Sci.* **500**, 218–241 (2002).
4. Chen, R. J. et al. Noncovalent functionalization of carbon nanotubes for highly specific electronic biosensors. *Proc. Natl Acad. Sci. USA* **100**, 4984–4989 (2003).
5. Kam, N. W. S., Jessop, T. C., Wender, P. A., and Dai, H. J. Nanotube molecular transporters: Internalization of carbon nanotube–protein conjugates into mammalian cells. *J. Am. Chem. Soc.* **126**, 6850–6851 (2004).
6. Pantarotto, D., Briand, J., Prato, M., and Bianco, A. Translocation of bioactive peptides across cell membranes by carbon nanotubes. *Chem. Commun.* 16–17 (2004).
7. Bianco, A., Kostarelos, K., Partidos, C. D., and Prato, M. Biomedical applications of functionalised carbon nanotubes. *Chem. Commun.* 571–577 (2005).
8. Cherukuri, P., Bachilo, S. M., Litovsky, S. H., and Weisman, R. B. Near-infrared fluorescence microscopy of single-walled carbon nanotubes in phagocytic cells. *J. Am. Chem. Soc.* **126**, 15638–15639 (2004).
9. Liu, Y. et al. Polyethylenimine-grafted multiwalled carbon nanotubes for secure noncovalent immobilization and efficient delivery of DNA. *Angew. Chem. Int. Edn Engl.* **44**, 4782 (2005).
10. Kam, N. W. S., Liu, Z., and Dai, H. Functionalization of carbon nanotubes via cleavable disulfide bonds for efficient intracellular delivery of siRNA and potent gene silencing. *J. Am. Chem. Soc.* **127**, 12492–12493 (2005).
11. Kam, N. W. S., O'Connell, M., Wisdom, J. A., and Dai, H. Carbon nanotubes as multifunctional biological transporters and near-infrared agents for selective cancer cell destruction. *Proc. Natl Acad. Sci. USA* **102**, 11600–11605 (2005).
12. Kam, N. W. S., Liu, Z., and Dai, H. J. Carbon nanotubes as intracellular transporters for proteins and DNA: An investigation of the uptake mechanism and pathway. *Angew. Chem. Int. Edn Engl.* **45**, 577–581 (2005).

13. Kam, N. W. S., and Dai, H. Carbon nanotubes as intracellular protein transporters: Generality and biological functionality. *J. Am. Chem. Soc.* **127**, 6021–6026 (2005).
14. Sayes, C. M. et al. Functionalization density dependence of single-walled carbon nanotubes cytotoxicity in vitro. *Toxicity Lett.* **161**, 135–142 (2006).
15. Chen, X. et al. Interfacing carbon nanotubes with living cells. *J. Am. Chem. Soc.* **128**, 6292–6293 (2006).
16. Dumortier, H. et al. Functionalized carbon nanotubes are non-cytotoxic and preserve the functionality of primary immune cells. *Nano Lett.* **6**, 1522–1528 (2006).
17. Wang, H. F. et al. Biodistribution of carbon single-wall carbon nanotubes in mice. *J. Nanosci. Nanotechnol.* **4**, 1019–1024 (2004).
18. Singh, R. et al. Tissue biodistribution and blood clearance rates of intravenously administered carbon nanotube radiotracers. *Proc. Natl Acad. Sci. USA* **103**, 3357–3362 (2006).
19. Moghimi, S. M., Hunter, A. C., and Murray, J. C. Long-circulating and target-specific nanoparticles: Theory to practice. *Pharmacol. Rev.* **53**, 283–318 (2001).
20. Moghimi, S. M., Hunter, A. C., and Murray, J. C. Nanomedicine: current status and future prospects. *FASEB J.* **19**, 311–330 (2005).
21. Mizejewski, G. J. Role of integrins in cancer: Survey of expression patterns. *Proc. Soc. Exp. Biol. Med.* **222**, 124–138 (1999).
22. Jin, H., and Varner, J. Integrins: Roles in cancer development and as treatment targets. *Br. J. Cancer* **90**, 561–565 (2004).
23. Xiong, J. P. et al. Crystal structure of the extracellular segment of integrin $\alpha_v\beta_3$ in complex with an Arg-Gly-Asp ligand. *Science* **296**, 151 (2002).
24. Cai, W. et al. Peptide-labeled near-infrared quantum dots for imaging tumor vasculature in living subjects. *Nano Lett.* **6**, 669–676 (2006).
25. Cai, W. et al. In vitro and in vivo characterization of ^{64}Cu-labeled Abegrin^{TM}, a humanized monoclonal antibody against integrin $\alpha_v\beta_3$. *Cancer Res.* **66**, 9673 (2006).
26. Chen, X. et al. Pegylated Arg-Gly-Asp peptide: ^{64}Cu labeling and PET imaging of brain tumor $\alpha_v\beta_3$-integrin expression. *J. Nucl. Med.* **45**, 1776–1783 (2004).
27. Heller, D. A., Baik, S., Eurell, T. E., and Strano, M. S. Single-walled carbon nanotube spectroscopy in live cells: Towards long-term labels and optical sensors. *Adv. Mater.* **17**, 2793–2799 (2005).

28. Jain, R. K. Vascular and interstitial barriers to delivery of therapeutic agents in tumors. *Cancer Metastasis Rev.* **9**, 253–266 (1990).
29. Wu, Y. et al. MicroPET imaging of glioma integrin $\alpha_v\beta_3$ expression using ^{64}Cu-labeled tetrameric RGD peptide. *J. Nucl. Med.* **46**, 1707–1718 (2005).
30. Cai, W., Zhang, X., Wu, Y., and Chen, X. A thiol-reactive ^{18}F-labeling agent, N-[2-(4-^{18}F-fluorobenzamido) ethyl]maleimide (^{18}F-FBEM), and the synthesis of RGD peptide-based tracer for PET imaging of $\alpha_v\beta_3$ integrin expression. *J. Nucl. Med.* **47**, 1172–1180. (2006).

Supplementary Information: Chapter 14

14.S1 Comparison of Aqueous Stability of SWNTs Functionalized by Various Methods

Phospholipid (PL)-PEG (PEG M.W. 2000 Da or 5400 Da) and sodium dodecyl sulphate (SDS) suspended SWNTs were made by sonicating raw Hipco SWNTs in aqueous solution of PL-PEG-NH_2 (1 mg/ml) or SDS (10 mg/ml) for 1 h followed by centrifugation at 24,000 g for 6 h to remove aggregates. Excess PL-PEG-NH_2 or SDS in the supernatant was removed by repeated filtration through 100 kDa filters and resuspension in water. Hydroxylated SWNTs (SWNTols) were prepared following the method reported by Wang et al. [1]. In brief, SWNTs were mixed with KOH and agitated vigorously with the help of small stainless steel balls for 1 h at room temperature (r.t.). Then the SWNTols were dissolved in distilled water and centrifuged at 12,000 g for 15 min. Excess KOH was removed by filtration using 100 kDa filters.

PL-PEG-NH_2 suspended SWNTs are the most stable without any visible aggregation in water and phosphate buffered saline (PBS) after removing excess amount of free PL-PEG-NH_2 molecules. The nanotubes are stable in PBS without aggregation even after heating at 70°C for one week (Fig. 14.S1a, first row), suggesting no detachment of the coating under such harsh condition. SDS, a commonly used surfactant, can also be used to solubilize SWNTs, giving a good suspension in water. However, SDS suspended SWNTs are not stable in the absence of free SDS (Fig. 14.S1a, second row), suggesting that the binding between SDS and SWNTs is not sufficiently strong. The in vivo biodistribution study of hydroxylated SWNTs (SWNTols) has been previously performed before by Wang et al. [1]. However, this type of SWNTs is not very stable in PBS buffer. Obvious aggregation of SWNTols was seen after incubation in PBS overnight at either room temperature or 70°C (Fig. 14.S1, third row).

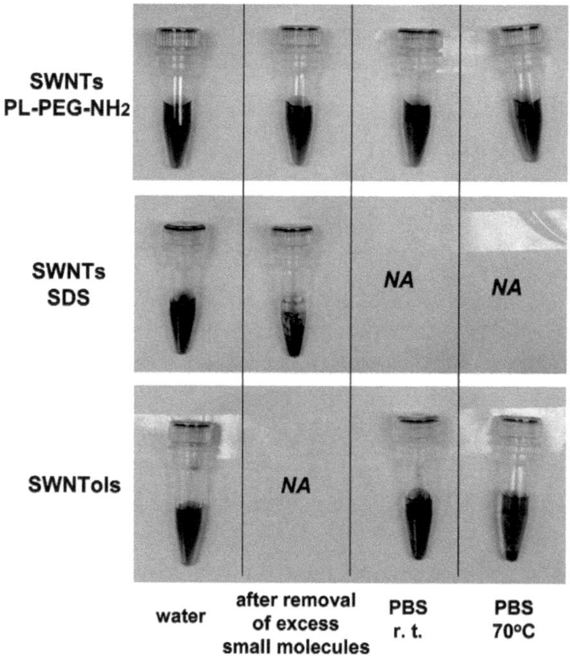

Figure 14.S1 Stability of SWNTs functionalized by various methods. First row, phospholipid-PEG (PL-PEG-NH_2) suspended SWNTs are stable at all conditions, even at 70°C for one week (the fourth photo). Second row, SDS suspended SWNTs are not stable after removal of excess SDS molecules in aqueous solution. Third row, hydroxylated SWNTs (SWNTols) aggregate in PBS buffer after overnight incubation at either room temperature or 70°C (the third and fourth photo, respectively).

14.S2 Measurement of SWNT Concentration by UV-VIS-NIR Spectroscopy

PL-PEG-NH_2 functionalized SWNTs exhibit high optical absorbance from ultraviolet (UV) to near infrared (NIR) regimes (Fig. 14.S2). The absorbance spectrum of a SWNT solution can be utilized to estimate the concentration of nanotubes [2]. The absorbance at 808 nm exhibits a molar extension coefficient of $7.9 \times 10^6\ M^{-1} \cdot cm^{-1}$ for SWNTs with average length of ~200 nm [2].

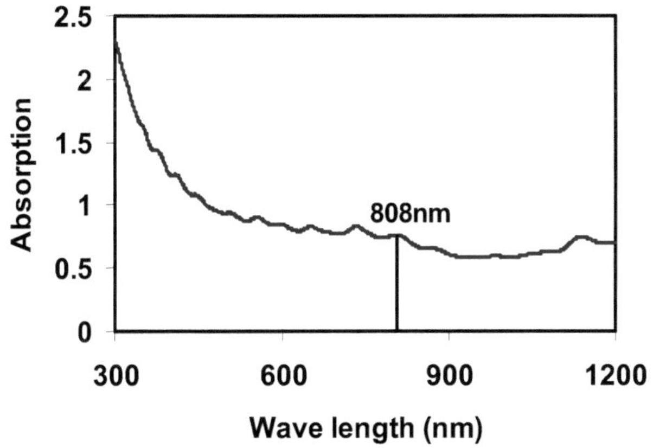

Figure 14.S2 A UV-VIS-NIR spectrum of SWNTs functionalized by PL-PEG-NH_2.

14.S3 Number of DOTA and RGD Peptides on the SWNTs

For SWNT-PEG_{2000} and SWNT-PEG_{5400} with and without conjugated RGD, the average number of DOTA chelators per SWNT was measured using a previously reported procedure with slight modifications [3]. Briefly, a defined amount of nonradioactive $CuCl_2$ (~150 fold excess of SWNT concentration) in 40 μl 0.1 M NaOAc buffer (pH = 6.5) was added to 0.2 mCi of $^{64}CuCl_2$ in 20 μl 0.1 M NaOAc buffer. Twenty μl of each SWNT conjugate in 100 μl 0.1 M NaOAc buffer was added to the above carrier-added $^{64}CuCl_2$ solution. The reaction mixtures were incubated with constant shaking at 40°C for 1 h. ^{64}Cu-labeled SWNTs were purified using 100 kDa filters and the radio-labeling yield was calculated by measuring the radio activity retained in the SWNTs solution. The number of DOTA per SWNT (average length ~150 nm) was determined as moles (Cu^{2+}) × yield/moles (SWNT).

Table 14.S1 DOTA number assay. Standard deviations (std.) were obtained from 4 parallel samples

	Average number of DOTA per SWNT	Std.
SWNT-PEG_{2000}	75.5	9.3
SWNT-PEG_{5400}	61.9	0.6
SWNT-PEG_{2000}-RGD	47.4	6.4
SWNT-PEG_{5400}-RGD	36.2	4.5

The measured numbers of DOTA per nanotube of SWNTs with RGD conjugation (SWNT-PEG_{2000}-RGD and SWNT-PEG_{5400}-RGD) are less than those without RGD (SWNT-PEG_{2000} and SWNT-PEG_{5400}). Assuming 100% yield of thiolated RGD conjugation (a large excess of thiolated RGD was used), the ratio between RGD and DOTA on nanotubes will be 0.59:1 for SWNT-PEG_{2000}-RGD and 0.71:1 for SWNTPEG$_{5400}$-RGD as calculated by the following equation:

$$\frac{\begin{bmatrix}\text{\#No. of RGD per}\\ \text{SWNT-PEG-RGD}\end{bmatrix}}{\begin{bmatrix}\text{\#No. of DOTA per}\\ \text{SWNT-PEG-RGD}\end{bmatrix}} = \frac{\begin{bmatrix}\text{\#No. of DOTA per}\\ \text{SWNT-PEG}\end{bmatrix} - \begin{bmatrix}\text{\# of DOTA per}\\ \text{SWNT-PEG-RGD}\end{bmatrix}}{\begin{bmatrix}\text{\#No. of DOTA per}\\ \text{SWNT-PEGRGD}\end{bmatrix}}$$

14.S4 Non-Specific Binding (NSB) of SWNTs Functionalized with Different Length PEGs

Ex vivo, we found that PEGylation by PL-PEG_{5400} imparted to SWNTs high hydrophilicity and resistance to protein NSB. PL-PEG_{2000} functionalization was insufficient to prevent protein NSB to SWNTs.

Four hundred nanoliter of mouse IgG (MIgG) solution was spotted on a polylysine coated glass slide (Electron Microscopy Sciences) and allowed to dry. The glass slide was then rinsed briefly with PBS at pH 7.4 and subsequently blocked with 0.5% tween20 and 3% fetal bovine serum in PBS overnight. After blocking, the slide was rinsed briefly with PBS and water followed by drying with blowing air. Forty nanomolar of SWNT-PEG_{2000} and SWNT-PEG_{5400} solution were dropped over the MIgG spots and allowed to incubate for 1 hr. After incubation, the slide was soaked in PBS for 30 min and then briefly rinsed with water before blow drying. Raman spectrum was taken at different positions inside the MIgG spot (~10 mW power, 50 × objective, laser spot size ~2 μm^2, 30 s scan) and the averaged spectrums were presented in Fig. 14.S3. The NSB level of SWNTs on substrates was determined based on the intensity of SWNT Raman G band peak.

SWNTs coated with PEG_{2000} exhibited appreciable NSB on MIgG spots on substrates, indicating the SWNTs sidewalls were not densely covered by hydrophilic PEG chains. In contrast, SWNTs coated with PEG_{5400} showed little NSB on MIgG spots, suggesting

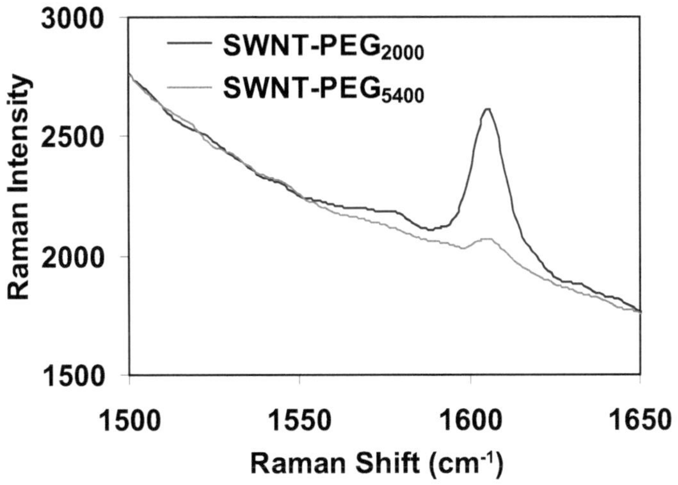

Figure 14.S3 Non-specific binding (NSB) of SWNTs with different PEG lengths. SWNT-PEG_{5400} has significantly lower NSB on mouse IgG protein substrate than SWNT-PEG_{2000}.

that the longer length of PEG_{5400} afforded sufficient hydrophilicity and biological inertness to SWNTs. The lower NSB of SWNTs coated with PEG_{5400} was consistent with the in-vivo behaviour of lower liver uptake and longer circulation half-life.

We suggest that the optimum length of PEG on SWNTs to impart high hydrophilicity and minimize NSB of proteins and antibodies (ex vivo and in vivo) will depend on the density of PEG chains on the tubes. For the particular phospholipid approach used in the current work, the packing of lipids on SWNT sidewalls may not be dense and long PEG_{5400} chains of the lipids are needed to impart sufficient hydrophilicity and NSB resistance. For denser PEGylation on SWNT sidewalls by other means, it is possible much shorter PEG can afford excellent inertness of SWNTs in vivo. Much remains to be done in developing such chemistry for nanotubes.

14.S5 Receptor Binding Assay of SWNT-PEG-RGD

The in vitro integrin $\alpha_v\beta_3$-binding affinity and specificity of SWNT-PEG_{2000}-RGD and SWNT-PEG_{5400}-RGD were assessed via competitive displacement cell-binding assays using ^{125}I-echistatin as the $\alpha_v\beta_3$ integrin-specific radio-ligand [4]. Experiments were performed on integrin $\alpha_v\beta_3$-positive U87MG human glioblastoma cells. The cells were harvested, washed twice with PBS and re-suspended (2×10^6 cells/ml) in binding buffer (20 mM Tris, pH 7.4, 150 mM NaCl, 2 mM $CaCl_2$, 1 mM $MgCl_2$, 1 mM $MnCl_2$, 0.1% BSA). Filter multiscreen DV plates (96-well, pore size: 0.65 μm, Millipore, Billerica, MA) were seeded with 1×10^5 cells per well and incubated with ^{125}I-echistatin (30,000 counts per minute (cpm)/well) in the presence of increasing concentrations of SWNTPEG$_{2000}$-RGD or SWNT-PEG_{5400}-RGD (0–100 nM). The total incubation volume was adjusted to 200 μl. After the cells were incubated for 2 h at room temperature, the plates were filtered through a multi-screen vacuum manifold and washed twice with cold binding buffer. The hydrophilic polyvinylidene fluoride (PVDF) filters were collected and the radioactivity was determined using a gamma counter (Packard, Meriden, CT). The half maximal inhibitory concentration (IC_{50}) values were calculated by fitting the data by nonlinear regression using GraphPad Prism™ (GraphPad

Software, Inc., San Diego, CA). Experiments were carried out twice with triplicate samples.

Echistatin is a viper venom disintegrin containing RGD loop which binds integrin $\alpha_v\beta_3$ with high affinity. The presence of SWNT-PEG-RGD can inhibit the binding between echistatin and integrin $\alpha_v\beta_3$ by competition, confirming the successful conjugation of RGD on SWNTs (Fig. 14.S3). Both SWNT-PEG_{2000}-RGD and SWNT-PEG_{5400}-RGD inhibited the binding of ^{125}I-echistatin to $\alpha_v\beta_3$ integrin on U87MG cells, with IC_{50} values being 4.1 and 11.1 nM of SWNTs, respectively. The enhanced integrin $\alpha_v\beta_3$ binding affinity for the SWNT-PEG-RGD conjugates as compared to the monomeric RGD peptide ($IC_{50} = 67.9 \pm 6.1$ nM) [4] is likely due to the cooperative or multivalent binding of multiple RGD units on the SWNT to multiple $\alpha_v\beta_3$ integrin, in accordance with our previous findings that multimeric RGD peptides exhibit more potent binding to integrin $\alpha_v\beta_3$ than the monomeric peptide c(RGDyK) [4].

14.S6 Immunofluorescence Staining

Frozen tumour sections (5 μm) were warmed to room temperature, fixed with ice-cold acetone for 10 min, and dried in the air for 30 min. The sections were blocked with 10 % donkey serum for 1 hour at r.t.. For CD31 and human integrin $\alpha_v\beta_3$ double staining, the sections were incubated with rat anti-mouse CD31 (1:100, BD BioSciences, San Jose, CA) and Abegrin™ (anti-human integrin $\alpha_v\beta_3$ monoclonal antibody) (100 μg/mL) for 1 hour at r.t.. After incubating with Cy3-conjugated donkey anti-rat secondary antibody (1:200, Jackson ImmunoResearch Laboratories, Inc.) and FITC-conjugated donkey antihuman secondary antibody (1:200, Jackson ImmunoResearch Laboratories, Inc.), the tumour sections were examined under the microscope (Carl Zeiss Axiovert 200M, Carl Zeiss USA, Thornwood, NY). For CD31 and mouse integrin β_3 double staining, hamster anti-mouse β_3 (1:100, BD BioSciences) and FITC-conjugated goat anti-hamster secondary antibody (1:400, Jackson ImmunoResearch Laboratories, Inc) were used.

CD31 is a transmembrane glycoprotein highly expressed on endothelial cells but not on tumour cells. CD31 staining was

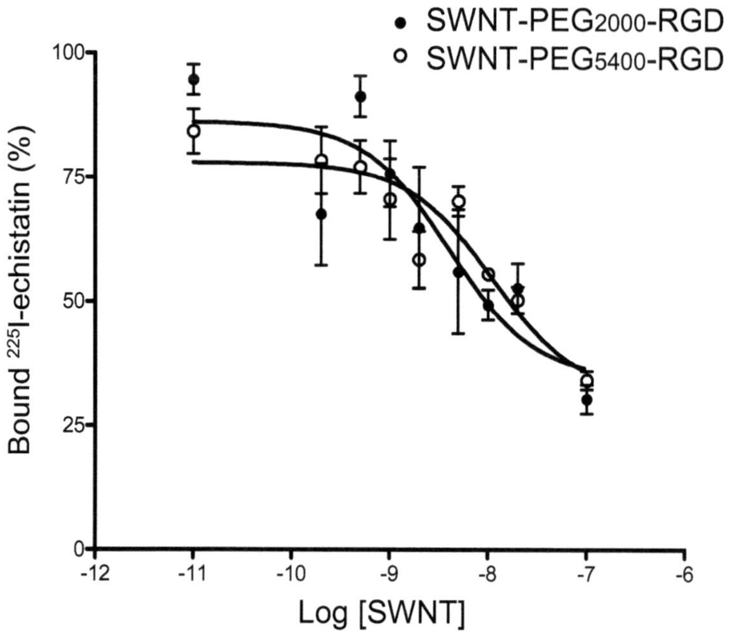

Figure 14.S4 Receptor binding assay. In vitro inhibition of ^{125}I-echistatin binding to $\alpha_v\beta_3$ integrin on U87MG human glioblastoma cells by SWNT-PEG_{2000}-RGD (•) and SWNT-PEG_{5400}-RGD (○), with IC_{50} values being 4.1 and 11.1 nM, respectively.

performed to visualize tumour vasculature. Human integrin $\alpha_v\beta_3$ is expressed on U87MG tumour cells (a human cancer cell line) while the tumour vasculature expresses mouse integrin $\alpha_v\beta_3$. As shown in Fig. 14.S4a, Abegri™ (antihuman integrin $\alpha_v\beta_3$ monoclonal antibody) staining afforded strong signal for U87MG tumour cells but not HT-29 tumour cells. No co-localization between Abegrin™ staining and CD31 staining was observed for either U87MG tumour or HT-29 tumour, suggesting the vessels in neither of the two tumours were stained by Abegrin™ as murine tumour vessels do not express human integrin $\alpha_v\beta_3$. In Fig. 14.S4b, anti-mouse β_3 staining showed strong vessel staining for U87MG tumour but weak staining for HT-29 tumour. The good co-localization of CD31 and mouse β_3 staining indicates that the tumour vasculature expresses

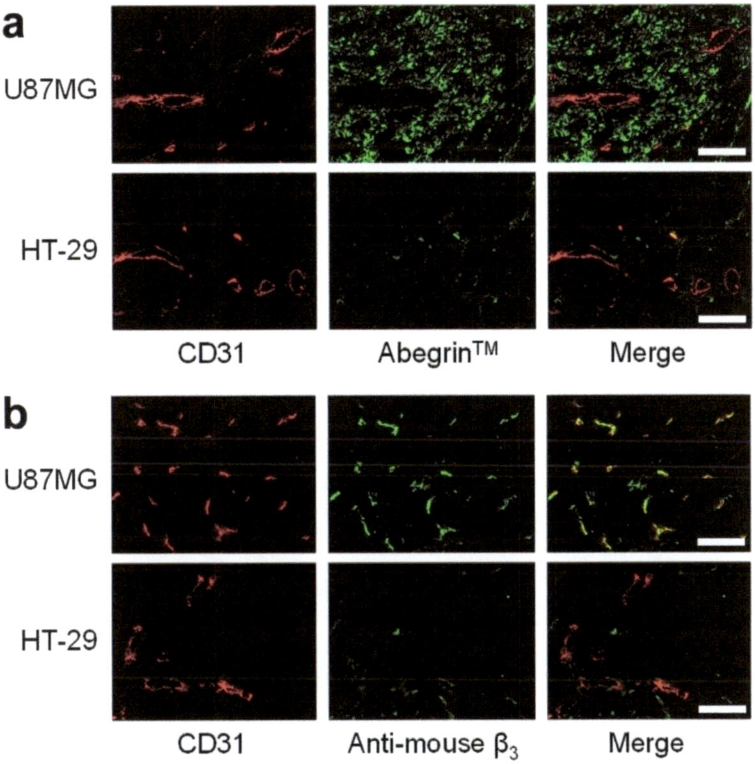

Figure 14.S5 Immunofluorescence staining of U87MG and HT-29 tumour sections. (a) CD31 and Abegri^{TM} (anti-human integrin $\alpha_v\beta_3$ monoclonal antibody) immunofluorescence staining showed that U87MG cells have much higher integrin $\alpha_v\beta_3$ expression than HT-29 cells. (b) CD31 and mouse β_3 staining of U87MG and HT-29 tumour sections revealed that U87MG tumour vessels also have higher integrin $\alpha_v\beta_3$ expression than HT-29 tumour vessels. Scale bar: 100 μm.

mouse integrin $\alpha_v\beta_3$, although the expression level is different in different tumours.

Taken together, Fig. 14.S4 shows that U87MG tumour has high human integrin $\alpha_v\beta_3$ expression in tumour cells and mouse integrin $\alpha_v\beta_3$ expression on the tumour vasculature while HT-29 tumour has low integrin $\alpha_v\beta_3$ expression in both. Note that both mouse integrin $\alpha_v\beta_3$ (expressed on tumour vasculature) and human integrin $\alpha_v\beta_3$ (expressed on U87MG cells) can be recognized by c(RGDyK).

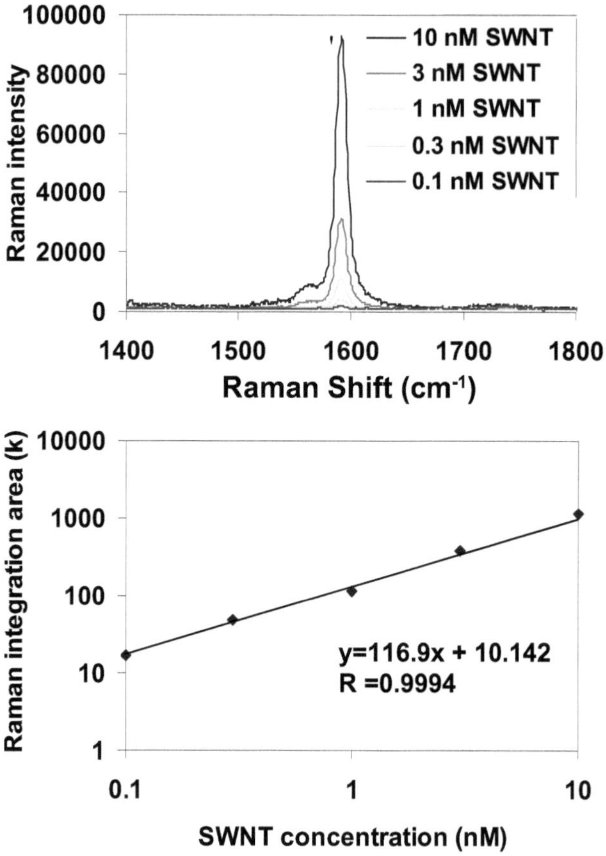

Figure 14.S6 Standard Raman calibration curve of SWNT solutions. (a) Raman spectra of SWNT solutions of different concentrations after subtracting background. (b) G-band intensity vs. SWNT concentration.

14.S7 Raman Measurement of SWNTs in Tissues

SWNT solutions with known concentrations in a capillary glass tube were measured by Renishaw micro-Raman instrument (laser excitation wavelength = 785 nm). A glass capillary tube filled with SWNT solution was placed under the objective of the Raman microscope. After focusing at the centre of the capillary, we recorded the Raman spectrum of the solution (100 mW power, 20 × objective, laser spot size $\sim 3 \times 60\ \mu m^2$, 10 s collection time). Spectra were also taken by slightly changing the focus. The G band peak area was integrated from 1570 cm^{-1} to 1620 cm^{-1} and averaged for multiple spectra and then plotted against the concentrations of SWNTs (Fig. 14.S6).

Raman spectra of tissue suspensions were collected similarly and the spectra were used to measure SWNT concentrations in the tissues against the standard calibration curve in Fig. 14.S6. The Raman spectroscopy based %ID/g of SWNTs in a specific tissue was calculated by the following equation:

$$\%\ \text{ID/g} = \frac{\begin{pmatrix}\text{[SWNT] in}\\ \text{tissue suspension}\end{pmatrix} \times \begin{pmatrix}\text{volume of tissue}\\ \text{suspension}\end{pmatrix}}{\begin{pmatrix}\text{[SWNT] in}\\ \text{injected solution}\end{pmatrix} \times \begin{pmatrix}\text{volume of}\\ \text{injected SWNT}\end{pmatrix} \times \begin{pmatrix}\text{wet weight}\\ \text{of tissue}\end{pmatrix}}$$

Supplementary References

1. Wang, H. F. et al. Biodistribution of carbon single-wall carbon nanotubes in mice. *J. Nanosci. Nanotechnol.* **4**, 1019 (2004).
2. Kam, N. W. S., O'Connell, M., Wisdom, J. A., and Dai, H. Carbon nanotubes as multifunctional biological transporters and near-infrared agents for selective cancer cell destruction. *Proc. Natl. Acad. Sci. USA* 102, 11600–11605 (**2005**).
3. Meares, C. F. et al. Conjugation of antibodies with bifunctional chelating agents isothiocyanate and bromoacetamide reagents methods of analysis and subsequent addition of metal ions. *Anal. Biochem.* **142**, 68 (1984).
4. Wu, Y., Cai, W., and Chen, X. Near-infrared fluorescence imaging of tumor integrin $\alpha_v\beta_3$ expression with Cy7-labeled RGD multimers. *Mol. Imaging Biol.* **8**, 226–236 (2006).

Chapter 15

A Pilot Study in Non-Human Primates Shows No Adverse Response to Intravenous Injection of Quantum Dots

Ling Ye,[a,*] Ken-Tye Yong,[b,c,*] Liwei Liu,[d] Indrajit Roy,[c] Rui Hu,[b,c] Jing Zhu,[c] Hongxing Cai,[d] Wing-Cheung Law,[c] Jianwei Liu,[a] Kai Wang,[a] Jing Liu,[a] Yaqian Liu,[a] Yazhuo Hu,[a] Xihe Zhang,[d] Mark T. Swihart,[c] and Paras N. Prasad[c,*]

[a] *Institute of Gerontology and Geriatrics, Laboratory Animal Center, Chinese PLA General Hospital, Beijing 100853, PR China*
[b] *School of Electrical and Electronic Engineering, Nanyang Technological University, Singapore 639798, Singapore*
[c] *Institute for Lasers, Photonics and Biophotonics, University at Buffalo, The State University of New York, Buffalo, New York 14260-4200, USA*
[d] *ChangChun University of Science and Technology, ChangChun, Jilin 130022, PR China*
ktyong@ntu.edu.sg, lye301@gmail.com, pnprasad@buffalo.edu

Quantum dots have been used in biomedical research for imaging [1, 2], diagnostics [3, 4] and sensing purposes [5, 6]. However, concerns over the cytotoxicity of their heavy metal constituents [7, 8] and conflicting results from in vitro [7, 9] and small animal

*These authors contributed equally to this work.
Reproduced with permission from *Nature Nanotechnology*, **7**, 453–458, 2012.

Nano-Enabled Medical Applications
Edited by Lajos P. Balogh

ISBN 978-981-4800-32-7 (Hardcover), 978-0-429-39903-9 (eBook)
www.jennystanford.com

[10–14] toxicity studies have limited their translation towards clinical applications. Here, we show in a pilot study that rhesus macaques injected with phospholipid micelle-encapsulated CdSe/CdS/ZnS quantum dots do not exhibit evidence of toxicity. Blood and biochemical markers remained within normal ranges following treatment, and histology of major organs after 90 days showed no abnormalities. Our results show that acute toxicity of these quantum dots in vivo can be minimal. However, chemical analysis revealed that most of the initial dose of cadmium remained in the liver, spleen and kidneys after 90 days. This means that the breakdown and clearance of quantum dots is quite slow, suggesting that longer-term studies will be required to determine the ultimate fate of these heavy metals and the impact of their persistence in primates.

CdSe nanocrystals are among the brightest, best-studied and most widely available quantum dots for bioimaging, but there are significant toxicity concerns because of their cadmium content [15–17]. Several reports have demonstrated quantum dot toxicity in cell culture studies [7, 9], but this toxicity has not been observed in vivo in small animals [10–14]. Thus, a significant discrepancy exists between the conclusions drawn from in vitro and in vivo studies. Factors that may trigger toxic effects from quantum dots include interactions with blood components, non-specific accumulation in major organs, and immune responses [11]. In small animals, systemically administered quantum dots accumulate in the reticuloendothelial system, including the liver, spleen and lymphatic system [7]. The excretion of quantum dots is mainly governed by their hydrodynamic size [18]. Current understanding of the toxicological and pharmacological effects of quantum dots is limited, and these effects are strongly dependent on their surface coating [19, 20]. An improved understanding of the toxicity of various quantum dot formulations is essential, and long-term toxicological and pharmacokinetic investigations involving the degradation, excretion, persistence and immune response of quantum dots will ultimately be needed for designing quantum-dot-based probes for imaging and therapy.

Through non-human primate studies, the toxicity and pharmacokinetics of quantum dot formulations can be investigated in greater detail and with lower risk, time and cost than in human clinical trials [21]. Although chimpanzees are genetically closest to humans, they are an endangered species and are difficult to maintain and support. Rhesus macaques, which diverged ~25 million years ago from the lineage that led to both chimpanzees and humans, are more plentiful and easily maintained [22]. Because of their close genetic and physiological relationship to humans, and also their availability, rhesus macaques provide an excellent non-human primate model (Supplementary Table 15.S1). In the present study, male rhesus macaques were studied for 90 days following intravenous injection of 25 mg kg^{-1} phospholipid micelle-encapsulated quantum dots (Fig. 15.1). Four treated animals and one control animal were then euthanized for ex vivo histological analysis, and a further two treated animals were observed for an additional year.

Body weights were monitored every other day (Supplementary Fig. 15.S1) and the fluctuations were no greater after treatment than before treatment, suggesting minimal systemic effects. A brief transient decrease in body weight on the day of injection is a normal result of fasting on the day before treatment. Similarly, no major fluctuation in body temperature was observed (Supplementary Fig. 15.S1). Behaviour was monitored and recorded twice each day. No abnormalities in eating, drinking, grooming, exploratory behaviour, activity, physical features, urination or neurological status were observed. No vomiting was observed before or after injection of the quantum dot formulation.

Because the micelles encapsulating the quantum dots are similar in size to viruses and large proteins, they may affect the immune system or induce an inflammatory response, which would be indicated by changes in haematological factors (for example, red and white blood cell count) [23]. Accordingly, standard haematological and biochemical markers were monitored. A complete blood count (CBC) was performed at regular intervals, and the results did not suggest any acute toxicity (Fig. 15.2a–j). Similarly, established serum biochemistry assays were performed. The eight important hepatic indicators—alkaline phosphatase (ALP), total protein (TPROT), albumin (ALB), total bilirubin (TBILI), direct bilirubin (DBIL), alanine transaminase (ALT), aspartate transaminase (AST) and

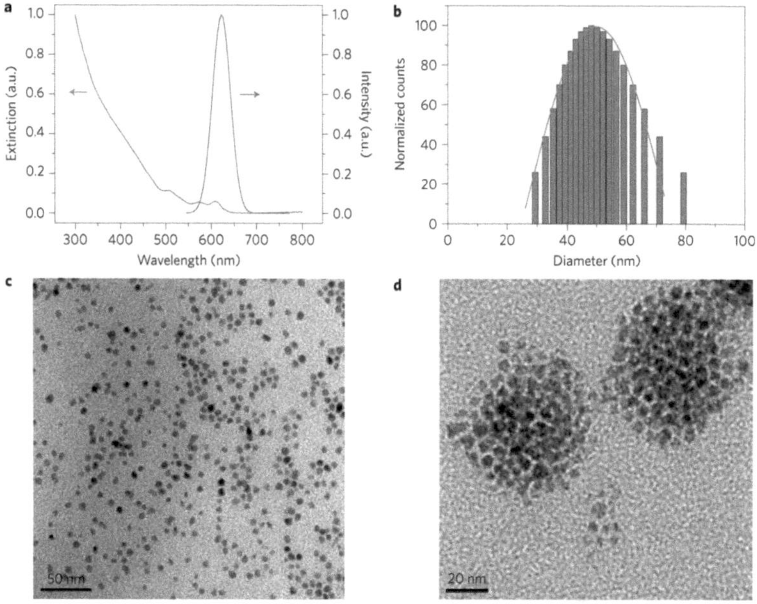

Figure 15.1 Characterization of phospholipid micelle-encapsulated CdSe/CdS/ZnS quantum dot formulation. (a) UV–vis absorbance and photoluminescence emission spectra of micelle-encapsulated quantum dots. The excitonic peaks near the absorption edge and the narrow emission peak show that the quantum dots are relatively monodisperse and defect free. (b) Dynamic light scattering data showing the distribution of hydrodynamic diameter of the micelles encapsulating the quantum dots. The blue line is a Gaussan fit with a mean of 52 nm and standard deviation of 12.7 nm. (c, d) Transmission electron microscopy (TEM) images showing quantum dots before (c) and after (d) micelle encapsulation. The average diameter of the quantum dots is 7–8 nm. TEM grids were prepared by drop-casting from chloroform (c) or water (d) dispersions and allowing the solvent to evaporate. After encapsulation, the particles remain clustered in groups that are representative of single micelles with multiple encapsulated quantum dots.

gamma glutamyl transaminase (γ-GT)—were measured weekly and showed no sign of liver injury (Fig. 15.2k–r). Indicators of kidney function—blood urea nitrogen (BUN) and creatinine (CRE)—were also normal (Fig. 15.2t,u). Triglycerides (TG), blood glucose (GLU) and cholesterol (CHO) were also in the healthy range throughout the examination period (Fig. 15.2v–x). Prothrombin time (PT) before treatment (Fig. 15.2s) was slightly above the normal range. Values after treatment were slightly lower than those before treatment, but still slightly above the normal range from the literature. Nonetheless, the international normalized ratio (INR, a ratio of PT to that of a normal control, scaled to account for assay sensitivity) ranged from 0.92 to 1.12, indicating normal blood-clotting behaviour. Urine analysis showed that urine specific gravity was 1.01–1.02, which lies in the normal range.

After 90 days, the concentrations in the tissues of the elemental constituents of the quantum dots were measured by inductively coupled plasma mass spectroscopy (ICP-MS, Fig. 15.3). This showed that quantum dots had predominantly accumulated in the liver, spleen and kidney, where the average cadmium concentrations were 18, 63 and 84 $\mu g\ g^{-1}$, respectively. Much lower levels (below 1 $\mu g\ g^{-1}$) were observed in the heart, lung, lymph, thymus, brain, muscle, skin and testis (Fig. 15.3). Nonetheless, cadmium was detected at levels exceeding those of the control animal in all tissues. Selenium uptake was also highest in the liver, spleen and kidney, with concentrations of 2.8, 12.5 and 1.9 $\mu g\ g^{-1}$, respectively. Less than 0.4 $\mu g\ g^{-1}$ of selenium was detected in the heart, lung, lymph, thymus, brain, muscle, skin and testis. Zinc concentrations were comparable to those of the control animal, but suggest some possible accumulation in the kidney and liver. The blood clearance profile was also investigated (Fig. 15.3d), revealing clearance times of less than 6 h. At 24 h, the concentrations of both cadmium and selenium in the blood decreased to 0.1 and 0.2 $\mu g\ ml^{-1}$. Negligible concentrations of both cadmium and selenium were detected after 72 h.

ICP-MS analysis of the quantum dot formulation (nominally 10 mg ml^{-1}) gave cadmium, selenium and zinc concentrations of 326 mg ml^{-1}, 75.0 mg ml^{-1} and 11.2 mg ml^{-1}, respectively. Thus, 25 mg kg^{-1} of the formulation corresponded to 0.81 mg kg^{-1} of cadmium. After 90 days, the cadmium levels in the kidneys, liver and spleen corresponded to ~35%, 58% and 6% of the initial dose, respectively. This sums to 99% of the injected dose. The total cadmium in the organs of the control animal, which represents

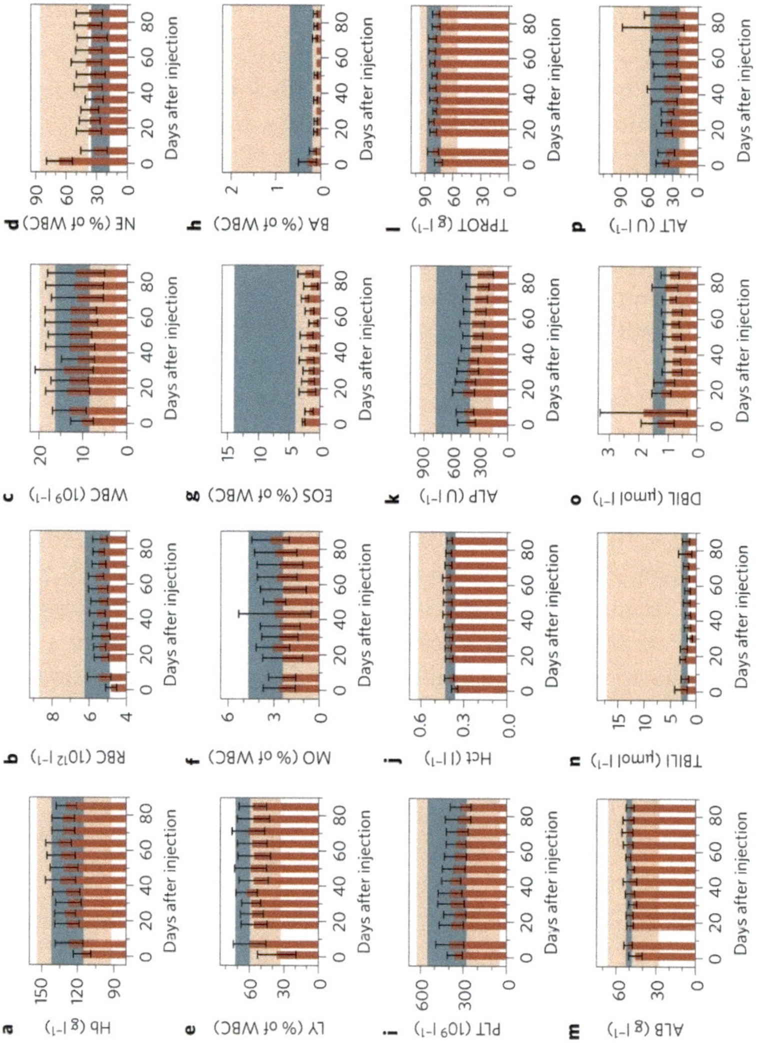
a
Hb (g l−1)
b
RBC (10^12 l−1)
c
WBC (10^9 l−1)
d
NE (% of WBC)
e
LY (% of WBC)
f
MO (% of WBC)
g
EOS (% of WBC)
h
BA (% of WBC)
i
PLT (10^9 l−1)
j
Hct (l l−1)
k
ALP (U l−1)
l
TPROT (g l−1)
m
ALB (g l−1)
n
TBILI (μmol l−1)
o
DBIL (μmol l−1)
p
ALT (U l−1)
Days after injection

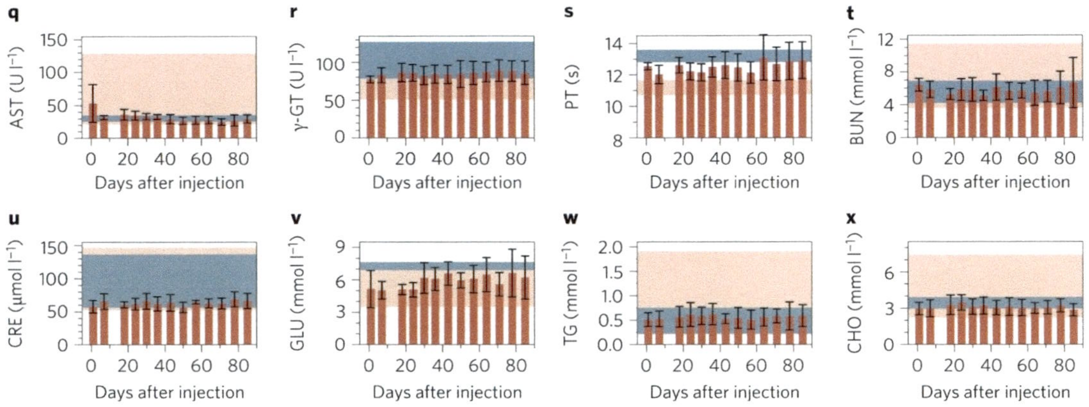

Figure 15.2 Blood test results for treated rhesus macaques. (a–x), The results ($n = 6$, measured weekly) show no abnormalities in immune response, kidney or liver function, blood clotting or blood chemistry over the 90 day period following treatment. The region rendered in pink represents the normal range from the literature, and the blue region represents the range observed in the test subjects before treatment. Error bars represent one standard deviation above or below the mean. Abbreviations: haemoglobin, Hb; red blood cell count, RBC; white blood cell count, WBC; neutrophil granulocyte, NE; lymphocyte, LY; monocyte, MO; eosinophil granulocyte, EOS; basophil granulocyte, BA; platelet count, PLT; haematocrit, Hct; alkaline phosphatase, ALP; total protein, TPROT; albumin, ALB; total bilirubin, TBILI; direct bilirubin, DBIL; alanine transaminase, ALT; aspartate transaminase, AST; gamma glutamyl transferase, γ-GT; prothrombin time, PT; blood urea nitrogen, BUN; creatinine, CRE; blood glucose, GLU; triglyceride, TG; total cholesterol, CHO.

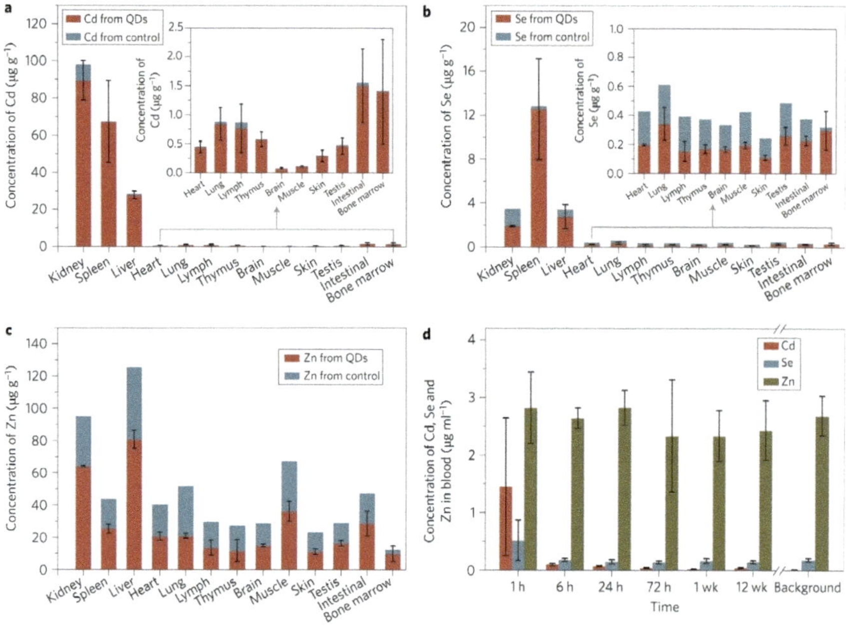

Figure 15.3 ICP-MS analysis of the major organs of treated ($n = 53$) and control ($n = 51$) rhesus macaques. Most of the injected dose remained in the kidneys, liver and spleen 90 days after injection. ICP-MS analysis of blood samples at select time points showed rapid clearance of quantum dots from the bloodstream. (a–c) In vivo biodistribution of cadmium (a), selenium (b) and zinc (c). (d) Blood clearance profile of quantum dots. Error bars indicate plus and minus one standard deviation of the measurements from the tissues of three treated animals.

a background level that should be subtracted from the levels measured in the treated animals, corresponded to ~5% of the dose injected in the treated animals. This suggests that more than 90% of the injected cadmium remains in the animals' organs after 90 days.

Histological analysis of tissues from four treated animals and one control animal (Fig. 15.4 and Supplementary Figs. 15.S2–15.S4) revealed no pathological changes in the heart, lung, kidney, liver or spleen. Hepatocytes in the liver samples were observed to be normal, and there were no signs of inflammatory response. No pulmonary fibrosis was detected in the lung samples. The glomerulus structure in the kidney section was observed without

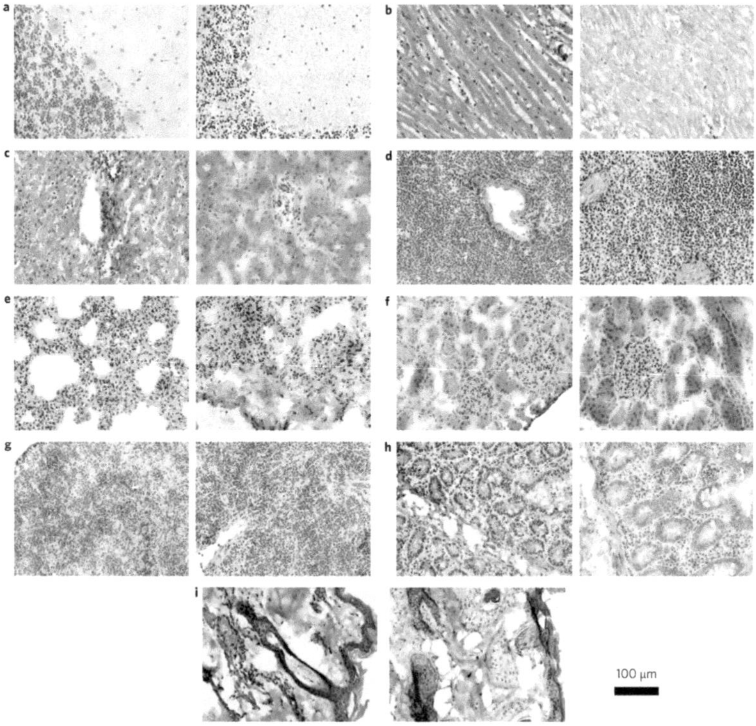

Figure 15.4 Histological images from the major organs of the rhesus macaques three months after intravenous injection of the QD formulation. Evaluations were carried out by pathologists and found no anomalies. In each pair, the left image is from the control animal and the right image is from a treated animal. (a–i) Tissues were collected from brain (a), heart (b), liver (c), spleen (d), lung (e), kidney (f), lymph (g), intestine (h) and skin (i). Images were taken at 40× magnification with standard haematoxylin and eosin staining. Histological images from three additional treated monkeys are provided in Supplementary Figs. 15.S2–S4.

difficulty. Necrosis was not found in any of the histological samples analysed. Two clinical pathologists, both unaware of the quantum dot treatment or of which tissues came from the control animal, analysed the tissue sections and found no abnormalities.

Several studies of quantum dot toxicity have shown that coating the cadmium-based core with another material, such as ZnS, or adding a protective polymer coating can reduce or eliminate toxicity in vitro and in vivo [7]. Studies using quantum dots with cadmium-based cores (for example, CdTe and CdSe) have demonstrated cytotoxic effects associated with leaching of heavy metal ions from them, which is thought to interfere with tissue and cell health and function [8]. If this release of metal ions is prevented, toxicity may be reduced. Our previous study in mice showed that if the quantum dots were excreted without being degraded, their toxicity could be reduced significantly [10]. Quantum dots have not been used in humans, and limited information is available on the stability of quantum dots in vivo. Recent reports based on in vivo experiments in mice argued that functionalized quantum dots might remain in the body for weeks to months [7, 24] and showed that the bioconjugated quantum dots were distributed in lymph nodes, kidneys, liver, lung, bone marrow and spleen [25, 26].

We found that in rhesus macaques the micelle-encapsulated quantum dots were mainly located in the kidneys, liver and spleen. The ICP-MS results strongly suggest that the quantum dot formulation is slowly degrading in the liver and spleen, releasing free cadmium ions that accumulate in the kidney. This would explain the much higher cadmium-to-selenium ratio in the kidney compared with the liver and spleen. Previously, others have used ICP-MS analysis to show that after intravenous injection of cadmium-based quantum dots in mice, over the course of 28 days the cadmium levels in the kidneys eventually reached nearly 10% of the injected dose (and 40% in the liver) [27]. The redistribution of the cadmium over time may signify the degradation of quantum dots in vivo, because the natural accumulation sites of Cd^{2+} ions are the liver and kidneys. Fluorescence imaging of organ tissues from the treated monkeys (Supplementary Fig. 15.S5) is consistent with this interpretation; quantum dot fluorescence was observed in the liver and spleen, but no quantum dot signal was observed in the kidney, despite the high concentration of the elemental constituents of the quantum dots present there.

Although previous studies of quantum dots in primates are not available, the present results can be considered in the context of earlier rodent studies. Ballou et al. reported that mice injected with 20 pmol per gram animal weight of amphiphilic polyacrylic acid and PEG-coated quantum dots were viable until necropsy (133 days) and showed no signs of necrosis or damage at the sites of tissue injection and no signs of quantum dot breakdown in vivo [12]. Larson et al. reported no adverse effects in mice injected with 20 nM and 1 μM CdSe/ZnS (ref. 28). Fitzpatrick et al. intravenously injected amphiphilic polymer-coated quantum dots into mice and found slow degradation of quantum dots over time in vivo, leading to alterations in quantum dot fluorescence, suggesting that toxicological parameters may change with residence time of the quantum dot in tissue [24]. Fischer et al. injected mercaptoundecanoic acid-, lysine- and BSA-coated CdSe/ZnS quantum dots into rats and found a differential distribution based on the coatings [13]. No quantum dots were found in the urine or faeces over the course of 10 days, suggesting persistence in the body. These few studies dramatically underscore the need for shortand long-term toxicity evaluations that examine multiple organ systems before quantum dot risk can be adequately assessed. So far, 6 nmol CdSe/ZnS (corresponding to $\sim$150 mg kg^{-1}) is the highest dosage used in mice [29]. However, we used the lower concentration range suggested by Larson's work, which is suitable for in vivo imaging applications [28]. This required a total of $\sim$2 g of quantum dots for this study, whereas the higher dosage would have required over 10 g.

Hauck et al. recently studied the toxicity and biodistribution of CdSe/ZnS core–shell quantum dots in healthy Sprague–Dawley rats [11]. They observed no appreciable toxicity, even after breakdown of the quantum dots in vivo over time (>80 days). The dose used by Hauck et al. corresponded to $\sim$1 mg kg^{-1} cadmium, which is almost identical to the dose used here. However, the biodistribution in the primates studied here showed much higher quantum dot concentrations in the liver, spleen and kidney compared with that observed by Hauck et al. in rats.

In conclusion, rhesus macaques intravenously injected with $\sim$25 mg kg^{-1} of a cadmium-based quantum dot formulation survived without any evidence of toxicity. All measured biochemical markers were in the normal range. Two treated monkeys underwent continued evaluation and showed no ill effects for one year. However, given the persistence of elevated cadmium and selenium levels in organs after three months, longer-term studies will be required to determine the ultimate fate of these heavy metals and any ill effects arising due to their presence. All data obtained in this pilot study suggest that the phospholipid micelle-encapsulated quantum dot formulation has very low toxicity on the timescale considered. This suggests a potential role for quantum dots in human clinical applications, such as image-guided surgery for tumour removal, where the advantages provided by quantum dots can outweigh the potential risk associated with accumulation of their constituent elements in the body.

15.1 Methods

15.1.1 Quantum Dot Synthesis

CdO (1.6 mmol), 3 mmol TDPA and 3 g TOPO were heated in a 100 ml three-necked flask to 290–300°C under argon. After 10–15 min, a clear solution was obtained. This was maintained at 300°C for 5 min, then 0.8 ml of 1 M TOP-Se was rapidly injected. The reaction was stopped after 2–3 min by removing the heating mantle. Quantum dots were collected by the addition of ethanol and centrifugation. The CdS/ZnS graded shell was formed following the method of Manna et al. [30]. Approximately 0.3 g CdSe quantum dots was dispersed in ~5 ml toluene. Separately, 1 mmol CdO, 4 mmol zinc acetate and 7 g TOPO were dissolved in 10 ml oleic acid and heated to 180°C for ~30 min under argon. The quantum dot dispersion was injected slowly into the hot mixture, which was then held at 180°C, with a needle outlet that released evaporating toluene. After ~15 min, the needle was removed, and the temperature was raised to 210°C. TOP-S (2 ml) was then added dropwise, and the mixture was held at ~210°C for 10–15 min. Aliquots were removed via syringe and injected into excess toluene to quench further growth of the quantum dots.

15.1.2 Phopholipid-micelle Encapsulation

Chloroform dispersions of quantum dots (~4 mgml^{-1}) and of DSPE-mPEG (Avanti Polar Lipids or Laysan Bio 1,2-diacyl-*sn*-glycero-3-phosphoethanolamine-*N*-[methoxy(polyethylene glycol)], ~10 mg

ml^{-1}) were mixed in a 1:4 mass ratio and gently stirred for 5–10 min. A Labconco rotary evaporator with a water bath at 25°C was used to evaporate the chloroform. The resulting lipidic film was hydrated with 3–5 ml of high-pressure liquid chromatography grade water and sonicated for 10–15 min using a bath sonicator. The resulting dispersion was filtered through a 0.45 or 0.2 μm membrane filter and kept at room temperature for further use. To remove excess phospholipid, the micelle-encapsulated quantum dots were centrifuged at 10,000 r.p.m. for 15 min. The precipitate was then redispersed in 1–2 ml of HPLC water.

15.1.3 Animal Studies

The study was conducted at the Laboratory Animals Center of Chinese PLA General Hospital (301 Hospital), Beijing, under a protocol that was reviewed and approved by the ethics committee of the hospital. Seven adult male rhesus macaques (six treated, one control) ranging from 3 to 4 years in age and from 4.8 to 7.7 kg in mass were purchased from the Institute of Beijing XIEERXIN Biology Resource. Animals were individually housed in stainless steel cages (20–22°C, 40–60% relative humidity, 12/12 hours light/dark cycle, 15 air changes per hour) and fed a commercial monkey diet. Water was available *ad libitum*. Research staff inspected the monkeys three times each day. Quantum dot (10 mg ml^{-1}) dispersions in 0.9% sodium chloride were filter-sterilized before injection. Animals were anaesthetized by subcutaneous injection of 50–75 mg ketamine (0.1 g 2 ml^{-1}). Encapsulated quantum dots (25 mg kg^{-1}) were administrated to six monkeys by intravenous drip. Urine and fasting blood samples were collected weekly. Body mass, temperature, appearance and exploratory behaviour were recorded at the same time. Before quantum dot injection, three monkeys were subject to urine and blood testing as the normal control. Venous blood samples (7–8 ml) were collected in evacuated tubes containing EDTA, sodium citrate anticoagulants and non-anticoagulant agents. The blood analysis (haematology, coagulation and chemistry) was carried out on Sysmex XS-800i, Roche STA-R Evolution and Cobas 6000-C501 units, respectively. Urine (3–5 ml) was taken for urinalysis on a Roche URISYS 2400 and Sysmex UF-100i.

15.1.4 Histological Analysis

Monkeys were killed by ketamine anaesthetic and 10% KCl. Heart, liver, spleen, lungs, kidneys, colon (section 10 cm in length), muscle (5 cm × 7 cm), lymph nodes and brain were collected, weighed and fixed with 10% buffered formalin. Bone marrow (from the femur) was prepared by smearing on glass slides and staining with Wright's stain. Haematoxylin and eosin-stained histological sections of the fixed organs were observed with an Olympus BX60 microscope at ×40 magnification. Tissue sections were examined by two independent clinical pathologists who were not told what treatment the monkeys had undergone or which samples were from the control monkey.

15.1.5 Inductively Coupled Plasma Mass Spectrometry (ICP-MS) Analysis and Fluorescence Imaging

Tissue samples for ICP-MS analysis and fluorescence imaging were collected from three animals and kept at -20°C. For ICP-MS, three fragments were sampled as a set from each organ of interest. Organs and sample sets were weighed before being frozen, with each set weighing ~1 g. Standard cryosectioning was performed for fluorescence imaging of the organs. Elemental concentrations were determined using a quadrupole-based ICP-MS system (Elan DRC-II, PerkinElmer SCIEX Instruments; resolution, ~0.7–0.9 AMU). Samples were digested with HNO_3 and HCl (HNO_3:HCl ratio of 10:1) at 60°C for analysis.

Acknowledgements

This work was supported by The John R. Oishei Foundation, Air Force Office of Scientific Research (grant no. FA95500610398), the Singapore Ministry of Education (Grants Tier 2 MOE2010-T2-2-010 (M4020020.040 ARC2/11) and Tier 1 M4010360.040 RG29/10), Nanyang Technological University (start-up grant no. M4080141.040), the Beijing Natural Science Foundation (no. 7092097) and the National Natural Science Foundation of China (no. 21071150). The authors thank A. Maitra of Johns Hopkins University for helpful discussions.

Author Contributions

K.T.Y. and L.Y. designed the research. K.T.Y, L.Y., R.H., L.L., J.Z., I.R.W.C.L., J.L., K.W., J.L., Y.L. and Y.H. performed the research. L.Y., K.T.Y., L.L., I.R., R.H., J.Z., H.C., W.C.L., J.L., K.W., J.L., Y.L., Y.H., X.Z., M.T.S. and P.N.P. analysed the data. K.T.Y., L.Y., I.R., M.T.S. and P.N.P. co-wrote the paper.

Additional Information

The authors declare no competing financial interests.

References

1. Medintz, I. L., Uyeda, H. T., Goldman, E. R., and Mattoussi, H. Quantum dot bioconjugates for imaging, labelling and sensing. *Nature Mater.* **4**, 435–446 (2005).
2. Yong, K-T. Mn-doped near-infrared quantum dots as multimodal targeted probes for pancreatic cancer imaging. *Nanotechnology* **20**, 015102 (2009).
3. Michalet, X. et al. Quantum dots for live cells, in vivo imaging, and diagnostics. *Science* **307**, 538–544 (2005).
4. Bruchez, M., Moronne, M., Gin, P., Weiss, S., and Alivisatos, A. P. Semiconductor nanocrystals as fluorescent biological labels. *Science* **281**, 2013–2016 (1998).
5. Chan, W. C. W., and Nie, S. Quantum dot bioconjugates for ultrasensitive nonisotopic detection. *Science* **281**, 2016–2018 (1998).
6. Mattoussi, H., Palui, G., and Na, H. B. Luminescent quantum dots as platforms for probing in vitro and in vivo biological processes. *Adv. Drug Deliv. Rev.* **64**, 138–166 (2012).
7. Derfus, A. M., Chan, W. C. W., and Bhatia, S. N. Probing the cytotoxicity of semiconductor quantum dots. *Nano Lett.* **4**, 11–18 (2004).

8. Choi, H. S. et al. Tissue- and organ-selective biodistribution of NIR fluorescent quantum dots. *Nano Lett.* **9**, 2354–2359 (2009).
9. Cho, S. J. et al. Long-term exposure to CdTe quantum dots causes functional impairments in live cells. *Langmuir* **23**, 1974–1980 (2007).
10. Yong, K-T., Roy, I., Ding, H., Bergey, E. J., and Prasad, P. N. Biocompatible nearinfrared quantum dots as ultrasensitive probes for long-term in vivo imaging applications. *Small* **5**, 1997–2004 (2009).
11. Hauck, T. S., Anderson, R. E., Fischer, H. C., Newbigging, S., and Chan, W. C. W. In vivo quantum-dot toxicity assessment. *Small* **6**, 138–144 (2010).
12. Ballou, B., Lagerholm, B. C., Ernst, L. A., Bruchez, M. P., and Waggoner, A. S. Noninvasive imaging of quantum dots in mice. *Bioconj. Chem.* **15**, 79–86 (2004).
13. Fischer, H. C., Liu, L., Pang, K. S., and Chan, W. C. W. Pharmacokinetics of nanoscale quantum dots: in vivo distribution, sequestration, and clearance in the rat. *Adv. Funct. Mater.* **16**, 1299–1305 (2006).
14. Dubertret, B. et al. In vivo imaging of quantum dots encapsulated in phospholipid micelles. *Science* **298**, 1759–1762 (2002).
15. Colvin, V. L. The potential environmental impact of engineered nanomaterials. *Nature Biotechnol.* **21**, 1166–1170 (2003).
16. Werlin, R. et al. Biomagnification of cadmium selenide quantum dots in a simple experimental microbial food chain. *Nature Nanotech.* **6**, 65–71 (2011).
17. Hardman, R. A toxicologic review of quantum dots: toxicity depends on physicochemical and environmental factors. Environ. *Health Perspect.* **114**, 165–172 (2006).
18. Choi, H. S. et al. Renal clearance of quantum dots. *Nature Biotechnol.* **25**, 1165–1170 (2007).
19. Hoshino, A. et al. Physicochemical properties and cellular toxicity of nanocrystal quantum dots depend on their surface modification. *Nano Lett.* **4**, 2163–2169 (2004).
20. Kirchner, C. et al. Cytotoxicity of colloidal CdSe and CdSe/ZnS nanoparticles. *Nano Lett.* **5**, 331–338 (2005).
21. Xia, H-J., Zhang, G-H., Wang, R-R., and Zheng, Y-T. The influence of age and sex on the cell counts of peripheral blood leukocyte subpopulations in Chinese rhesus macaques. *Cell Mol. Immunol.* **6**, 433–440 (2009).
22. Xia, H., Liu, H., Zhang, G., and Zheng, Y. Phenotype and function of monocytederived dendritic cells from Chinese rhesus macaques. *Cell Mol. Immunol.* **6**, 159–165 (2009).

23. Ho, C-C. et al. Quantum dot 705, a cadmium-based nanoparticle, induces persistent inflammation and granuloma formation in the mouse lung. Nanotoxicology http://dx.doi.org/10.3109/17435390.2011.635814 (2011).
24. Fitzpatrick, J. A. J. et al. Long-term persistence and spectral blue shifting of quantum dots in vivo. *Nano Lett.* **9**, 2736–2741 (2009).
25. Schipper, M. L. et al. Particle size, surface coating, and PEGylation influence the biodistribution of quantum dots in living mice. *Small* **5**, 126–134 (2009).
26. Al-Jamal, W. T., Al-Jamal, K. T., Cakebread, A., Halket, J. M., and Kostarelos, K. Blood circulation and tissue biodistribution of lipid–quantum dot (L-QD) hybrid vesicles intravenously administered in mice. *Bioconj. Chem.* **20**, 1696–1702 (2009).
27. Yang, R. S. H. et al. Persistent tissue kinetics and redistribution of nanoparticles, quantum dot 705, in mice: ICP-MS quantitative assessment. *Environ. Health Perspect.* **115**, 1339–1343 (2007).
28. Larson, D. R. et al. Water-soluble quantum dots for multiphoton fluorescence imaging in vivo. *Science* **300**, 1434–1436 (2003).
29. Gao, X., Cui, Y., Levenson, R. M., Chung, L. W. K., and Nie, S. In vivo cancer targeting and imaging with semiconductor quantum dots. *Nature Biotechnol.* **22**, 969–976 (2004).
30. Manna, L., Scher, E. C., Li, L-S., and Alivisatos, A. P. Epitaxial growth and photochemical annealing of graded CdS/ZnS shells on colloidal CdSe nanorods. *J. Am. Chem. Soc.* **124**, 7136–7145 (2002).

Supplementary Information: Chapter 15

15.S1 Detailed Materials and Methods

A. Synthesis of CdSe/CdS/ZnS Quantum Dots

The following protocol was adopted from our previous report [1] to obtain CdSe QD. 1.6 mmol cadmium oxide, 3 mmol tetradecylphosphonic acid (TDPA), and 3 g tri-n-octylphosphine oxide (TOPO) were loaded into a 100 ml three-necked flask. Next, the reaction mixture was slowly heated under an argon atmosphere to 290–300°C. After 10 to 15 min of heating, a clear homogeneous solution was obtained. The reaction mixture was maintained at 300°C for another 5 min, then 0.8 mL of 1 M (tri-n-octylphosphine) TOP-Se was rapidly injected. The reaction was stopped after 2–3 min by removing the heating mantle. The QDs were separated from the surfactant solution by addition of ethanol and centrifugation. The reddish QD precipitate could be readily redispersed in various organic solvents (hexane, toluene, and chloroform). The method of forming a CdS/ZnS graded shell on CdSe QDs was adapted from Manna et al. [2] A CdSe QD dispersion was prepared in advance by dispersing ~0.3 g of CdSe QDs in ~5 mL of toluene. Separately, 1 mmol of cadmium oxide, 4 mmol of zinc acetate, and 7 g of TOPO were dissolved in 10 mL of oleic acid. This mixture was heated to 180°C for ~30 min under an argon flow, and then the CdSe QD dispersion was injected into the continuously stirred hot reaction mixture. The reaction mixture was held at 180°C, with a needle outlet that allowed the toluene to evaporate. After ~15 min of heating, the needle was removed, and the reaction temperature was raised to 210°C. Upon reaching the desired temperature, 2 mL of TOP-S was added drop wise into the reaction mixture. The reaction mixture was then held at ~210°C for 10 to 15 min, and then an aliquot was removed via syringe and was injected into a large volume of toluene at room temperature, thereby quenching any further growth of the QDs. The QDs were separated from the toluene solution by addition of ethanol and centrifugation.

B. Preparation of phospholipid-micelle encapsulated CdSe/CdS/ZnS Quantum Dots

The as-prepared organic-dispersible QDs were collected by addition of ethanol (volume ratio of QDs to ethanol 1:3) and centrifugation (12,000 rpm for 15 min). The precipitate was collected and dried under vacuum and then re-dispersed in chloroform. Next, QDs dispersed in chloroform (~4 mg/mL), and DSPE-mPEG (1,2-Diacyl-*sn*-Glycero-3-Phosphoethanolamine-*N*-[Methoxy(Polyethylene glycol)] with average molecular weight of 5000, in chloroform solution (~10 mg/mL) were mixed together at a weight ratio of 1:4. The PEGylated phospholipid was purchased from Avanti polar lipids, Inc. and Laysan Bio Inc. Each mixture was gently stirred for 5 to 10 min. A Labconco rotary evaporator with a water bath of 25°C was used to evaporate the organic solvent. The lipidic film, deposited on the reaction vial, was hydrated with 3 to 5 mL of HPLC water and subjected to ultrasonication for 10 to 15 min using a bath sonicator. The resulting dispersion was filtered through a 0.45 or 0.2 μm membrane filter and kept at room temperature for further use. To remove the excess phospholipid from the QD dispersion, the micelle-encapsulated QDs were further purified using centrifugation at 10,000 rpm for 15 min. The QD precipitate was then re-dispersed in 1 to 2 mL of HPLC water (the water was filtered with 0.2 μm membrane filter).

C. Animal studies

Seven adult male rhesus macaques were used in this study (six treated, one control). Their ages ranged from 3 to 4 years and their body mass was between 4.8 and 7.7 kg. They were purchased from the Institute of Beijing XIEERXIN Biology Resource. During the study, animals were individually housed in stainless steel cages. They were fed a commercial monkey diet from local supplies in Beijing and water was available ad libitum. Climatic conditions were controlled at 20–22°C, 40–60% relative humidity, 12/12 h light/dark cycle, and 15 air changes per hour. Environmental enrichment was provided as determined by standard procedures. Experimental investigation was performed over 90 days. The research staff members at the Animals Center regularly inspected the monkeys three times per day. The study was conducted at Laboratory Animals Center of Chinese PLA General Hospital (301 Hospital), Beijing. All procedures were conducted under a protocol that was reviewed and approved by the ethics committee of Chinese PLA General Hospital. For QD injection, 10 mg/mL QD was dispersed in 0.9% sodium chloride and filter sterilized. Animals were anesthetized with 50–75 mg ketamine (0.1 g/2 mL) by subcutaneous injection. 25 mg/kg QDs was administrated to six of the monkeys by intravenous drip. In all cases, the QD injection was finished within 30 min for each monkey. Urine and fasting blood samples were collected once per week. Body mass, temperature, appearance, and exploratory behaviour were recorded at the same time. Before QD injection, three monkeys were subject to urine and blood testing as the normal control. 7–8 mL venous blood samples were collected into evacuated tubes containing EDTA, sodium citrate anticoagulants, and non-anticoagulant agents. The blood analysis (haematology, coagulation and chemistry) was carried out on Sysmex XS-800i, Roche STA-R Evolution and Cobas 6000-C501, respectively. 3–5 mL of urine was taken for urinalysis on Roche URISYS 2400 and Sysmex UF-100i.

D. Histological analysis

Four monkeys were sacrificed for tissue analysis. The monkeys were sacrificed by ketamine anaesthetic and 10% KCl. Heart, liver, spleen, lungs, kidneys, colon (section 10 cm in length), muscle (5 cm × 7 cm), lymph nodes, and brain were collected, weighed and fixed with 10% buffered formalin following PBS rinsing. Bone marrow (from the femur) was taken and prepared for light microscopy by smearing on glass slides and staining with Wright's stain. Haematoxylin and eosin-stained histological sections of the fixed organs were observed with an Olympus BX60 microscope at 40× magnification. All the experiments were performed following the rules, guidelines, and protocols at Chinese PLA General Hospital, P. R. China. The other two monkeys were retained for further evaluation.

E. ICP-MS analysis and fluorescent imaging

After sacrifice of the animals, organs were harvested and weighted. Tissue samples for inductively coupled plasma mass spectrometry (ICP-MS) analysis and fluorescent imaging were collected from 3 animals and kept at –20°C. For ICP-MS, three fragments were sampled as a set from each organ of interest. Organs and samples were weighed before freezing, with each set weighing around 1 g in total. Standard cryosection procedures were performed for fluorescent imaging of the organs of interests. The fluorescent images were taken from a typical inverted microscope based fluorescent imaging system. The concentration of cadmium and other materials in the tissues and blood samples was determined using a quadrupole based ICP-MS system (Elan DRC-II, PerkinElmer SCIEX Instruments, USA). The samples were digested with HNO_3 and HCl (HNO_3:HCl ratio of 10:1) at 60°C before the measurements. The ICP-MS operates in a peak hopping scan mode, with nebulizer gas flow at 0.96 L/min, auxiliary gas flow at 1.80 L/min, and plasma gas flow at 15.0 L/min. The equipment has a resolution of 0.7 ~ 0.9 amu.

Supplementary References

1. Qian, J. et al. Imaging pancreatic cancer using surface-functionalized quantum dots. *J. Phys. Chem. B.* **111**, 6969–6972 (2007).
2. Manna, L., Scher, E. C., Li, L.-S., and Alivisatos, A. P. Epitaxial growth and photochemical annealing of graded CdS/ZnS shells on colloidal CdSe nanorods. *J. Am. Chem. Soc.* **124**, 7136–7145 (2002).

Figures and Table

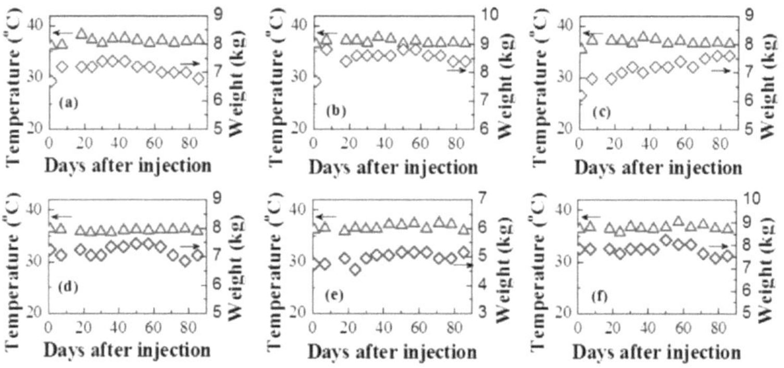

Figure 15.S1 Body mass and body temperature of the six treated monkeys over the course of the study. Red triangles correspond to the body temperature, on the left axis of each plot. Blue diamonds correspond to the body mass, on the right axis of each plot.

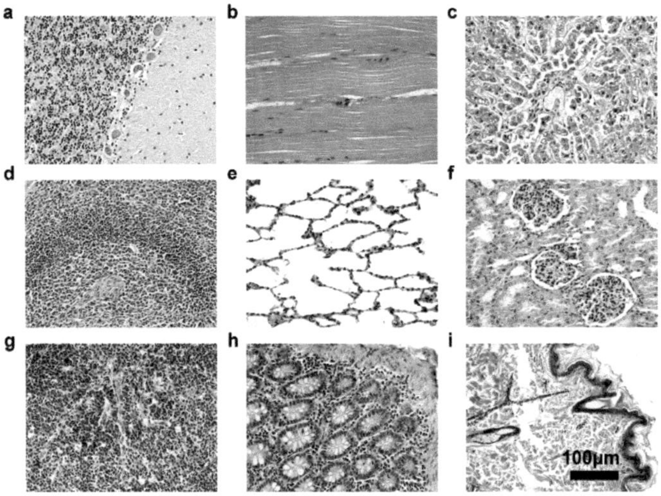

Figure 15.S2 Histological sections of the major organs of rhesus macaque number 2 (treated with QD formulation) 90 days after injection. No anomalies were observed in the tissues. Tissues were harvested from (a) brain, (b) heart, (c) liver, (d) spleen, (e) lung, (f) kidney, (g) lymph, (h) intestine, and (i) skin. The images were taken at 40× magnification.

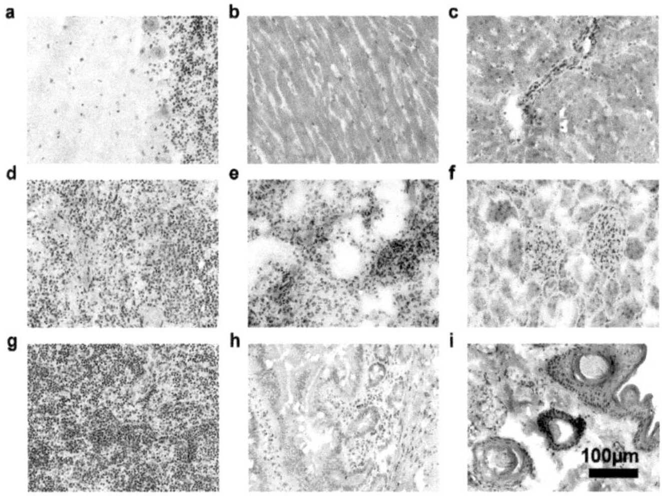

Figure 15.S3 Histological sections of the major organs of rhesus macaque number 3 (treated with QD formulation) 90 days after injection. No anomalies were observed in the tissues. Tissues were harvested from (a) brain, (b) heart, (c) liver, (d) spleen, (e) lung, (f) kidney, (g) lymph, (h) intestine, and (i) skin. The images were taken at 40× magnification.

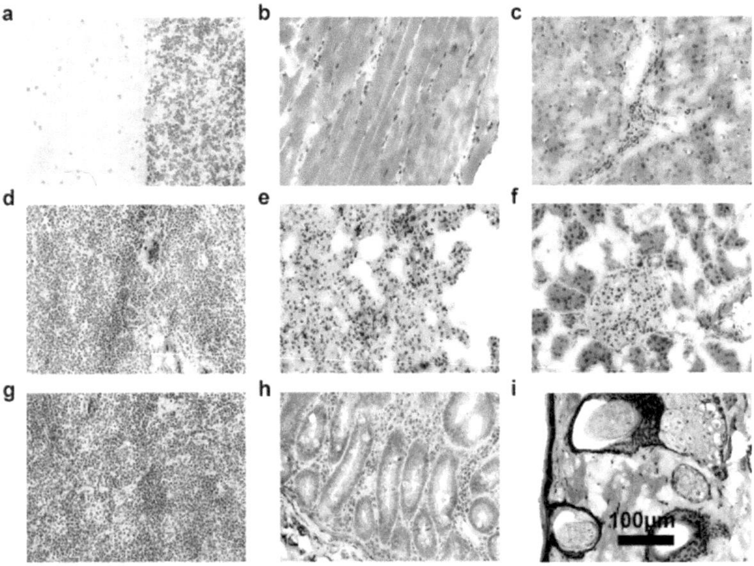

Figure 15.S4 Histological sections of the major organs of rhesus macaque number 4 (treated with QD formulation) 90 days after injection. No anomalies were observed in the tissues. Tissues were harvested from (a) brain, (b) heart, (c) liver, (d) spleen, (e) lung, (f) kidney, (g) lymph, (h) intestine, and (i) skin. The images were taken at 40× magnification.

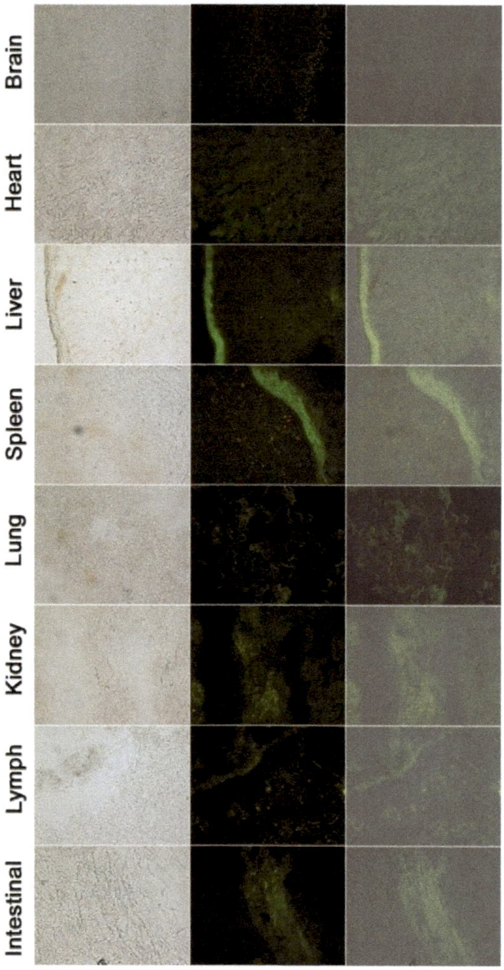

Figure 15.S5 Fluorescence imaging of tissue sections from monkey 90 days after injection with micelle-encapsulated QDs. In all cases, green represents emission from the tissue and red represents emission from QDs. Left, centre, and right panels are transmission, fluorescence, and overlaid images, respectively.

Table 15.S1 Standard Normal ranges for blood chemistry parameters for humans and Rhesus macaques

	Human	Monkey		Human	Monkey		Human	Monkey
WBC (10^9/L)	3.5–10.0	2.4–20.0	ALB (g/L)	35–50	27.8–66.1	GLU (mmol/L)	3.4–6.2	1.4–6.7
RBC (10^{12}/L)	3.5–5.5	5.24–8.70	ALT (U/L)	0–40	15–99	TG (mmol/L)	0.4–1.7	0.21–1.91
Hct (L/L)	0.35–0.52	0.359–0.621	AST (U/L)	0–40	21–129	CHO (mmol/L)	3.1–5.7	2.22–7.45
PLT (10^9/L)	100–300	46–627	ALP (U/L)	0–130	156–951	PT (s)	12–16	6.95–19.15
NE	0.5–0.7	0.38–0.86	TPROT (g/L)	55–80	55–95.7	PTA (%)	60–120	NA
LY	0.2–0.4	0.33–0.61	DBIL (μmol/L)	0–8.6	0.22–2.95	INR	0.95–1.5	0.5–1.74
MO	0.03–0.08	0–0.03	TBILI (μmol/L)	0–21	2.1–17.1			
EOS	0.01–0.05	0–0.06	γ-GT (U/L)	0–50	50.66–93.5			
BA	0–0.01	0–0.02	BUN (μmol/L)	1.8–7.5	3.6–11.5			
Hb (g/L)	110–176	92–154	CRE (μmol/L)	30–110	52.5–147.1			

Abbreviations: Hb, haemoglobin; WBC, white blood cells; RBC, red blood cell; Hct, haematocrit; PLT, platelet; NE, neutrophil; LY, lymphocyte; MO, monocyte; EOS, eosinophil; BA, basophil; ALB, alanine albumin; ALT, alanine transaminase; AST, aspartate transaminase; ALP, alkaline phosphatise; TPROT, total protein; DBIL, serum direct bilirubin; TBILI, serum total bilirubin; γ-GT, serum gamma glutamyl transferas; BUN, blood urea nitrogen; CRE, creatinine; TG, triglyceride; GLU, blood glucose; CHO, total cholesterol; PT, prothrombin time; PTA, plasma thromboplastin antecedent.

Chapter 16

Nanomedicines for Ocular NSAIDs: State-of-the-Art Update of the Safety on Drug Delivery

Joana R. Campos,[a] Joana Araújo,[b] Elisabet Gonzalez-Mira,[c,d] Maria A. Egea,[b] Elena Sanchez-Lopez,[b] Marta Espina,[b] Selma B. Souto,[e] Maria L. Garcia,[b,f] and Eliana B. Souto[a]

[a] *Department of Pharmaceutical Technology, Faculty of Pharmacy, University of Coimbra, Pólo das Ciências da Saúde, Azinhaga de Santa Comba, 3000-548 Coimbra, Portugal*
[b] *Department of Pharmacy, Pharmaceutical Technology and Physical Chemistry, Faculty of Pharmacy and Food Sciences, University of Barcelona, Av. Joan XXIII 27-31, 08028 Barcelona, Spain*
[c] *Institut de Ciència de Materials de Barcelona, Campus Universitari de Bellaterra, 08193, Cerdanyola del Vallès, Spain*
[d] *Centro de Investigación Biomédica en Red-Bioingeniería Biomateriales y Nanomedicina, Spain*
[e] *Department of Endocrinology, Braga Hospital, Sete Fontes, 4710-243 São Victor Braga, Portugal*
[f] *Institute of Nanoscience and Nanotechnology, Faculty of Pharmacy, University of Barcelona, Barcelona 08028, Spain*
ebsouto@ff.uc.pt, ebsouto@ebsouto.pt

The eyes are among the most readily accessible organs in terms of location in the body, nevertheless drug delivery to eye tissues is particularly problematic. Poor bioavailability of drugs from ocular

Nano-Enabled Medical Applications
Edited by Lajos P. Balogh

ISBN 978-981-4800-32-7 (Hardcover), 978-0-429-39903-9 (eBook)
www.jennystanford.com

dosage forms is mainly due to pre-corneal loss factors, e.g. tear dynamics, non-productive absorption, transient residence time in the cul-de-sac, and relative impermeability of the corneal epithelial membrane. There is a clear need for effective topical formulations able of promoting drug penetration and maintaining therapeutic levels with a reasonable frequency of application, strategy that can also result in enhancement of side effects that probably would not be acceptable. Delivery of a drug via nanotechnology-based product fulfils mainly three objectives, i.e. enhancement of drug permeation, controlled release, and targeting. The inflammatory response of the ocular tissues is a common side effect associated to ophthalmic surgery. Together with steroidal agents, non-steroidal anti-inflammatory drugs are used in eye surgery. In this work, the state-of-art on the safety of drug delivery to the eye has been revised and updated, focusing on the most promising nanomedicines incorporating anti-inflammatory drugs for ophthalmic application.

16.1 Introduction

Ocular diseases directly affect human vision and quality of life. Significant achievements have been made in the discovery of ocular pathological mechanisms and management of ocular disease [1]. Ocular drug delivery (ODD) is one of the most interesting and challenging endeavours facing the pharmaceutical scientists [2]. ODD includes multiple pharmaceutical strategies to reach a therapeutic target in the eye through a variety of different routes [3]. Efficient drug delivery to the ocular tissues faces a number of challenges due to the anatomy, physiology and biochemistry of the eye and the blood–ocular barriers (including blood-aqueous and blood-retinal barrier) provide a unique structure that restricts the entry of drug molecules at the required site of action [4–6]. From a drug delivery perspective, the eye can be divided in two parts: the anterior segment comprising the cornea, the iris, the ciliary body and the lens, and the posterior segment constituted by the choroid and the retina [2, 7]. Moreover, there are two ocular annexes (eyelids and conjunctiva) and three different fluids (tears, aqueous humour and vitreous humour) also involved in ocular drug penetration [2, 7].

A suitable ocular formulation should release the drug overcoming the protective barriers of the eye without causing permanent tissue damage. For drug delivery on the anterior segment, common routes of administration are topical instillation and subconjunctival injection, while for the posterior segment, common routes include systemic dosing, periocular and intravitreal injections, and topical dosing [8].

Topical instillation is the conventional and preferred route to achieve the therapeutic goal and the majority (nearly 90%) of ocular diseases are treated with topical application of drug solutions (eye drops), particularly to the anterior segment of the eye. However, this route typically has a low clinical success rate. This is attributed in part to the unfavourable physicochemical properties, i.e. low water solubility, of the active substances such as anti-infective agents, anti-inflammatories, and immunosuppressants, which are typically included in these formulations [3, 9]. The topical ocular administration of drugs has two different purposes: to treat superficial eye diseases, such as infections (i.e. conjunctivitis, blepharitis, keratitis sicca), and to provide intra-ocular treatment through the cornea for diseases such as glaucoma or uveitis. For their favourable cost advantage, the greater simplicity of formulation development and production, and the good patients' acceptability, more than 90% of the marketed ophthalmic formulations are in the form of eye drops for soluble drugs, but these conventional systems cannot be considered optimal in the treatment of vision threatening ocular diseases since most of the drugs are washed off from the eye by various mechanisms (lachrymation, tear dilution and tear turnover) [3, 10, 11]. Static barriers, including tight junctions of the conjunctiva, the alternating hydrophobic corneal epithelium, hydrophilic corneal stroma, and hydrophobic endothelium, and dynamic barriers, including the rapid tear turnover and the vasculature and lymphatics of the conjunctiva, all contribute to the highly impenetrable anterior surface the entry of foreign substances [9, 10]. The tear drainage rate constant from the pre-corneal area is 1.45 min^{-1} which results in 500 to 700 times greater drug loss from the ocular surface in comparison to the drug absorption rate into the anterior chamber

[10]. As a result, effective intraocular drug delivery represents a major challenge due to the presence of various elimination mechanisms and physiological barriers that result in less than 5% of administered drug penetrates the cornea and reaches intra-ocular tissues [9–11]. Efforts to enhance ocular bioavailability from topical instillations have been accomplished using novel formulations, such as hydrogels, polymeric micelles, nanosuspensions, and lipid-based nanocarriers [9, 10].

The cornea is an important route for drug absorption after topical application. It consists of five different layers, namely epithelium, Bowman's membrane, stroma, Descemet's membrane, and endothelium [12]. The corneal epithelium, lipophilic in nature, acts as a selective barrier for small molecules and prevents the diffusion of macromolecules via the paracellular route, it plays a major role in limiting trans-corneal drug absorption with a drug permeability rate of only 10^{-7}–10^{-5} cm s^{-1} [12–14]. While small lipophilic drugs are passively transported via the transcellular pathway, hydrophilic drugs undergo restricted permeation through tight junctions via the paracellular pathway [10]. The stroma, located below the epithelium, is a highly hydrophilic layer making up 90% of the cornea. The corneal stroma being hydrophilic impedes transport of lipophilic drug molecules while hydrophilic molecules generally diffuse freely through the stroma [10]. The corneal endothelium consists of a single layer of flattened epithelium-like cells and it is responsible for maintaining normal corneal hydration. Since the cornea is characterized by lipophilic and hydrophilic structures, it represents an effective barrier to the absorption of both hydrophilic and lipophilic molecules [15, 16].

After instillation of an ophthalmic drug solution, the drug is first mixed with the lachrymal fluid and remains in contact with the ocular mucosa for a very short period of time, typically 1–2 min, because of the permanent production of lachrymal fluid (0.5–2.2 μL/min). Drainage through the upper and lower canaliculus into the lachrymal sac, which opens in the nasolacrimal duct, induces a rapid elimination of conventional dosage forms during blinking. The conjunctiva and sclera are more permeable than cornea for drugs topically applied into the eye, but the circulation removes the drugs before it can be absorbed by inner ocular tissues.

Both transconjunctival penetration and transnasal absorption after drainage are generally undesirable, not only because of the loss of active ingredient but also because of possible severe systemic side effects.

Drugs penetrate across the corneal epithelium via the transcellular pathway (mainly for lipophilic drugs) or paracellular pathway (hydrophilic drugs). The transcorneal penetration appears to be hindered by the binding of the drug to the corneal tissues, acting these ones as drug reservoirs [17]. In the case of ionizable drugs (weak acids and weak bases), their permeation depends on the chemical equilibrium between the ionized and non-ionized forms in the eye drop and eventually in the lachrymal fluid. It is well known that the lipid membranes are more permeable to the non-ionized form of the drug than the ionized form. Besides the degree of ionization, other relevant issue affecting the corneal penetration is the charge of the ionized molecule. Hydrophilic charged cationic compounds permeate more easily through the cornea than anionic forms, because the corneal epithelium is negatively charged [16]. Also the pH and buffering capacity of the instilled solution can have a significant effect on ophthalmic drug absorption [18].

To overcome the disadvantages of eye drops, several ophthalmic drug delivery systems have been investigated, which can control the release of the drug and sustain therapeutic levels over an extended period of time. The efforts have been concentrated on the development of the ideal formulation for topical ODD would be able to (i) increase the drug residence time at the ocular surface, (ii) protect the drug from degradation occurring in the presence of tear enzymes, (iii) provide targeted drug delivery to the site of action, increasing bioavailability and minimizing the non-productive conjunctival absorption, and (iv) promote drug penetration through the cornea, enabling it to act as a controlled release system, thus reducing the dosing frequency. Clearly the materials used in any ODD system must be biocompatible, biodegradable, and non-irritating [3]. Several approaches have been proposed with, e.g. viscosity enhancers, penetration enhancers, pro-drugs and drug delivery devices. In the case of viscosity enhancers, studies indicate that these substances have a limited value, because the formulations are liquid and will be easily removed from the eye by its defensive mechanisms [19]. Concerning the penetration enhancers, although several are being applied, the higher sensitiveness of the eye tissues imposes great caution in their selection [15, 20]. For drug delivery devices, since they must remain in the eye for several hours, irritation of local tissue can be more problematic than for eye drops, which are eliminated within seconds.

An optimal ocular drug delivery system should be developed so that it could be administered in the form of eye drops, causing no blurred vision or irritability and would need no more than one or two instillations a day.

Nanotechnology, a term derived from the Greek word "nano", meaning dwarf, is the engineering of functional systems at molecular or submicron scale. The past decades have witnessed a significant progress in the development of nanosized ocular drug delivery systems. Such increasing interest in nanomedicine may be attributed to the tremendous advances in nanotechnology, polymer chemistry and chemical engineering. Nevertheless, additional research is required in the area of ocular drug delivery, especially with regards to the delivery of hydrophobic compounds, nucleic acids and proteins in order to improve their therapeutic outcomes and consequently the quality of life of the patients [10]. In the quest of improving efficacy, nanotechnology offers advantages that allow a more targeted drug delivery and controllable release of the therapeutic compound with the aim to manage better drug pharmacokinetics, pharmacodynamics, non-specific toxicity, immunogenicity and biorecognition of systems. Nanotechnology-based delivery systems, depending on their particle charge, surface properties and relative hydrophobicity, can be designed to be successfully used in over-coming retinal barriers and providing protection for the encapsulated drug, prolonging exposure by controlled release. It has also led to the solution of various solubility-related problems of poorly soluble drugs [21] and can be used to target mononuclear phagocyte systems to allow regional specific delivery and minimize side effects in other organs [22]. The drug delivery systems based on nanotechnology with an appropriate particle size and narrow size range, ensuring low irritation, adequate bioavailability and compatibility with ocular tissues may prove to be the best delivery tools for ocular diseases and there should be sought one of these systems to every suspended drug.

This review will focus on recent attempts of nanotechnology-based systems for treating and managing various ocular ailments. In this review, we discuss the use of different types of nanomedicines, which have been suggested as suitable carriers for ocular drug delivery highlighting the therapeutic significance of each drug delivery system for clinical translation. These include nanoparticles, liposomes, niosomes, dendrimers, and microemulsions, as well as microparticles (microcapsules and microspheres), and hybrid systems (combination of different strategies).

16.2 Ocular Anti-Inflammatory Drug Therapy

Being the most sensory organs of the body, diseases affecting eye makes the coloured world into blurred or dark in nature [23]. Ocular discomforts involve anterior/posterior-segment diseases, symptomatic distress, associated inflammation and severe retinal disorders [24]. The inflammation refers to a typical reaction of cells/tissue in response to injury or any other disease condition. It is the kind of reciprocal attempt of the body's immune system to eradicate foreign bodies or microbes to retard further injury and can be resulted into swelling, redness, heat, and pain. Inflammation is the outcome of imbalance of biochemical homeostasis of the eyes [25]. Inflammatory diseases of the eye such as iritis, uveitis and eye inflammation caused due to an eye injury, allergens, chemical exposure like acids, pesticides or any other harmful chemicals from manufacturing processes in industries. These may further lead to complications like clouding in the retina, the formation of cataracts, an increase of intraocular pressure such as glaucoma and swelling or detachment of the retina [24]. Uveitis is an inflammatory ailment of the eye, which is caused by autoimmunity, infectious organisms, toxic material, or tumours. It could lead to severe pain and increased light sensitivity in the affected patients. Although inflammation is a natural defensive phenomenon, it results in discomfort in the eye and also makes the eye vulnerable to other diseases [26]. Ocular inflammation is also a common result of cataract surgery, producing pain and photophobia in many patients and potentially leading to serious complications including increased intraocular pressure (IOP), posterior capsule opacification, cystoids macular oedema (CME) and decreased visual acuity [27]. This mechanism leads to the stimulation of some membrane protein such as phospholipase A_2, following tissue injury/damage, breaks down cell membrane phospholipids to arachidonic acid, which forms the substrate for further reactions mainly by the cyclo-oxygenase and the lipoxygenase pathways. The resulting arachidonic acid metabolites, along with other chemical mediators, interact to cause inflammation.

The main products of the cyclo-oxygenase (COX) pathway are prostaglandins (PGs), probably derived from the iris ciliary body [28]. PGs cause miosis, increase in vascular permeability of the blood–ocular barriers and changes in intraocular pressure. Furthermore, they possess chemokinetic activity and can serve as mediators in both humoral and cellular phases of the inflammatory response [24, 25]. The main products of lipoxygenase (LO) pathway are leukotrienes (LT) such as LT B_4 and peptidoleukotrienes. LT B_4 causes aggregation of polymorphonuclear cells, exudation of plasma, translocation of calcium and stimulation of phospholipase A_2 [24, 29]. Peptidoleukotrienes result in bronchoconstriction, secretion of mucus, plasma leakage from post capillary venules, vasoconstriction of arterioles and miosis.

Chronic ocular inflammation may lead to disruption to ocular tissues and blood vasculatures surrounding the eyes. This condition is regarded as complicated by ophthalmologists that have been found to result in impaired or complete vision loss [30]. For the treatment of ocular inflammation there are two important categories of molecules available which are non-steroidal anti-inflammatory drugs (NSAIDs) and corticosteroids [24]. NSAIDs are a class of anti-inflammatory drugs that are used in ophthalmologic surgery; they constitute a heterogeneous group of compounds with different structural classes, which do not include a steroid nucleus derived biosynthetically from cholesterol in their chemical structure. Also known as cyclo-oxigenase inhibitors based on their mode of action, NSAIDs are important modulators of ocular inflammatory reactions [28]. NSAIDs inhibit the biosynthesis of

PGs, thus inhibiting intraoperative miosis during cataract surgery, reducing the vascular permeability of the blood–ocular barrier and modifying inflammation. However, they are not capable of inhibiting the actions of PGs once formed [30].

Glucocorticoids inhibit the phospholipase A_2 enzyme, thus inhibiting the biosynthesis of both PGs and LTs. Other effects include effects on white cells (inhibition of migration and reduction of the number of cells), inhibition of degranulation, lysosome stabilization, suppression of lymphokines, reduction of capillary permeability and suppression of vasodilatation [31].

Although steroidal agents have been the standard treatment for ocular inflammation in the past, the use of NSAIDs has increased over the past two decades. The main advantage of using topical NSAIDs is the avoidance of undesirable effects of steroidal agents, which include the decreased immunological response to infection, cataract formation, steroid-induced raised intraocular pressure (IOP) and inhibition of re-epithelialization following epithelial denudation. Clinical evidence suggested that the combined use of NSAIDs and steroids is synergistic [32] and in fact, it has become the standard of care to use a regimen of NSAIDs and steroids before and after cataract surgery [33].

NSAIDs are effective drugs compared to placebos for the relief of anterior chamber inflammation [34]. The goals of topical prophylactic NSAIDs treatment include the prevention of intraoperative miosis [35], management of postoperative inflammation, prevention or treatment of CME [36] and reduction of ocular pain [37]. There are four NSAIDs currently approved by the U.S. Food and Drug Administration (FDA) for the treatment of post operative inflammation after cataract surgery [38], namely kerotolac, bromfenac, diclofenac, and nepafenac. This latter is the only prodrug NSAID and thus requires conversion to its more active state, amfenac, through intraocular enzymatic hydrolysis [39]. A study compared aqueous humour concentrations of these four NSAIDs after administration in patients having cataract surgery and the cyclooxigenase-1 (COX-1) and COX-2 inhibitory activity was also determined, via in vitro measurement of PG inhibition to rank order the potency of the molecules, showing nepafenac significantly greater ocular bioavailability than the others and amfenac greater potency at COX-2 inhibition [38]. Furthermore, diclofenac, nepafenac, ketorolac, and bromfenac have relatively greater significant effects than other topical NSAIDs. NSAIDs were first approved by the FDA to prevent surgically induced miosis [40]. Newer NSAIDs are being investigated for their ability to reduce the incidence of CME after cataract surgery. CME is a major complication after cataract surgery and remains the primary cause of surgical visual disorders [41]. Whether NSAIDs can effectively prevent the development of CME is still controversial. Recent comprehensive analyses investigated the ability of NSAIDs to reduce the incidence of CME after cataract surgery, but a positive effect was not observed [40]. Although ambiguity surrounds NSAIDs regarding CME prevention, NSAIDs play an important role in cataract surgery.

NSAID ophthalmic preparations have very similar anti-inflammatory mechanisms, yet their therapeutic efficacies differ. NSAIDs represent a class of drugs that are effective for reducing anterior chamber inflammation and relieving ocular pain. NSAIDs also show fewer treatment-related/serious ocular adverse events.

16.3 Nanomedicines for Ocular Therapy

As mentioned before, nanotechnology-based novel ocular nanosystems (ONS) are being explored for last two decades to the development of ocular drug delivery systems that would prolong the drug retention, allowing the drug to remain in contact with the cornea for longer duration and thus increasing bioavailability and decreasing the elimination process of the eye [42]. Nanoparticulate technology is advocated as an ophthalmic drug delivery approach that may enhance dosage form acceptability while providing sustained release in the ocular milieu. This was achieved incorporating the drugs into the solid-lipid nanoparticles (SLN) [43], nanoemulsion [44, 45], liposomes [46], niosomes [47], cubosomes [48], spanlastics [49], nanomicelles [10], and nanostructured lipid-carriers (NLC) [50], which have been proved to be superior to conventional ones regarding pharmacokinetic and pharmacodynamics perspective. Numerous ONS have been developed into non-biodegradable systemic vehicles, but now advancement came into the field employing biodegradable materials [51]. To date, many efforts have been made on both carrier design and exploring the mechanisms of their biological actions. Meanwhile, much attention is being focused on the fabrication and modification of multi-functional nanocarriers for ocular target therapy [52]. Examples of drugs loaded in ONS are listed in Table 16.1.

Most of the anti-inflammatory drugs are hydrophobic and with low corneal permeability. ONS enable efficacious and safe delivery of these drugs through the ocular surface for which novel approaches are employed [24]. The physicochemical and pharmaceutical aspects related to the drug, such as pH, temperature, tonicity and osmolality of tear fluid, tear turnover, surface charge and construction of cornea structure, pore size of cornea and conjunctiva, differential lipophilicity of ocular layers, ocular blood flow, ocular metabolism, sensitivity of ocular surface, desired site of action, desired drug release pattern, and requisite residence/contact time are a major issue considered by formulation scientists when formulating dispersed systems especially those intended for parenteral or ocular administration. Very small particles as nanoparticles are well tolerated and possess adhesive properties, which could prolong the residence time of the drug in the cul-de-sac, prevent tear wash out (due to tear dynamics), and increase ocular bioavailability. Once sufficient local drug concentration prevails in the anterior chamber, anti-inflammatory drugs can target the corresponding receptors/phenomenon to diminish the effect of inflammatory manifestations. This leads to relief from pain and discomfort caused by inflammation [9]. Other potential advantages of nanoscaled drug delivery systems in ocular therapy are (i) the possibility of self-administration by patients as eye drops, (ii) no impairment of sight because of small dimensions of the delivery systems, (iii) protection against metabolic enzymes (such as peptidases and nucleases), (iv) possible uptake into corneal cells, (v) prolonged drug release, reducing the need for repeated instillation or injection, and (vi) and targeting towards affected tissues, reducing possible side effects and required dose. For precise targeting of inflammation, it is very necessary to understand its aetiology, molecular interactions in biological phenomena, resultant products/incidences and their manifestations [24].

Table 16.1 Nanomedicines for the treatment of eye diseases (adapted from [24])

Type of nanomedicine	Administration route/therapeutic use	Loaded drug	Composition
Liposomes	Topical	Ganciclovir	Phosphatidyl choline, cholesterol, sodium deoxycholate
		Bromfenac	Distearoylphosphatidylcholine, dicetyl phosphate, cholesterol
		Ciprofloxacin	Cholesterol, phosphatidyl choline
		Diclofenac	Distearoylphosphatidylcholine, cholesterol, polyvinyl alcohol
	Topical inflammation	Prednisolone acetate	Phosphatidylcholine, lactose monohydrate
	Topical, ocular mycoses	Fluconazole	Cholesterol, phosphatidyl choline
		Ketaconazole	Span 60, Tween 80
Cubosomes	Topical, inflammation	Furbiprofen	Glyceryl mono-oleate, Poloxamer 407
		Dexamethasone	Monolein, Poloxamer 407
	Topical	Ketorolac	Glyceryl mono-oleate, Poloxamer 407
		Cyclosporine A	Glyceryl mono-oleate, Poloxamer 407
Niosomes	Topical, glaucoma	Acetazolamide	Span 60, cholesterol
	Topical, inflammation	Prednisolone acetate	Span 60, cholesterol
Spanlastic	Topical, ocular mycoses	Fluconazole	Span 60, Tween 80
		Itraconazole	Span 60, Tween 80

	Topical, uveitis	Triamcinolone acetonide	PLGA
		Cyclosporine A	PLA
		Oleanolic acid, Ursolic acid	PLGA, Poloxamer 188, Transcutol P
		Indomethacin	Polycaprolactone, poly L-Lysine (PLL), chitosan, Poloxamer 188
		Timolol maleate	Poly(isobutylcyanoacrylate), Pluronic F68, Tween 80, Dextran T40 and T70
	Topical, inflammation	Pilocarpine	Gelatin
Nanoparticle		Hydrocortisone	Albumin
		Cyclosporine A	PLGA, Eudragit RL, Carbopo
		Dexamethasone	Hyaluronic acid-coated chitosan nanoparticles
		Diclofenac	N-trimethyl chitosan
		Bromfenac	Chondroitin sulphate, Chitosan
		Dexibuprofen	PEG, PLGA
		Indomethacin	Chitosan, Poloxamer 188
	Topical, inflammation after ocular surgery	Flurbiprofen	PLGA, Poloxamer 188
	Topical, glaucoma	Metipranolol	Polypsiloncaprolactone, isobutylcyanoacrylate
	Topical	Dexamethasone	RPβ-CD

(*Continued*)

Table 16.1 (*Continued*)

Type of nanomedicine	Administration route/therapeutic use	Loaded drug	Composition
Solid Lipid Nanoparticle	Topical	Valcyclovir	Tristearin, stearic acid, Poloxamer 188, sodium taurocholate
	Topical, inflammation	Indomethacin	Comprito ATO888, Tween 80, Poloxamer 188
		Cyclosporin A	Comprito ATO888, Tween 80, Poloxamer 188
			Tripalmitin, Compritol ATO888, Tween 80
		Diclofenac	Goat fat, Phospholipon G, Tween 80
Nanostructured Lipid Carrier	Topical, inflammation	Ibuprofen	Gelucire 44/14, Transcutol P, Cremophor EL
		Cyclosporine A	Precifac, Miglyol, Polyethylene glycol monostearate
		Flurbiprofen	Compritol ATO888, iglyol 812N, Solutol HS15 Tween 80, chitosan
		Flurbiprofen	Stearic acid, castor oil, Tween 80
Nanosuspension/ Nanoemulsion	Topical, inflammation	Ibuprofen	Eudragit RS100
		Diclofenac	PLGA, Isopropyl, myristate, Tween 80, glycerin
		Difluprednate	Castor oil, Tween 80
	Topical, uveitis	Piroxicam	Eudragit, RS100, isopropyl
		Dexamethasone	Isopropyl, myristate, Tween 80, propylene, glycol, chitosan

Nanomicelles	Topical	Rapamycin	Vitamin E TPGS, Octoxynol 40
		Dexamethasone	Sodium taurocholate, egg lecithin
		Metipranolol	Pluronic F127, PVA, chitosan
		Cyclosporin A	Dimethoxy poly(ethylene glycol) 500, Solutol
			Poly (L-lactide)-b-poly(methacrylic acid co-3-acrylamidophenylboronic acid
		Dexamethasone	Dithiol polyethyleneglycol- poly (D,L-lactide)
		Biotin-12 Hydroxy stearic acid Acyclovir	Vitamin E TPGS, Octoxynol 40
		α-Lipoic acid	Soluplus®
		Netilmicin, Dexamethasone alcohol, Dexamethasone phosphate	Polyhydroxyethylaspartamide polyethylene glycol-hexadecylamine
	Topical, posterior uveitis	Dexamethasone	Polyoxy 40 stearate, Polysorbate 80
	Topical, inflammation	Lonorxicam	Poly(ethylene oxide), poly (propylene oxide), Synperonic® PE/P84, Synperonic® PE/F127, Tetronic® T701
		Dexamethasone	Copolymer of N-isopropylacrylamate vinyl pyrollidine and acrylic acid

16.3.1 Polymeric Hydrophilic Particles

Nanoparticles are biodegradable and polymeric colloidal nanosized (from 10 to 1000 nm) particles that are mainly made of lipid, protein, and natural or synthetic polymer composition [53]. Depending upon the process used for the preparation, nanospheres or nanocapsules can be obtained. Nanospheres have a matrix-like structure, where active compounds can be firmly adsorbed onto their surface, entrapped or dissolved in the matrix. Nanocapsules have a polymeric shell and an inner core. In this case, the active substances are not only dissolved in the core but may also be adsorbed onto the nanocapsule surface [54]. The increasing use of nanoparticle formulations to uplift the drugs residence time inside cornea with high permeation rate and ocular delivery is a concern. Nanoparticles stabilized themselves as liquid colloidal systems for viable ONS; however, the limitations posed by their stability issue, particle size uniformity and release kinetics and bulk production issues should also considered [24].

Different studies have shown the long-lasting bioavailability, good biocompatibility, and biodegradability of nanoparticles in the body. The first studies addressing ocular disease treatment have been reported by Ticho et al. using Piloplex systems consisting of pilocarpine-loaded nanospheres produced with poly(methylmethacrylate-acrylic acid) copolymer [55]. Although in clinical trials Piloplex lowered the intra-ocular pressure, those systems have not been accepted for further commercial development because of their non-biodegradability, local toxicity and difficulty in the large scale manufacture of sterile preparations.

The materials used in the preparation of the sustained release systems must be biodegradable and transparent improving the patient's comfort when compared to the implanted systems. In addition, they must fulfil the requirements of controlled degradation time and minimum inflammatory or toxicological potential [56]. Ocular bioavailability will depend on the polymer's bioadhesion properties, which are affected by its swelling properties, hydration time, molecular weight, and degree of cross-linking. Nanoparticles composed of various biodegradable polymers, e.g. poly(lactic acid) (PLA) [57], poly(alkyl cyanoacrylate) (PACA) [58], poly(lactic-co-glycolic acid) (PLGA) [59–66], poly(epsilon-caprolactone) (PCL) [67], natural polymers, such as chitosan [68], gelatin [69], sodium alginate and albumin [70], have demonstrated promising results over the last decade in the use for efficient drug delivery to the ocular tissues.

The physicochemical characteristics of the drug also influence the release rate and, consequently, the therapeutic effects. In the case of nanoparticles, both the surface charge and the binding of the drug to the particles were found to be more important than the drug loading [20]. The active molecules are confined to polymeric matrices by relatively strong non-covalent interactions such as ionic, hydrogen-bonding, hydrophobic or dipole [8]. Upon administration onto the eye, the particles are maintained at the delivery site and the drug is released from their matrix through diffusion, chemical reaction, polymer degradation, or ion-exchange mechanisms.

Without bioadhesion, nanoparticles are eliminated from the precorneal site almost as quickly as aqueous solutions, so the use of mucoadhesive polymers to the negatively charged cornea was investigated. The cationic polymer chitosan, a deacetylated chitin, has attracted a great deal of attention because of its unique properties, such as acceptable biocompatibility, biodegradability, and ability to enhance the paracellular transport of drugs [68]. Calvo et al. showed that the bioavailability of encapsulated indomethacin doubled when PCL nanoparticles were coated with chitosan [71]. The different behaviour of chitosan-coated systems, as compared to the non-coated ones, led to inferring that the surface modification of a colloidal drug carrier might be an approach to modulate its interaction with the ocular mucosa.

Other cationic polymers used for nanosphere production are synthetic acrylic copolymers, such as Eudragit®. Pignatello et al. formulated nanoparticles with Eudragit RL and RS using a modified quasi-emulsion solvent diffusion technique and solvent evaporation method [72]. Administration of these nanoparticles in rabbits, sustained release, and increased absorption of the incorporated NSAIDs (ibuprofen and flurbiprofen) were observed. Furthermore, no inflammation or discomfort was detected in the rabbits' eyes, suggesting a local tolerance of these nanoparticles [73, 74].

Adibkia et al. also studied a NSAID-loaded Eudragit nanoparticle formulation [75]. The nanoparticles composed of Eudragit RS100 were loaded with piroxicam applying the solvent evaporation/extraction technique. These nanoparticles were used to control inflammatory symptoms in rabbits with endotoxin-induced uveitis (EIU), being compared with a microsuspension of the piroxicam alone. The in vivo examinations revealed that the inflammation could be inhibited by the nanoparticle formulation more significantly than the pure piroxicam microsuspension, improving ocular delivery for locally inhibition of inflammation.

PACA nanoparticles became also very popular in ocular anti-inflammatory therapy, for the delivery of e.g. dexamethasone [76], triamcinolone [77], revealing improved encapsulation efficiencies for the anti-inflammatory drugs and acceptable physicochemical characteristics for ocular administration. In another study with PLGA, Vega et al. applied an experimental design to optimize a formulation of flurbiprofen-loaded nanoparticles [59, 64, 66, 78]. The particle surface was coated with Poloxamer 188, resulting in zeta potential values of approximately −25 mV and a particle size of 230 nm. Drug entrapment efficiencies of over 90% were obtained and in vitro release studies showed a burst release, followed by a slower release, completed after 90 min. In vivo studies performed in rabbits demonstrated that the formulations did not induced toxicity or irritation. Nanoparticle formulations were compared with the commercial eye drops (Ocoflur^{TM}) after induction of inflammation by instillation of sodium arachidonate. The results indicated a very good anti-inflammatory efficacy for flurbiprofen-loaded PLGA nanoparticles providing controlled and continuous drug delivery. The conjugation with cell-penetrating peptides improved lurbiprofen delivery from poly(lactic-co-glycolic acid)-polyethylene glycol nanoparticles [63].

In another study, Badawi et al. [79] produced indomethacin-loaded chitosan nanoparticles by modified ionic gelation of the polymer with tripolyphosphate. Transmission electron microscopy revealed regular well-identified spherical shape. The nanoparticles had a mean size of 280 nm, with a zeta potential of +17 mV and a high loading efficiency of 84.8%. In vitro release studies revealed a slight initial burst release during the first hour followed by slow and gradual indomethacin release of 76% during a 24 h period. In vivo studies showed that the nanoparticles were able to contact intimately with the cornea providing slow gradual indomethacin release with long-term drug level, thereby increasing delivery to both external and internal ocular tissues [79].

De Campos et al. reported cyclosporine A-loaded nanoparticles for topical administration. Chitosan is able to interact with ocular mucins and also open tight junctions of the epithelial cells, promoting the transcorneal flux. This system improved the interaction of the drug with corneal and conjunctival tissues, finding therapeutic levels of cyclosporin in ocular surface tissues [80]. Chaiyasan et al. used chitosandextran sulphate nanoparticles as a topical ocular delivery system with lutein as a model drug. The positively charged nanoparticles exhibited mucoadhesive properties measured by electrostatic interactions when examined in vitro using a membrane coated with mucin to simulate the ocular surface, being considered suitable for topical ophthalmic drug delivery [81]. In other study, brimonidine-loaded nanoparticles were prepared by the ionic gelation technique to improve the bioavailability and to prolong the residence time of the formulation in the eye. Nanoparticles showed an entrapment efficiency of the drug. In vivo studies showed a sustained effect of the formulation in comparison with conventional eye drops [82]. Liu et al. showed that the ocular delivery of baicalin (an anti-inflammatory and analgesic agent) could be enhanced by the use of solid lipid nanoparticles. These nanoparticles enhanced the corneal penetration of the drug in permeation studies performed in excised rabbit corneas. A significantly higher bioavailability of tobramycin was shown when 100-nm drug-loaded solid lipid nanoparticles were administered on the ocular surface in rabbit eyes [83]. Although not an anti-inflammatory drug, a higher therapeutic efficacy and a more prolonged effect were observed for methazolamide-loaded solid lipid nanoparticles when compared to a commercial formulation of this anti-glaucoma drug [84]. Hao et al. prepared solid lipid nanoparticles within the colloidal range, dispersed in poloxamer-based hydrogels with thermosensitive in situ gelling properties. In vitro corneal penetration revealed a nearly steady sustained drug release. Toxicity tests in the hen's egg test-chorioallantoic membrane showed no irritation, suggesting these systems as potential vehicles for ocular application [85]. Wang et al. prepared diblock elastin-like polypeptide-based nanoparticles with a multivalent presentation of lacritin at the surface. The in vivo efficacy was explored in wounds on the ocular surface of female non-obese diabetic mice. Two doses of nanoparticles after surgery produced significantly faster wound healing than controls. Histology analysis revealed that no significant corneal inflammation was observed. This report provided the first in vivo verification of lacritin's wound healing potential and can be further applied to rationally bioengineer other peptide therapeutics into self-assembling nanoparticles for treating visual impairments [86]. Kao et al. used a long-lasting pilocarpine-loaded chitosan/carbopol nanoparticle ophthalmic formulation that showed a slower release profile and the most significant long-lasting decrease in the pupil diameters of the rabbits, in comparison with those obtained with other systems such as a pilocarpine solution, a gel, and liposomes [87]. The bioadhesive polymer hyaluronic acid has also been combined with dorzolamide- and timolol-loaded nanoparticles, which was efficient in reducing intraocular pressure in rabbits in comparison with the same dose of a conventional solution of both hypotensive drugs [88]. The in vivo efficacy of cyclosporine A was also enhanced with these systems [89]. Cationic nanoparticles were used to enhance the bioavailability of drugs by topical administration, through electrostatic interactions between the negatively charged ocular surfaces and the positively charged formulation [82]. Başaran et al. incorporated cyclosporine A into three different types of cationic chitosan nanoparticles. They used low, medium, and high molecular weight chitosan to prepare nanoparticles. Medium-molecular-weight chitosan nanoparticles showed a more uniform release pattern and were used for the in vivo studies. Nanoparticles cyclosporine A was applied in sheep eyes and the results showed a prolonged drug release from chitosan-based nanoparticles [90].

The major developmental issues for nanoparticles include formulation stability, control of particle size, control of drug release rate, and large-scale production of sterile formulations. Studies show the potential of colloidal systems for either hydrophilic or hydrophobic drugs, and since they are liquid dosage forms, do not interfere with the vision function, so they have great potential as drug delivery systems for ocular tissues.

16.3.2 Microemulsions

Microemulsions are clear low-viscosity liquid systems for topical drug delivery. Conventional ophthalmic dosage forms tend to be either simple solutions of water-soluble drugs or suspensions of water-insoluble drugs. Unfortunately, as these delivery vehicles generally result in poor levels of drug absorption across the cornea, most of the applied drug does not reach its intended site of action.

Microemulsions are colloidal dispersions composed of an oil phase, aqueous phase, surfactant and co-surfactant at appropriate ratios to stabilize the system [82]. These systems are based on low interfacial tension, unlike coarse emulsions micronized with external energy. This is achieved by the presence of a co-surfactant, which leads to spontaneous formation of thermodynamically stable systems. Microemulsions are transparent liquids, because their droplet size of the dispersed phase is very small, usually bellow 150 nm [91, 92]. They possess optimal characteristics to be administered by topical route. Depending on their charge and viscosity, the retention time onto the ocular surface can be increased allowing the delivery and permeation of drugs across the ocular surface [82]. These systems should offer a possible solution to the problem of poor delivery to the cornea, by sustaining the release of the drug, as well as by providing a higher penetration of drug into the deeper layers of the eye [93, 94]. In addition, they offer the potential of increasing the solubility of the drug in the ophthalmic delivery vehicle [95]. For the production of microemulsions, usually high concentrations of surfactants and co-surfactants are required to develop pseudo-ternary phase diagrams and thus obtain the microemulsion region [96, 97]. Therefore, it is important to assess the ocular tolerance of these systems [98]. Several parameters influence the ocular tolerance, including the viscosity, osmolarity and tonicity, surface tension and charge.

Klang et al. evaluated in vivo the corneal penetration of indomethacin from a commercial hydro-poly(ethylene glycol) (PEG) ocular solution in comparison to a negatively and to a positively charged microemulsion [99]. The microemulsion formulations provided significantly higher indomethacin levels than the commercially available eye drop solution in the aqueous humour and sclera-retina. Furthermore, the spreading coefficient of the positively charged microemulsion on cornea was four fold higher than that of the negatively charged microemulsion, revealing better wettability properties on the cornea of the former in comparison to either commercial solution or to the negatively charged microemulsion. These results were attributed to the enhanced residence time of the drop on the epithelial layer provided by the positive charges, thus enabling a better drug penetration through the cornea to the internal tissues of the eye. The use of bioadhesive polymers in the aqueous phase of the microemulsions has demonstrated to enhance the drug bioavailability as well. Chauhan et al. reported timolol microemulsions in 2-hydroxyethyl methacrylate (HEMA) to extend the release rate of the drug. Gels that had timolol-loaded microemulsions exhibited a fast release in PBS and in saline due to higher solubility of timolol in these solutions compared to that in water. While 100% of the drug had diffused in water, only 55% was released in PBS. Although these systems have limited utility for ophthalmic drug delivery of hypotensive drugs due to the release rate results, the drug loading of timolol was very high, so it must be possible to find applications for delivery of other type of molecules to the eye [100]. Phase-transition microemulsions have also been studied for ophthalmic drug delivery purposes. Here, there is a phase change after their instillation that produces a viscosity increment and a flow rheological change, enhancing the ocular residence time of formulations on the ocular surface [101–104].

A range of other drugs have been formulated in a microemulsion in an attempt to sustain release, and in general, these studies showed that it was possible to delay the effect of drug incorporated in a microemulsion, thereby improving bioavailability. The proposed mechanism of the delayed action is that microemulsion droplets act as drug reservoirs, and are not eliminated by the lachrymal drainage. As expected, microemulsions provide higher loading for lipophilic drugs (e.g. dexamethasone [105]) rather than hydrophilic drugs (e.g. pilocarpine hydrochloride [96]).

16.3.3 Liposomes

Liposomes are membrane-like vesicles consisting of one or more concentric bilayers, produced from natural non-toxic phospholipids and cholesterol, alternating aqueous and lipophilic compartments [106]. The phospholipids include phosphatidylcholine, phosphatidylserine, phosphatidylethanolamine and soya phosphatidylcholine, which are biocompatible and help to bind to the ocular surface and permeate through the pores due to lipidic and surfactant-like properties. The cholesterol is incorporated during the preparation of liposomes to maintain the rigidity of lipid bilayer as well as to prevent the leakage of drug from the formulation [24]. Because of their size, amphiphilic properties and biocompatibility, liposomes are promising systems for drug delivery. The hydrophilic drugs can be encapsulated into core region, while the hydrophobic drugs could entrapped in the lipid bilayers [107]. They can be classified according to structural features as multilamellar vesicles (MLVs), small unilamellar vesicles (SUVs), and large unilamellar vesicles (LUVs), depending on their size and the number of lipid bilayers. With respect to their electrical charge, liposomes can have a positive, negative, or neutral surface charge, depending on their chemical composition (type of phospholipids, drug substances and other lipid molecules). They can accommodate both hydrophilic and lipophilic drugs, respectively into the aqueous compartment or the lipid bilayer [54]. The various methods to the preparation of liposomes are thin film hydration, French press, extrusion method, reverse phase evaporation, microencapsulation vesicle method and ethanol injection. The properties of a liposome that needs to be optimized during formulation include the changes with different composition of lipids, the size of vesicles, surface charge and method of preparations [108]. Liposomes can permeate through the ocular surface by several biological mechanisms, such as fusion, adsorption, endocytosis and lipid exchange. As showed in various investigations, they usually elicit zero order kinetic of drug release pattern with characteristic rapid release accompanied by sustained release [109].

Liposome-based ocular drug delivery systems were first being introduced in 1981 [42]. Modified release liposome formulations have been investigated to improve the treatment of posterior segment by maintaining the drug concentration, enhancing direct delivery, and minimizing intraocular toxicity [110]. For ophthalmic delivery, liposomes are ideal and effective drug carriers due to its biocompatibility, the structural similarity with the cell membrane and ability to provide a platform for both, water-soluble and insoluble drugs. The drug can also be delivered to the posterior region of the eye by crossing the permeability barrier easily. Many anti-inflammatory drugs have been formulated basing liposomal approach for ocular use [24]. Besides providing controlled action at the corneal surface, vesicular systems also help in providing controlled ocular delivery by preventing the drug metabolism from the enzymes present at the tear/corneal epithelial surface.

Liposomes offer advantages over most ophthalmic delivery systems since they are completely biodegradable, biocompatible, relatively non-toxic, have the ability to provide an intimate contact with the corneal surface, are easily to prepare, being versatile in their physical characteristics. However, they have some disadvantages slightly difficult to overcome. Owing to their structure, liposomes are less stable compared with polymeric and lipid nanoparticles, which are characterized by the presence of a solid matrix, raising concerns on liposome long-term shelf life. A change in lipid composition may lead to more stable liposomes, e.g. by increasing the amount of cholesterol, nevertheless one should keep in mind that this could, at the same time, jeopardize the controlled/prolonged drug release profile. Their lower stability also results in a generally faster release of the drug compared with nanoparticle formulations, turning liposomes less suitable for intraocular use, where release over very long time periods is often desirable.

The binding affinity of liposomes to the cornea suggests that the interaction is probably electrostatic in nature [16] and those with a positive surface charge provide more stable adsorption because the corneal epithelium is thinly coated with negatively charged mucin. Besides charge and size, the use of bioadhesive polymers (e.g. PACA, chitosan, hyaluronic acid) to prolong the residence time of an ocular preparation in the precorneal region is another approach, which can further improve liposomal drug delivery [111].

As topical steroids poorly penetrate the cornea with intact epithelium, Al-Muhammed et al. followed for several hours the concentrations of a radiolabelled cationic liposome formulation containing dexamethasone in rabbit cornea, aqueous humour, iris, ciliary body, and sclera [112]. The results depicted high dexamethasone levels in the conjunctiva up to 6 h, suggesting that this formulation might create a drug reservoir for subsequent ocular penetration. Such results were also attributed to the presence of saturated phospholipids in liposome composition, which may protect them from tear lysozyme and esterase activity, resulting in greater drug bioavailability. Encapsulation of idoxuridine in liposomes results in an increase in the penetration of the drug into the cornea [113]. The treatment of uveitis has been achieved with fewer side effects by using tacrolimus encapsulated in liposomes in comparison to non-encapsulated ones [114]. Positively charged liposomes, composed by cationic lipids, like dioleoyl-3-trimethyleammonium propane chloride (DOTAP) or stearylamine (SA), interact with the anionic mucins present on the ocular surface, which allows the formation of a completely coated layer, increasing the contact time of the formulation and the drug bioavailability [115]. Schaeffer et al. observed an increase of penicillin G bioavailability with the use of drug-loaded liposomes composed by phosphatidilcholine (PC), cholesterol (Cht) and SA, in comparison with the equivalent concentration of free drug [116]. Law et al. reported an increased residence time of acyclovir-loaded liposomes with the same components that promoted an enhanced bioavailability of the drug, in comparison with anionic liposomes, which were drained fast because they were not able to provide an electrostatic interaction with the ocular surface [117]. Saettone et al. compared a positively charged liposomal acyclovir formulation (PC, Cht, and SA) with a commercialized ointment. Liposomes increased the acyclovir concentration in the aqueous humour of rabbits when compared with the ointment at the same drug concentration [118]. Acetazolamide- and pilocarpine-loaded liposomes were investigated for the treatment of ocular hypertension, promoting a more pronounced and sustained reduction in intraocular pressure than conventional hypotensive solutions [119]. Li et al. investigated the potential of liposomes to enhance the local anti-glaucomatous effect of brinzolamide [120]. In vitro release studies of brinzolamide-loaded liposomes displayed a biphasic release pattern with an initial burst release, followed by a second sustained release period. In Franz type diffusion cells, corneal permeability increase in comparison with the commercially available formulation. In vivo studies in rabbits showed a more sustained and effective IOP reduction for drug-loaded liposomes.

The bioavailability of the drug in the tear fluid and the uptake into the cornea were increased after instillation of liposomes in rabbits, being less cytotoxic than distamycin A alone in rabbit corneal epithelial cells. Various studies have shown that hybrid systems (liposomes combined with hydrogels) can increase the residence time of the formulations, thus prolonging the drug release rate [14]. The presence of the hydrogel increased the amount of drug permeated through the cornea in comparison to the drug aqueous solution at the same concentration. The dispersion of drug-loaded liposomes in collagen shields is one of interesting technological strategy to achieve higher concentrations of the therapeutic agent in ocular tissues. Pleyer et al. used it in rabbit eyes. They found that cyclosporine levels in cornea, sclera, aqueous humour, and vitreous were significantly higher in these animals, in comparison with conventional cyclosporine eye drops [121]. The combination of liposomes and bioadhesive polymers such as carboxymethyl cellulose or hyaluronic acid increased the hypotensive effect of a melatonin analog 5-MCA-NAT in rabbit eyes compared to other liposomes formulated without polymers. This hybrid formulation was able to increase the bioavailability of the therapeutic agent, being soft with the ocular surface. As the novel formulation simulates tear film composition, it might relieve symptoms of secondary diseases affecting the ocular surface derived from the chronic use of anti-glaucoma topical medications [122–125]. Dong et al. designed and prepared ibuprofen-loaded liposomes coated by regenerated silk fibroins, which are novel adhesive excipients [126]. No detectable toxicity was observed in cells after of exposure to silk fibroins. The novel liposomes showed sustained release and in vitro corneal permeation of ibuprofen as compared to a drug solution and liposomes without silk fibroins. Liposomes are able to improve drug stability in biologic fluids and can reduce retinal toxicity [127]. Various drugs with known high toxicity have been successfully formulated as liposomal formulations. As amphotericin B, which produced less toxic effects than a marketed solution after a single intravitreal injection in healthy pigmented rabbits [128, 129]. An acyclovir pro-drug called HDP-P-GCV was entrapped into liposomes and injected into the vitreous in an animal model of cytomegalovirus retinitis, being able to protect the treated eyes against the virus [130]. Bochot et al. developed PEGylated liposomes for the intravitreal delivery of an oligonucleotide model, pdT16. The liposomal formulation (PC:Cht:PEG-Polyethylene glycol-distearoyl ethyl phosphocoline) protected the active substance against degradation in the vitreous and prolonged the pdT16 release into the vitreous, retina, and choroid [131]. Lajavardi et al. used PEGylated liposomes to enhance intravitreal delivery of a proteic substance, vasointestinal peptide (VIP), for the treatment of uveitis. Labelled rhodamine (Rh) liposomes (PC:PG:Cht:PEG-Polyethylene glycol-distearoyl ethyl phosphocoline:PE-Rh) dispersed in hyaluronic acid were able to protect the protein from degradation in the vitreous and to maintain a sustained release of VIP [132]. Asteriti et al. investigated the potential of liposomes to deliver functional proteins to retinal photoreceptors and modulate their physiological response. Vesicles, prepared with L-α-phosphatidylcholine and cholesterol, were loaded with rhodamine B, recombinant myristoylated recoverin (an endogenous protein regulating the duration of the phototransduction cascade in photoreceptor cells), or antibodies against recoverin [133]. Liposomes have been extensively studied for these reasons for encapsulation of ocular therapeutics.

16.3.4 Lipid Nanoparticles

Lipid nanoparticles consist of solid lipid spheres, with average diameters between 50 to 1000 nm, dispersed in aqueous medium [134–136]. These particles are similar to those described as polymeric nanoparticles, being in this case composed of a solid lipid core matrix, stabilized by surfactants. Solid lipidic nanoparticles (SLN) and Nanostructured Lipid Carriers (NLC) are constituted from biodegradable lipids corresponds to nanoparticles that are formed primarily of polymers or polymeric derivatives [137]. As such, these lipid nanoparticles will have high encapsulation efficiency and loading capacity for lipophilic molecules. SLN consist of lipids that are in the solid state at room temperature and surfactant or mixture of surfactants, while in the case of NLC the inclusion of liquid lipid is an additional component [24].

Surfactants used for their production include biological membrane lipids (lecithin, pure phospholipids), bile salts (e.g. sodium taurocholate), biocompatible non-ionic molecules (e.g. ethylene oxide/propylene oxide copolymers, sorbitan esters, fatty acid ethoxylates), and mixtures thereof [138]. Lipid nanoparticles combine several advantages, such as consistent formulation, stable structure, small size, easy of fabrication, monodispersity, larger drug payload than metallic and polymeric nanoparticles with an addition of non-toxicity due to reactive surface properties [24]. Thus, they avoid some of the disadvantages previously pointed out for other colloidal carrier systems, such as polymer nanoparticles, microemulsions and liposomes. When optimized, lipid nanoparticles exhibit high physical stability, protection of incorporated labile drugs against degradation and excellent in vivo tolerability. However, these systems generally exhibit a low drug pay-load capacity and drug expulsion during storage due to the transition of highly ordered lipid particles [139, 140]. These disadvantages can be remedied by using structured lipid matrices in SLN formulation and surface modification of the particle [141].

Attama et al. prepared sodium diclofenac-loaded lipid nanoparticles combining the homolipid from goat (goat fat) and a phospholipid, with high encapsulation efficiency applying the hot high-pressure-homogenization technique (above 90%) [142]. Administration of this formulation in bioengineered human cornea depicted a sustained release of the analgesic drug. Furthermore, permeation of sodium diclofenac through the cornea construct was improved by surfacing nanoparticles with phospholipid, which showed the performance of this formulation for ocular application. Several anti-fungal and anti-viral drugs have also been delivered to an anterior segment of the eye via SLN and NLC carrier system and found to be effective in delivering ocular therapeutics [143]. Ibuprofen loaded-NLC were developed using Compritol 888 ATO as the main lipid core combined with Gelucire 44/14, Miglyol 812, stearylamine (charge-inducing agent), Transcutol P as corneal permeability enhancer, and Cremphor EL. These particles showed prolonged controlled release profiles. NLC containing stearylamine were more stable and had a longer pre-ocular retention time due to its cationic surface charge. Gelucire 44/14 and Transcutol P could enhance the corneal permeability by mainly due to lipophilic properties of Gelucire, and to mechanism involving structural changes in the epithelium as a result of micelle formation induced by Transcutol P [144]. Flurbiprofen, another NSAIDs aryl propionic derivate, avoids the production of prostaglandins, by blocking the action of cyclo-oxygenase (COX), reducing inflammation followed ocular surgery or laser treatment [145]. Gonzalez-Mira et al. prepared by an ultrasound technique, and optimized by factorial design approaches, to improve flurbiprofen bioavailability and to manage inflammation induced by ocular surgery. Stearic acid and castor oil (50:50 wt%) were the solid and liquid lipids selected as NLC matrix suitable for ocular administration. The obtained results suggest that, when the oil content increased, a more stable flurbiprofen-loaded NLC was obtained. The ocular tolerance was assessed in vitro and in vivo by the Eytex and Draize test. The developed systems were physicochemically stable showing a good tolerance after topical instillation [146]. Another formulation of FB-loaded SLN, containing stearic acid (SA), a fatty acid, as the solid lipid and a mixture of Miglyol 812 and castor oil (CO) as the liquid lipids, was also developed. The optimal NLC formulation showed an appropriate average size for ophthalmic administration with a narrow size distribution, negatively charged surface and high encapsulation efficiency [147]. The in vitro experiments proved that sustained release of FB was achieved using NLC as drug carriers. Optimal NLC formulation did not show toxicity in ocular tissues [148]. Luo et al. analysed the potential of chitosan oligosaccharides (COS)-coated NLC for ocular flurbiprofen delivery. In these particles, containing Compritol 888 ATO, Gelucire 44/14, Solutol HS-15, and Tween 80, the ocular bioadhesion test revealed that the clearance of the formulations was significantly delayed in the presence of COS. On the other hand, COS-coated NLC could improve corneal penetration [149]. Indomethacin was successfully encapsulated in SLN (In-SLN). A screening of lipids and surfactants has been carried out, to produce a stable formulation composed of Compritol 888 ATO (2% w/v), Tween 80 (0.75% w/v), Poloxamer 188 (0.25% w/v), and glycerin (2.25% w/v). The encapsulation efficiency of In-SLN, particle size, and surface electrical charge were considered suitable for ocular administration. A considerable increase in the chemical stability and in in vitro corneal permeability of indomethacin was observed with the In-SLN formulation against indomethacin solution and indomethacin hydroxypropyl-beta-cyclodextrin-based formulations [150].

SLN and NLC are safe and flexible nanocarriers system for the delivery of the drug, gene, and nucleic acids. New technologies have been developed for SLN and NLC production, and are currently under investigation to obtain the optimum encapsulation of different drug categories and to deliver the bioactive within the ocular sites [5, 6, 24, 62].

16.3.5 Drug Nanocrystals

By definition drug nanocrystals are nanoparticles being composed of 100% drug without any matrix material, typically with a size range between 200 and 500 nm [151]. They are typically produced in the form of nanosuspensions, which are submicron (colloidal) dispersions of drug particles, stabilized by surfactants, polymers, or a mixture of both. A formulation should have a volume median diameter (D50) < 1 mm and a volume diameter 90% undersize (D90) < 2.5 mm to be classified as a nanosuspension. The term nanocrystals has been extended to describe nanosized suspensions of partially crystalline or even amorphous drugs formed because of changes from the crystalline to the amorphous form during processing. Nanocrystals are regarded as an important nanoformulation approach exhibiting advantages of increased dissolution and saturation solubility with chemical stability and low toxicity [152].

There are several methods used to reduce the particle size of a drug, i.e. nanocrystals are produced in the form of nanosuspensions using top-down (e.g. wet milling or high pressure homogenization) and bottom-up methods (e.g. anti-solvent precipitation) [152]. The bottom up technologies start from the molecules which are dissolved and precipitate them by adding the solvent to a non-solvent [153], although it uses relatively simple, low cost equipment, it needs to avoid particle growth to microcrystals, which represents a problem. The top-down technologies are disintegration methods like jet milling, pearl milling [154] and high-pressure homogenization (HPH) [155]. A general problem of pearl mills is potential erosion of material from the milling pearls leading to product contamination. Milling by high pressure homogenization can be done by two principles: microfluidization and piston-gap homogenisers [151]. The obtained product is a suspension of drug nanocrystals in a liquid stabilised by a surfactant or polymer, showing an increased dissolution velocity and higher saturation solubility in comparison to micronized drugs. The physicochemical properties are dependent on the type of stabilizer, i.e. non-ionic surfactants (e.g. polysorbates, poloxamines and poloxamers) and ionic surfactants (e.g. bile salts and alkyl-sulphonates) [156]. In order to obtain a controlled release dosage form with better physical stability the nanosuspension has to be transformed into a solid, dry multi-particulate product, by means of lyophilization or spray drying procedures. Lyophilization requires a cryoprotective agent added to the solution to avoid particle aggregation after reconstitution of the system. Spray drying usually applies high temperatures in the process which is not suitable for thermo-labile drugs.

Several drugs (e.g. cinnarizine, griseofulvin, indomethacin, itraconazole, loviride, mebendazole, naproxen, phenylbutazone, and phenytoin) have been formulated as nanocrystals for the evaluation of the Ostwald ripening effect as a function of the storage temperature [157]. Except for indomethacin, after 3 months of storage at room temperature, Ostwald ripening occurred in all drugs. Lower temperatures were able to slow down the ripening whereas higher temperatures accelerated the effect. Lyophilization and spray-drying also affected particle agglomeration in particular for drugs with more hydrophobic surfaces, which were harder to disintegrate, and thus dissolution was compromised upon drying. The same was found for compounds having higher log P values. Drug nanocrystals can be considered as a universal formulation approach for poorly soluble drugs. The striking advantages the advantages of nanocrystals for drug delivery to the eye are: improved ocular safety, increased retention of the formulation in cul-de-sac, enhanced corneal permeability across the corneal and conjunctival epithelium, enhanced ocular bioavailability, dual drug release profile in the eye, and increased tolerability [152].

According to Sharma et al., the dual drug release profile of nanocrystals in the eye means that they exhibit both immediate and sustained drug release profiles after their topical administration [158]. The immediate drug release can be linked to the increased saturation solubility and dissolution of the nanocrystals, resulting in initial higher concentrations available for absorption and, thus, rapid onset of action. Contrariety, the prolonged drug release derives from the high surface area of the nanocrystals, which facilitates interactions with biological membranes. The increased interactions with the ocular mucosa provide nanocrystals with mucoadhesive properties, increasing their retention time in the cul-de-sac region and, thus, prolonged drug action is achieved. Enhancing the viscosity of nanosuspensions or inclusion of nanocrystals into an in situ gelling system can further increase the retention time, which prolong the release profile of the drug [159]. Tuomela et al. prepared nanosuspensions of the poorly water-soluble drug brinzolamide as ocular formulations for the treatment of glaucoma. From the polymers and/or surfactants that were screened as stabilizers during wet media milling, hydroxypropyl methylcellulose it was verified to be the stabilizer of choice because it was capable of maintaining the reduced particle size of the nanosuspensions (460 nm) [160]. Currently there are few studies about NSAIDs in the form of nanocrystals for ophthalmic application, since the major pre-requisite for nanocrystal formulation is the hardness of the drug crystals.

16.3.6 Nanoemulsions

Nanoemulsions (NE) possess low surface tension and include water, oil, and amphiphilic surfactants as main formulation composition. Nevertheless, co-surfactant can also be added sometimes to further increase the solubility and stability of the NE. The formation of droplets via the action of lipid and water inside nanosystems is the basis to react with the environment of biological membrane. Available as water-in-oil (w/o) and oil-in-water (o/w) system, they provide choice to the formulation scientist. However, the o/w is of particular interest in nano-ophthalmics considering the lipophilic nature of the corneal membrane [161].

Considering the use of surfactants and co-surfactants in relatively greater amount (compared to liposomes), their ocular tolerance is the major criteria that must be considered judiciously for acceptance. Other parameters to take into account carefully for the development of an acceptable ONS include viscosity, pH, tonicity, surface charge and size of NE [24].

There was reported the use of NE to deliver the poorly water-soluble drugs into corneal segment giving delayed and sustained release, but this can ascribed to their droplets that act as drug reservoir and remains unaffected by the lachrymal fluid. Nevertheless, a formulation having positive surface charge will be able to bind more specifically to the corneal membrane through electrostatic interaction [44]. There is a promising and successful utilization of nanoemulsion-based ONS. The selection of suitable cationic material for the effective formulation composition is necessary when targeting to ocular sites [162].

The most important advantage of nanoemulsion to note is replacing the conventional eye drop formulation, because they can be easily instilled and in comparison, they able to produce sustained delivery and higher retention of drug molecules into eye sites for better clinical outcomes [24].

16.3.7 Niosomes

Niosomes have gained popularity in ocular drug delivery because they can provide prolonged duration of action at the corneal surface and increase the ocular bioavailability of ophthalmic therapeutics. They are non-ionic surfactant vesicular systems formed from the self-assembly of non-ionic amphiphilic compounds in aqueous media, resulting in bilayer vesicles of 10–0.5 μm in size [82].

The use of niosomes has shown promising results for the treatment of ocular hypertension associated to glaucoma. Aggarwal et al. reported that acetazolamide-loaded cationic niosomes provided good entrapment efficiency and corneal permeability, being effective for the reduction of intraocular pressure in experimental animals.

However, positively charged niosomes, composed of sorbitan ester, cholesterol, and SA, resulted toxic for corneal cells. It was prepared a bioadhesive niosomal formulation of acetazolamide by coating niosomes with Carbopol®, to reduce corneal toxicity of niosomes, a bioadhesive niosomal formulation of acetazolamide. The bioadhesive formulation showed a increment in the duration of action in comparison with a commercial formulation and it was less toxic than non-coated vesicles [163]. In other study, authors observed that the niosomal formulation coated with 0.5% w/v chitosan and containing a 0.25 w/v solution of timolol maleate extended the release of the drug in vitro up to 10 h. At same way, a longer prolonged effect was observed when chitosan-coated formulations where instilled on rabbit eyes, in comparison to the administration of a conventional hypotensive solution (8 and 2 h) [164]. To demonstrate that compounds can influence the entrapment efficacy as well as the size and shape of niosomes Abdelkader et al. developed naltrexone hydrochloride-loaded niosomes and tested the incorporation of different additives to the formulation [165]. Ex vivo transcorneal permeation studies made in excised cow corneas proved that niosomes were capable of controlling drug permeation and enhancing its corneal permeability. These formulations resulted practically non-irritant [166]. And they observed that niosomal preparations were able to protect the encapsulated drug from the photo-induced oxidation up to 3-fold in comparison to free naltrexone solutions [167]. Li et al. developed proniosome-derived niosomes for topical ophthalmic delivery of tacrolimus. Freshly excised rabbit corneas were employed to analyse the in vitro permeation of tacrolimus from the vesicles and the drug retention in the cornea. They observed a significant increase in tacrolimus permeation as compared with commercial ointments. The in vivo ocular irritation test of 0.1% tacrolimus-loaded niosomes showed no irritation and showed also good biocompatibility with cornea after 21 consecutive days of instillations (four times/day). The in vivo anti-allograft rejection assessment was performed in a rat corneal xenotransplantation model. With these tests, they observed that tacrolimus-loaded niosomes delayed the occurrence of corneal allograft rejection and significantly prolonged the median survival time of corneal allografts, in comparison to cyclosporine eye drops, drug-free niosomes, or untreated eyes [168]. Abu et al. prepared and characterized an ocular niosomal/hydrogel formulation containing 0.5% atenolol for the treatment of ocular hypertension associated to glaucoma. Niosomes were prepared with span 60 and cholesterol at different molar ratios. The size of niosomal formulations ranged and the results showed a high retention of atenolol inside vesicles. Examination of rabbit eyes did not show irritation signs. Atenolol niosome/carbopol gel showed slower and sustained in vitro release of atenolol in comparison to the rest of formulations and a maximum IOP reduction of 49.80% after 5.33 h of drug administration [169].

16.3.8 Cubosomes

Cubosomes are liquid crystalline particles in nano size range (100–300 nm), usually composed of amphiphilic lipids (Monoolein, and phytantriol) and with or without stabilizer/surfactant (Poloxomer 407) [170]. Their formation entails the formation of self-assembly of lipid molecules in cubic structure owing to their molecular orientation and structural symmetry. The structure of cubosomes entails curved bicontinuous lipid bilayers describing three-dimensional honey combed arrangement. As their name indicates, they are cubic structures with numerous water channels within themselves. Once their make-up of amphiphilic lipids and possessing aqueous channels, they can resourcefully accommodate hydrophilic, lipophilic and amphiphilic molecules with high loading efficiency [24]. During the loading process, the hydrophilic drug gets accommodated in the water channels and also at the surface of the polar head of amphiphilic lipids while lipophilic drug can be incorporated in lipidic architectures [171, 172]. Surfactants and lipids play an important role to stabilize the cubosomes against agglomeration in the bulk cubic phase (colloidal stability). Due to cubic phase, cubosomes tend to re-coalesce and stabilizers like poloxamer 407 (PEO99–PPO67–PEO99 tri-block copolymer) stabilize such a colloidal system by complexing with its surface which exposes them to surrounding water phase [24].

They have as properties low viscosity, high stability, large surface area and they are stable with a wide range of temperature at any dilution levels due to its unique structural features. The hydrotrope concentration and drug to stabilizer ratio also significantly affect the formulation of cubosomes [24].

Cubosomes are promising ONS due to their biocompatibility and they have bioadhesive properties. They are composed of unique lipids like monolein, glyceryl monooleate which are highly biocompatible [171]. Many cubosome-based nanocarriers have been reported as safe for therapeutic application [170]. The optimized cubosomal ophthalmic nanocarrier system might be a promising and a future hope for greater retention of drugs in ocular sites due to high bioadhesivness. Cubosomes enhance the transcorneal permeation and bioavailability of drugs and therefore they are indicative successful implication in ophthalmic delivery. They established themselves as effective ONS due to their cubic phase and curved like structure which leads to better stabilization of system under aqueous environment [24].

16.3.9 Nanomicelles

Nano-sized micelles has its vital importance and attention due to its ability to target both anterior segments related inflammation as well as a posterior segment of the eye [10]. Nanomicelles are composed of a hydrophobic core and a hydrophilic shell and they are spherical vesicles highly acceptable for passive drug delivery of hydrophobic compounds. Polymeric micelles (10–200 nm) are amphiphilic molecules or block copolymers which can generally self-assemble into organized core–shell/supramolecular structures in aqueous media at concentrations exceeding their critical micellar concentrations (CMC) [173, 174]. Low-molecular weight surfactant-based micelles exhibit higher CMC in contrast to polymeric micelles, which lead to diminished stability and potential side effects. The potential of polymeric micelles to solubilize and stabilize hydrophobic compounds leads to their prolonged retention which improving therapeutic outcomes [175].

However, micelle-based nanomedicines suffer from a myriad of problems when employed in in vivo systems. Dilution in the blood stream and interactions with blood components (including plasma proteins such as albumin, fibrinogen, prothrombin, and gamma globulins) lead to their premature disintegration in the systemic circulation. This proves the primary reason for the limited improvements in therapeutic efficacy of micellar nanomedicines in clinical trials when administered systemically [176, 177], while such issue would be minimal for topical ocular administration. To prevent the premature disintegration and release of therapeutic cargo, core cross-linked polymeric micelles have evolved as a promising strategy. The cross-linking both in the core and in the corona of micelles can be based on covalent bonds, hydrogen bonding or π-π stacking. The covalent attachment in core cross-linked polymeric micelles allows stable drug retention and enables tailored release kinetics [10].

Over past couple of years, there was an increasing number of cross-linked micellar-based nanomedicines due to the growing interest for the development of core cross-linked polymeric micelles for various disease conditions including ocular indications. Promising in vivo results have entailed these cross-linked polymeric micelles to enter early stage clinical evaluations [178, 179].

Polymeric micelles are highly effective delivery systems for intraocular drug delivery, facilitating the formulation and entrapment of highly hydrophobic drugs, aiding drug administration, enabling controlled drug release kinetics, enhancing therapeutic efficacy, reducing side effects and improving the in vivo stability [10]. It may be noted that the hydrophobicity of a drug limits the possibility to formulate its clear solution in a concentration sufficient to reach therapeutic levels in the ocular tissues [174]. This limitation can be solved employing the nanomicelles due to their ability to entrap the lipophilic drug in its hydrophobic core which gives a clear solution suitable for ocular administration. Most widely used surfactants for the preparation of nanomicelles included sodium dodecyl sulphate, n-dodecyl tetra (ethylene oxide), Vitamin E TPGS, octoxynol-40, dodecyl trimethyl ammonium bromide and dioctanoyl phosphatidylcholine (C8 lecithin). Regarding the formulation aspect of nanomicelles, they often protect the drug from degradation and enable its ocular permeation. An ocular product with high tolerance and greater clinical output is the need of the current ophthalmic market. It can be considered that nanomiceller formulation is a safe alternative to conventional eye drops for treating of corneal diseases [180]. Nanomicelles carrying dexamethasone to the anterior segment of the eye showed higher bioavailability in comparison to the dexamethasone suspension [181]. Beside, stable nanomicelles have been formulated to deliver therapeutic agents to the posterior segment of the eye [182]. Nanomicelles loaded with different drugs such as rapamycin have been developed to treat different eye diseases [183]. Nanomicelles provides an opportunity to ocular biologists and formulation scientists to reduce the instillation frequency and increase the clinical outcomes due to its sustained release action and high tolerability with inside eye environment [24].

16.3.10 Spanlastics

Based on the vulnerability of inflamed eye and anionic barrier property of the ocular surface, the need of non-ionic, non-irritant ONS experienced for which noisome and its recent variant, spanlastics were regarded as better alternatives [184]. Spanlastics are surfactant-based novel elastic nanovesicular systems consist of non-ionic surfactants and spans (sorbitan esters). They lack cholesterol. The inclusion of non-ionic surfactant with the exclusion of cholesterol give spanlastics its tenets of ultradeformability [185]. Comparing with niosomes, these are more deformable and multilamellar leading to enhanced entrapment efficiency. Their size is generally larger than the pore size of biomembranes but due to their surfactant nature, they temporarily increase the size of pore so that the bigger vesicles can permeate or penetrate [186]. They consist of edge activators like tweens which lower the interfacial tension of vesicles and enhance their fluidity and deformability by interfering with the packing characteristics. Their drug delivery aptitude and penetration enhancer activity are of special importance to ophthalmic nanomedicines [187].

16.3.11 Hydrogels

Hydrogels are 3D crosslinked polymeric networks capable of absorbing large quantities of water. Their diffusion and permeability properties can be facilely tuned by varying processing parameters, such as crosslinking time, temperature, type of crosslinker, ratio of polymer and crosslinker which make varying the crosslinking density. Hydrogels are usually formed at ambient temperatures, without the need for organic solvents or harsh processing conditions, which may denature proteins or antibodies. And combined with the presence of an aqueous internal environment, makes hydrogels attractive candidates for entrapment of fragile entities such as proteins, antibodies, peptides, cells, and oligonucleotides. In ocular drug delivery, hydrogels have been explored for sustained delivery of drugs, intravitreally as well as periocularly [188].

16.3.12 Chitosan Nanoparticles

Chitosan is the most frequently available natural polymer after cellulose [189]. It is a polysaccharide derived from the natural compound called chitin which is found mainly in the exoskeleton of crustaceans. Chitosan nanoparticles are mainly used to treat the disorders of the anterior segment of the eye and rarely used for the posterior segment of eye, because chitosan has mucoadhesive properties due to the surface positive charge. The hydroxyl group increases the mucoadhesive properties and presence of surface positive charge chitosan enhanced binding to the negatively charged corneal surface, increasing their residence time and decreased clearance [190].

16.3.13 Dendrimers

Poly(amidoamine) (PAMAM) is highly branched, repetitively, tree-like nanostructured polymers referred to as dendrimers that can be used for ocular drug delivery purposes. Dendrimers have well-defined size, tailorable structure, and appropriate biodistribution in both posterior and anterior segments of the eye, which make them one of the most applicable drug delivery systems in nanoscales [191]. Some dendrimers have the favourable potential to be used as surface coating agents and drug carriers and have also antimicrobial properties [42]. In vitro studies of the interactions between PAMAM dendrimers and negatively charged ocular mucins revealed strong interactions between the mucins and the dendrimers in the tear film [192]. It is also reported that encapsulating the fluoroquinolones, such as prulifloxacin and nadifloxacin, within PAMAM dendrimers enhances drugs' solubility and improves their anti-bacterial potency compared to free drugs [193]. Dendrimers can potentially be used as topical eye drops with the capacity to form a gel layer on the surface of the cornea and provide sustained transportation of anti-microbial agents without affecting the visual power. PAMAM dendrimers have also been used to effectively carry pilocarpine nitrate, tropicamide, and glucosamine to the rabbit eyes [42]. Despite this, selective localization of dendrimers in the retinal ganglion cells may lead to inflammation and develop into the toxicity or even exacerbate the possible toxicity of dendrimers [194].

16.3.14 Microparticles

Microparticles consist of micronized drug particles for intravitreal injection, but also consist of micronized polymer/drug matrix-like structures (microspheres) or polymer-coated microparticles (microcapsules), that are applied topically to the eye surface. Microparticles are solid drug carriers (1–1000 μm) capable of providing sustained and controlled release of loaded active agents. According to their structure, microparticles can receive the name of microcapsules (reservoir structure) and microspheres (matrix structure). They can be too large for intravitreal injections where nanoparticles with diameters of less than 200 nm have been shown to result in more effective drug delivery to the retina than larger particles [82, 195]. The use of microparticles for topical therapies is not very common due to their short contact time on the ocular surface, and it is also important to know that the size of microparticles has to be controlled to be less than 10 μm to avoid possible eye irritation [196, 197]. The advantage of microparticles for intraocular administrations is that they can be injected in the form of suspensions by using conventional needles. Microparticles are typically dispersed in a physiological vehicle (phosphate buffer or balanced salt solutions), but sometimes viscous vehicles, such as hyaluronic acid or hydroxypropylmethyl cellulose, are used to improve the injectability of the suspensions [198].

Many authors have tested the use of microparticles as drug carriers over the last decades. Posterior segment diseases such as proliferative vitreoretinopathy, age-related macular degeneration, diabetic retinopathy, uveitis, macular oedema, cytomegalovirus, glaucoma, and retinitis pigmentosa have been treated with drug-loaded microspheres with a relatively high rate of success [199]. Addo et al. observed that the use of tetracaine-loaded microparticles enhanced the anaesthetic effect of the drug in rabbits, in comparison with a commercialized ophthalmic solution [200]. Other authors have studied the hypothesis that mucoadhesive microparticles may offer a means for slow drug release from the particles that remain adherent to the ocular surface for an extended period of time. Sensoy et al. showed the utility of mucoadhesive polymers for the preparation of bioadhesive microspheres loaded with sulphacetamide for the treatment of keratitis in rabbits [201]. Choy et al. observed that the use of a mucoadhesion promoter such as polyethyleneglycol (PEG) increased the microparticles' retention time on the ocular surface of rabbits [202, 203].

A large variety of active substances have been included in microspheres (i.e. anti-proliferatives, anti-inflammatories, immuno-suppressants, antibiotics, and even biotechnological therapeutic agents for intraocular administration. Microparticulate systems have been prepared from biodegradable polymers like gelatin, albumin, polyorthoesters, polyanhydrides, and polyesters. The PLA and poly(glycolic) acid (PGA) polymers and their co-polymers PLGA are the most used. For ocular application, and especially for the treatment of diseases affecting the back of the eye, these erodible polymers have been employed to prepare different devices (implants, scleral plugs, pellets, discs, films, rods, nanoparticles, and microparticles) [204]. Park et al. also utilised PEG as additive with poly(lactic-co-glycolic acid) (PLGA) microspheres loaded with brimonidine. The mucoadhesive microparticles increased drug bioavailability of the hypotensive agent compared to the marketed brimonidine formulation [205]. Moritera et al. showed the administration of adryamicin-loaded microspheres in an animal model of proliferative vitreoretinopathy. The retinal detachment was decreased from 50% to 10% in the rabbits after the intravitreal injection and a significant decrease in the retinal toxicity was produced, when comparing a single injection of microspheres with the administration of the same amount of drug in solution [206, 207]. Veloso et al. evaluated the anti-viral effect of ganciclovir-loaded microspheres in infected rabbits' eyes. In treated eyes, vitritis, retinitis, and optic neuritis decreased in contrast to control eyes. No adverse tissue reaction was clinically or histopathologically observed in the injected eyes [208]. Conti et al. developed PLA and PLGA microparticles loaded with acyclovir. After 14 days, the active substance was detected in the vitreous of rabbits receiving D,L-PLA microparticles [209]. Duvvuri et al. dispersed ganciclovir-loaded microspheres in a thermogellin PLGAPEG-PLGA solution that was then injected in rabbit eyes. Drug levels in the vitreous were significantly higher and maintained for a prolonged time in comparison with a drug solution [210].

The treatment of intraocular inflammation, uveitis, or intraocular immune disorders with microparticles has also been extensively investigated. He et al. prepared Cyclosporine-loaded PLGA 75:25 microparticles which produced a increment in the residence time of the drug in comparison with a cyclosporine A solution [211]. Microspheres were also able to maintain therapeutic concentrations in the disease-related tissues such as the choroid–retina and iris–ciliary body. Dexamethasone-loaded PLGA 50:50 microspheres were tested for the prevention of ocular inflammation in an animal model of uveitis provoked by lipopolysaccharide injection [212]. Authors proved that the intraocular inflammation was significantly reduced in animals receiving drug-loaded microspheres in comparison to control. Some studies performed with microspheres loaded with anti-inflammatory drugs have already been performed in humans. Cardillo and co-workers showed human studies of the intravitreal injection of triamcinolone-loaded PLGA microspheres. The eyes treated with triamcinolone microspheres showed a marked decrease in retinal thickness as well as an improved visual acuity [213].

The development of new bioengineered products like peptides, proteins, or nucleic acids for the treatment of ophthalmic diseases has made it necessary to design new microparticulate systems able to effectively administrate such molecules to the target site. This strategy has been used for the treatment of glaucoma.

The therapeutic approach focuses on promoting the survival of retinal ganglion cells (RGCs) [82]. Andrieu-Soler et al. developed PLGA microspheres loaded with GDNF (a glial-cell-linederived neurothropic factor) which was efficient in preventing the retinal degeneration in a retinosis (rd1) [214]. The efficacy of GDNF-loaded microspheres has also been tested in a glaucoma rat model based on the intraocular pressure elevation with an episcleral injection of a hypertonic NaCl solution, reporting a significant increase of the survival of retinal ganglion cells and their axons [215]. ChecaCasalengua et al. developed PLGA microparticles loaded with an anti-oxidant (vitamin E) and GDNF. Authors reported that a single injection of these microspheres in a rat model of glaucoma based on intraocular pressure elevation was sufficient to promote a retinal ganglion cell's survival, in contrast to RGC survival observed in eyes treated with GDNF in solution, vitamin E and blank microspheres [216]. Andrieu-Soler et al. showed the efficacy of PLGA microspheres loaded with recombinant human glial cell line-derived neurotrophic factor (rhGDNF) in preventing retinal degeneration in an animal model (mice) of rd1. They observed a significant delay of rod photoreceptor degeneration in the animals receiving the rhGDNF-loaded microspheres in comparison to the control group or the group receiving blank microspheres. Retrobulbar administration of PLGA microspheres loaded with the IKK2 inhibitor, 2-[(aminocarbonyl)amino]-5-(4-fluorophenyl)-3-thiophenecarboxamide (TPCA-1), promoted a significant reduction of laser-induced choroidal neovascularization (CNV) [211]. Zhang et al. reported PLGA microspheres for the release of ranibizumab by intraocular administration. The proposed encapsulation method provided minimal loss of bioactivity of the anti-vascular endothelial growth factor. After intravitreal injection of the microparticles in a chick model, the active molecule was released in a controlled fashion and no significant reaction or cell death was observed [217]. The utility of microparticles in rare ocular diseases, such as retinitis pigmentosa, has also been evaluated. PLGA microspheres loaded with tauroursodeoxycholic acid (TUDCA) and phosphate solution (control) were intravitreally injected in rats. The authors suggested the potential of these formulations to delay vision loss in retinitis pigmentosa [82].

16.4 Cytotoxicity and Toxicological Concerns

Due to the relative safety and convenience of ocular application, as well as the relatively low risk of systemic side effects in comparison with other routes of administration, topical administration of ophthalmic drugs is usually the first choice. Several ophthalmic formulations of NSAIDs are marketed for topical use to manage pain and inflammation. Corticosteroids, which exert a profound anti-inflammatory effect, have no direct influence on pain. In contrast to the corticosteroids, NSAIDs do not cause raised ocular tension, secondary cataract formation, or reactivation of herpes simplex virus infection upon application to the eye. Frequently reported adverse reactions of NSAIDs include transient burning, stinging and minor signs of ocular irritation and instillation [218]. Hypersensitivity reactions with itching, reddening, photosensitivity and keratitis punctata may occur. Rarely, topically applied NSAIDs may cause systemic symptoms. Sitenga et al. described a single case of acute exacerbation of asthma after inadvertent application of kerotolac eye drops in a 44-year-old woman with chronic asthma, rhinosinusitis, and nasal polyps [219].

Stroobants et al. investigated the effect of six NSAID eye drops on the normal corneal epithelium of rabbits [220]. The corneal surface damage was assessed by scanning electron microscopy, and all NSAIDs caused changes in the epithelium cells with severity increasing from the peripheral to the central corneas. It was revealed an increased number of dark cells in the corneal epithelium, development of craters, tendency for cells to lose their hexagonal pattern, and some alterations in the membranes, such as surface roughness, projections, cell wrinkling, and deformations of the nuclei.

For ophthalmic drug delivery, a suitable carrier system should fulfil not only suitable biopharmaceutical properties but also ocular tolerability. The developed ONS must have the capability to permeate/penetrate through the biological membrane easily, even they can accumulate in intracellular structures, and they can sustain at the site for the longer period. Due to their higher surface-to-volume ratio, they have different physicochemical properties than the same bulk material. It is necessary to assess the toxicity of ONS even though the utilized material/matrix is considered safe in its bulk form [24]. Without adequate regulatory guidance for ocular drugs, novel formulations can pose unique challenges to those involved in designing non-clinical programs and choosing species, strains, and ocular toxicity parameters.

Although nanomaterials offer several advantages helping to deliver the drug effectively at the site of action, their toxicity profile is of great concern. Thereby in vivo toxicity studies of formulated nanomaterials are an important parameter to be considered. For any delivery system to be clinically adoptable, it is desirable that they should be safe, tolerable, non-irritant, and efficacious. Their concentration should be estimated carefully as very high concentration may cause irritation and corneal damage. Ionic surfactants are toxic while non-ionic ones can be used but with proper screening of suitable concentration required [221, 222]. Suitable animal model and toxicity tests are mandatory to evaluate the nanotoxicity. Draize test implemented by US FDA in 1940 is used for the prediction of toxicity considering rabbit eye. Rabbit is a suitable animal model to perform the toxicity study, because its eye is more sensitive and susceptible for ONS [223]. However, it suffers from many drawbacks, such as that it takes large amount of time to give the results, high dose requirement and most importantly over prediction of human responses. To circumvent the disadvantages of Draize test several strategies have been demonstrated, such as Bovine Cornea Opacity Permeability assay (very fast method, gives results of permeability in 24 h), Low-volume eye-irritation test (less dose required; 0.01 ml/0.01 g), and ocular organotypic models (able to retain normal physiology of enucleated eye or cornea) [24]. In our laboratory, it was designed and developed a corneal device to study possible modifications in the corneal structures after application of nanomedicines or free drugs to the eye [224]. The habitual conditions for the permeation studies were modified to reproduce the behaviour when eye drops are administered to the eye, and for this reason, a new tetra-compartmental pharmacokinetic model was developed. In order to determine possible modifications of the corneal structures after the permeation studies, these corneas can be submitted to scanning microscopy.

Chitosan was the most studied polymer, with a number of reports that claim its low toxicity and good biocompatibility following topical administration to rabbits and also cell cultures [68, 79, 110, 225]. With regard to NSAIDs-loaded nanoparticles, there are few studies about in vivo toxicology and no clinical assays have been performed yet. Flurbiprofen-loaded nanoparticles of both polymers, PLGA [59] and Eudragit [74], were applied to rabbit eyes according to the Draize test [226], showing no sign of toxicity or irritation to ocular tissues. Diclofenac-loaded nanoparticles produced with PLGA, PACA or PCL were studied by a modified Draize test, with the in vivo results showing no sign of irritation or damaging effects to ocular tissues in rabbit eyes for as long as 24 h after application [227]. In an in vivo study using rabbit eye, Eudragit nanosuspension containing 0.1% ibuprofen showed a very good ocular tolerability [73, 228]. Moreover, in vitro cell viability studies and in vivo Draize test show non-irritant properties of Dexibuprofen PLGA-PEG nanospheres [229].

Considering the highly increasing interest in developing ocular nanomedicines, there is a need for in vitro and in vivo toxicity evaluation tests for precise understanding of the interaction of delivery system with the biological system at molecular, cellular, and tissue level. These interactions will be helpful in the estimation of biocompatibility, stability, biological performance and side effects of ONS [230]. It is useful to understand anatomical differences between species for interpreting toxicological and pathological responses to the eye and for human risk assessment of these new therapies.

16.5 Conclusions and Future Trends

Ocular diseases include various anterior and posterior segment diseases. The efficient ocular drug delivery is a great challenge to researchers and pharmacologists due to the unique anatomy and physiology of the eye. Although there are conventional non-invasive and invasive treatments, such as eye drops, injections and implants, the current treatments suffer from either low bioavailability or severe adverse ocular effects. The emerging nanoscience and nanotechnology are playing an important role in the development of novel strategies for ocular disease therapy, emerging as great advantage in ocular disease treatment since last two decades [24, 52].

In the past years, a large amount of research has been published dealing with nanomedicines for ocular NSAIDs delivery, but little work in vivo has been accomplished and no clinical studies have been performed yet. It is therefore required to establish correlations between human and animal studies. In general, the most used animal species in ocular field is the rabbit, but although the rabbit eye is comparable to the human eye in terms of size, it has differences such as higher surface sensitivity, lower tear production, lower blinking frequency and higher mucous production. All these factors could likely improve the mucoadhesion and reduce clearance from the ocular surface, which can increase the ocular residence time of systems and that could not be super-imposable in humans.

Several approaches have been proposed for ocular NSAIDs delivery namely, polymeric and lipid nanoparticles, liposomes, microemulsions and drug nanocrystals. Liposomes and cubosomes are very effective and biocompatible; however, they lack in physical stability. NE consists of a large amount of surfactant and co-surfactant which impart irritancy and intolerance. Niosomes and spanlastics are made of non-ionic surfactants, but still may pose the problem of ocular incompatibility. Nanoparticles and surface modified nanoparticles can permeate and retain in intracellular structures, which in the long term may be hazardous, cause irregularity in cell functioning, generate toxic metabolites or cause mutation. Nanotoxicity is the major restriction for usage of nanocarrier in ophthalmics. In these nanoformulations each carrier system has its unique properties. Lipid-based formulations were found to increase the loading and entrapment of active substance of either nature (hydrophilic or hydrophobic). Polymeric particles are suitable for the long-term ophthalmic therapy as they could sustain the release as well as action of loaded therapeutic bioactive. Pharmaceutical, biopharmaceutical and pharmacological characteristics of these nanocarriers are in contrast relies on formulation properties, such as particle size, particle surface charge, effective surface area, bio adhesion and stability that governs the overall biological performance of these carriers under the most dynamic ocular environment. A vibrant research is going on towards optimizing these nanocarrier systems for maximizing their ocular permeation efficiency to impart most effective ocular therapy. All the ONS have the potential to permeate the corneal layers; however, some of the drug is loss due to poor retention and nasolacrimal drainage. Thus, another hurdle in ocular therapy is how one can enhance the corneal permeability [24].

There is need of great understanding of design and development, pharmacokinetic and pharmacodynamics, toxicity, safety, and efficacy of ONS. There is great improvement needed in this aspect. Many researchers are working towards the development of safe and efficient ONS hoping for the future of nanotechnology-based ocular therapeutics [24]. Research focusing on in vivo and clinical studies is still a demand in the near future in order to bring successful clinical use for these nanomedicines.

Acknowledgements

The authors acknowledge the financial support received from Portuguese Science and Technology Foundation (FCT/MCT), Spanish Ministry of Science and Innovation (MAT 2014-59134-R) and from European Funds (PRODER/COMPETE) under the project reference M-ERA-NET/0004/2015-PAIRED, co-financed by FEDER, under the Partnership Agreement PT2020.

References

1. Y. Weng, J. Liu, S. Jin, W. Guo, X. Liang, Z. Hu, Nanotechnology-based strategies for treatment of ocular disease, *Acta Pharm. Sin. B*, 7 (2017) 281–291.
2. B. Hai-Zhi, J. G. Hovhannes, G. Lance, L. Xiao-Jing, X. Cathie, K. Tatiana, Ocular disposition, pharmacokinetics, efficacy and safety of nanoparticle-formulated ophthalmic drugs, *Curr. Drug Metab.*, 8 (2007) 91–107.
3. J. Alvarez-Trabado, Y. Diebold, A. Sanchez, Designing lipid nanoparticles for topical ocular drug delivery, *Int. J. Pharm.*, 532(1) (2017) 204–217.

4. J. F. Fangueiro, F. Veiga, A. M. Silva, E. B. Souto, Ocular drug delivery: New strategies for targeting anterior and posterior segments of the eye, *Curr. Pharm. Des.*, 22 (2016) 1135–1146.
5. E. Sanchez-Lopez, M. Espina, S. Doktorovova, E. B. Souto, M. L. Garcia, Lipid nanoparticles (SLN, NLC): Overcoming the anatomical and physiological barriers of the eye—Part II—Ocular drug-loaded lipid nanoparticles, *Eur. J. Pharm. Biopharm.*, 110 (2017) 58–69.
6. E. Sanchez-Lopez, M. Espina, S. Doktorovova, E. B. Souto, M. L. Garcia, Lipid nanoparticles (SLN, NLC): Overcoming the anatomical and physiological barriers of the eye—Part I—Barriers and determining factors in ocular delivery, *Eur. J. Pharm. Biopharm.*, 110 (2017) 70–75.
7. A. Patel, K. Cholkar, V. Agrahari, A. K. Mitra, Ocular drug delivery systems: An overview, *World J. Pharmacol.*, 2 (2013) 47–64.
8. H. Z. Bu, H. J. Gukasyan, L. Goulet, X. J. Lou, C. Xiang, T. Koudriakova, Ocular disposition, pharmacokinetics, efficacy and safety of nanoparticle-formulated ophthalmic drugs, *Curr. Drug Metab.*, 8 (2007) 91–107.
9. M. Li, M. Xin, G. Chuanlong, G. Lin, X. Wu, New nanomicelle curcumin formulation for ocular delivery: Improved stability, solubility, and ocular anti-inflammatory treatment, *Drug. Dev. Ind. Pharm.*, 43(11) (2017) 1846–1857.
10. A. Mandal, R. Bisht, I. D. Rupenthal, A. K. Mitra, Polymeric micelles for ocular drug delivery: From structural frameworks to recent preclinical studies, *J. Control. Release*, 248 (2017) 96–116.
11. J. C. Lang, Ocular drug delivery conventional ocular formulations, *Adv. Drug Deliv. Rev.*, 16 (1995) 39–43.
12. S. Reichl, C. Kölln, M. Hahne, J. Verstraelen, In vitro cell culture models to study the corneal drug absorption, *Expert Opin. Drug Metab. Toxicol.*, 7 (2011) 559–578.
13. I. P. Kaur, A. Garg, A. K. Singla, D. Aggarwal, Vesicular systems in ocular drug delivery: An overview, *Int. J. Pharm.*, 269 (2004) 1–14.
14. C. Le Bourlais, L. Acar, H. Zia, P. A. Sado, T. Needham, R. Leverge, Ophthalmic drug delivery systems—Recent advances, *Prog. Retin. Eye Res.*, 17 (1998) 33–58.
15. I. P. Kaur, A. Garg, A. K. Singla, D. Aggarwal, Vesicular systems in ocular drug delivery: An overview, *Int. J. Pharm.*, 269 (2004) 1–14.
16. C. L. Bourlais, L. Acar, H. Zia, P. A. Sado, T. Needham, R. Leverge, Ophthalmic drug delivery systems: Recent advances, *Prog. Retin. Eye. Res.*, 17 (1998) 33–58.

17. L. Romanelli, P. Valeri, L. A. Morrone, G. Pimpinella, G. Graziani, B. Tita, Ocular absorption and distribution of bendazac after topical administration to rabbits with different vehicles, *Life Sci.*, 54 (1994) 877–885.
18. T. Loftssona, T. Jarvinen, Cyclodextrins in ophthalmic drug delivery, *Adv. Drug Deliv. Rev.*, 36 (1999) 59–79.
19. I. P. Kaur, M. Kanwar, Ocular preparations: The formulation approach, *Drug. Dev. Ind. Pharm.*, 28 (2002) 473–493.
20. S. Ding, Recent development in ophthalmic drug delivery, *Pharm. Sci. Tech. Today*, 1 (1998) 328–335.
21. O. Kayser, A. Lemke, N. Hernandez-Trejo, The impact of nanobiotechnology on the development of new drug delivery systems, *Curr. Pharm. Biotechnol.*, 6 (2005) 3–5.
22. I. Brigger, C. Dubernet, P. Couvreur, Nanoparticles in cancer therapy and diagnosis, *Adv. Drug Deliv. Rev.*, 54 (2002) 631–651.
23. E. Addo, O. A. Bamiro, R. Siwale, Anatomy of the eye and common diseases affecting the eye, in: R. T. Addo (ed.) *Ocular Drug Delivery: Advances, Challenges and Applications*, Springer International Publishing, Cham, 2016, pp. 11–25.
24. L. Lalu, V. Tambe, D. Pradhan, K. Nayak, S. Bagchi, R. Maheshwari, K. Kalia, R. K. Tekade, Novel nanosystems for the treatment of ocular inflammation: Current paradigms and future research directions, *J. Control. Release*, 268 (2017) 19–39.
25. V. L. Perez, R. R. Caspi, Immune mechanisms in inflammatory and degenerative eye disease, *Trends Immunol.*, 36 (2015) 354–363.
26. P.-S. Launay, É. Reboussin, H. Liang, K. Kessal, D. Godefroy, W. Rostene, J. Sahel, C. Baudouin, S. Melik Parsadaniantz, A. Reaux Le Goazigo, Ocular inflammation induces trigeminal pain, peripheral and central neuroinflammatory mechanisms, *Neurobiol. Dis.*, 88 (2016) 16–28.
27. B. D. Srinivasan, P. S. Kulkarni, Inhibitors of the arachidonic acid cascade in the management of ocular inflammation, *Prog. Clin. Biol. Res.*, 312 (1989) 229–249.
28. P. Bhattacherjee, The role of arachidonate metabolites in ocular inflammation, *Prog. Clin. Biol. Res.*, 312 (1989) 211–227.
29. R. W. Lee, L. B. Nicholson, H. N. Sen, C.-C. Chan, L. Wei, R. B. Nussenblatt, A. D. Dick, Autoimmune and autoinflammatory mechanisms in uveitis, *Semin. Immunopathol.*, 36 (2014) 581–594.

30. A. T. Giannico, L. Lima, G. C. Shaw, H. H. A. Russ, T. R. Froes, F. Montiani-Ferreira, Effects of prostaglandin analogs on blood flow velocity and resistance in the ophthalmic artery of rabbits, *Arquivos Brasileiros de Oftalmologia*, 79 (2016) 33–36.
31. A. J. Flach, Cyclo-oxygenase inhibitors in ophthalmology, *Surv. Ophthalmol.*, 36 (1992) 259–284.
32. J. S. Heier, T. M. Topping, W. Baumann, M. S. Dirks, S. Chern, Ketorolac versus prednisolone versus combination therapy in the treatment of acute pseudophakic cystoid macular edema, *Ophthalmology*, 107 (2000) 2034–2038;discussion 2039.
33. T. P. O'Brien, Emerging guidelines for use of NSAID therapy to optimize cataract surgery patient care, *Curr. Med. Res. Opin.*, 21 (2005) 1131–1137.
34. K.-H. Ruan, V. Cervantes, S.-P. So, Engineering of a novel hybrid enzyme: An anti-inflammatory drug target with triple catalytic activities directly converting arachidonic acid into the inflammatory prostaglandin E(2), *Protein Eng. Des. Sel.*, 22 (2009) 733–740.
35. A. J. Flach, Topical nonsteroidal antiinflammatory drugs in ophthalmology, *Int. Ophthalmol. Clin.*, 42 (2002) 1–11.
36. K. Miyake, K. Masuda, S. Shirato, T. Oshika, K. Eguchi, H. Hoshi, Y. Majima, W. Kimura, F. Hayashi, Comparison of diclofenac and fluorometholone in preventing cystoid macular edema after small incision cataract surgery: A multicentered prospective trial, *Jpn. J. Ophthalmol.*, 44 (2000) 58–67.
37. M. O. Price, F. W. Price, Efficacy of topical ketorolac tromethamine 0.4% for control of pain or discomfort associated with cataract surgery, *Curr. Med. Res. Opin.*, 20 (2004) 2015–2019.
38. T. Walters, M. Raizman, P. Ernest, J. Gayton, R. Lehmann, In vivo pharmacokinetics and in vitro pharmacodynamics of nepafenac, amfenac, ketorolac, and bromfenac, *J. Cataract. Refract. Surg.*, 33 (2007) 1539–1545.
39. D. A. Gamache, G. Graff, M. T. Brady, J. M. Spellman, J. M. Yanni, Nepafenac, a unique nonsteroidal prodrug with potential utility in the treatment of trauma-induced ocular inflammation: I. Assessment of anti-inflammatory efficacy, *Inflammation*, 24 (2000) 357–370.
40. P. Duan, Y. Liu, J. Li, The comparative efficacy and safety of topical non-steroidal anti-inflammatory drugs for the treatment of anterior chamber inflammation after cataract surgery: A systematic review

and network meta-analysis, *Graefes Arch. Clin. Exp. Ophthalmol.*, 255 (2017) 639–649.

41. A. Rothova, Inflammatory cystoid macular edema, *Curr. Opin. Ophthalmol.*, 18 (2007) 487–492.
42. M. A. Kamaleddin, Nano-ophthalmology: Applications and considerations, *Nanomed. Nanotechnol. Biol. Med.*, 13 (2017) 1459–1472.
43. N. Soni, N. Soni, H. Pandey, R. Maheshwari, P. Kesharwani, R. K. Tekade, Augmented delivery of gemcitabine in lung cancer cells exploring mannose anchored solid lipid nanoparticles, *J. Colloid Interface Sci.*, 481 (2016) 107–116.
44. N. Morsi, M. Ibrahim, H. Refai, H. El Sorogy, Nanoemulsion-based electrolyte triggered in situ gel for ocular delivery of acetazolamide, *Eur. J. Pharm. Sci.*, 104 (2017) 302–314.
45. B. Gorain, H. Choudhury, R. K. Tekade, S. Karan, P. Jaisankar, T. K. Pal, Comparative biodistribution and safety profiling of olmesartan medoxomil oil-in-water oral nanoemulsion, *Regul. Toxicol. Pharmacol.*, 82 (2016) 20–31.
46. Y. Huang, Q. Tao, D. Hou, S. Hu, S. Tian, Y. Chen, R. Gui, L. Yang, Y. Wang, A novel ion-exchange carrier based upon liposome-encapsulated montmorillonite for ophthalmic delivery of betaxolol hydrochloride, *Int. J. Nanomed.*, 12 (2017) 1731–1745.
47. R. G. S. Maheshwari, S. Thakur, S. Singhal, R. P. Patel, M. Tekade, R. K. Tekade, Chitosan encrusted nonionic surfactant based vesicular formulation for topical administration of ofloxacin, *Sci. Adv. Mater.*, 7 (2015) 1163–1176.
48. J. Huang, T. Peng, Y. Li, Z. Zhan, Y. Zeng, Y. Huang, X. Pan, C.-Y. Wu, C. Wu, Ocular cubosome drug delivery system for timolol maleate: Preparation, characterization, cytotoxicity, ex vivo, and in vivo evaluation, *AAPS PharmSciTech*, 18 (2017) 2919–2926.
49. A. M. Al-Mahallawi, O. M. Khowessah, R. A. Shoukri, Enhanced non invasive trans-tympanic delivery of ciprofloxacin through encapsulation into nano-spanlastic vesicles: Fabrication, in-vitro characterization, and comparative ex-vivo permeation studies, *Int. J. Pharm.*, 522 (2017) 157–164.
50. H. Almeida, P. Lobão, C. Frigerio, J. Fonseca, R. Silva, J. M. Sousa Lobo, M. H. Amaral, Preparation, characterization and biocompatibility studies of thermoresponsive eyedrops based on the combination of nanostructured lipid carriers (NLC) and the polymer Pluronic F-127

for controlled delivery of ibuprofen, *Pharm. Dev. Technol.*, 22 (2017) 336–349.

51. H. Chi, Y. Gu, T. Xu, F. Cao, Multifunctional organic–inorganic hybrid nanoparticles and nanosheets based on chitosan derivative and layered double hydroxide: Cellular uptake mechanism and application for topical ocular drug delivery, *Int. J. Nanomed.*, 12 (2017) 1607–1620.
52. Y. Weng, J. Liu, S. Jin, W. Guo, X. Liang, Z. Hu, Nanotechnology-based strategies for treatment of ocular disease, *Acta Pharm Sin B.*, 7(3) (2017) 281–291.
53. R. Ghanghoria, R. K. Tekade, A. K. Mishra, K. Chuttani, N. K. Jain, Luteinizing hormone-releasing hormone peptide tethered nanoparticulate system for enhanced antitumoral efficacy of paclitaxel, *Nanomedicine*, 11 (2016) 797–816.
54. R. M. Mainardes, L. P. Silva, Drug delivery systems: Past, present, and future, *Curr. Drug Targets*, 5 (2004) 449–455.
55. U. Ticho, M. Blumenthal, S. Zonis, A. Gal, I. Blank, Z. W. Mazor, A clinical trial with Piloplex: A new long-acting pilocarpine compound: Preliminary report, *Ann. Ophthalmol.*, 11 (1979) 555–561.
56. G. G. Giordano, P. Chevez-Barrios, M. F. Refojo, C. A. Garcia, Biodegradation and tissue reaction to intravitreous biodegradable poly(D,L-lactic-co-glycolic)acid microspheres, *Curr. Eye Res.*, 14 (1995) 761–768.
57. C. Giannavola, C. Bucolo, A. Maltese, D. Paolino, M. A. Vandelli, G. Puglisi, V. H. Lee, M. Fresta, Influence of preparation conditions on acyclovir-loaded poly-d,l-lactic acid nanospheres and effect of PEG coating on ocular drug bioavailability, *Pharm. Res.*, 20 (2003) 584–590.
58. S. K. Das, I. G. Tucker, D. J. Hill, N. Ganguly, Evaluation of poly(isobutylcyanoacrylate) nanoparticles for mucoadhesive ocular drug delivery. I. Effect of formulation variables on physicochemical characteristics of nanoparticles, *Pharm. Res.*, 12 (1995) 534–540.
59. E. Vega, M. A. Egea, O. Valls, M. Espina, M. L. Garcia, Flurbiprofen loaded biodegradable nanoparticles for ophtalmic administration, *J. Pharm. Sci.*, 95 (2006) 2393–2405.
60. C. Canadas, H. Alvarado, A. C. Calpena, A. M. Silva, E. B. Souto, M. L. Garcia, G. Abrego, In vitro, ex vivo and in vivo characterization of PLGA nanoparticles loading pranoprofen for ocular administration, *Int. J. Pharm.*, 511 (2016) 719–727.
61. G. Abrego, H. Alvarado, E. B. Souto, B. Guevara, L. H. Bellowa, A. Parra, A. Calpena, M. L. Garcia, Biopharmaceutical profile of

pranoprofen-loaded PLGA nanoparticles containing hydrogels for ocular administration, *Eur. J. Pharm. Biopharm.*, 95 (2015) 261–270.

62. E. Sanchez-Lopez, M. A. Egea, B. M. Davis, L. Guo, M. Espina, A. M. Silva, A. C. Calpena, E. M. B. Souto, N. Ravindran, M. Ettcheto, A. Camins, M. L. Garcia, M. F. Cordeiro, Memantine-loaded PEGylated biodegradable nanoparticles for the treatment of glaucoma, *Small*, 14 (2018) 1701808.
63. A. Vasconcelos, E. Vega, Y. Perez, M. J. Gomara, M. L. Garcia, I. Haro, Conjugation of cell-penetrating peptides with poly(lactic-co-glycolic acid)-polyethylene glycol nanoparticles improves ocular drug delivery, *Int. J. Nanomed.*, 10 (2015) 609–631.
64. E. Vega, M. A. Egea, M. L. Garduno-Ramirez, M. L. Garcia, E. Sanchez, M. Espina, A. C. Calpena, Flurbiprofen PLGA-PEG nanospheres: Role of hydroxy-beta-cyclodextrin on ex vivo human skin permeation and in vivo topical anti-inflammatory efficacy, *Colloids Surf., B*, 110 (2013) 339–346.
65. E. Vega, M. A. Egea, A. C. Calpena, M. Espina, M. L. Garcia, Role of hydroxypropyl-beta-cyclodextrin on freeze-dried and gamma-irradiated PLGA and PLGA-PEG diblock copolymer nanospheres for ophthalmic flurbiprofen delivery, *Int. J. Nanomed.*, 7 (2012) 1357–1371.
66. E. Vega, F. Gamisans, M. L. Garcia, A. Chauvet, F. Lacoulonche, M. A. Egea, PLGA nanospheres for the ocular delivery of flurbiprofen: Drug release and interactions, *J. Pharm. Sci.*, 97 (2008) 5306–5317.
67. C. Losa, L. Marchal-Heussler, F. Orallo, J. L. Vila Jato, M. J. Alonso, Design of new formulations for topical ocular administration: Polymeric nanocapsules containing metipranolol, *Pharm. Res.*, 10 (1993) 80–87.
68. M. J. Alonso, A. Sanchez, The potential of chitosan in ocular drug delivery, *J. Pharm. Pharmacol.*, 55 (2003) 1451–1463.
69. J. Vandervoort, A. Ludwig, Preparation and evaluation of drug-loaded gelatin nanoparticles for topical ophthalmic use, *Eur. J. Pharm. Biopharm.*, 57 (2004) 251–261.
70. J. M. Irache, M. Merodio, A. Arnedo, M. A. Camapanero, M. Mirshahi, S. Espuelas, Albumin nanoparticles for the intravitreal delivery of anticytomegaloviral drugs, *Mini Rev. Med. Chem.*, 5 (2005) 293–305.
71. P. Calvo, J. L. Vila-Jato, M. J. Alonso, Comparative in vitro evaluation of several colloidal systems, nanoparticles, nanocapsules, and nanoemulsions, as ocular drug carriers, *J. Pharm. Sci.*, 85 (1996) 530–536.

72. R. Pignatello, C. Bucolo, G. Puglisi, Ocular tolerability of Eudragit RS100 and RL100 nanosuspensions as carriers for ophthalmic controlled drug delivery, *J. Pharm. Sci.*, 91 (2002) 2636–2641.

73. R. Pignatello, C. Bucolo, P. Ferrara, A. Maltese, A. Puleo, G. Puglisi, Eudragit RS100 nanosuspensions for the ophthalmic controlled delivery of ibuprofen, *Eur. J. Pharm. Sci.*, 16 (2002) 53–61.

74. R. Pignatello, C. Bucolo, G. Spedalieri, A. Maltese, G. Puglisi, Flurbiprofen-loaded acrylate polymer nanosuspensions for ophthalmic application, *Biomaterials*, 23 (2002) 3247–3255.

75. K. Adibkia, M. R. Siahi Shadbad, A. Nokhodchi, A. Javadzedeh, M. Barzegar-Jalali, J. Barar, G. Mohammadi, Y. Omidi, Piroxicam nanoparticles for ocular delivery: Physicochemical characterization and implementation in endotoxin-induced uveitis, *J. Drug Targeting*, 15 (2007) 407–416.

76. C. Gomez-Gaete, N. Tsapis, M. Besnard, A. Bochot, E. Fattal, Encapsulation of dexamethasone into biodegradable polymeric nanoparticles, *Int. J. Pharm.*, 331 (2007) 153–159.

77. J. Valero, M. A. Egea, M. Espina, F. Gamisans, M. L. Garcia, Effect of polymerization coadjuvants on nancapsule elaboration and triamcinolone entrapment., *Drug. Dev. Ind. Pharm.*, 22 (1996) 167–173.

78. J. Araujo, E. Vega, C. Lopes, M. A. Egea, M. L. Garcia, E. B. Souto, Effect of polymer viscosity on physicochemical properties and ocular tolerance of FB-loaded PLGA nanospheres, *Colloids Surf., B*, 72 (2009) 48–56.

79. A. A. Badawi, H. M. El-Laithy, R. K. El Qidra, H. El Mofty, M. El Dally, Chitosan based nanocarriers for indomethacin ocular delivery, *Arch. Pharm. Res.*, 31 (2008) 1040–1049.

80. F. Lallemand, O. Felt-Baeyens, K. Besseghir, F. Behar-Cohen, R. Gurny, Cyclosporine A delivery to the eye: A pharmaceutical challenge, *Eur. J. Pharm. Biopharm.*, 56 (2003) 307–318.

81. W. Chaiyasan, S. P. Srinivas, W. Tiyaboonchai, Crosslinked chitosan-dextran sulfate nanoparticle for improved topical ocular drug delivery, *Mol. Vision*, 21 (2015) 1224–1234.

82. I. Bravo-Osuna, V. Andrés-Guerrero, P. Pastoriza Abal, I. T. Molina-Martínez, R. Herrero-Vanrell, Pharmaceutical microscale and nanoscale approaches for efficient treatment of ocular diseases, *Drug Deliv. Transl. Res.*, 6 (2016) 686–707.

83. Z. Liu, X. Zhang, H. Wu, J. Li, L. Shu, R. Liu, L. Li, N. Li, Preparation and evaluation of solid lipid nanoparticles of baicalin for ocular drug

delivery system in vitro and in vivo, *Drug Dev. Ind. Pharm.*, 37 (2011) 475–481.

84. R. Li, S. Jiang, D. Liu, X. Bi, F. Wang, Q. Zhang, Q. Xu, A potential new therapeutic system for glaucoma: Solid lipid nanoparticles containing methazolamide, *J. Microencapsulation*, 28 (2011) 134–141.
85. J. Hao, X. Wang, Y. Bi, Y. Teng, J. Wang, F. Li, Q. Li, J. Zhang, F. Guo, J. Liu, Fabrication of a composite system combining solid lipid nanoparticles and thermosensitive hydrogel for challenging ophthalmic drug delivery, *Colloids Surf., B*, 114 (2014) 111–120.
86. W. Wang, J. Despanie, P. Shi, M. C. Edman-Woolcott, Y.-A. Lin, H. Cui, J. M. Heur, M. E. Fini, S. F. Hamm-Alvarez, J. A. MacKay, Lacritin-mediated regeneration of the corneal epithelia by protein polymer nanoparticles, *J. Mater. Chem. B*, 2 (2014) 8131–8141.
87. A. Trapani, E. De Giglio, D. Cafagna, N. Denora, G. Agrimi, T. Cassano, S. Gaetani, V. Cuomo, G. Trapani, Characterization and evaluation of chitosan nanoparticles for dopamine brain delivery, *Int. J. Pharm.*, 419 (2011) 296–307.
88. S. Wadhwa, R. Paliwal, S. R. Paliwal, S. P. Vyas, Hyaluronic acid modified chitosan nanoparticles for effective management of glaucoma: Development, characterization, and evaluation, *J. Drug Targeting*, 18 (2010) 292–302.
89. E. H. Gökçe, G. Sandri, S. Eğrilmez, M. C. Bonferoni, T. Güneri, C. Caramella, Cyclosporine A-loaded solid lipid nanoparticles: Ocular tolerance and in vivo drug release in rabbit eyes, *Curr. Eye Res.*, 34 (2009) 996–1003.
90. E. Başaran, E. Yenilmez, M. S. Berkman, G. Büyükköroğlu, Y. Yazan, Chitosan nanoparticles for ocular delivery of cyclosporine A, *J. Microencapsulation*, 31 (2014) 49–57.
91. M. Kreilgaard, Dermal pharmacokinetics of microemulsion formulations determined by in vivo microdialysis, *Pharm. Res.*, 18 (2001) 367–373.
92. M. Kreilgaard, Influence of microemulsions on cutaneous drug delivery, *Adv. Drug Deliv. Rev.*, 54 Suppl 1 (2002) S77–98.
93. R. Gaudana, J. Jwala, S. H. Boddu, A. K. Mitra, Recent perspectives in ocular drug delivery, *Pharm. Res.*, 26 (2009) 1197–1216..
94. V. B. Patravale, A. A. Date, R. M. Kulkarni, Nanosuspensions: a promising drug delivery strategy, *J. Pharm. Pharmacol.*, 56 (2004) 827–840.
95. W. Warisnoicharoen, A. B. Lansley, M. J. Lawrence, Nonionic oil-in-water microemulsions: the effect of oil type on phase behaviour, *Int. J. Pharm.*, 198 (2000) 7–27.

96. J. Chan, G. M. Maghraby, J. P. Craig, R. G. Alany, Phase transition water-in-oil microemulsions as ocular drug delivery systems: in vitro and in vivo evaluation, *Int. J. Pharm.*, 328 (2007) 65–71.
97. F. F. Lv, L. Q. Zheng, C. H. Tung, Phase behavior of the microemulsions and the stability of the chloramphenicol in the microemulsion-based ocular drug delivery system, *Int. J. Pharm.*, 301 (2005) 237–246.
98. L. Baydoun, P. Furrer, R. Gurny, C. C. Muller-Goymann, New surface-active polymers for ophthalmic formulations: Evaluation of ocular tolerance, *Eur. J. Pharm. Biopharm.*, 58 (2004) 169–175.
99. S. Klang, M. Abdulrazik, S. Benita, Influence of emulsion droplet surface charge on indomethacin ocular tissue distribution, *Pharm. Dev. Technol.*, 5 (2000) 521–532.
100. C.-C. Li, M. Abrahamson, Y. Kapoor, A. Chauhan, Timolol transport from microemulsions trapped in HEMA gels, *J. Colloid Interface Sci.*, 315 (2007) 297–306.
101. M. J. Lawrence, G. D. Rees, Microemulsion-based media as novel drug delivery systems, *Adv. Drug Deliv. Rev.*, 45 (2000) 89–121.
102. J. Chan, G. M. E. Maghraby, J. P. Craig, R. G. Alany, Effect of water-in-oil microemulsions and lamellar liquid crystalline systems on the precorneal tear film of albino New Zealand rabbits, *Clin. Ophthalmol.*, 2 (2008) 129–138.
103. J. Chan, G. M. M. E. Maghraby, J. P. Craig, R. G. Alany, Phase transition water-in-oil microemulsions as ocular drug delivery systems: In vitro and in vivo evaluation, *Int. J. Pharm.*, 328 (2007) 65–71.
104. R. G. Alany, T. Rades, J. Nicoll, I. G. Tucker, N. M. Davies, W/O microemulsions for ocular delivery: Evaluation of ocular irritation and precorneal retention, *J. Control. Release*, 111 (2006) 145–152.
105. S. L. Fialho, A. da Silva-Cunha, New vehicle based on a microemulsion for topical ocular administration of dexamethasone, *Clin. Exp. Ophthalmol.*, 32 (2004) 626–632.
106. M. C. Teixeira, C. Carbone, E. B. Souto, Beyond liposomes: Recent advances on lipid based nanostructures for poorly soluble/poorly permeable drug delivery, *Prog. Lipid Res.*, 68 (2017) 1–11.
107. Y. Fan, Q. Zhang, Development of liposomal formulations: From concept to clinical investigations, *Asian J. Pharm. Sci.*, 8 (2013) 81–87.
108. J. Zhang, X. Liang, X. Li, Z. Guan, Z. Liao, Y. Luo, Y. Luo, Ocular delivery of cyanidin-3-glycoside in liposomes and its prevention of selenite-induced oxidative stress, *Drug Dev. Ind. Pharm.*, 42 (2016) 546–553.

109. M. Kanamala, W. R. Wilson, M. Yang, B. D. Palmer, Z. Wu, Mechanisms and biomaterials in pH-responsive tumour targeted drug delivery: A review, *Biomaterials*, 85 (2016) 152–167.
110. Y. Diebold, M. Jarrin, V. Saez, E. L. Carvalho, M. Orea, M. Calonge, B. Seijo, M. J. Alonso, Ocular drug delivery by liposome-chitosan nanoparticle complexes (LCS-NP), *Biomaterials*, 28 (2007) 1553–1564.
111. A. Zimmer, E. Mutschler, G. Lambrecht, D. Mayer, J. Kreuter, Pharmacokinetic and pharmacodynamic aspects of an ophthalmic pilocarpine nanoparticle-delivery-system, *Pharm. Res.*, 11 (1994) 1435–1442.
112. J. Al-Muhammed, A. Y. Ozer, M. T. Ercan, A. A. Hincal, In-vivo studies on dexamethasone sodium phosphate liposomes, *J. Microencapsulation*, 13 (1996) 293–306.
113. S. K Dharma, P. H Fishman, G. Peyman, A preliminary study of corneal penetration of 125I-labelled idoxuridine liposome, *Acta Ophthalmol. (Copenh.)*, 64 (1986) 298–301..
114. R. Zhang, R. He, J. Qian, J. Guo, K. Xue, Y.-F. Yuan, Treatment of experimental autoimmune uveoretinitis with intravitreal injection of tacrolimus (FK506) encapsulated in liposomes, *Invest. Ophthalmol. Visual Sci.*, 51 (2010) 3575–3582.
115. I. K. Gipson, Distribution of mucins at the ocular surface, *Exp. Eye Res.*, 78 (2004) 379–388.
116. H. E. Schaeffer, D. L. Krohn, Liposomes in topical drug delivery, *Invest. Ophthalmol. Visual Sci.*, 22 (1982) 220–227.
117. S. L. Law, H. Y. Hung, Properties of acyclovir-containing liposomes for potential ocular delivery, *Int. J. Pharm.*, 161 (1998) 253–259.
118. P. Chetoni, S. Rossi, S. Burgalassi, D. Monti, S. Mariotti, M. F. Saettone, Comparison of liposome-encapsulated acyclovir with acyclovir ointment: Ocular pharmacokinetics in rabbits, *J. Ocul. Pharmacol. Ther.*, 20 (2004) 169–177.
119. O. N. El-Gazayerly, A. H. Hikal, Preparation and evaluation of acetazolamide liposomes as an ocular delivery system, *Int. J. Pharm.*, 158 (1997) 121–127.
120. H. Li, Y. Liu, Y. Zhang, D. Fang, B. Xu, L. Zhang, T. Chen, K. Ren, Y. Nie, S. Yao, X. Song, Liposomes as a novel ocular delivery system for brinzolamide: In vitro and in vivo studies, *AAPS PharmSciTech*, 17 (2016) 710–717.
121. U. Pleyer, B. Elkins, D. Rückert, S. Lutz, J. Grammer, J. Chou, K.-H. Schmidt, B. J. Mondino, Ocular absorption of cyclosporine A from

liposomes incorporated into collagen shields, *Curr. Eye Res.*, 13 (1994) 177–181.

122. D. Quinteros, M. Vicario-de-la-Torre, V. Andrés-Guerrero, S. Palma, D. Allemandi, R. Herrero-Vanrell, I. T. Molina-Martínez, Hybrid formulations of liposomes and bioadhesive polymers improve the hypotensive effect of the melatonin analogue 5-MCA-NAT in rabbit eyes, *PLoS One*, 9 (2014) e110344.
123. D. C. Broadway, I. Grierson, C. O'Brien, R. A. Hitchings, Adverse effects of topical antiglaucoma medication: I. The conjunctival cell profile, *Arch. Ophthalmol.*, 112 (1994) 1437–1445.
124. I. R. Schwab, J. V. Linberg, V. M. Gioia, W. H. Benson, G. M. Chao, Foreshortening of the inferior conjunctival fornix associated with chronic glaucoma medications, *Ophthalmology*, 99 197–202.
125. R. Nuzzi, C. Finazzo, A. Vercelli, C. Cracco, Conjunctiva and subconjunctival tissue in primary open-angle glaucoma after long-term topical treatment: An immunohistochemical and ultrastructural study, *Graefes Arch. Clin. Exp. Ophthalmol.*, 233 (1995) 154–162.
126. Y. Dong, P. Dong, D. Huang, L. Mei, Y. Xia, Z. Wang, X. Pan, G. Li, C. Wu, Fabrication and characterization of silk fibroin-coated liposomes for ocular drug delivery, *Eur. J. Pharm. Biopharm.*, 91 (2015) 82–90.
127. A. Bochot, E. Fattal, Liposomes for intravitreal drug delivery: A state of the art, *J. Control. Release*, 161 (2012) 628–634.
128. J. P. Cannon, R. Fiscella, S. Pattharachayakul, K. W. Garey, F. De Alba, S. Piscitelli, D. P. Edward, L. H. Danziger, Comparative toxicity and concentrations of intravitreal amphotericin B formulations in a rabbit model, *Invest. Ophthalmol. Visual Sci.*, 44 (2003) 2112–2117.
129. K. R. Liu, G. A. Peyman, B. Khoobehi, Efficacy of liposome-bound amphotericin B for the treatment of experimental fungal endophthalmitis in rabbits, *Invest. Ophthalmol. Visual Sci.*, 30 (1989) 1527–1534.
130. L. Cheng, K. Y. Hostetler, S. Chaidhawangul, M. F. Gardner, J. R. Beadle, K. S. Keefe, G. Bergeron–Lynn, G. M. Severson, K. A. Soules, A. J. Mueller, W. R. Freeman, Intravitreal toxicology and duration of efficacy of a novel antiviral lipid prodrug of ganciclovir in liposome formulation, *Invest. Ophthalmol. Visual Sci.*, 41 (2000) 1523–1532.
131. A. l. Bochot, E. Fattal, V. R. Boutet, J. R. Deverre, J. C. Jeanny, H. l. N. Chacun, P. Couvreur, Intravitreal delivery of oligonucleotides by sterically stabilized liposomes, *Invest. Ophthalmol. Visual Sci.*, 43 (2002) 253–259.

132. L. Lajavardi, S. Camelo, F. Agnely, W. Luo, B. Goldenberg, M.-C. Naud, F. Behar-Cohen, Y. de Kozak, A. Bochot, New formulation of vasoactive intestinal peptide using liposomes in hyaluronic acid gel for uveitis, *J. Control. Release*, 139 (2009) 22–30.
133. S. Asteriti, G. Dal Cortivo, V. Pontelli, L. Cangiano, M. Buffelli, D. Dell'Orco, Effective delivery of recombinant proteins to rod photoreceptors via lipid nanovesicles, *Biochem. Biophys. Res. Commun.*, 461 (2015) 665–670.
134. R. H. Muller, K. Mader, S. Gohla, Solid lipid nanoparticles (SLN) for controlled drug delivery: A review of the state of the art, *Eur. J. Pharm. Biopharm.*, 50 (2000) 161–177.
135. R. H. Muller, R. D. Petersen, A. Hommoss, J. Pardeike, Nanostructured lipid carriers (NLC) in cosmetic dermal products, *Adv. Drug Deliv. Rev.*, 59 (2007) 522–530.
136. R. H. Muller, M. Radtke, S. A. Wissing, Solid lipid nanoparticles (SLN) and nanostructured lipid carriers (NLC) in cosmetic and dermatological preparations, *Adv. Drug Deliv. Rev.*, 54 Suppl 1 (2002) S131–155.
137. R. Arora, S. Katiyar, V. Kushwah, S. Jain, Solid lipid nanoparticles and nanostructured lipid carrier-based nanotherapeutics in treatment of psoriasis: A comparative study, *Expert Opin. Drug Deliv.*, 14 (2017) 165–177.
138. G. Sandri, S. Motta, M. C. Bonferoni, P. Brocca, S. Rossi, F. Ferrari, V. Rondelli, L. Cantù, C. Caramella, E. Del Favero, Chitosan-coupled solid lipid nanoparticles: Tuning nanostructure and mucoadhesion, *Eur. J. Pharm. Biopharm.*, 110 (2017) 13–18.
139. E. B. Souto, W. Mehnert, R. H. Muller, Polymorphic behaviour of Compritol888 ATO as bulk lipid and as SLN and NLC, *J. Microencapsulation*, 23 (2006) 417–433.
140. E. B. Souto, R. H. Muller, Investigation of the factors influencing the incorporation of clotrimazole in SLN and NLC prepared by hot high-pressure homogenization, *J. Microencapsulation*, 23 (2006) 377–388.
141. A. A. Attama, B. C. Schicke, C. C. Muller-Goymann, Further characterization of theobroma oil-beeswax admixtures as lipid matrices for improved drug delivery systems, *Eur. J. Pharm. Biopharm.*, 64 (2006) 294–306.
142. A. A. Attama, S. Reichl, C. C. Muller-Goymann, Diclofenac sodium delivery to the eye: In vitro evaluation of novel solid lipid nanoparticle

formulation using human cornea construct, *Int. J. Pharm.*, 355 (2008) 307–313.

143. A. Beloqui, M. Á. Solinís, A. Rodríguez-Gascón, A. J. Almeida, V. Préat, Nanostructured lipid carriers: Promising drug delivery systems for future clinics, *Nanomed. Nanotechnol. Biol. Med.*, 12 143–161.
144. X. Li, S.-f. Nie, J. Kong, N. Li, C.-y. Ju, W.-s. Pan, A controlled-release ocular delivery system for ibuprofen based on nanostructured lipid carriers, *Int. J. Pharm.*, 363 (2008) 177–182.
145. E. Sánchez-López, M. Espina, S. Doktorovova, E. B. Souto, M. L. García, Lipid nanoparticles (SLN, NLC): Overcoming the anatomical and physiological barriers of the eye – Part II - Ocular drug-loaded lipid nanoparticles, *Eur. J. Pharm. Biopharm.*, 110 (2017) 58–69.
146. E. Gonzalez-Mira, M. A. Egea, M. L. Garcia, E. B. Souto, Design and ocular tolerance of flurbiprofen loaded ultrasound-engineered NLC, *Colloids Surf., B*, 81 (2010) 412–421.
147. E. Gonzalez-Mira, M. A. Egea, E. B. Souto, A. C. Calpena, M. L. García, Optimizing flurbiprofen-loaded NLC by central composite factorial design for ocular delivery, *Nanotechnology*, 22 (2011) 045101.
148. E. Gonzalez-Mira, S. Nikolić, A. C. Calpena, M. A. Egea, E. B. Souto, M. L. García, Improved and safe transcorneal delivery of flurbiprofen by NLC and NLC-based hydrogels, *J. Pharm. Sci.*, 101 707–725.
149. Q. Luo, J. Zhao, X. Zhang, W. Pan, Nanostructured lipid carrier (NLC) coated with chitosan oligosaccharides and its potential use in ocular drug delivery system, *Int. J. Pharm.*, 403 (2011) 185–191.
150. K. Hippalgaonkar, G. R. Adelli, K. Hippalgaonkar, M. A. Repka, S. Majumdar, Indomethacin-loaded solid lipid nanoparticles for ocular delivery: Development, characterization, and in vitro evaluation, *J. Ocul. Pharmacol. Ther.*, 29 (2013) 216–228.
151. C. M. Keck, R. H. Muller, Drug nanocrystals of poorly soluble drugs produced by high pressure homogenisation, *Eur. J. Pharm. Biopharm.*, 62 (2006) 3–16.
152. M. Malamatari, K. M. G. Taylor, S. Malamataris, D. Douroumis, K. Kachrimanis, Pharmaceutical nanocrystals: Production by wet milling and applications, *Drug Discovery Today*, 23 (2018) 534–547.
153. M. Trotta, M. Gallarate, F. Pattarino, S. Morel, Emulsions containing partially water-miscible solvents for the preparation of drug nanosuspensions, *J. Control. Release*, 76 (2001) 119–128.

154. E. Merisko-Liversidge, G. G. Liversidge, E. R. Cooper, Nanosizing: A formulation approach for poorly-water-soluble compounds, *Eur. J. Pharm. Sci.*, 18 (2003) 113–120.
155. R. H. Muller, C. Jacobs, O. Kayser, Nanosuspensions as particulate drug formulations in therapy. Rationale for development and what we can expect for the future, *Adv. Drug Deliv. Rev.*, 47 (2001) 3–19.
156. J. E. Kipp, The role of solid nanoparticle technology in the parenteral delivery of poorly water-soluble drugs, *Int. J. Pharm.*, 284 (2004) 109–122.
157. B. Van Eerdenbrugh, L. Froyen, J. Van Humbeeck, J. A. Martens, P. Augustijns, G. Van den Mooter, Drying of crystalline drug nanosuspensions-the importance of surface hydrophobicity on dissolution behavior upon redispersion, *Eur. J. Pharm. Sci.*, 35 (2008) 127–135.
158. O. P. Sharma, V. Patel, T. Mehta, Nanocrystal for ocular drug delivery: Hope or hype, *Drug Deliv. Transl. Res.*, 6 (2016) 399–413.
159. S. Gupta, M. K. Samanta, A. M. Raichur, Dual-drug delivery system based on in situ gel-forming nanosuspension of forskolin to enhance antiglaucoma efficacy, *AAPS PharmSciTech*, 11 (2010) 322–335.
160. A. Tuomela, P. Liu, J. Puranen, S. Rönkkö, T. Laaksonen, G. Kalesnykas, O. Oksala, J. Ilkka, J. Laru, K. Järvinen, J. Hirvonen, L. Peltonen, Brinzolamide nanocrystal formulations for ophthalmic delivery: Reduction of elevated intraocular pressure in vivo, *Int. J. Pharm.*, 467 (2014) 34–41.
161. Y. Megumi Nishitani, K. Edna Tomiko Myiake, L. Raimar, B.-C. Nadia Araci, Challenges and future prospects of nanoemulsion as a drug delivery system, *Curr. Pharm. Des.*, 23 (2017) 495–508.
162. P. Daull, M. Amrane, J.-S. Garrigue, Novasorb® cationic nanoemulsion and latanoprost: The ideal combination for glaucoma management?, *Glauc. Open Access*, 2 (2017) 107.
163. D. Aggarwal, A. Garg, I. P. Kaur, Development of a topical niosomal preparation of acetazolamide: Preparation and evaluation, *J. Pharm. Pharmacol.*, 56 (2004) 1509–1517.
164. I. P. Kaur, D. Aggarwal, H. Singh, S. Kakkar, Improved ocular absorption kinetics of timolol maleate loaded into a bioadhesive niosomal delivery system, *Graefes Arch. Clin. Exp. Ophthalmol.*, 248 (2010) 1467–1472.
165. H. Abdelkader, S. Ismail, A. Kamal, R. G. Alany, Design and evaluation of controlled-release niosomes and discomes for naltrexone hydrochloride ocular delivery, *J. Pharm. Sci.*, 100 (2011) 1833–1846.

166. H. Abdelkader, S. Ismail, A. Hussein, Z. Wu, R. Al-Kassas, R. G. Alany, Conjunctival and corneal tolerability assessment of ocular naltrexone niosomes and their ingredients on the hen's egg chorioallantoic membrane and excised bovine cornea models, *Int. J. Pharm.*, 432 (2012) 1–10.
167. H. Abdelkader, Z. Wu, R. Al-Kassas, R. G. Alany, Niosomes and discomes for ocular delivery of naltrexone hydrochloride: Morphological, rheological, spreading properties and photo-protective effects, *Int. J. Pharm.*, 433 (2012) 142–148.
168. Q. Li, Z. Li, W. Zeng, S. Ge, H. Lu, C. Wu, L. Ge, D. Liang, Y. Xu, Proniosome-derived niosomes for tacrolimus topical ocular delivery: In vitro cornea permeation, ocular irritation, and in vivo anti-allograft rejection, *Eur. J. Pharm. Sci.*, 62 (2014) 115–123.
169. I. I. Abu Hashim, M. S. El-dahan, R. M. Yusif, A.-E. H. Abd-ElGawad, H. Arima, Potential use of niosomal hydrogel as an ocular delivery system for atenolol, *Biol. Pharm. Bull.*, 37 (2014) 541–551.
170. S. D. Aindrilla, M. C. Harsiddhi, R. J. Kisan, J. K. Vilasrao, Cubosomes: Innovative nanostructures for drug delivery, *Curr. Drug Deliv.*, 13 (2016) 482–493.
171. Z. Karami, M. Hamidi, Cubosomes: Remarkable drug delivery potential, *Drug Discovery Today*, 21 (2016) 789–801.
172. N. Garti, D. Libster, A. Aserin, Lipid polymorphism in lyotropic liquid crystals for triggered release of bioactives, *Food Funct.*, 3 (2012) 700–713.
173. Y. Lu, K. Park, Polymeric micelles and alternative nanonized delivery vehicles for poorly soluble drugs, *Int. J. Pharm.*, 453 (2013) 198–214.
174. A. Zafar, J. Ahmad, R. T. Addo, S. Akhter, Progress of controlled drug delivery systems in topical ophthalmology: Focus on nano and micro drug carriers, in: R. T. Addo (ed.) *Ocular Drug Delivery: Advances, Challenges and Applications*, Springer International Publishing, Cham, 2016, pp. 131–163.
175. T. Arnarson, P. H. Elworthy, Effects of structural variations of non-ionic surfactants on micellar properties and solubilization: Variation of oxyethylene content on properties of C22 monoethers, *J. Pharm. Pharmacol.*, 34 (1982) 87–89.
176. R. Trivedi, U. B. Kompella, Nanomicellar formulations for sustained drug delivery: Strategies and underlying principles, *Nanomedicine (Lond.)*, 5 (2010) 485–505.

177. J. Rios-Doria, A. Carie, T. Costich, B. Burke, H. Skaff, R. Panicucci, K. Sill, A versatile polymer micelle drug delivery system for encapsulation and in vivo stabilization of hydrophobic anticancer drugs, *J. Drug Deliv.*, 2012 (2012) 8.

178. M. Talelli, M. Barz, C. J. Rijcken, F. Kiessling, W. E. Hennink, T. Lammers, Core-crosslinked polymeric micelles: Principles, preparation, biomedical applications and clinical translation, *Nano Today*, 10 (2015) 93–117.

179. J.-H. Chang, N. K. Garg, E. Lunde, K.-Y. Han, S. Jain, D. T. Azar, Corneal neovascularization: An anti-VEGF therapy review, *Surv. Ophthalmol.*, 57 (2012) 415–429.

180. F. Alvarez-Rivera, D. Fernández-Villanueva, A. Concheiro, C. Alvarez-Lorenzo, αLipoic acid in Soluplus® polymeric nanomicelles for ocular treatment of diabetes-associated corneal diseases, *J. Pharm. Sci.*, 105 (2016) 2855–2863.

181. C. Civiale, M. Licciardi, G. Cavallaro, G. Giammona, M. G. Mazzone, Polyhydroxyethylaspartamide-based micelles for ocular drug delivery, *Int. J. Pharm.*, 378 (2009) 177–186.

182. S. Patel, C. Garapati, P. Chowdhury, H. Gupta, J. Nesamony, S. Nauli, S. H. S. Boddu, Development and evaluation of dexamethasone nanomicelles with potential for treating posterior uveitis after topical application, *J. Ocul. Pharmacol. Ther.*, 31 (2015) 215–227.

183. K. Cholkar, S. Gunda, R. Earla, D. Pal, A. K. Mitra, Nanomicellar topical aqueous drop formulation of rapamycin for back-of-the-eye delivery, *AAPS PharmSciTech*, 16 (2015) 610–622.

184. Y. Pathak, V. Sutariya, A. Hirani, *Nano-Biomaterials For Ophthalmic Drug Delivery*, Springer, Switzerland, 2016.

185. D. A. Farghaly, A. A. Aboelwafa, M. Y. Hamza, M. I. Mohamed, Topical delivery of fenoprofen calcium via elastic nano-vesicular spanlastics: Optimization using experimental design and in vivo evaluation, *AAPS PharmSciTech*, 18 (2017) 2898–2909.

186. S. Kakkar, I. P. Kaur, Spanlastics: A novel nanovesicular carrier system for ocular delivery, *Int. J. Pharm.*, 413 (2011) 202–210.

187. P. Deol, I. P. Kaur, G. Sharma, S. Sandhu, M. Singh, M. Yadav, Potential of nanomaterials as movers and packers for drug molecules, *Solid State Phenom.*, 222, (2015) 159–178.

188. A. P. Catalina, P. Marcel, S. Corina, F. P. Radu, O. Lacramioara, Modern drug delivery systems for targeting the posterior segment of the eye, *Curr. Pharm. Des.*, 21 (2015) 6055–6069.

189. A. Khatkar, S. Sigroha, Chitosan- A naturally derived antioxidant polymer with diverse applications, *Curr. Org. Chem.*, 21 (2017) 333–341.

190. A. Madni, M. A. Rahem, N. Tahir, M. Sarfraz, A. Jabar, M. Rehman, P. M. Kashif, S. F. Badshah, K. U. Khan, H. A. Santos, Non-invasive strategies for targeting the posterior segment of eye, *Int. J. Pharm.*, 530 (2017) 326–345.

191. S. P. Kambhampati, R. M. Kannan, Dendrimer nanoparticles for ocular drug delivery, *J. Ocul. Pharmacol. Ther.*, 29 (2013) 151–165.

192. I. Bravo-Osuna, M. Noiray, E. Briand, A. M. Woodward, P. Argüeso, I. T. Molina Martínez, R. Herrero-Vanrell, G. Ponchel, Interfacial interaction between transmembrane ocular mucins and adhesive polymers and dendrimers analyzed by surface plasmon resonance, *Pharm. Res.*, 29 (2012) 10.1007/s11095-11012-10761-11091.

193. Y. Cheng, H. Qu, M. Ma, Z. Xu, P. Xu, Y. Fang, T. Xu, Polyamidoamine (PAMAM) dendrimers as biocompatible carriers of quinolone antimicrobials: An in vitro study, *Eur. J. Med. Chem.*, 42 (2007) 1032–1038.

194. S. Kannan, H. Dai, R. S. Navath, B. Balakrishnan, A. Jyoti, J. Janisse, R. Romero, R. M. Kannan, Dendrimer-based postnatal therapy for neuroinflammation and cerebral palsy in a rabbit model, *Sci. Transl. Med.*, 4 (2012) 130ra146–130ra146.

195. E. Sakurai, H. Ozeki, N. Kuno, Y. Ogura, Effect of particle size of polymeric nanospheres on intravitreal kinetics, *Ophthalmic. Res.*, 33 (2001) 31–36

196. S. Ebrahim, G. A. Peyman, P. J. Lee, Applications of liposomes in ophthalmology, *Surv. Ophthalmol.*, 50 (2005) 167–182.

197. J. F Szoka, and, D. Papahadjopoulos, Comparative properties and methods of preparation of lipid vesicles (liposomes), *Annu. Rev. Biophys. Bioeng.*, 9 (1980) 467–508.

198. J. A. A. Veloso, Q. Zhu, R. Herrero-Vanrell, M. F. Refojo, Ganciclovir-loaded polymer microspheres in rabbit eyes inoculated with human cytomegalovirus, *Invest. Ophthalmol. Visual Sci.*, 38 (1997) 665–675.

199. R. Herrero-Vanrell, I. Bravo-Osuna, V. Andrés-Guerrero, M. Vicario-de-la-Torre, I. T. Molina-Martínez, The potential of using biodegradable microspheres in retinal diseases and other intraocular pathologies, *Prog. Retin. Eye Res.*, 42 (2014) 27–43.

200. R. T. Addo, A. Siddig, R. Siwale, N. J. Patel, J. Akande, A. N. Uddin, M. J. D'Souza, Formulation, characterization and testing of tetracaine

hydrochloride-loaded albumin-chitosan microparticles for ocular drug delivery, *J. Microencapsulation*, 27 (2010) 95–104.

201. D. Sensoy, E. Cevher, A. Sarıcı, M. Yılmaz, A. Özdamar, N. Bergişadi, Bioadhesive sulfacetamide sodium microspheres: Evaluation of their effectiveness in the treatment of bacterial keratitis caused by Staphylococcus aureus and Pseudomonas aeruginosa in a rabbit model, *Eur. J. Pharm. Biopharm.*, 72 (2009) 487–495.
202. Y. Bin Choy, J.-H. Park, M. R. Prausnitz, Mucoadhesive microparticles engineered for ophthalmic drug delivery, *J. Phys. Chem. Solids*, 69 (2008) 1533–1536.
203. Y. B. Choy, J.-H. Park, B. E. McCarey, H. F. Edelhauser, M. R. Prausnitz, Mucoadhesive microdiscs engineered for ophthalmic drug delivery: Effect of particle geometry and formulation on preocular residence time, *Invest. Ophthalmol. Visual Sci.*, 49 (2008) 4808–4815.
204. Y. Tsutomu, T. Yasuhiko, K. Hideya, O. Yuichiro, Recent advances in intraocular drug delivery systems, *Recent Pat. Drug Deliv. Formul.*, 5 (2011) 1–10.
205. C. G. Park, Y. K. Kim, M. J. Kim, M. Park, M. H. Kim, S. H. Lee, S. Y. Choi, W. S. Lee, Y. J. Chung, Y. E. Jung, K. H. Park, Y. B. Choy, Mucoadhesive microparticles with a nanostructured surface for enhanced bioavailability of glaucoma drug, *J. Control. Release*, 220 (2015) 180–188.
206. T. Moritera, Y. Ogura, Y. Honda, R. Wada, S. H. Hyon, Y. Ikada, Microspheres of biodegradable polymers as a drug-delivery system in the vitreous, *Invest. Ophthalmol. Visual Sci.*, 32 (1991) 1785–1790.
207. T. Moritera, Y. Ogura, N. Yoshimura, Y. Honda, R. Wada, S. H. Hyon, Y. Ikada, Biodegradable microspheres containing adriamycin in the treatment of proliferative vitreoretinopathy, *Invest. Ophthalmol. Visual Sci.*, 33 (1992) 3125–3130.
208. T. R. Thrimawithana, S. Young, C. R. Bunt, C. Green, R. G. Alany, Drug delivery to the posterior segment of the eye, *Drug Discovery Today*, 16 (2011) 270–277.
209. B. Conti, C. Bucolo, C. Giannavola, G. Puglisi, P. Giunchedi, U. Conte, Biodegradable microspheres for the intravitreal administration of acyclovir: In vitro/in vivo evaluation, *Eur. J. Pharm. Sci.*, 5 (1997) 287–293.
210. S. Duvvuri, K. G. Janoria, D. Pal, A. K. Mitra, Controlled delivery of ganciclovir to the retina with drug-loaded poly(D,L-lactide-co-glycolide) (PLGA) microspheres dispersed in plga-peg-plga gel: A

novel intravitreal delivery system for the treatment of cytomegalovirus retinitis, *J. Ocul. Pharmacol. Ther.*, 23 (2007) 264–274.

211. Y. He, Y. Liu, Y. Liu, J. Wang, X. Zhang, W. Lu, Z. Ma, X. Zhu, Q. Zhang, Cyclosporine-loaded microspheres for treatment of uveitis: In vitro characterization and in vivo pharmacokinetic study, *Invest. Ophthalmol. Visual Sci.*, 47 (2006) 3983–3988.

212. E. Barcia, R. Herrero-Vanrell, A. Díez, C. Alvarez-Santiago, I. López, M. Calonge, Downregulation of endotoxin-induced uveitis by intravitreal injection of polylactic-glycolic acid (PLGA) microspheres loaded with dexamethasone, *Exp. Eye Res.*, 89 (2009) 238–245.

213. J. A. Cardillo, F. Paganelli, R. A. Costa, A. A. Silva, A. G. Oliveira, A. A. S. Lima–Filho, M. E. Farah, R. B. Jr, B. D. Kuppermann, Intravitreal bioerudivel sustained–release triamcinolone microspheres system (RETAAC): Safety and pharmacokinetic in the rabbit eye and a preliminary report of its potential usefulness for the treatment diabetic macular edema, *Invest. Ophthalmol. Visual Sci.*, 47 (2006) 3886–3886.

214. C. Andrieu-Soler, A. Aubert-Pouëssel, M. Doat, S. Picaud, M. Halhal, M. Simonutti, M.-C. Venier-Julienne, J.-P. Benoit, F. Behar-Cohen, Intravitreous injection of PLGA microspheres encapsulating GDNF promotes the survival of photoreceptors in the rd1/rd1 mouse, *Mol. Vis.*, 11 (2005) 1002–1011.

215. M. S. Ward, A. Khoobehi, E. B. Lavik, R. Langer, M. J. Young, Neuroprotection of retinal ganglion cells in DBA/2J mice with GDNF-loaded biodegradable microspheres, *J. Pharm. Sci.*, 96 (2007) 558–568.

216. P. Checa-Casalengua, C. Jiang, I. Bravo-Osuna, B. A. Tucker, I. T. Molina-Martínez, M. J. Young, R. Herrero-Vanrell, Retinal ganglion cells survival in a glaucoma model by GDNF/Vit E PLGA microspheres prepared according to a novel microencapsulation procedure, *J. Control. Release*, 156 (2011) 92–100.

217. L. Zhang, T. Si, A. J. Fischer, A. Letson, S. Yuan, C. J. Roberts, R. X. Xu, Coaxial electrospray of ranibizumab-loaded microparticles for sustained release of anti-VEGF therapies, *PLoS One*, 10 (2015) e0135608.

218. M. Joseph Titi, A critical look at ocular allergy drugs, *Am. Fam. Physician*, 53 (1996) 2637–2642, 2645–2636.

219. G. L. Sitenga, E. B. Ing, R. G. Van Dellen, B. R. Younge, J. A. Leavitt, Asthma caused by topical application of ketorolac, *Ophthalmology*, 103 (1996) 890–892.

220. A. Stroobants, K. Fabre, P. C. Maudgal, Effect of non-steroidal anti-inflammatory drugs (NSAID) on the rabbit corneal epithelium studied by scanning electron microscopy, *Bull. Soc. Belge. Ophtalmol.*, 276 (2000) 73–81.
221. T. W. Prow, Toxicity of nanomaterials to the eye, *Wiley Interdiscip. Rev. Nanomed. Nanobiotechnol.*, 2 (2010) 317–333.
222. L. Hongjun, B. Qian, C. Xiaobo, Z. Ying-Lan, Current Methods and Research Progress in Nanomaterials Risk Assessment, *Curr. Drug Metab.*,13 (2012) 354–363.
223. M. K. Prinsen, C. F. M. Hendriksen, C. A. M. Krul, R. A. Woutersen, The Isolated Chicken Eye test to replace the Draize test in rabbits, *Regul. Toxicol. Pharmacol.*, 85 (2017) 132–149.
224. R. Valls, E. Vega, M. L. Garcia, M. A. Egea, O. J. Valls, Transcorneal permeation in a corneal device of non-steroidal anti-inflammatory drug delivery systems, *Open Med. Chem. J.*, 2 (2008) 66–71.
225. S. S. Suri, H. Fenniri, B. Singh, Nanotechnology-based drug delivery systems, *J. Occup. Med. Toxicol.*, 2 (2007) 16.
226. J. H. Kay, J. K. Calandra, Interpretation of eye irritation test, *J. Soc. Cosmet. Chem.*, 13 (1962) 281–289.
227. S. M. Agnihotri, P. R. Vavia, Diclofenac-loaded biopolymeric nanosuspensions for ophthalmic application, *Nanomedicine*, 5 (2009) 90–95.
228. C. Bucolo, A. Maltese, G. Puglisi, R. Pignatello, Enhanced ocular anti-inflammatory activity of ibuprofen carried by an Eudragit RS100 nanoparticle suspension, *Ophthalmic Res.*, 34 (2002) 319–323.
229. E. Sanchez-Lopez, M. A. Egea, A. Cano, M. Espina, A. C. Calpena, M. Ettcheto, A. Camins, E. B. Souto, A. M. Silva, M. L. Garcia, PEGylated PLGA nanospheres optimized by design of experiments for ocular administration of dexibuprofen-in vitro, ex vivo and in vivo characterization, *Colloids Surf., B*, 145 (2016) 241–250.
230. N. K. Mehra, D. Cai, L. Kuo, T. Hein, S. Palakurthi, Safety and toxicity of nanomaterials for ocular drug delivery applications, *Nanotoxicology*, 10 (2016) 836–860.

Chapter 17

Rare Earth Nanoparticles Prevent Retinal Degeneration Induced by Intracellular Peroxides

Junping Chen,[a,b] Swanand Patil,[c] Sudipta Seal,[c] and James F. McGinnis[a,b,d,e]

[a] *Oklahoma Center for Neuroscience, University of Oklahoma Health Sciences Center, Oklahoma City, Oklahoma 73104, USA*
[b] *Dean A. McGee Eye Institute, University of Oklahoma Health Sciences Center, Oklahoma City, Oklahoma 73104, USA*
[c] *Surface Engineering and Nanotechnology Facility, Advanced Materials Processing Analysis Center, Mechanical Materials Aerospace Engineering, Nanoscience and Technology Center, University of Central Florida, 4000 Central Florida Blvd, Orlando, Florida 32816, USA*
[d] *Department of Ophthalmology, University of Oklahoma Health Sciences Center, Oklahoma City, Oklahoma 73104, USA*
[e] *Department of Cell Biology, University of Oklahoma Health Sciences Center, Oklahoma City, Oklahoma 73104, USA*
james-mcginnis@ouhsc.edu

Photoreceptor cells are incessantly bombarded with photons of light, which, along with the cells' high rate of oxygen metabolism, continuously exposes them to elevated levels of toxic reactive oxygen intermediates (ROIs). Vacancy-engineered mixed-valence-state cerium oxide nanoparticles (nanoceria particles) scavenge

Reproduced with permission from *Nature Nanotechnology*, **1**, 142–150, 2006.

Nano-Enabled Medical Applications
Edited by Lajos P. Balogh

ISBN 978-981-4800-32-7 (Hardcover), 978-0-429-39903-9 (eBook)
www.jennystanford.com

ROIs. Our data show that nanoceria particles prevent increases in the intracellular concentrations of ROIs in primary cell cultures of rat retina and, in vivo, prevent loss of vision due to light-induced degeneration of photoreceptor cells. These data indicate that the nanoceria particles may be effective in inhibiting the progression of ROI-induced cell death, which is thought to be involved in macular degeneration, retinitis pigmentosa and other blinding diseases, as well as the ROI-induced death of other cell types in diabetes, Alzheimer's disease, atherosclerosis, stroke and so on. The use of nanoceria particles as a direct therapy for multiple diseases represents a novel strategy and suggests that they may represent a unique platform technology.

17.1 Introduction

Photoreceptor cells have the highest rate of oxygen metabolism of any cells in the body and are continuously exposed to the deleterious effects of oxidative stress and to constant bombardment by photons of light. In all types of blindness, including those caused by inherited retinal degeneration, diabetic retinopathy, macular degeneration and retinal detachment, irrespective of the initiating defect, the intracellular concentration of reactive oxygen intermediates (ROIs) is thought to rise chronically or acutely and to activate a cell death pathway [1–3]. The term 'oxidative stress' refers to the damage to cells and cellular components caused by exposure to the highly unstable small molecules called reactive oxygen intermediates [4]. These ROIs, including hydrogen peroxide, hypochlorite ions, hydroxyl radicals, hydroxyl ions and superoxide anions [5], react with almost any nearby DNA, RNA, lipid, carbohydrate or protein. They are produced primarily by the normal oxygen metabolism that occurs in the mitochondrial respiratory chain, although they are also made by activated leukocytes. Such ROIs are also present in the air (if it contains cigarette smoke, radon or ozone) and are generated by the Sun's ultraviolet rays.

Cerium is a rare earth element of the lanthanide series and cerium oxide (CeO_2) is an inorganic compound that is insoluble in water, which has routinely been used in polishing glass and jewellery, and in catalytic converters for automobile exhaust systems and other commercial applications. Although most of the rare earth elements exist in the trivalent state, cerium also occurs in a +4 state and may flip-flop between the two in a redox reaction [6–8]. Cerium oxides are excellent oxygen buffers because of their redox capacity [9]. As a result of alterations in the cerium oxidation state, CeO_2 forms oxygen vacancies or defects in the lattice structure by loss of oxygen and/or its electrons. The valence and defect structure of CeO_2 is dynamic and may change spontaneously or in response to physical parameters such as temperature, the presence of other ions, and oxygen partial pressure [7, 8, 10]. Earlier studies have shown that, with a decrease in particle size, nanoceria particles demonstrate the formation of more oxygen vacancies [11, 12]. The increased surface-area-to-volume ratio that exists in a nanoparticle of ~5 nm diameter enables CeO_2 to regenerate its activity and thereby act catalytically.

We hypothesized that these engineered nanoceria particles would be able to scavenge ROIs within retinal neurons. To test this, we used an in vitro cell culture system and an in vivo albino rat light-damage model. The photoreceptor cells in the albino rat are very sensitive to light and can be induced to degenerate by varying the exposure time or the intensity of the environmental lighting [13]. This sensitivity requires rhodopsin, the photoreceptor-cell visual pigment [14]. The nanoceria particles were effective in both cases and most importantly prevented light-induced loss of functional vision in vivo. In the light-damage model, retinal function, as determined by electroretinography (ERG), is preserved in a dose-dependent manner by the intravitreal injection of nanoceria particles prior to application of bright light, but they are also effective even when injected after light damage has been initiated. Our use of vacancy-engineered nanoceria particles represents a completely new therapeutic strategy, which we think has a very high probability of also being successful in preventing a variety of degenerative diseases. The nanoceria particles will not eliminate the cause, but should inhibit the progressive degenerative response to the ROIs.

17.2 Results

17.2.1 Nanoceria Particles Inhibit the Intracellular Accumulation of Peroxides

As an initial test of the ability of nanoceria particles to protect retinal neurons, primary cell cultures of dissociated rat retinas were grown for 7 days, as previously described [15], and incubated with 1, 3, 5, 10 or 20 nM nanoceria particles for 0.5, 12, 24 or 96 h. After incubation, the cells were washed with phosphate buffer saline three times and incubated with 10 μM 2,7-dichlorofluorescein diacetate (DCFH-DA) at 37°C for 30 min. The cells were then washed again and incubated with 1 mM H_2O_2 at 37°C for 30 min. The cells were harvested and the intensity of fluorescence was determined by flow cytometry. The ROI level was calculated as a ratio (mean intensity of experimental cells/mean intensity of control cells) and plotted as shown in Fig. 17.1. The data demonstrate that the nanoceria particles, when present for 12, 24 and 96 h, are effective in inhibiting the H_2O_2-induced rise of intracellular ROIs at all concentrations of nanoceria particles tested. The addition of 5 nM nanoceria particles at the time of plating also resulted in fewer apoptotic cells when assayed at 1, 2, 3 and 4 weeks later (data not shown).

17.2.2 In vivo Protection of Photoreceptor Cells

The cell culture data prompted us to test the efficacy of the nanoceria particles in vivo. To do this we used a light-damage animal model in which albino rats are exposed to 2,700 lux of light for 6 h, conditions that result in the death of about 50–60% of the photoreceptor cells throughout the retina. For the experimental paradigm, rats were injected with 2 μl of saline or a 0.1, 0.3 or 1.0 μM suspension of the nanoceria particles in saline into the vitreous of both eyes on day 0. Three days later, the animals were exposed to 2,700 lux of light for 6 h and then returned to normal cyclic lighting (12 h dark, 12 h light at 100 lux) for 7 days. The rats were then killed, the eyes enucleated, and fixed, processed, paraffin-embedded sections cut and stained with haematoxylin and eosin (H&E). Morphological evaluation of photoreceptor cell protection was performed using

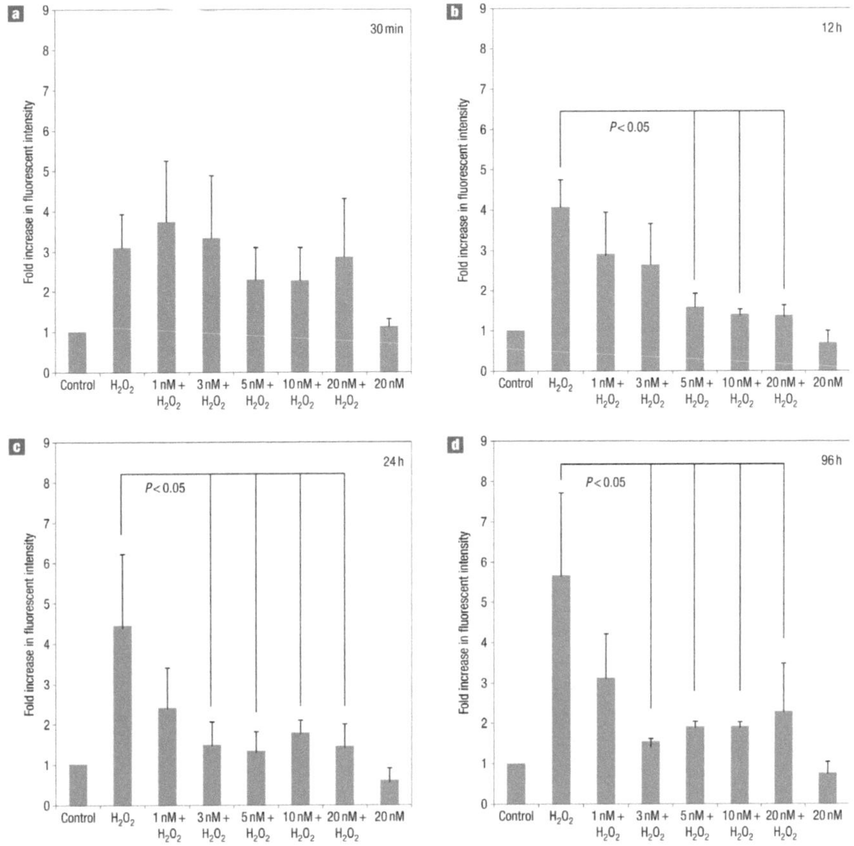

Figure 17.1 Inhibition of ROIs by nanoceria particles. Pretreatment of cultured retinal neurons with nanoceria particles (1, 3, 5, 10 and 20 nM) inhibits the intracellular accumulation of ROIs in response to exposure to 1 mM H_2O_2 for 30 min. (a–d) Incubation with nanoceria particles for 0.5 (a), 12 (b), 24 (c) and 96 h (d). The protection is dose-and time-dependent. (Statistical analysis was done by one-way ANOVA and Newman–Keuls test for post hoc analysis. Data are shown as mean + s.d., $n = 3$, $P < 0.05$.)

bright field microscopy. Exposure to the 2,700 lux of light reduced the thickness of the outer nuclear layer (ONL) of photoreceptor cell nuclei from the 10–13 rows (Fig. 17.2a, control) in normal rats to 1–2 rows in the most degenerated region of the retinas in light-exposed (LE) rats (Fig. 17.2b) or vehicle- (0.9% NaCl) injected

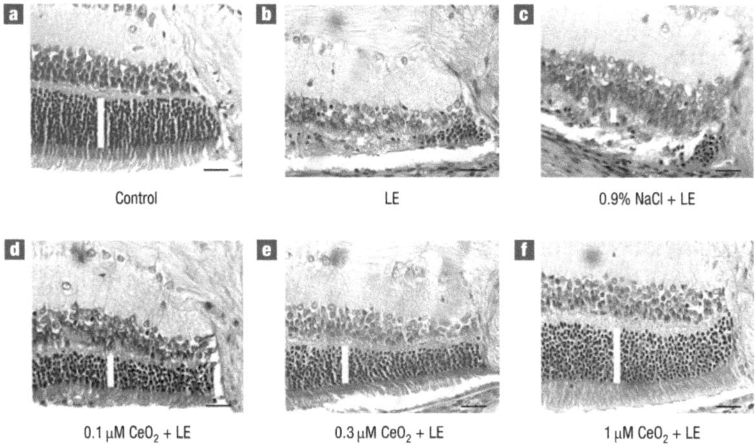

Figure 17.2 Intravitreal injection of nanoceria particles protects rat retina photoreceptor cells from light-induced degeneration. Representative images of photomicrographs of H&E stained sections adjacent to the optic nerve are shown. The white bars indicate the thickness of the layer of nuclei of rods and cones (the outer nuclear layer, ONL). Injections were given on day 0, rats were exposed to light on day 3, and the experiment ended on day 10. (a) No light exposure (LE), no injection. (b–f) Rats were exposed to 6 h of 2,700 lux white light. No injection (b), injections of saline (c), 0.1 μM CeO_2 (d), 0.3 μM CeO_2 (e) and 1.0 μM CeO_2 (f) 3 days before LE. The scale bars in the lower right corner represent 25 μm.

rats (Fig. 17.2c). However, in rats treated with nanoceria particles, there was significant dose-dependent protection of photoreceptors, resulting in the ONL having ~5–6 rows (Fig. 17.2d, 0.1 μM CeO_2), ~7–8 rows (Fig. 17.2e, 0.3 μM CeO_2) or ~11–12 rows (Fig. 17.2f, 1 μM CeO_2) of nuclei. In addition, the outer and inner segments of treated retinas were highly preserved compared to the LE and vehicle-injected retinas. The data shown are representative of those obtained in three separate experiments and are from the right eye only as there was no difference between right and left eyes of individual rats (data not shown).

17.2.3 Quantitative Histology

To obtain an overall view of the light damage and protection throughout the retina [16], the thickness of the ONL was measured

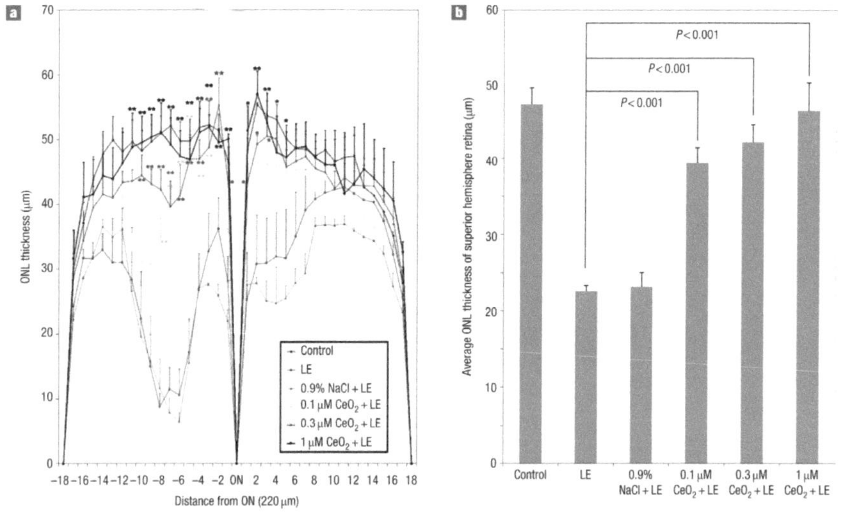

Figure 17.3 Nanoceria particles provide pan-retinal protection against light damage. The thickness of the ONL of the retina was measured along the vertical meridian every 220 μm, starting at the optic nerve (ON). (a) The ONL thickness of control retinas (no LE, no injection), LE, and LE and vehicle-treated nanoceriaparticle-treated retinas are plotted against each of the successive 220 μm distances from the optic nerve (ON). The negative x-axis numbers on the left side of the ON refer to the points on the superior hemisphere and those on the right side refer to the inferior hemisphere. Statistical analysis was done by one-way ANOVA and Newman–Keuls test for post hoc analysis. The results are expressed as mean ONL thickness + s.e.m. ($n = 6$ for each point; $^{*}P < 0.05$; $^{**}P < 0.001$ versus light exposure (LE) group). (b) Summation of the ONL thickness across the superior hemisphere provides a quantitative assessment of the protective effects of the nanoceria particles in the most sensitive hemisphere. Statistical analysis was done by one-way ANOVA and Newman–Keuls test for post hoc analysis. The results are expressed as mean ONL thickness + s.e.m. ($n = 6$ for each group).

every 220 μm, starting from the optic nerve and proceeding in the superior and inferior directions to where the retina ends. The thickness of the ONL is directly proportional to the number of photoreceptor cell nuclei (~5 μm diameter) at each point and a plot of the thickness versus each 220 μm distance from the optic nerve is presented in Fig. 17.3a. These data demonstrate that the superior retina (left of the optic nerve) is more sensitive

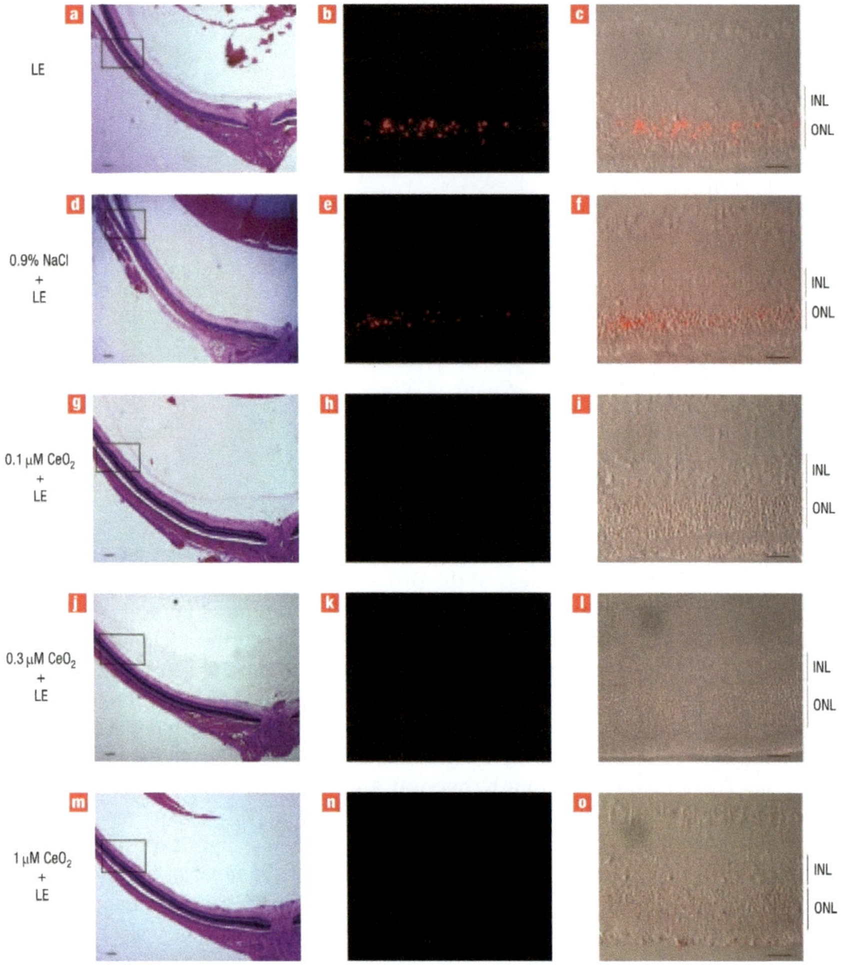

Figure 17.4 Nanoceria particles prevent the appearance of TUNEL-positive photoreceptor cells, which occurs days after exposure to damaging light. The eyes were injected at day 0, exposed to light on day 3, and the experiment ended on day 8. (a, d, g, j, m) Sections stained with H&E. The black rectangles indicate the sites from which the images in the second and third columns were taken. (b, e, h, k, n) Sections stained with anti-digoxigenin-rhodamine for the TUNEL assay. (c, f, i, l, o) The same sections visualized using Nomarsky optics. INL, inner nuclear layer where bipolar cell nuclei are located; ONL, outer nuclear layer where photoreceptor cell nuclei are located. The scale bars in (a, d, g, j, m) represent 100 μm; the scale bars in (c, f, i, l, o) represent 25 μm.

to light damage than the inferior retina (right of optic nerve) and that the nanoceria particles are effective in protecting the photoreceptor cells in both regions. Injections of 0.1 μM and 0.3 μM suspensions are partially protective, but the data obtained for a 1.0 μM suspension demonstrate almost complete protection and are essentially indistinguishable from the control animal, which was not exposed to light. These conclusions are especially obvious when the data from the average thickness of the most sensitive region, the superior retina, are plotted on a bar graph (Fig. 17.3b).

17.2.4 Protection against Long-Term Light Damage

Exposure to bright light results in the rapid death of numerous photoreceptor cells, but it also damages some cells less extensively and they die over a longer time period [17]. To visualize these damaged cells, animals were killed 5 days after light exposure and their eyes processed for histological analysis using the TUNEL (terminal deoxynucleotidyl transferase dUTP nick end labelling) assay. This fluorescent assay reveals those cells that have been damaged by the light and have progressed along an apoptotic cell death pathway. The results shown in Fig. 17.4 are representative of such experiments and demonstrate that exposure to bright light results in TUNEL-positive cells in those retinas from eyes that were either uninjected (Fig. 17.4b,c) or injected with saline (Fig. 17.4e,f). However, all three concentrations (0.1, 0.3, 1.0 μM) of nanoceria particles protected the retinas and prevented the appearance of TUNEL-positive cells (Fig. 17.4h,k,n), suggesting that even at lower concentrations, nanoceria particles are effective in inhibiting long-term light-induced injury to photoreceptor cells.

17.2.5 Nanoceria Particles Protect Retinal Function

The morphological data clearly indicate that the nanoceria particles preserved the photoreceptor cells, but the best criterion for showing protection of vision from light damage is to demonstrate the preservation of retinal function. To address this, the same experimental paradigm was used, but prior to ending the experiment, ERG was performed on day 7. Representative ERG data of those

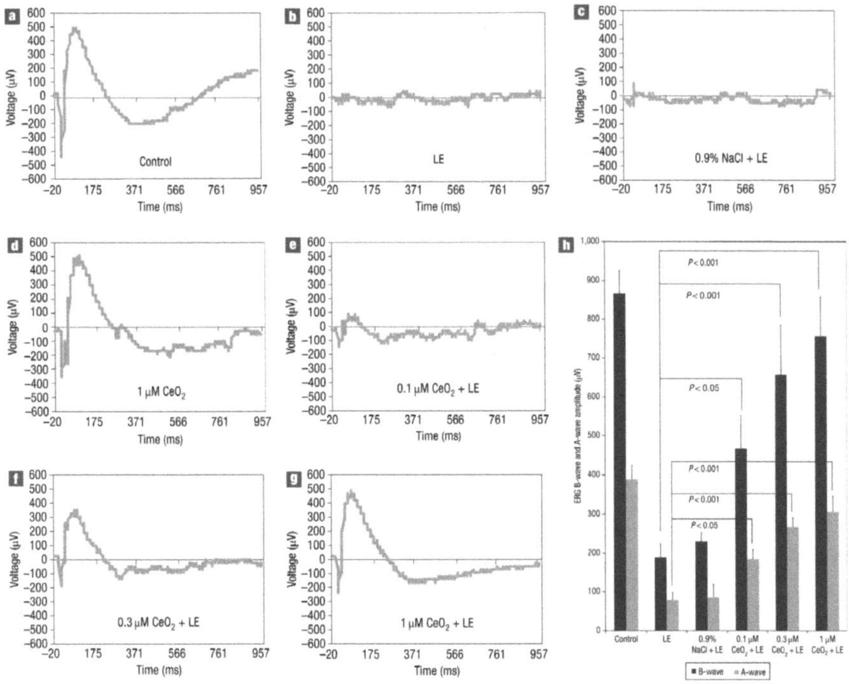

Figure 17.5 Retinal function is protected by the nanoceria particles in a dose-dependent manner. (a–g) Representative ERGs of each group are plotted as voltage (μV) against time (ms). (h) Mean of the ERG B-wave and A-wave amplitudes are shown for each group of animals. Statistical analysis was done by one-way ANOVA and Newman–Keuls test for post hoc analysis. The results are expressed as mean amplitude + s.d. ($n = 6$ for each point). Control animals had no injections and were not exposed to bright light. LE animals had no injections and were exposed to bright light.

experiments are shown in Fig. 17.5. The control animals, not exposed to light, exhibit a strong ERG (Fig. 17.5a), whereas those animals that were exposed to light and were either uninjected (Fig. 17.5b) or injected with saline (Fig. 17.5c) had almost flat ERGs, indicating the absence of retinal function. The ERG data from the rats that were injected with nanoceria particles but not exposed to the bright light (Fig. 17.5d) are essentially identical to those obtained with rats that were not injected and not exposed to the bright light, indicating that the nanoceria particles were not simply preventing the passage of light into the eye. However, rats injected with nanoceria particles that were exposed to light showed ERGs that reflected a progressive improvement in retinal function that paralleled the progressive increase in the concentration of nanoceria particles from 0.1 to 1.0 μM. The quantification of ERG A- and B-wave amplitudes (Fig. 17.5h) clearly demonstrates the functional improvement of retinas following treatment with nanoceria particles. This is especially evident in the eyes pretreated with 1 μM nanoceria particles, which had 79% and 87% of the control A- and B-wave amplitudes, compared with 22% and 26% for the retinas with 0.9% NaCl vehicle treatment.

17.2.6 Nanoceria Particles Can Rescue Retina Function After Light Damage

The data presented show that nanoceria particles were very effective in protecting retina function when administered 3 days before light exposure. To determine if the nanoceria particles had any ability to rescue photoreceptor cells after they had been exposed to damaging light, rats were subjected to 6 h of 2,700 lux light and were injected intravitreally, 2 h later, with 2 μl of a 1 μM nanoceria particle suspension. The retinal function of the animals was determined 7 days later using ERG and quantitative histology applied to the retinal sections. A summary of the histology data from such an experiment is presented in Fig. 17.6a, which indicates that, even after light damage, there is protection of photoreceptor cells throughout the retina. Similarly, a summary of the ERG data (Fig. 17.6b) demonstrates that a significant amount of retinal function is rescued by post-treatment with the nanoceria particles.

17.3 Discussion

Vision is dependent on rod and cone photoreceptors, and human diseases that cause blindness do so by a variety of primary events that ultimately produce two subsequent effects in the photoreceptor cells. The first effect is thought to be the production of either an acute or a chronic rise in the intracellular concentration of

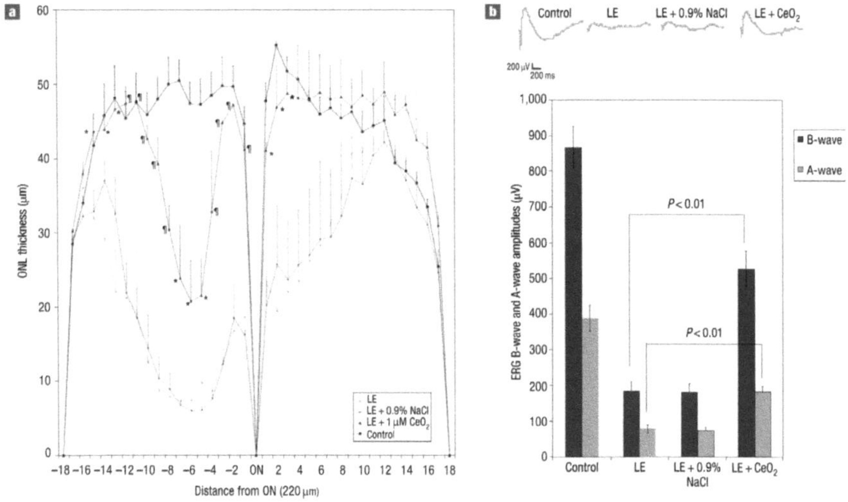

Figure 17.6 Nanoceria particles protect photoreceptor cells even when administered after LE. Rats were exposed to 2,700 lux for 6 h, and 2 h later were either uninjected or injected intravitreally with 2 μl of saline or a 1 μM suspension of CeO_2 in saline. (a) The thickness of the ONL is plotted against the site of measurement along the vertical meridian starting at the optic nerve (ON) for each group of animals. Control: no injections, no light exposure; LE: light exposure, no injections; LE+ 0.9% NaCl: light exposure then saline injection; LE + CeO_2: light exposure then 1 μM CeO_2 injection. Statistical analysis was done by one-way ANOVA and Newman–Keuls test for post hoc analysis. The results are expressed as the mean ONL thickness ± s.e.m. ($n = 4$ for each point; $*P < 0.05$, $P < 0.001$ versus the LE group). (b) Summary of the functional protection (ERG) provided by post-treatment nanoceria particle injection. These data are from the same animals as in a. Each column represents the mean of the ERG B- and A-wave amplitudes. Statistical analysis was done by one-way ANOVA and Newman–Keuls test for post hoc analysis. The results are expressed as mean amplitude ± s.d. ($n = 4$; $P < 0.01$ versus LE group). The insets are representative ERG waveforms from four independent experiments.

ROIs in the photoreceptor cells, which in turn induces the second effect, the activation of an apoptotic signal transduction pathway and the eventual death of the photoreceptor cells. The intracellular ROIs are especially damaging because they can react with almost every type of organic molecule that exists within the cell. Most of the ROIs within the body are generated by respiration, the use of oxygen to produce energy from the metabolism of organic molecules. The retina of the eye is exposed to large amounts of such toxic compounds, generated by normal oxidative reactions as well as those produced by the constant absorption of photons of light. The retina of the albino rat is extremely sensitive to photon-induced damage, which results in the production of excess ROIs, the subsequent degeneration of the photoreceptor cells, and the loss of visual function [18]. We have used retinal neurons in culture and the light-damage model in vivo to demonstrate and characterize the ability of a suspension of vacancy-engineered nanoceria particles to prevent ROI-induced injury to the photoreceptor cells and to preserve the visual function of the retina.

Investigations of nanocrystalline CeO_2 have revealed that its lattice constant increases with decreasing nanoparticle size [11, 12]. This has been attributed to an increase in oxygen vacancies in the crystal structure [11, 12, 19]. It is implied that the migration enthalpy of the oxygen vacancy in CeO_2 is smaller at the nanoscale [8, 20]. Additionally, at the nanoscale, the surface area of CeO_2 particles is dramatically enlarged in relation to its volume to increase oxygen exchange and redox reactions. Thus, oxygen vacancies are likely to form more readily at the nanoscale. We hypothesized that nanoceria particles, owing to their chemical and physical structure, protect cells from free-radical-induced damage. The defects in the nanoceria particles can act as chemical spin traps similar to nitrosone compounds, currently used as biological antioxidants. However, one CeO_2 particle may offer many sites of spin-trap activity, where current pharmacological agents offer only a few per molecule. Additionally, the lattice defects in nanoceria particles possess the potential for regeneration and do not require repetitive dosage as seen with the use of vitamins C and E. We propose that nanoceria particles act as free-radical scavengers by switching between the +3 and +4 valence states via various surface chemical reactions. The radical-scavenging mechanism can be given by the following set of chemical reactions.

$$Ce^{3+} \rightleftharpoons Ce^{4+} + e^-$$
$$Ce^{3+} + OH^\bullet \rightarrow Ce^{4+} + OH^-$$
$$Ce^{4+} + O_2^{\bullet -} \rightarrow Ce^{3+} + O_2$$

This regenerative property makes nanoceria particles a very attractive proposition for treating ROI-mediated cellular damage and diseases, as most of the present free-radical scavengers do not have such a property and need a repetitive dosage.

Initially, we tested the ability of nanoceria particles to protect retinal neurons in culture and found that their presence at 5 nM was sufficient to prevent the increase in apoptotic cells seen in cultures grown in the absence of the nanoceria particles. Their protective effects were further demonstrated by showing that the nanoceria particles prevented the ROI-induced apoptosis that occurred in cultured retinal neurons exposed to exogenously added H_2O_2. The addition of the nanoceria particles to such cultures was also shown to prevent the intracellular accumulation of ROIs that was observed in retinal neurons to which H_2O_2 was added. To determine if the nanoceria particles would act in a similar manner in vivo, a light-damage paradigm was established and the number of photoreceptor cells was determined by measuring the thickness of the ONL. The histological evaluation of individual sections supported the conclusion that the nanoceria particles provided protection from light damage, even at very low concentrations (2 μl of 0.1, 0.3 and 1.0 μM). The light-dependent degeneration, although more severe in the superior retina, occurs throughout the retina, and the protective effect of the nanoceria particles is not limited to the area of the injection but was also shown to occur across the entire retina. The pan-retinal dose-dependent protection was further documented using ERG, to show that pretreatment with the nanoceria particles protected retinal function.

Injection of the nanoceria particles 3 days prior to light exposure prevented the subsequent appearance of TUNEL-positive cells seen in the retinas of uninjected or saline-injected control animals. We did not observe any dead, damaged or TUNEL-positive retinal pigment epithelium (RPE) cells in our experiments and the ERG data indicate that the RPE are fully functional in the nanoceria-particle-treated cells. This is contrary to the observations made with the ZUR SIV albino rat [21]. Even when injected after light damage, the nanoceria particles were still partially effective, suggesting that, as long as the cells have not died, they may be prevented from either entering or proceeding down an apoptotic pathway. We do not yet know how the nanoceria particles are taken up into photoreceptor cells, nor the process by which they are eliminated. However, because ROIs are formed within the photoreceptor cells in response to light damage, and because ROIs react over very short distances, we think that the nanoceria particles must actually enter the photoreceptor cells. Similarly, inhibition of the appearance of TUNEL-positive cells days after the light insult suggests a relatively long active lifespan in the retina. Because our data demonstrated that the nanoceria particles prevent the peroxide-induced increase in the intracellular concentration of ROIs in cultured retinal cells, we think the nanoceria particles function by the same mechanism in vivo. However, it may be that they regulate the concentration of redox couples such as glutathione disulphide/glutathione [22] or induce scavenging enzymes [23, 24], although we think it is more likely that they inhibit apoptosis by directly preventing the rise in ROI levels.

Our in vivo results were obtained by direct injection of the nanoceria particles into the vitreous of the eye and it is likely that other routes of administration would also be effective. Neurodegeneration within the retina is not unlike neurodegeneration within other areas of the central nervous system. It is not generally possible to directly analyse human tissue in vivo, but the preponderance of data suggests that Alzheimer's disease [25], Parkinson's disease [1] and Huntington's disease [26] are caused by different initiating events that result in an increase in ROI levels within the populations of neurons undergoing degeneration. This suggests that the nanoceria particles can also be effective in preventing the ROI-mediated death or loss of function of the neurons uniquely affected in each of these illnesses. Similarly, there are many diverse diseases initiated by ROI, including atherosclerosis [27] and stroke [28], and inductive reasoning leads to the conclusion that the vacancy-engineered nanoceria particles can be beneficial against any of the diseases in which ROIs are involved.

17.4 Material and Methods

17.4.1 Animals

Sprague–Dawley albino rats were bred and maintained in a regular cyclic light environment (12 h on, 12 h off, with less than 100 lux of in-cage luminance) and used for experimentation between two and three months of age. Animals were cared for and handled according to the Association for Research in Vision and Ophthalmology (ARVO) statement for the use of animals in ophthalmic and vision research, and the IACUC-approved animal use protocols, which comply with the University of Oklahoma Faculty of Medicine guidelines for use of animals in research.

17.4.2 Primary Culture of Retinal Neurons

Primary dissociated cell cultures of rat retinas were established from 0- to 2-day-old rat pups as described [15]. The medium and culture conditions do not appear to support the growth of glial cells because no glial cell markers (glutamine synthetase [Muller cells] [29], glial fibrillary acidic protein [astrocytes] [30] and glycerol phosphate dehydrogenase [oligodendrocytes] [31]) are detected from day 7 to day 28.

17.4.3 Measurement of Intracellular ROI by Flow Cytometry

Intracelluar ROI production was measured by flow cytometry using DCFH-DA (Sigma), an oxidant-sensitive fluorescent probe. In the presence of intracellular peroxides, H_2DCF is oxidized to a highly fluorescent compound, 2,7-dichlorofluorescein [32]. The retinal cells were exposed to nanoceria particles for the indicated times, washed, and incubated with 10 μM DCFH-DA (dissolved in DMSO, filter sterilized) at 37°C for 30 min. The cells were then incubated with 1 mM H_2O_2 at 37°C for 30 min after the excess DCFH-DA was washed off with phosphate buffer saline. The cells were then harvested with trypsin. The intensity of fluorescence was detected by flow cytometry with an excitation filter of 485 nm. The ROI level was calculated as a ratio = mean intensity of experimental cells/mean intensity of control cells.

17.4.4 Intravitreal Injection and Light-Induced Photoreceptor Degeneration

Rats were anaesthetized, pupils dilated, a topical anaesthetic applied to the cornea, and 2 μl of 0.1, 0.3 or 1 μM nanoceria particles in 0.9% NaCl were injected. Controls were injected with 2 μl of 0.9% NaCl. Because we do not know if the nanoparticles can diffuse throughout the animal, both eyes of each animal were injected with the same suspension. Control animals were injected with the vehicle, saline, in both eyes. For light-induced photoreceptor degeneration, the rats were exposed to 2,700 lux (measured with a photometer) of constant light for 6 h. During light exposure, rats were maintained in transparent polycarbonate cages (one or two rats per cage) with stainless-steel wire covers. A water bottle was kept at the side of the cage and food was placed in the bottom of the cage on the bedding. They were returned to cyclic light for 5–7 days before the experiment was ended.

17.4.5 Morphologic Evaluation by Quantitative Histology

After ERG recordings, the rats were killed by an overdose of carbon dioxide, the eyes enucleated, fixed in Perfix, embedded in paraffin, and 5 μm-thick sections were cut along the vertical meridian so that the superior and inferior hemispheres were separated by the optic nerve. To evaluate quantitatively the morphologic changes using H&E stained sections [16], we measured the ONL thickness at 220 μm intervals, starting at the optic nerve head and moving along the vertical meridian toward the superior or inferior ora serrata. This produced 18 data points in each hemisphere. The mean ONL thickness of each point was then calculated from the retinas of at least four eyes.

17.4.6 Functional Rescue of Photoreceptor Cells as Evaluated by Electroretinography

The ERG experiments were performed as described previously [33]. The animals used in each set of experiments were all the same sex (males) and were taken from the same litter. They were maintained in the same room and treated identically. All the ERG recordings were performed at the same time of day, from 10 a.m. to 12 p.m. Animals were kept in total darkness overnight before ERG were recorded. For quantitative analysis, the A-wave amplitude was measured as the difference between baseline and the peak of the A-wave, and the B-wave amplitude was measured as the difference between the peaks of the A- and B-waves.

17.4.7 Photoreceptor Cell Apoptosis Evaluation by TUNEL Assay

Apoptotic cells in sections were visualized using a Nikon Eclipse 500 Fluorescent microscope with a 594 nm filter using an apoptosis detection kit (ApopTag Rhodamin In Situ Apoptosis Detection, Intergen) based on the TUNEL assay exactly as described in the accompanying instructions.

17.4.8 Statistical Analysis

Statistical analysis was performed by including all sets of animals in a one-way ANOVA analysis. The Newman–Keuls test for post hoc analysis was done to compare control animals with animals injected with nanoceria particles. This representation evaluates the ability of the nanoceria particles to rescue photoreceptor cells.

Acknowledgements

The authors thank M. Wu and M. Ramsey for technical assistance, J. Ash for advice on construction of data images and M. Dittmar for help with the care and use of the animals. This work was supported by National Institutes of Health grant EY13050, an NEI core grant EY12190, P20 RR017703 from the COBRE program of the National Center for Research Resources, general funds from Presbyterian Health Foundation, and by an unrestricted grant from Research to Prevent Blindness to the Department of Ophthalmology, Dean McGee Eye Institute. S.S. was supported by the National Science Foundation (NSF BES: 0510270).

Author Contributions

J.C. and J.F.M. conceived and designed the experiments, J.C. performed the experiments, J.F.M. and J.C. analysed the data, S.P. and S.S. generated the nanoceria particles, and J.C., S.P., S.S. and J.F.M. cowrote the paper.

Competing Financial Interests

The authors declare that they have no competing financial interests.

References

1. Emerit, J., Edeas, M., and Bricaire, F. Neurodegenerative diseases and oxidative stress. *Biomed. Pharmacother.* **58**, 39–46 (2004).
2. Caldwell, R. B. et al. Vascular endothelial growth factor and diabetic retinopathy: pathophysiological mechanisms and treatment perspectives. *Diabetes Metab. Res. Rev.* **19**, 442–455 (2003).
3. Lewis, G. P., Erickson, P. A., Anderson, D. H., and Fisher, S. K. Opsin distribution and protein incorporation in photoreceptors after experimental retinal detachment. *Exp. Eye Res.* **53**, 629–640 (1991).
4. van Reyk, D. M., Gillies, M. C., and Davies, M. J. The retina: oxidative stress and diabetes. *Redox. Rep.* **8**, 187–192 (2003).
5. Beatty, S., Koh, H., Phil, M., Henson, D., and Boulton, M. The role of oxidative stress in the pathogenesis of age-related macular degeneration. *Surv. Ophthalmol.* **45**, 115–134 (2000).

6. Suzuki, T., Kosacki, I., Anderson, H. U., and Colomban, P. Electrical conductivity and lattice defects in nanocrystalline cerium oxide thin films. *J. Am. Ceram. Soc.* **84**, 2007–2014 (2001).
7. Herman, G. S. Characterization of surface defects on epitaxial CeO_2(001) films. *Surf. Sci.* **437**, 207–214 (1999).
8. Conesa, J. C. Computer modeling of surfaces and defects on cerium dioxide. *Surf. Sci.* **339**, 337–352 (1995).
9. Land, P. L. Defect equilibria for extended point-defects, with application to nonstoichiometric ceria. *J. Phys. Chem. Solids* **34**, 1839–1845 (1973).
10. Mamontov, E., Egami, T., Brezay, R., Koranne, M., and Tyagi, S. lattice defects and oxygen storage capacity of nanocrystalline ceria and ceria-zirconia. *J. Phys. Chem. A* **104**, 11110–11116 (2000).
11. Deshpande et al. Size dependency variation in lattice parameter and valency states in nanocrystalline cerium oxide. *Appl. Phys. Lett.* **87**, 1–3 (2005).
12. Tsunekawa, S., Sahara, R., Kawazoe, Y., and Ishikawa, K. Lattice relaxation of monosize CeO_{2-x} nanocrystalline particles. *Appl. Surf. Sci.* **152**, 53–56 (1999).
13. Williams, T. P., and Howell, W. L. Action spectrum of retinal light-damage in albino rats. Invest. Ophthalmol. *Vis. Sci.* **24**, 285–287 (1983).
14. Grimm, C. et al. Protection of Rpe65-deficient mice identifies rhodopsin as a mediator of light-induced retinal degeneration. *Nat. Genet.* **25**, 63–66 (2000).
15. McGinnis, J. F. et al. Unique retina cell phenotypes revealed by immunological analysis of recoverin expression in rat retina cells. *J. Neurosci. Res.* **55**, 252–260 (1999).
16. Rapp, L. M., and Williams, T. P. A parametric study of retinal light damage in pigmented and albino rats, in *The Effects of Constant Light on Visual Processes* (eds Williams, T. P., and Baker, B. N.) 135–159 (Plenum Press, New York, 1980).
17. Specht, S., Organisciak, D. T., Darrow, R. M., and Leffak, M. Continuing damage to rat retinal DNA during darkness following light exposure. *Photochem. Photobiol.* **71**, 559–566 (2000).
18. Wenzel, A., Grimm, C., Samardzija, M., and Reme, C. E. Molecular mechanisms of light-induced photoreceptor apoptosis and neuroprotection for retinal degeneration. *Prog. Retin. Eye Res.* **24**, 275–306 (2005).

19. Patil, S., Seal, S., Guo, Y., Schulte, A., and Norwood, J. Role of trivalent La and Nd dopants in lattice distortion and oxygen vacancy generation in cerium oxide nanoparticles. *Appl. Phys. Lett.* **88**, 1–3 (2006).
20. Kosacki, I., Suzuki, T., Anderson, H. U., and Colomban, P. Raman scattering and lattice defects in nanocrystalline CeO_2 thin films. *Solid State Ionics* **149**, 99–105 (2002).
21. Hafezi, F., Marti, A., Munz, K., and Reme, C. E. Light-induced apoptosis: differential timing in the retina and pigment epithelium. *Exp. Eye Res.* **64**, 963–970 (1997).
22. Schafer, F. Q., and Buettner, G. R. Redox environment of the cell as viewed through the redox state of the glutathione disulfide/glutathione couple. *Free Radic. Biol. Med.* **30**, 1191–1212 (2001).
23. Zoete, V., Rougee, M., nkova-Kostova, A. T., Talalay, P., and Bensasson, R. V. Redox ranking of inducers of a cancer-protective enzyme via the energy of their highest occupied molecular orbital. *Free Radic. Biol. Med.* **36**, 1418–1423 (2004).
24. Gao, X., and Talalay, P. Induction of phase 2 genes by sulforaphane protects retinal pigment epithelial cells against photooxidative damage. *Proc. Natl. Acad. Sci. USA* **101**, 10446–10451 (2004).
25. Huber, A., Stuchbury, G., Burkle, A., Burnell, J., and Munch, G. Neuroprotective therapies for Alzheimer's disease. *Curr. Pharm. Des.* **12**, 705–717 (2006).
26. Butterfield, D. A., and Kanski, J. Brain protein oxidation in age-related neurodegenerative disorders that are associated with aggregated proteins. *Mech. Ageing Dev.* **122**, 945–962 (2001).
27. Singh, U., and Jialal, I. Oxidative stress and atherosclerosis. *Pathophysiology* **13**, 129–142 (2006).
28. Khalil, A. A., Aziz, F. A., and Hall, J. C. Reperfusion injury. *Plast. Reconstr. Surg.* **117**, 1024–1033 (2006).
29. Linser, P. J., and Moscona, A. A. Induction of glutamine synthetase in embryonic neural retina: its suppression by the gliatoxic agent alpha-aminoadipic acid. *Brain Res.* **227**, 103–119 (1981).
30. Vernadakis, A., Mangoura, D., Sakellaridis, N., and Linderholm, S. Glial cells dissociated from newborn and aged mouse brain. *J. Neurosci. Res.* **11**, 253–262 (1984).
31. Leveille, P. J., McGinnis, J. F., Maxwell, D. S., and de Vellis, J. Immunocytochemical localization of glycerol-3-phosphate dehydrogenase in rat oligodendrocytes. *Brain Res.* **196**, 287–305 (1980).

32. LeBel, C. P., Ischiropoulos, H., and Bondy, S. C. Evaluation of the probe 2′,7′-dichlorofluorescin as an indicator of reactive oxygen species formation and oxidative stress. *Chem. Res. Toxicol.* **5**, 227–231 (1992).
33. Cao, W. et al. In vivo protection of photoreceptors from light damage by pigment epithelium-derived factor. *Invest Ophthalmol. Vis. Sci.* **42**, 1646–1652 (2001).

Chapter 18

Nanomechanical Analysis of Cells from Cancer Patients

Sarah E. Cross,[a,b,*] Yu-Sheng Jin,[c,*] Jianyu Rao,[c,*] and James K. Gimzewski[a,b,*]

[a] *Department of Chemistry and Biochemistry, University of California, Los Angeles, California 90095, USA*

[b] *California NanoSystems Institute, University of California, Los Angeles, California 90095, USA*

[c] *Department of Pathology and Laboratory Medicine, University of California, Los Angeles, California 90095, USA*

gimzewski@cnsi.ucla.edu, jrao@mednet.ucla.edu

Change in cell stiffness is a new characteristic of cancer cells that affects the way they spread [1, 2]. Despite several studies on architectural changes in cultured cell lines [1, 3], no ex vivo mechanical analyses of cancer cells obtained from patients have been reported. Using atomic force microscopy, we report the stiffness of live metastatic cancer cells taken from the body (pleural) fluids of patients with suspected lung, breast and pancreas cancer. Within the same sample, we find that the cell stiffness of metastatic cancer cells is more than 70% softer, with a standard deviation over five times narrower, than the benign cells that line the body

*The authors contributed equally to this work.

Reproduced with permission from *Nature Nanotechnology*, **2**, 780–783, 2007.

Nano-Enabled Medical Applications
Edited by Lajos P. Balogh

ISBN 978-981-4800-32-7 (Hardcover), 978-0-429-39903-9 (eBook)
www.jennystanford.com

Table 18.1 Patient characteristics and cytological diagnosis versus mechanical measurements (Young's modulus, E)

Case no.	Age/sex	Clinical history	Cytological diagnosis of pleural fluid*	Stiffness (kPa): 'Tumour'	Stiffness (kPa): 'Normal'
1	52/Female	Non-small cell carcinoma of the lung	Positive for metastatic malignant cells	0.56 ± 0.09	2.10 ± 0.79
2	60/Female	Non-small cell carcinoma of the lung	Positive for metastatic malignant cells	0.52 ± 0.12	2.05 ± 0.87
3	49/Female	Breast ductal adenocarcinoma	Positive for metastatic malignant cells	0.50 ± 0.08	1.93 ± 0.50
4	85/Male	Pancreatic adenocarcinoma	Positive for metastatic malignant cells	0.54 ± 0.08	0.54 ± 0.12
5	40/Male	Liver cirrhosis	Negative for malignant cells	—	1.86 ± 0.50
6	47/Male	Fever and hepatic failure	Negative for malignant cells	—	1.75 ± 0.61
7	92/Female	Anasarca peripheral oedema	Negative for malignant cells	—	2.09 ± 0.98

*Cytomorphological diagnosis was made based on morphological analysis combined with immunohistochemical analysis (see Supplementary Information). Stiffness values (E) represent mean $\pm$ s.d.

cavity. Different cancer types were found to display a common stiffness. Our work shows that mechanical analysis can distinguish cancerous cells from normal ones even when they show similar shapes. These results show that nanomechanical analysis correlates well with immunohistochemical testing currently used for detecting cancer.

Recent progress in the study of cancer cell motility and invasion has generated a greater understanding of the mechanical properties involved in malignant transformation. In particular, the dynamic reorganization of the cytoskeleton has become a specific point of interest regarding changes in cell morphology, motility, adhesion and invasion [4, 5]. Techniques for measuring cell adhesion and motility have been established in the laboratory for in vitro analysis of cellular phenotypic events associated with tumour cells. Current cytoskeleton analysis techniques include biochemical purifications, gene expression and polymerization assays, immunofluorescent labelling, and time-lapse and electron microscopies [6, 7]. Recently, a change in the physical properties, in particular cell elasticity, of tissue cells has been recognized as an indication of disease [2, 8, 9] and has emerged as a marker for cellular phenotypic events associated with cell adhesion and cytoskeletal organization [2, 10–12]. In particular, several studies have shown a reduction in stiffness with increasing metastatic efficiency in human cancer cell lines using several different in vitro biomechanical assays [13–15].

In this chapter, we report on the associated cell stiffness (elasticity) of pathologically defined human metastatic cancer cells and benign reactive mesothelial cells in human body (pleural) fluid samples (as approved by the Institutional Review Board; Table 18.1) using atomic force microscopy (AFM) [16]. AFM has been used to probe a number of properties inherent to microbial cells, mammalian cells and biomolecules [10, 13–15, 17–22], including analysis of cellular mechanical strain and elasticity, due to the precise application of low forces to cells with minimal disruption [18, 19]. Although AFM has previously been used to probe cellular mechanics under physiological conditions [13–15, 21, 23, 24], it has not been used for cancer cell analysis in a clinical setting, which is the purpose of this chapter.

Analyses of body fluid samples, rather than primary tumour samples, were chosen here, because tumour cells in body fluid are all metastatic in nature and thus provide a clonal population of metastatic cells for analysis. Additionally, the co-existence of both benign and metastatic cells in a single specimen provides a native internal control. Body cavities, including pleural, pericardial and peritoneal cavities, are covered by serous membranes consisting of a single row of flat mesothelial cells on the surface and an underlying submesothelial layer, which cover a very large surface area in close contact with every major organ of the body. Because of their continuity with the lymphatic system, these cavities are commonly the seat of metastasis. Metastatic malignant effusions constitute an unequivocal sign of widespread cancer. Current cancer cell detection relies on qualitative morphological analyses of shape change resulting from biochemical alterations, such as cytoskeletal remodelling [25]. However, morphological analysis of cells recovered from an effusion is often difficult to diagnose because of the notorious reactivity of mesothelial cells in mimicking metastatic cancer cells morphologically, featuring enlarged nuclei, increased nuclear and cytoplasmic ratios, among other cytomorphological features.

We measured the nanomechanical properties associated with the observed shape changes inherent to metastatic adenocarcinoma cells and benign mesothelial cells in pleural effusions obtained from body cavity fluid samples collected from patients with suspected metastatic adenocarcinoma, using standard protocols (see Methods). Samples were centrifuged, and cell pellets were resuspended in culture medium for 12 h, based on time-culture experiments to establish the optimum incubation time for cell-substrate adherence for mechanical analysis while minimizing artifacts resulting from in vitro culture. The 12 h incubation also allowed the differentiation of benign and malignant cells based on their ex vivo growth and morphological characteristics (that is, 'tumour' cells display anchorage-independent growth patterns, such as rounding of cells, whereas 'normal' mesothelial cells show a large, flat morphology; Fig. 18.1), so that AFM analysis could be performed separately on these two cell populations. These analyses were later confirmed by immunofluorescence analysis

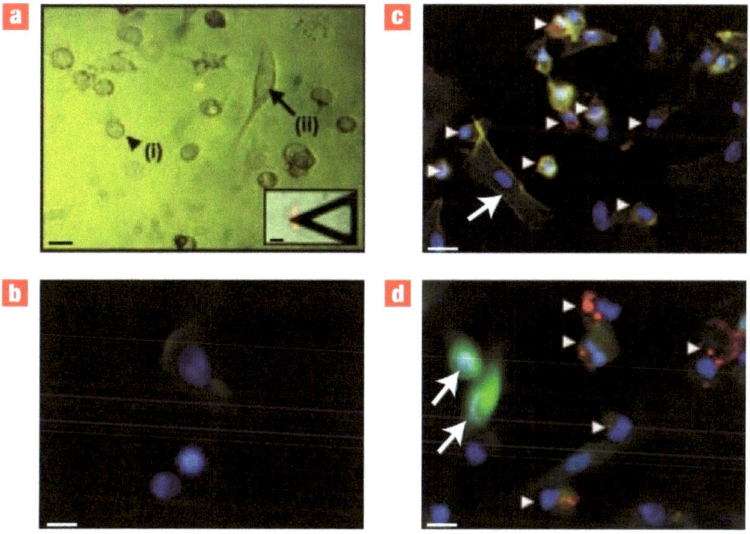

Figure 18.1 Optical and fluorescence images of a cytological sample. (a) Optical image demonstrating the round, balled morphology of visually assigned tumour cells (i) and the large, flat morphology of presumed benign mesothelial (normal) cells (ii). The inset shows alignment of an AFM tip over the central (nuclear) region of a cell. Scale bar = 50 μm. (b–d) Immunofluorescence analysis of a clinical pleural effusion displaying two cell populations (as shown in a): fluorescence image of the negative control (omitting primary antibodies for Ber-EP4 and Calretinin) (b); corresponding positive immunofluorescence triple labelling assay for DNA (blue), F-actin (green) and Ber-EP4 (red) (c) and for DNA, Ber-EP4 and Calretinin (DNA = blue, Ber-EP4 = red, Calretinin = green) (d). Arrowheads = tumour, arrow = mesothelial cells. Scale bars (main panels a–d) = 10 μm.

using biomarkers specific for metastatic cancer and mesothelial cells. Immunofluorescence analysis was performed independently, although on the same population of cells from the same pleural effusion sample, to confirm that cells subjected to AFM analysis based on growth characteristics indeed represented either tumour or mesothelial cells. For each sample, eight probable mesothelial ('normal') cells and eight probable malignant ('tumour') cells (except in the negative cases where only benign cells were present) were selected from their culture dish for ex vivo AFM analysis, which was conducted in an alternate fashion to ensure similar conditions were applied for both populations, without prior knowledge of the cytomorphological analysis. Cytomorphological (with immunohistochemical confirmatory analysis), AFM and immunofluorescence analyses were performed independently. Cytomorphological analysis was performed on its own population of cells, but the immunofluorescence and AFM analyses were made on the same population of cells; however, all of the cell populations were obtained from the same pleural effusion. Mechanical measurements were obtained at 37°C at a rate of 1 Hz (see Methods). Force–displacement curves were recorded on each cell to determine the relative cell stiffness (Young's modulus, E) of individual cells, yielding values of E for each cell type per clinical specimen (Table 18.1).

To confirm the selected cell populations actually represent tumour and mesothelial cells, immunofluorescence triple labelling assays of the samples were performed (see Methods; Fig. 18.1b–d). Labelling for DNA/Ber-EP4/F-actin (Fig. 18.1c) and DNA/Ber-EP4/Calretinin (Fig. 18.1d) both showed staining of the small, round cells for Ber-EP4 (red), a marker for metastatic adenocarcinoma cells, thus confirming that the round, balled cells (shown optically in Fig. 18.1a) were indeed metastatic adenocarcinoma cells. Moreover, the large, flat cells were positive for Calretinin staining (Fig. 18.1d), indicative of normal mesothelial cells and consistent with the optical morphology (Fig. 18.1a). Immunofluorescence analysis showed that the small round cells were not apoptotic, dead or mitotic, as they have intact DNA (Fig. 18.1b–d, blue fluorescence).

Cell elasticity measurements (Young's modulus, E) taken on cytological samples collected from patients with suspected metastatic adenocarcinoma are shown in Fig. 18.2. Data collected from seven different clinical samples (positive for metastatic tumour, $n = 40$; negative for metastatic tumour, $n = 48$) yielded average E values (mean $\pm$ s.d.) for all tumour and benign mesothelial cells of 0.53 ± 0.10 kPa and 1.97 ± 0.70 kPa, respectively (Fig. 18.2b,c; $P = 8.72 \times 10^{-22}$). A two-sample independent t-test conducted on the tumour and mesothelial cell populations showed that the population means were significantly different from each other at the 95% confidence level ($P = 8.72 \times 10^{-22}$). In particular, tumour cell elasticity measurements from a single clinical sample were found to have an average cell stiffness (mean ± s.d.) of 0.56 ± 0.09 kPa (Fig. 18.2e); however, the average cell stiffness for the morphologically determined benign mesothelial cells in the sample expressed a significantly increased average cellular elasticity (mean ± s.d.), with a value of 2.10 ± 0.79 kPa (Fig. 18.2f; $P = 7.77 \times 10^{-5}$). Interestingly, the tumour and benign mesothelial cells exhibited significantly different trends; elasticity measurements for the metastatic cancer cells and the benign mesothelial cells were best fit by Gaussian (normal) and log-normal fits, respectively. Tumour cells displayed a narrow, spiked peak with little spread, whereas benign mesothelial cells displayed a broad peak. Similarly, elasticity measurements collected on a single negative pleural fluid sample yielded an average cell stiffness (mean ± s.d.) of 2.09 ± 0.98 kPa (Fig. 18.2g; $n = 8$). The data clearly show that the cell stiffness of metastatic cancer cells is ~73 ± 11% less stiff than the benign mesothelial cells in the same sample and when compared to samples collected from other patients.

Figure 18.2h,i shows Gaussian fits of Young's modulus data collected on an individual clinical sample (Table 18.1, Case 4) containing only one population of cells, which were difficult to determine as either benign mesothelial cells or metastatic malignant cells, based on cytomorphology. However, following ex vivo culture there still appeared to be two populations of cells, one cell type that was larger and more flat, suggestive of benign mesothelial cells, and the other type, which was more round and balled up, suggestive of tumour cells. Despite the difference of morphology in ex vivo culture, nanomechanical analysis concluded that these were all tumour cells. Average values (mean ± s.d.) for both the tumour and morphologically classified benign mesothelial cells were 0.54 ± 0.08 kPa and 0.54 ± 0.12 kPa, respectively (Fig. 18.2h,i; $n = 8$). These population means were not statistically different, and indicated the presence of a single population of malignant cells. Subsequent immunohistochemical staining (IHC) with a battery of markers including Calretinin, B72.3 and Ber-EP4 (see Supplementary Information) confirmed that the sample contained predominantly one cell population, which was indeed malignant based on IHC findings (negative for Calretinin and positive for Ber-EP4 and B72.3). Additionally, this particular case demonstrates

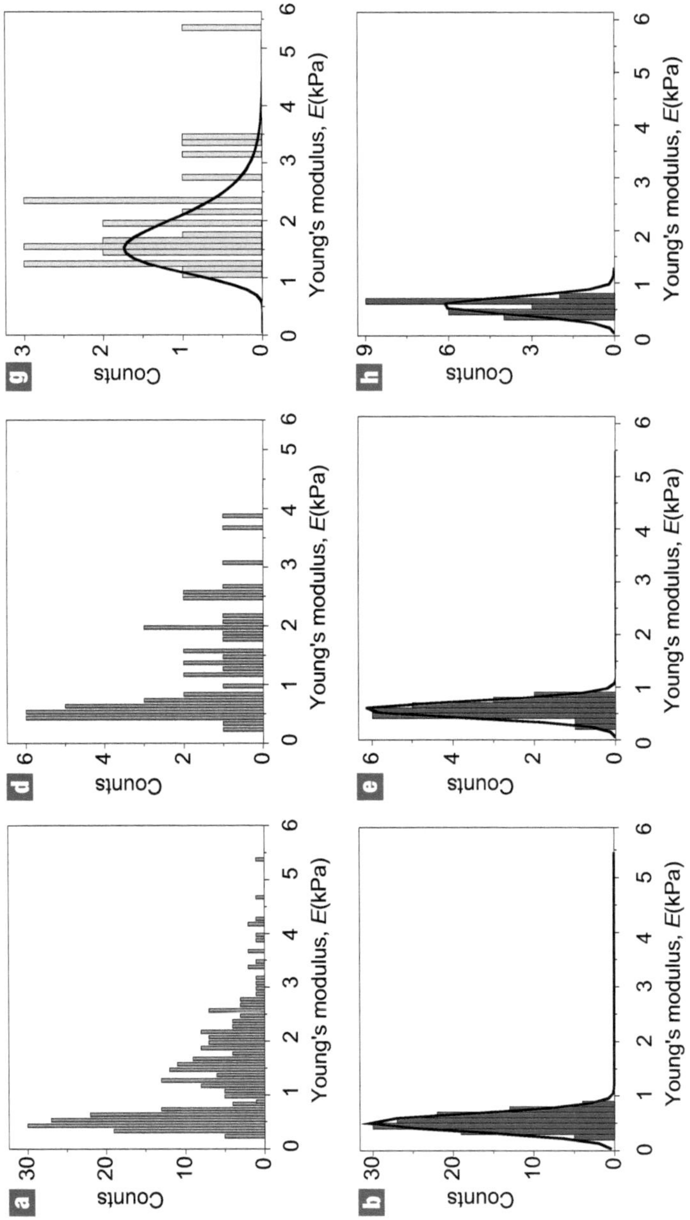
a
b
d
e
g
h
Counts
Young's modulus, E(kPa)

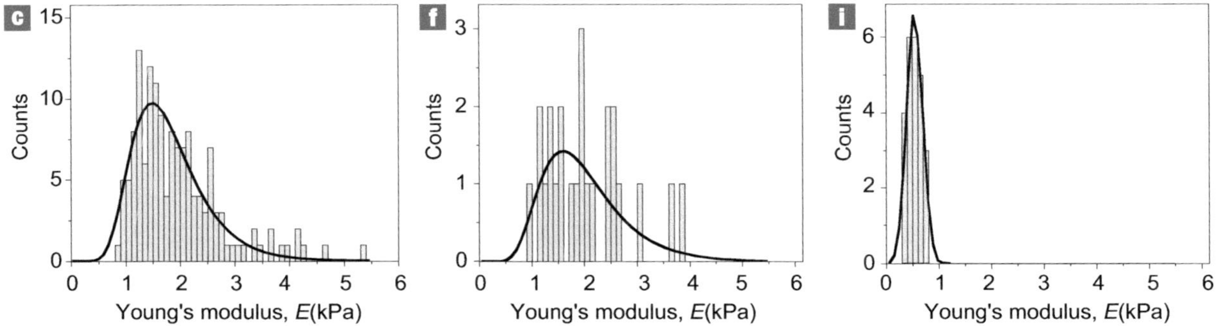

Figure 18.2 Histograms of the associated Young's modulus E for cytological samples collected from patients with suspected metastatic cancer. (a) Histogram of E for all data collected from seven different clinical samples (tumour, $n = 40$; normal, $n = 48$). (b) Gaussian fit for all tumour data from the seven samples. (c) Log-normal fit for all normal data from the samples. (d–f) Data from a single clinical sample: values of E for all cells (d), Gaussian fit of tumour cells (e), and lognormal fit of normal cells (f) for this sample. (g) Histogram of E with log-normal fit for a single negative sample. (h, i) Gaussian fits of visually diagnosed metastatic (tumour, h) and mesothelial (normal, i) cells in an individual clinical sample (Table 18.1, Case 4).

no correlation between ex vivo cultured cell morphology and expected measured stiffness. This lack of correlation suggests that it is unlikely that the ex vivo cultured morphological differences influence measured elasticity.

For each individual patient effusion, a new cantilever was used, to avoid any contribution of possible artifacts. The similarity in the data from patient to patient and cell to cell is strongly suggestive of a parallel between patient effusions and cell type. To further confirm that the measured differences between tumour and benign cells were not an artifact of ex vitro culturing, we used a modified cytocentrifugation procedure without ex vivo culturing, which yielded morphologically indistinguishable cell types. The resulting Young's modulus values for tumour and benign mesothelial cells were analogous to those obtained using the 12 h ex vivo culture method (see Supplementary Information for more details). This correlation of cell stiffness for the tumour and normal cells in the pleural effusions experimentally demonstrates that the mechanical analysis is representative of cell type, even when considering different preparation methods and the use of different substrates.

The need for biomarkers for cancer detection and analysis is critical due to the complexity of the disease. This study shows important implications for the combined use of imaging analysis with nanomechanical measurements as a novel biomarker for evaluating and sensing changes in tumour cells, and which is easily translated into clinical settings. Our results suggest that a shift in biologically driven biomechanical properties towards a decrease in cell stiffness correlates with an increase in metastatic potential. Under analogous conditions, cell stiffness of metastatic cancer cells is >70% lower than normal reactive mesothelial cells in the same sample, when compared to other pleural effusions, or even for patients with different clinical histories (Table 18.1). This finding is consistent with previous studies, which found a decrease in stiffness of human cancer cell lines with increasing metastatic efficiency using different single-cell in vitro biomechanical assays [13–15]. Despite morphological overlap between tumour and normal cell types, which often poses a significant problem for cytomorphological and immunohistochemical diagnosis of malignant cells, mechanical analysis provides the ability to distinguish cancer cells. These findings clearly demonstrate the ability of nanomechanical-based functional analysis to detect metastatic tumour cells in body fluid. Likewise, the distribution of measured cell stiffness for tumour cells is over five times narrower than the corresponding distribution for benign mesothelial cells. We found that tumour cell stiffness fits with a normal distribution, whereas the measured benign mesothelial cell stiffness values better fit a log-normal distribution. The finding that no quantitative overlap was observed in the measured cell elasticity values collected for the metastatic tumour and reactive mesothelial cells is of major diagnostic significance. Cancer is extremely biochemically diverse, yet mechanically a common modulus for each cell type is exhibited even for different tumour types and patient effusions (Table 18.1). The correlation of cytomechanical measurements with immunohistochemical analysis suggests that nanomechanical measurements of cancer cells has potential for the detection of cancer and may aid in personalized selection of medication and drug screening, especially in body cavity effusions where accurate diagnosis based solely on morphology has to date been challenging.

18.1 Methods

18.1.1 Cytological Sample Collection

The body cavity fluid samples were collected and processed using a standard protocol in cytology labs for conventional cytological analysis including Papanicolaou stain, Gimsa stain and cellblock preparation. An aliquot of the remaining sample (10 ml) was centrifuged at 500 g for 10 min. Cell pellets were resuspended with MEM-F12 culture medium and incubated for 12 h at 37°C in 5% CO_2 and 95% air. The culture medium was changed immediately before AFM analysis to wash off any dead and untouched cells.

18.1.2 Nanomechanical Measurements with AFM

All studies were conducted using a Nanoscope IV Bioscope (Veeco Digital Instruments) with a combined inverted optical microscope

(Nikon eclipse TE200). This combination permitted lateral positioning of the AFM tip over the nuclear region of the cell with micrometre precision (Fig. 18.1a, inset). The scan size for all measurements was set to 0 nm to maintain a constant position over the cell, and using the AFM software the tip was brought into contact with the central region of the cell [26]. Mechanical measurements were collected at 37°C using sharpened silicon nitride cantilevers with experimentally determined [27] spring constants of 0.02 N m^{-1} and a tip radius of <20 nm.

Force–displacement curves were recorded at 1 Hz for determination of Young's modulus E. E was calculated by converting the force curves into force–indentation curves [28] and fitting with the Hertz model, which describes the indentation of an elastic sample using a stiff conical indenter, as described elsewhere [10, 29]. The half opening angle of the AFM tip was 36° and the Poisson ratio of the cell was taken to be 0.5, as is typical for soft biological materials [28]. To prevent damage to the cell surface and to reduce any possible substrate-induced effects, measurements were analysed in force ranges resulting in shallow indentations of the cell (<400 nm) [30].

18.1.3 Immunofluorescence Triple Labelling

Two types of triple labelling assays were performed: One was DNA/F-actin/Ber- EP4 and the other DNA/Calretinin/Ber-EP4. For both experiments, cells were fixed first with 3.7% formaldehyde for 30 min at room temperature and then washed with 1× PBS three times, then incubated with 1% BSA in PBS pH 7.4 for 30 min. For DNA/F-actin/Ber-EP4 labelling, cells were first incubated with mouse anti-human Ber-EP4 (DAKO) at 1:300 dilution for 1 h, followed by Cy3-conjugated AffiniPure goat anti-mouse IgG(H + L) (Jackson ImmunoResearch Lab) at 1:200 dilution for 30 min, then with BODIPY FL phallacidin F-actin (Molecular Probes) at 1:40 dilution for 30 min. Finally, cells were incubated with 1:10,000 DAPI for 5 min. For DNA/Calretinin/Ber-EP4 labelling, cells were first incubated with mouse anti-human Ber-EP4 (DAKO) at 1:300 dilution for 1 h, followed by Cy3-conjugated AffiniPure goat anti-mouse IgG(H + L) (Jackson ImmunoResearch Lab) at 1:200 dilution for 30 min. Cells were then further incubated with 1:600 diluted rabbit anti-human Calretinin antibody (Zymed) for 1 h then with FITC-conjugated AffiniPure goat anti-rabbit IgG(H + L) (Jackson ImmunoResearch Lab) at 1:50 dilution for 30 min, followed by 1:10,000 DAPI for 5 min. All incubations were performed at room temperature and there were three PBS washing steps in between. Cells were covered with mounting medium for fluorescence microscopic examination. Images were taken using an Olympus BX-40 microscope with a × 40 objective.

18.1.4 Statistical Analysis

Data were expressed as mean ± s.d., and the statistical significance of differences in mean values was assessed using a two-sample independent Student's *t*-test at the 95% confidence level. Differences among means are reported using exact *P* values.

Acknowledgements

S.E.C. and J.K.G. acknowledge partial support from National Institutes of Health research grant no. 5 R21 GM074509 and from the Institute for Cell Mimetic Space Exploration, a National Aeronautics and Space Administration University Research Engineering Technology Institute. J.R. and Y.J. acknowledge partial support from National Institutes of Health research grant no. U01CA96116 and Alper grant, Johnsson Comprehensive Cancer Center.

References

1. Bhadriraju, K., and Hansen, L. K. Extracellular matrix- and cytoskeleton-dependent changes in cell shape and stiffness. *Exp. Cell Res.* **278**, 92–100 (2002).
2. Discher, D., Janmey, P., and Wang, Y. Tissue cells feel and respond to the stiffness of their substrate. *Science* **310**, 1139–1143 (2005).
3. Radmacher, M. Measuring the elastic properties of biological samples with the AFM. *IEEE Eng. Med. Biol. Mag.* **16**, 47–57 (1997).
4. Yamazaki, D., Kurisu, S., and Takenawa, T. Regulation of cancer cell motility through actin reorganization. *Cancer Sci.* **96**, 379–386 (2005).

5. Rao, J., and Li, N. Microfilament actin remodeling as a potential target for cancer drug development. *Curr. Cancer Drug Targ.* **4**, 267–283 (2004).
6. Motherby, H. et al. Pleural carcinosis confirmed by adjuvant cytological methods: A case report. *Diag. Cytopathol.* **19**, 370–374 (1998).
7. Osterheld, M., Liette, C., and Anca, M. M. D. Image cytometry: an aid for cytological diagnosis of pleural effusions. *Diag. Cytopathol.* **32**, 173–176 (2005).
8. McKnight, A. L. et al. MR elastography of breast cancer: preliminary results. *AJR Am. J. Roentgenol.* **178**, 1411–1417 (2002).
9. Bercoff, J. et al. In vivo breast tumor detection using transient elastography. *Ultrasound Med. Biol.* **29**, 1387–1396 (2003).
10. Rotsch, C., and Radmacher, M. Drug-induced changes of cytoskeletal structure and mechanics in fibroblasts: an atomic force microscopy study. *Biophys. J.* **78**, 520–535 (2000).
11. Lee, J., Ishihara, A., and Jacobson, K. How do cells move along surfaces? *Trends Cell Biol.* **3**, 366–370 (1993).
12. Stossel, T. P. On the crawling of animal cells. *Science* **260**, 1086–1094 (1993).
13. Suresh, S. et al. Connections between single-cell biomechanics and human disease states: gastrointestinal cancer and malaria. *Acta Biomaterialia* **1**, 15–30 (2005).
14. Guck, J. et al. Optical deformability as an inherent cell marker for testing malignant transformation and metastatic competence. *Biophys J.* **88**, 3689–3698 (2005).
15. Suresh, S. Biomechanics and biophysics of cancer cell. *Acta Biomaterialia* **3**, 413–438 (2007).
16. Binnig, G., Quate, C., and Gerber, C. Atomic force microscope. *Phys. Rev. Lett.* **56**, 930–933 (1986).
17. Rotsch, C., Braet, F., Wisse, E., and Radmacher, M. AFM imaging and elasticity measurements on living rat liver macrophages. *Cell Biol. Int.* **21**, 685–696 (1997).
18. Charras, G. T., and Horton, M. A. Single cell mechanotransduction and its modulation analyzed by atomic force microscopy indentation. *Biophys. J.* **82**, 2970–2981 (2002).
19. Dufrene, Y. F. Atomic force microscopy, a powerful tool in microbiology. *J. Bacteriol.* **184**, 5205–5213 (2002).

20. Pelling, A. E., Li, Y., Shi, W., and Gimzewski, J. K. Nanoscale visualization and characterization of *Myxococcus xanthus* cells with atomic force microscopy. *Proc. Natl Acad. Sci. USA* **102**, 6484–6489 (2005).
21. Pelling, A. E., Sehati, S., Gralla, E. B., Valentine, J. S., and Gimzewski, J. K. Local nanomechanical motion of the cell wall of *Saccharomyces cerevisiae*. *Science* **305**, 1147–1150 (2004).
22. Rief, M., Gautel, M., Oesterhelt, F., Fernandez, J. M., and Gaub, H. E. Reversible unfolding of individual titin immunoglobulin domains by AFM. *Science* **276**, 1109–1112 (1997).
23. Kasas, S., Gotzos, V., and Celio, M. R. Observation of living cells using the atomic force microscope. *Biophys. J.* **64**, 539–544 (1993).
24. Hansma, P. K. et al. Tapping mode atomic force microscopy in liquids. *Appl. Phys. Lett.* **64**, 1738–1740 (1994).
25. Salgia, R. et al. Expression of the focal adhesion protein paxillin in lung cancer and its relation to cell motility. *Oncogene* **18**, 67–77 (1999).
26. Wu, H. W., Kuhn, T., and Moy, V. T. Mechanical properties of 1929 cells measured by atomic force microscopy: effects of anticytoskeletal drugs and membrane crosslinking. *Scanning* **20**, 389–397 (1998).
27. Levy, R., and Maaloum, M. Measuring the spring constant of atomic force microscope cantilevers: thermal fluctuations and other methods. *Nanotechnology* **13**, 33–37 (2002).
28. Touhami, A., Nysten, B., and Dufrene, Y. F. Nanoscale mapping of the elasticity of microbial cells by atomic force microscopy. *Langmuir* **19**, 4539–4543 (2003).
29. Matzke, R., Jacobson, K., and Radmacher, M. Direct, high-resolution measurement of furrow stiffening during division of adherent cells. *Nature Cell Biol.* **3**, 607–610 (2001).
30. Stolz, M. et al. Dynamic elastic modulus of porcine articular cartilage determined at two different levels of tissue organization by indentation-type atomic force microscopy. *Biophys. J.* **86**, 3269–3283 (2004).

Supplementary Information: Chapter 18

Supplementary Discussion

Need for strong diagnostic marker for body cavity fluid samples

In general, histopathological examination of surgically removed tissue sample is the gold standard for the diagnosis of malignancy in any organ system, but this is not always possible for body fluid samples. Previous studies using such a standard showed cytomorphological analysis alone (without help of any ancillary testing, for example imuunohistochemistry) has the accuracy of about 50 to 70% to diagnose cancer in body fluid (Carlos WM Bedrossian MD in Malignant Effusions A Multimodal Approach to Cytologic Diagnosis. Igaku-Shoin. 1994; 121). Due to the tremendous burden associated with the diagnosis of metastatic carcinoma in body cavity fluid samples, various ancillary techniques have been utilized to assist the diagnosis of metastatic adenocarcinoma, mainly for the distinction of malignant epithelial cells from reactive mesothelial cells. Table 18.S1 lists the most commonly used markers up to date. A positive Calretinin staining has a sensitivity of 100% and a specificity of 80% in detecting mesothelial cells, whereas the sensitivity and specificity for detecting adenocarcinoma with BerEP4 is around 80% and 70%, respectively. A combination of immunohistochemical markers including Calretinin (positive in mesothelial cells but negative in metastatic adenocarcinoma) and BerEP4 and B72.3 (positive in metastatic adenocarcinoma but negative in mesothelial cells) performed on cell block sections of fluid samples the accuracy may be increased to 80% for separating benign from metastatic adenocarcinoma *(Politi, E., Kandraraki C., Apostolopoulou C., et al. Immunocytochemical panel for distinguishing between carcinoma and reactive mesothelial cells in body cavity fluids. Diagn Cytopathol. 32:151–5, 2005)* (Supplementary Fig. 18.1S). However, it is not 100% accurate, and often the cellblock may not be available. Calretinin helps separate benign mesothelial cells from

Table 18.S1 Ancillary tests for the diagnosis of metastatic adenocarcinoma

Histochemical
PAS with diastase digestion
Mucicarmine (neutral mucin)
Alcian blue
pH 1.5 (acid mucin)
Immunocytochemical
Calretinin
CEA
EMA
Leu M1
BerEp4
B72.3
Moc31
Ultrastructural
Surface microvilli
Intracytoplasmic lumina
Terminal bars
Secretory granules

metastatic malignant cells but it is not very specific. Tumor cells (especially mesothelioma) may occasionally be positive with Calretinin, whereas mesothelial cells may often appear negative. Thus, additional more accurate markers, such as the AFM measurements described in this manuscript, will be needed to further increase the accuracy of the diagnosis on cytological body fluid samples.

Whereas the current tests listed in Supplementary Table 18.S1 may be able to distinguish metastatic adenocarcinoma from mesothelial cells, such tests may not be utilized to distinguish other types of metastatic cancers such as squamous cell carcinoma, melanoma, and sarcoma. Furthermore, there is no test available so far to accurately diagnose malignant mesothelioma on a body cavity fluid cytology specimen, which by cytomorphology alone may be extremely challenging. This again argues for additional tests, such as nanomechanical analysis using AFM, which can be used to aid the diagnosis in body cavity fluid samples.

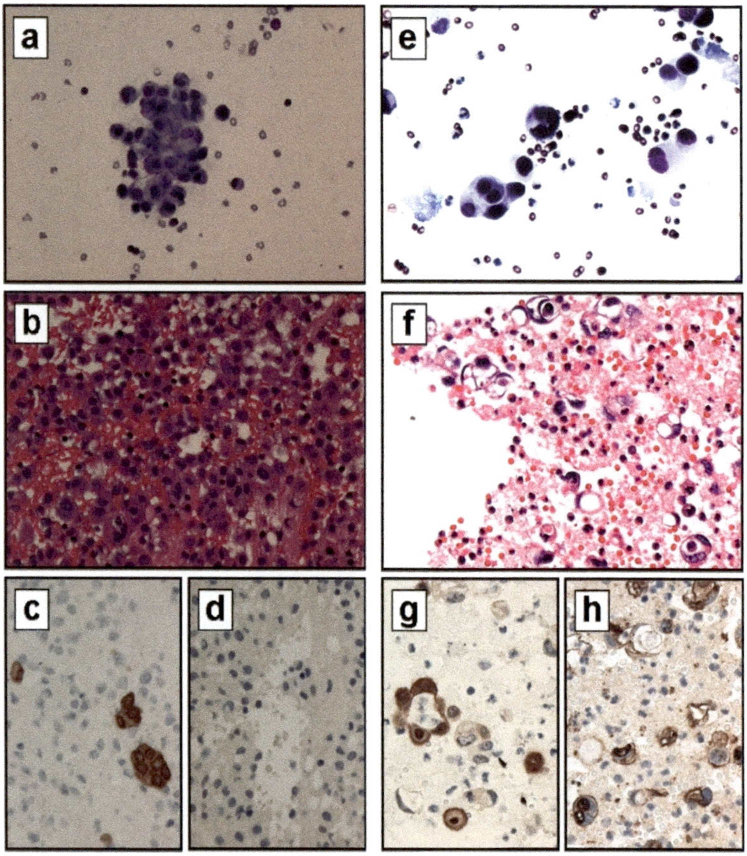

Figure 18.S1 Cytomorphological images of tested tumor-negative (a–d) and tumor-positive (e–h) samples. The Giemsa stained cytological smear (a, e) and Hematoxylin & Eosin stained tissue block section (H&E, b, f) demonstrated two populations of large cells, one representing tumor cells and the other, reactive mesothelial cells. It might be difficult, even for a trained cytopathologist, to separate these two cell types by morphology alone. There were also small inflammatory cells (lymphocytes and neutrophils, etc.) in the background, which can be readily distinguishable by morphology. To aid in the distinction of metastatic malignant adenocarcinoma cells from benign mesothelial cells, immunohistochemical staining (IHC) for various markers was performed on cellblock sections, as shown here. With the IHC staining, the two cell types were identified separately. Benign mesothelial cells (c, g) were positive for Calretinin and malignant cells associated with the tumor-positive sample (e–h) were positive for Ber-EP4 (h). All images are taken using a 40× objective.

Nanomechanical analysis and cell morphology

It has been well established that when tumor and normal cells are cultured, the tumor cells have some specific growth and morphological features distinctive from the normal cells, for example, anchorage independent growth pattern and pill up. In this study, we used these features to initially separate the tumor from normal cells for AFM measurements. But such difference of cultured cells may not be totally reliable. Furthermore, some samples may only have one population of cells (e.g., Case #4, Table 18.1 in the chapter). Therefore, immunofluorescence labeling of tumor and normal mesothelial cell specific markers (BerEp4 and Calretinin, respectively) were performed to confirm the selection.

To eliminate the concern that different morphology relating to "ex vivo" culturing condition, rather than the difference of cancer and normal cells, is the source of difference of measured stiffness between the two cell types we developed a preparation procedure to perform AFM analysis without the ex vivo culture. This procedure was originally used by Motherby, H. et al. Diag. Cytopathology 19, 370–374 (1998) but with a slight modification. In this procedure, cells were gently centrifuged on a slide (2 min for 500 g). The slide was then covered with same culture medium (90% D-MEM/F-12 (Ham) medium). AFM analysis was performed within 2 hrs of sample processing to preserve cell viability. The speed and time of centrifugation were determined to allow initial cellsubstrate adhesion with minimal alteration of the mechanical properties of the cells. Without ex vivo culture, the two cell types show similar morphological features. All cells, both "tumor" and "normal", appeared rounded and were indistinguishable by morphology alone. Below is the preliminary data from three clinical samples (2 positive samples):

"Tumor" ($n = 38$ measurements):

Young's modulus (E) $= 0.50 \pm 0.26$ kPa (mean $\pm$ s.d.)

"Normal" ($n = 27$ measurements):

Young's modulus (E) $= 2.73 \pm 1.30$ kPa (mean $\pm$ s.d.)

These average stiffness measurements, obtained using a completely different preparation procedure, are analogous to those obtained using the 12 h ex vivo culture method (average Young's Modulus (E) values (mean $\pm$ s.d.) for all "tumor" and "normal" cells using the ex vivo culture method as described in the manuscript were found to be 0.54 ± 0.14 kPa and 2.0 ± 0.82 kPa, respectively). This preliminary analysis shows that AFM measurements of cells directly centrifuged on slides without ex vivo culturing yielded similar findings as the cultured cells. Since morphologically the centrifuged tumor and normal mesothelial cells show similar features (e.g. both "tumor" and "normal" have a rounded morphology and are indistinguishable by shape), it is unlikely that morphology alone is sufficient to explain the difference of AFM measurement in these two cell types.

Chapter 19

Multi-Electrode Array Technologies for Neuroscience and Cardiology

Micha E. Spira and Aviad Hai
The Alexander Silberman Life Sciences Institute,
and the Harvey M. Kruger Family Center for Nanoscience,
The Hebrew University of Jerusalem, Jerusalem 91904, Israel
spira@cc.huji.ac.il

At present, the prime methodology for studying neuronal circuit-connectivity, physiology and pathology under in vitro or in vivo conditions is by using substrate-integrated microelectrode arrays. Although this methodology permits simultaneous, cell-non-invasive, long-term recordings of extracellular field potentials generated by action potentials, it is 'blind' to subthreshold synaptic potentials generated by single cells. On the other hand, intracellular recordings of the full electrophysiological repertoire (subthreshold synaptic potentials, membrane oscillations and action potentials) are, at present, obtained only by sharp or patch microelectrodes. These, however, are limited to single cells at a time and for short durations. Recently a number of laboratories began to merge the advantages of extracellular microelectrode arrays and intracellular microelectrodes. This Review describes the novel approaches,

Reproduced with permission from *Nature Nanotechnology*, **8**, 83–94, 2013.

Nano-Enabled Medical Applications
Edited by Lajos P. Balogh

ISBN 978-981-4800-32-7 (Hardcover), 978-0-429-39903-9 (eBook)
www.jennystanford.com

identifying their strengths and limitations from the point of view of the end users—with the intention to help steer the bioengineering efforts towards the needs of brain-circuit research.

A central goal of contemporary neuroscience is to understand the relationships between the functional connectivity-map of neuronal circuits and their physiological or pathological functions. At present, this goal seems impossible to achieve as available electrophysiological technologies only allow for extracellular recordings of large populations of neurons and are not suitable for providing simultaneous intracellular recordings of neural activity from hundreds of individual neurons. Furthermore, because of their mechanical instability, intracellular glass electrodes cannot be used to monitor long-term electrophysiological correlates of plasticity and learning. An ideal multiunit readout system should provide information that covers the entire repertoire of electrophysiological parameters from the individually recorded neurons. These include action potentials (APs), subthreshold excitatory- and inhibitory-postsynaptic potentials (EPSPs and IPSPs, respectively), and subthreshold membrane oscillations. Moreover, it should be possible to modulate the activity of individual neurons within the network by current application.

The available methodologies for the recording of neural activity include: (a) intracellular recordings and stimulation by sharp or patch electrodes, (b) extracellular recordings and stimulation by substrate-integrated microelectrode arrays (MEAs), (c) optical imaging and stimulation technologies of extrinsic fluorescent indicators or genetically encoded molecular probes, and (d) other methods such as functional magnetic resonance imaging, electroencephalography, electrocorticography and magnetoencephalography, designed to record activity from very-large-scale neural populations, which are not suitable for single-neuron resolution.

The invention of intracellular recording and stimulation technologies were hallmark developments that enabled the biophysical 'language' by which individual neurons transmit electrical information, communicate and 'compute' subthreshold synaptic information to be deciphered [1–4]. The power of intracellular recording systems is that they exhibit very good electrical coupling with the cell and provide accurate readout of the entire dynamic range of voltages generated by cells without distorting the readout over time. Yet, the use of sharp or patch microelectrodes is limited to individual neurons as steering of the electrode tips into target cells requires the use of bulky micromanipulators and the duration of intracellular recording sessions is limited by mechanical and biophysical instabilities.

In contrast, whereas the use of cell-non-invasive extracellular MEAs for in vitro recordings and polytrodes for in vivo recordings largely attenuate and temporally filter the electrical signals, it enables the simultaneous recording and stimulation of large populations of excitable cells for days and months without inflicting mechanical damage to the neuron's plasma membrane [5–10]. Extracellular field potential recordings (ambiguously referred to as local field potentials (LFPs) or field potentials (FPs)) reflect the spike activity of individual neurons or the superposition of fast APs, synaptic potentials and slow glial potentials in both time and space. As the physical processes that underlie the generation of FPs are understood it is theoretically possible to reconstruct their sources. Even though a great deal of information can be gained by using polytrodes, the information harboured in spike-pattern fingerprints is limited [11]. For example, extensive spike sorting cannot provide information as to whether the firing of an individual neuron is triggered by endogenous mechanisms, a barrage of incoming excitatory inputs or the cessation of inhibition (Box 19.1). What terminates the firing of a given neuron? Is it a barrage of inhibitory synaptic inputs, cessation of excitatory inputs or hyperpolarization of the membrane potential by endogenous mechanisms? Neurons that do not fire APs during a recording session are not 'visible' to extracellular electrodes (referred to as 'dark neurons'; Box 19.1). In some brain areas, 90% of the neurons are not spiking or are firing occasionally at very low rates of <0.16 spikes per second (for review see [12]). Intracellular recordings of synaptic potentials from such neurons would disclose a great deal of information as to the role of this 'silent majority' in information processing and the importance of individual neurons to the circuit behaviour. A great deal of neuroplasticity is associated with changes in the amplitude of synaptic potentials [13]. Unless these changes

Box 19.1 Illustration of some of the limitations of extracellular recordings with respect to intracellular recordings

Endogenous membrane properties as well as excitatory and inhibitory synaptic inputs regulate the firing patterns of individual neurons. This is depicted in the schematic of a neuron (blue) that receives an excitatory and an inhibitory synaptic input in (a) Subthreshold and supra-threshold electrophysiological activity of the neuron is recorded by an intracellular (upper orange electrode) and an extracellular (lower orange electrode) electrode. The amplifiers are depicted in yellow. The intracellular recordings are shown in the left panels of (b–e) and the corresponding extracellular recordings are shown in the right panels (green background). In (b), a neuron endogenously generates a train of APs (of approximately Δ100 mV) by depolarization of the membrane potential from the resting value of approximately −80 mV (bottom dashed line) reaching a threshold level to fire APs (middle dashed line) at about −50 mV, and then the membrane potential endogenously repolarizes. The extracellular electrode picks up the FPs generated by the APs (marked by vertical lines and green background). Note that the recorded FP amplitudes range between 0.01 and 1 mV, and are not drawn to scale. The attenuation factor (1/100 to 1/1000) is so large that subthreshold potentials generated by individual neurons cannot be recorded. Thus, the extracellular electrode is practically 'blind' to the subthreshold events (grey background, below the red dashed line). In (c) and (d), the very same pattern of APs firing is generated by excitatory (c) and inhibitory (d) synaptic inputs. Whereas in (c) summation of excitatory synaptic potentials depolarizes the neuron to reach the firing level, and the neuron stops firing when the barrage of the excitatory inputs stops (leading to membrane repolarization), in (d), the train of APs is generated by dis-inhibition (the cessation of the barrage of inhibitory synaptic inputs). The significant differences in these mechanisms (b–d) cannot be detected by the extracellular electrode. Furthermore, unless an

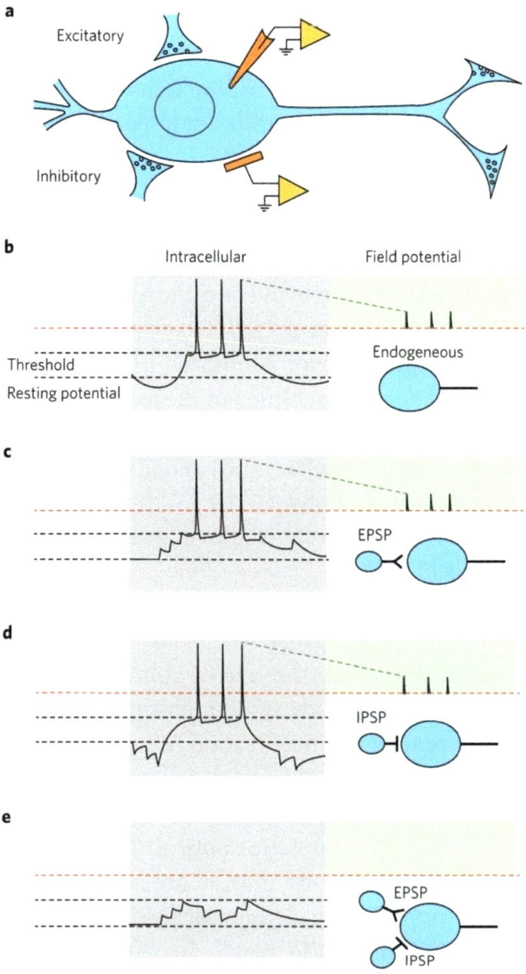

individual neuron is firing APs, synaptic inputs are not 'visible' to the extracellular electrodes at all (e). In this example, the extracellular electrode does not detect the presence of a neuron that receives a barrage of excitatory and inhibitory synaptic inputs. These inputs may be of significant importance to the functioning of the neuronal circuit.

reach firing threshold, extracellular recording systems are 'blind' to these critical events. It is conceivable that significant signalling between neurons is mediated by subthreshold potentials (chemical or electrical synapses) and is thus undetectable by conventional extracellular electrodes.

Since the development of the first MEA [14–17], technological efforts improved the quality of information gained by extracellular recordings mainly by increasing the density and the number of the electrodes that can be constructed and addressed over a single MEA. In vitro MEAs may contain over 10,000 electrodes [5, 7, 8, 18, 19]. In vivo polytrodes may have over a hundred [9, 20]. Nevertheless, the recording and stimulation qualities of these platforms (reflected by the electrical coupling coefficient between single neurons and the device, and the signal-to-noise ratio) remained poor. Typically the amplitudes of FPs range between 10 μV to 1 mV and a great deal of computational power is required to extract data and sort out the recorded signals [21, 22].

In parallel to the development of the extracellular MEA, efforts to develop optical imaging approaches began. These included imaging of membrane potentials by the use of voltage-sensitive dyes [23–27], imaging of neuronal activity by monitoring the changes in the free intracellular calcium concentration [28–30] and the monitoring of intrinsic signals [31]. This was followed by very powerful methods to optically excite or inhibit individual neurons or neuronal ensembles [32, 33]. The powerful optical imaging methods suffer from a number of limitations that at present prevent them from replacing electrophysiological approaches for studying neuronal circuits (for discussion see [34]).

Using nano- and micro-technologies, a number of laboratories began to merge the advantages of substrate-integrated extracellular MEA technologies with the critical advantages of intracellular electrodes. Namely, the construction of nano- or microdevices that enable simultaneous, long-term, multisite, intracellular recording and stimulation from many neurons under in vitro conditions. Further development and implementations of these technologies are expected to revolutionize basic and applied neuroscience.

In this Review we describe recent developments, expected benefits from their use and the foreseen limitations of the different approaches. We review these developments from the end-user point of view, rather than from the technological point of view. By considering the needs of contemporary 'circuit neuroscience' on the one hand and the recent technological developments in MEA fabrication on the other, we hope to shift the focus of upcoming technical developments from the habitual engineering optimizations to the pressing needs of neuroscience.

To objectively evaluate the different approaches, we examine the principal achievements in relation to a list of biophysical parameters that are needed to decipher the functional connectivity map of a neuronal network. An ideal imaginary device would allow the user to: (a) simultaneously record and stimulate hundreds of individual neurons intracellularly, (b) maintain a stable contact with the neurons for recording and stimulation for days and months, (c) monitor the transmembrane potential in the relevant cell-physiological range of -80 to $+30$ mV, (d) detect subthreshold potentials such as excitatory and inhibitory synaptic potentials with amplitudes in the range of ± 0.5–10 mV with a rise time of <1 ms and a slow decay time of 100–1,000 ms, and to record membrane oscillations in the range of ± 5 mV at frequencies of 1–50 Hz, (e) record APs with amplitudes of ~ 100 mV and duration of 1–500 ms (long APs for recording from cardiomyocytes).

To acquaint the reader with the basic terminology used in the field and as a technical introduction we begin the discussion by describing the structural and electrical relationships formed between neurons and substrate-integrated MEAs and briefly explain the contribution of the various parameters to the electrical coupling between excitable cells (neurons and cardiomyocytes) and MEAs. We then review the micro- and nanoengineering approaches recently used in attempts to merge the benefits of extracellular MEAs with those of the sharp glass electrodes, as well as the critical problems that would have to be addressed in order for the new approaches to be incorporated in basic research and clinical applications. On the basis of this discussion we propose a tentative recipe that we believe might provide an optimal approach to construct an MEA that can provide simultaneous, long-term, multisite, non-destructive intracellular recording and stimulation of neurons.

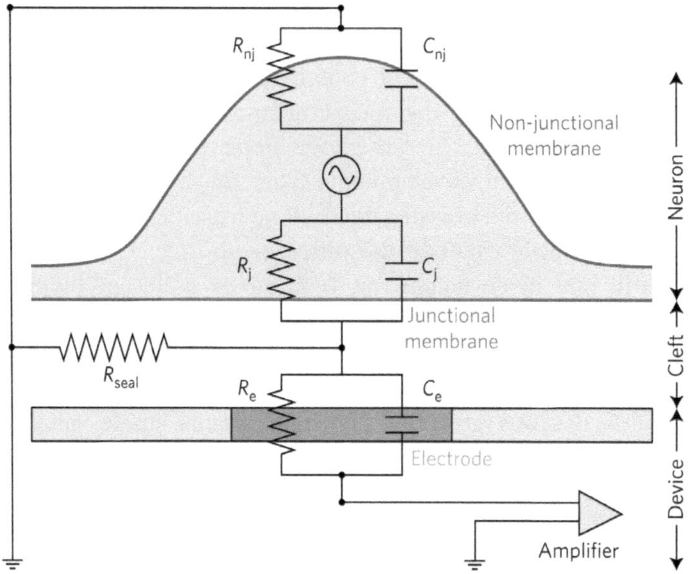

Figure 19.1 Schematic layout depicting the spatial relationships between a neuron and a substrate-integrated electrode and the analogue passive electrical circuit. The cell body of a neuron (light blue) resides on a sensing electrode (orange) integrated in the culture substrate (yellow). The electrode is coupled to an amplifier (yellow). A cleft filled by the culturing media (ionic solution) interposes between the cell membrane and the electrode–substrate. The neurons plasma membrane is subdivided into two: the part that faces the electrode (blue) is defined as the junctional membrane and is represented by the junctional membrane resistance (R_j) and the junctional membrane conductance (C_j). The rest of the membrane, defined as the non-junctional membrane (red), faces the bathing solution and the culture substrate. This part of the membrane is represented by the non-junctional resistance (R_{nj}) and the non-junctional capacitance (C_{nj}). The physiological solution within the cleft generates the seal resistance (R_{seal}) to ground. The electrode (orange) impedance is represented by the electrode resistance and capacitance (R_e and C_e, respectively). The electrode can be a passive element or a transistor. For simulation purposes of APs or intracellular current injections, current can be injected into the analogue cell-circuit in-between R_{nj} and R_j. Under physiological conditions current is generated by transient changes in the membrane conductances. The colour coding shown here is used in Fig. 19.4 to depict the different components of the analogue electrical circuit.

19.1 Electrical Circuit Analogue of the Neuron/Electrode Interface

The structural relationships between a neuron and a substrate-integrated planar electrode along with the analogue electrical circuit are schematically depicted in Fig. 19.1. The neuroelectronic hybrid is composed of three components (a) a neuron, (b) a cleft formed between the neuron and the substrate surface, and (c) the electrode.

Typically, differentiated neurons are non-isopotential structures with a cell body from which neurites in the form of a dendritic tree and a single long cylindrical axon emerge. These neuronal compartments adhere to MEA substrates by electrostatic or chemical interactions between adhesion molecules that protrude from the lipid membrane of the neurons and molecules deposited on the MEA platforms by the experimentalist [18, 35]. The cleft formed between the cell membrane and the MEA substrate is filled with the ionic solution.

For the simplified model depicted in Fig. 19.1, the neuron surface area is subdivided into: a junctional membrane that faces the sensing pad(s) (with R_j as the junctional resistance), and the non-junctional membrane (with R_{nj} as the non-junctional resistance) that faces the bathing solution and the substrate. Propagating APs or synaptic potentials produce complex extracellular current flow between activated compartments and other parts of the neuron. Fractions of these extracellular currents flow between the non-junctional and the junctional membranes. The cleft, formed between the neuron and the sensing element, generates a resistance that is referred to as the seal resistance (R_{seal}). The voltage formed over R_{seal} directly modulates the gate voltage of a field-effect transistor (FET), or the charge dispersal across a passive metal electrode [36–38].

19.2 Neuron–Device Electrical Coupling

The electrical coupling between a neuron or a cardiomyocyte and a sensing pad is defined here as the ratio between the maximal voltages recorded by the device in response to the maximal

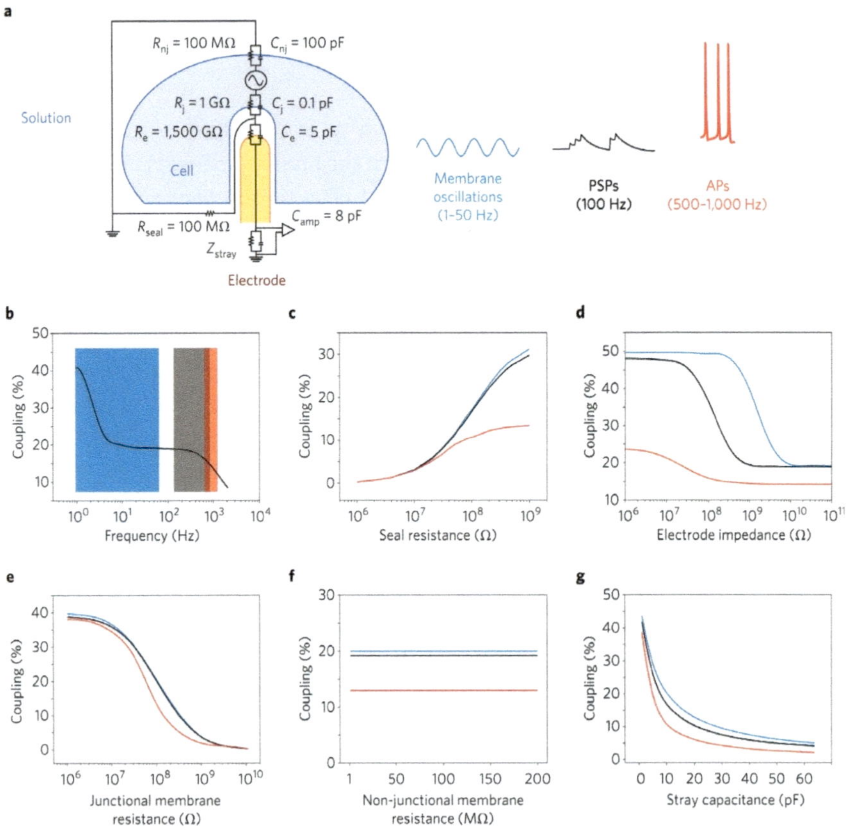

Figure 19.2 Simulation of the contribution of individual electrical components to the neuron-electrode electrical coupling coefficient. (a) Left: the analogue passive electrical circuit and default parameters used for the simulation parameters of a representative passive analogue electrical circuit. Right: the simulations were conducted for low-frequency signals depicting membrane oscillations, medium frequencies depicting postsynaptic potentials (PSPs) and fast frequencies depicting APs. C_{amp}, input capacitance of the amplifier; Z_{stray}, stray impedence. For the simulation, all default parameters were kept constant as shown in (a) while the tested parameter was varied in the range indicated by the horizontal axis. (b–g) The coupling coefficient as a function of: voltage pulse frequency (b), the seal resistance (c), the sensing pad impedance (d), the junctional membrane resistance (e), the non-juctional membrane resistance (f) and the stray capacitance (g). Simulations based on refs. [64, 65].

voltage generated by an excitable cell. Figure 19.2 illustrates how individual parameters of a passive analogue electrical circuit contribute to the coupling coefficient of slow membrane oscillations, medium-frequency synaptic potentials, and fast APs (Fig. 19.2a). The parameters used for the simulation approximate the passive physical properties of an excitable cell cultured on a substrate-integrated electrode (Fig. 19.2a).

Simulation of voltaic events of various frequencies illustrates that the coupling coefficient for higher-frequency APs is attenuated more strongly compared with postsynaptic potentials and membrane oscillations (Figs. 19.2b and 19.3).

Theoretical and experimental considerations revealed that reductively, the amplitude and shape of the FPs are determined by the multiplication of the R_{seal} value by the current that flows across it (Fig. 19.2c). For these reasons, intensive efforts were devoted to increase the value of R_{seal} [35, 37, 39]. Most of these research efforts yielded only limited improvements in the signal-to-noise ratio of the various devices. Studies of cell/electrode interfaces showed typical cleft thicknesses between 40–100 nm [40–44]. For most cell types this would correspond to R_{seal} in the range of 1–2 MΩ and FP recordings in the range of a few tens to few hundreds of microvolts.

Another factor that largely affects the electrical coupling coefficient between cells and an MEA is the input impedance of the sensing pad (Figs. 19.1 and 19.2d). The currents of living cells and electronic devices are fundamentally different: the former are formed by ions in solution whereas the latter, by electrons in mostly solid-state metals and semiconductors [37, 45]. The ramification of this difference is the effect of the impedance of the device on the electrical coupling, reflected by the sensing pad geometry and the material it is composed of. Typically the impedance of the sensing pad, either constructed from noble metals or insulated semiconductors, is attributed to the 'blocking' ion bilayer formed at the device active region and the ionic solution in which the neurons reside. For example, standard planar gold electrodes with radius of 30 μm have impedances of 50 KΩ at 1 KHz in electrolyte solution. Reducing the surface area of individual sensing pads to match the dimensions of individual neurons enables the density

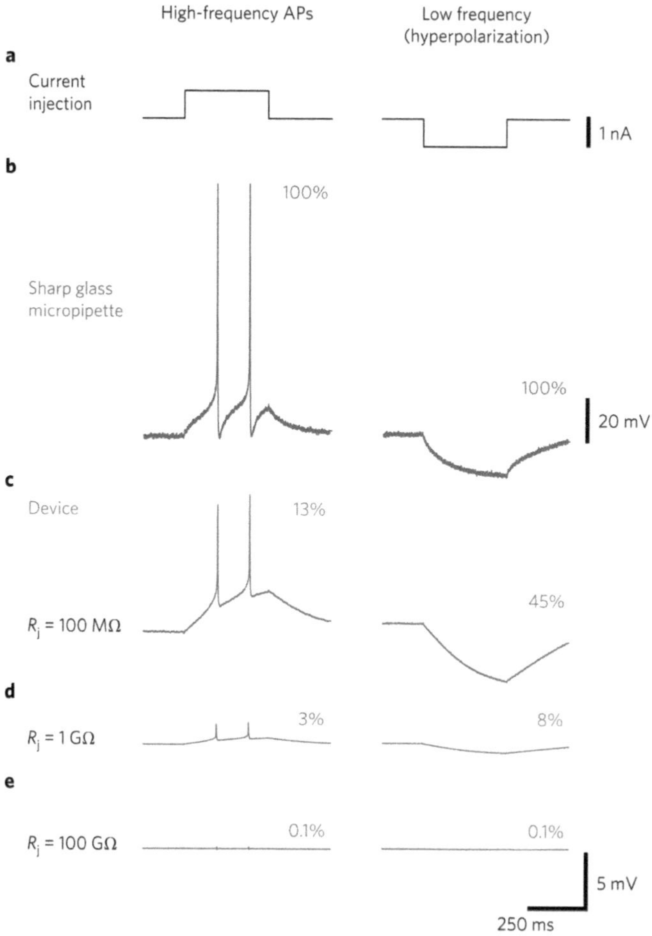

Figure 19.3 Dependency of the electrical coupling on the junctional membrane resistance and pulse duration. Shown is a simulation of the cell–device coupling of APs and a long hyperpolarizing pulse at three different values of R_j (100 MΩ, 1 GΩ and 100 GΩ). (a) Schematic illustrations of the depolarizing (left) and hyperpolarizing current pulse (right) delivered to generate two APs and membrane hyperpolarization, respectively. (b) Simulation of the ensuing intracellular potentials recorded by an intracellular electrode (red). (c–e) The recorded potentials by an extracellular-located electrode (as shown in Figs. 19.1 and 19.2) under different junctional membrane values (blue). Note that the amplitude of the extracellularly recorded APs (short pulses) is reduced faster than the voltage generated by the long pulse. That is, the coupling coefficient of the AP is more sensitive to the value of R_j than that of the long pulse. The values used for the simulation are identical to those shown in Fig. 19.2 although R_j is altered as indicated.

of the MEA and its spatial resolution to be increased [5, 7, 8, 46, 47]. This however, is reflected by the reduction of the FP amplitudes as a result of the increase in impedance and consequent reduction of the signal-to-noise ratio. Thus, the electrodes geometry and the ensuing impedance both place constraints on reducing the electrode size. Increasing the surface area by using nanostructures such as spongy platinum black or Ti_3N_4 [48–50], gold nanoflakes and nanopillars [51, 52], or carbon nanotubes [53, 54] is used to compensate for the dimensions of the electrode surface 'visible' to the cell. Although effective in reducing the impedance values up to 95% at approximately 1 KHz [54], in practice the recorded FPs are still in the range of hundreds of microvolts. This is most likely due to averaging of the complex positive and negative currents concomitantly generated by a number of sources, over fractions of the large surface area of the electrode. This 'averaging' usually results in reduced amplitude of the electrical readout. It should be noted, however, that reduction of the electrode impedance may be very effective in improving the readout signals when applied under conditions in which a single cell 'covers', engulfs or internalizes a single electrode such as those described in this chapter (Fig. 19.4).

The junctional membrane is defined as the resistance and capacitance of a membrane patch that faces the sensing pad or the gate of an FET (Fig. 19.1). The surface area of the junctional membrane can be anywhere between a very small fraction of the cell surface area, up to approximately 50% in cells that flatten while adhering strongly to substrate-integrated sensing pads. This variable depends on the geometry of the sensing pad and the morphology and adhesion characteristics of the specific cell. The junctional membrane can thus be of very high resistance and low capacitance. This implies that only a small fraction of the current generated across the neuron's membrane, flows through the junctional membrane. Reduction of the junctional membrane resistance would be very effective in improving the electrical coupling coefficient between a neuron and an electrode (Figs. 19.2e and 19.3). This is in fact the approach used by the classical methods of sharp electrodes, whole-cell patch electrodes or the perforated patch configuration [3, 55] (Fig. 19.4a,b). Attempts to improve the electrical coupling coefficient between cultured cells and planar

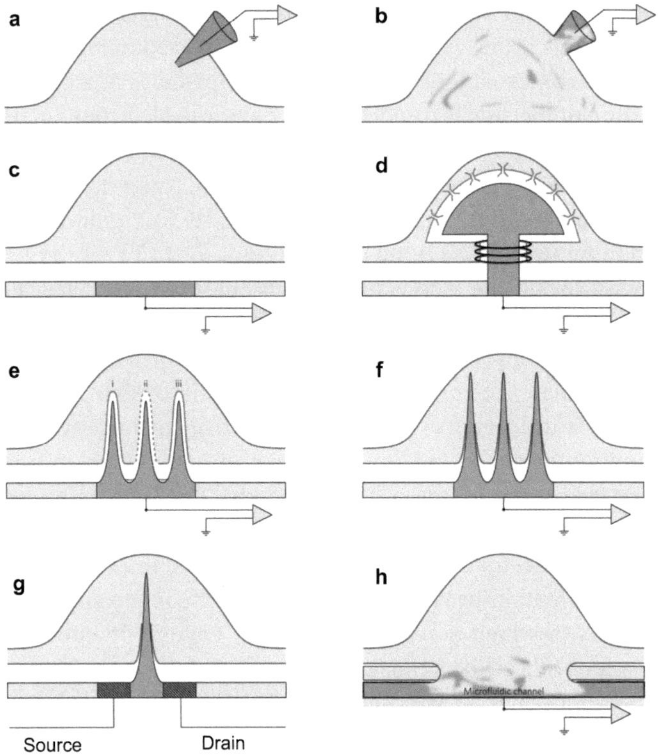

Figure 19.4 Different forms of the electrode/neuron interface configuration. The colour code represented here for the different components is the same as for Fig. 19.2. (a) A sharp glass intracellular microelectrode. (b) Whole-cell patch-electrode configuration. The 'mixing' of orange and blue schematically illustrates the perfusion of the cytosol by the electrodes content. (c) A neuron cultured on a substrate-integrated planar extracellular electrode. Note the cleft (white) separating the junctional membrane and the electrode. (d) A neuron engulfing a gold mushroom-shaped protruding microelectrode. Note actin rings surrounding the mushrooms stalk stabilizing the configuration. (e) Nanopillar electrodes extending into a cultured cardiomyocyte but that do not penetrate the plasma membrane (i). After the application of an electroporating pulse (ii) the nanopillar gains access to the cytoplasm. The electroporation is transient and the junctional membrane resistance recovers to control level within minutes (iii). (f) An array of nanopillars that penetrate the plasma membrane forming direct physical contact with the cytosol. (g) A nanopillar that serves as the gate for a nano-FET penetrates the cell's membrane. (h) Patch clamping of cultured neurons. The mixing of the ionic solution of the microfluidic system with the cytosol is depicted. For more details, see main text.

MEAs by expression of ion channels in the plasma membrane provided experimental demonstrations of the contribution of the conductance of the junctional membrane [37]. Nevertheless, experimental manipulation to improve the electrical coupling by expression of ion channels in neurons should not be used in studies of neuronal networks as such a manipulation changes the electroanatomy of the cells and their excitability and thus alters the functioning of the network being studied. In some recent studies, local increases in the junctional membrane conductance by localized electroporation have been used to transiently increase the coupling coefficient [56–60] (Fig. 19.4e).

Unavoidable stray capacitance along the conducting lines, together with the input impedance of the amplifying circuitry, further attenuates the recorded signals (Fig. 19.2g).

Recently, a number of laboratories began to merge the advantages of extracellular MEAs and intracellular microelectrodes. In the following paragraphs we describe the novel approaches, identifying their strengths and limitations from the point of view of the end users.

19.3 Neurons Actively Engulf Protruding Electrodes

The first series of studies reporting on successful multisite, non-invasive, intracellular recording and stimulation by MEAs were published by our laboratory between 2007 and 2010 [61–65]. In these studies we increased the neuron–microelectrode electrical coupling coefficient from approximately 0.1% as recorded by a planar extracellular MEA to approximately 50% by the use of a chemically functionalized micrometre-size mushroom-shaped gold protrusion as the sensing electrode (Figs. 19.4d and 19.5a). The increased coupling coefficient was associated with an intracellular recording of a monophasic positive attenuated intracellular AP instead of a typical biphasic FP. The unique neuron–electrode configuration used in this work, made it possible, for the first time, to record with a MEA action potentials as well as synaptic potentials

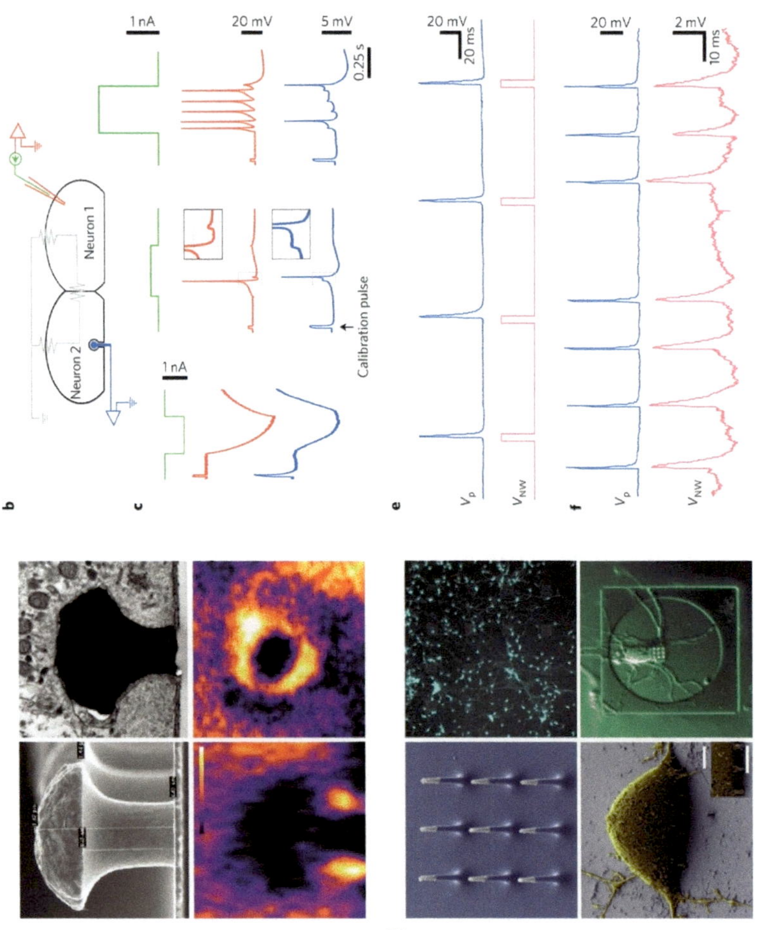
a
b
c
d
e
f
Neuron 1
Neuron 2
1 nA
1 nA
20 mV
5 mV
0.25 s
Calibration pulse
20 mV
20 ms
20 mV
2 mV
10 ms
V_p
V_{NW}
V_p
V_{NW}

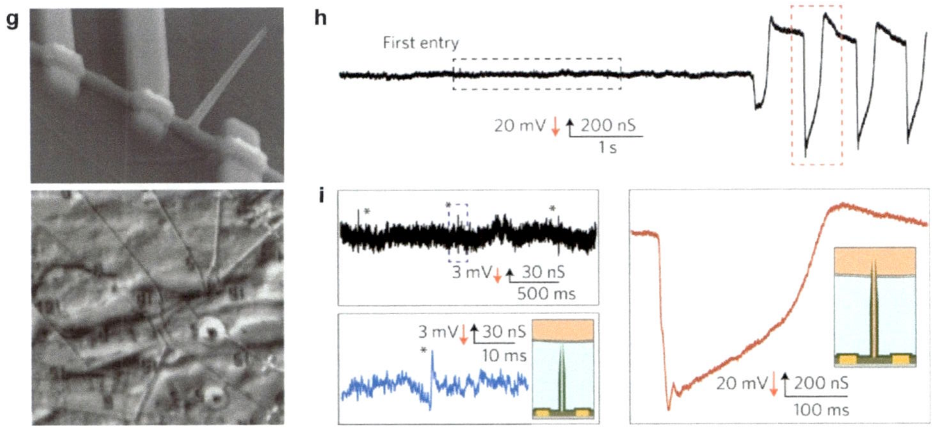

Figure 19.5 Recently developed MEA devices record intracellular potential from excitable cells. (a) Gold mushroom-shaped microelectrodes (gMμEs) functionalized with RGD-based peptide (top left, scanning electron microscopy (SEM) image) are engulfed by *Aplysia* neurons (top right, transmission electron microscopy image), which induces cytoskeletal reorganization around the structure (bottom, confocal microscopy images). The gMμE is 1.42 μm high. (b, c) Stimulation and recording of APs and subthreshold synaptic potentials is achieved by the gMμEs with a signal-to-noise ratio similar to sharp-glass and patch-clamp micropipettes. Green, current injection into the neuron; red, intracellular recording by a glass microelectrode; blue, in-cell recording by an extracellular gold mushroom microelectrode. (d) SEM (left) and optical (right) images of vertical nanowire electrode arrays (VNEAs). Scale bars, 2.5 μm. (e, f) Vertical nanowire electrode arrays (VNEA) can stimulate (e) and record (f) from rat cortical neurons monophasic APs. V_p, patch clamp voltage; V_{NW}, nanowire recordings voltage. (g–i) A phospholipid-functionalized silicon nanotube as the gate electrode of an FET device records APs from cardiomyocytes with a signal-to-noise ratio similar to glass electrodes. (g) SEM image of the device (top) and of cardiomyocytes grown on it (bottom). (h) Recording from a nanotube before penetration (left) of the cell's plasma membrane and after penetration (right). (i) Left: FPs recorded by the nanotube before penetration of the cell. Right: a single intracellularly recorded potential. Figure reproduced with permission from: (a) Top: [62], © 2009 RSC; Bottom: [63], © 2009 IOP; (b,c) [64], © 2010 Am. Physiological Soc.; (d–f) [71], © 2012 NPG; (g–i) [78], © 2012 NPG.

(Fig. 19.5c). The key to the multi-electrode-array 'in-cell recording' approach developed by us is the outcome of three converging cell biological principals: (a) the activation of endocytotic-like mechanisms in which cultured *Aplysia* neurons are induced to actively engulf gold mushroom-shaped microelectrodes (gMμE) that protrude from a flat substrate, (b) the generation of high R_{seal} between the cell's membrane and the engulfed gMμE, and (c) the increased junctional membrane conductance.

The neuron/gMμE interface was generated by chemically 'luring' the neurons to engulf the protruding gMμE by a highly conserved cell biological mechanism—endocytosis (which is a cell biological mechanism that underlies the internalization of particles into the cells [66]). The shape and the dimension of the gMμE were selected to mimic the geometry and dimensions of dendritic spines [67]. To facilitate the engulfment, the gMμEs were chemically functionalized by an RGD-based peptide [35]. This is one of a group of well-known molecular recognition motifs that trigger adhesion and engulfment mechanisms [68]. The localized presentation of the peptide by the gMμE (and possibly the electrode geometry itself) led to active engulfment of the electrodes by *Aplysia* neurons and a number of cell lines [61–65]. The engulfment of the microelectrodes is generated by molecular cascades that include the restructuring of the cytoskeleton to form an actin ring around the stalk of the 'gold mushroom' (Figs. 19.4d and 19.5a, left panel).

Using biophysical parameters obtained by direct measurements of the non-junctional resistance, non-junctional capacitance, electrode resistance, electrode capacitance and R_{seal}, as well as the estimated value of the junctional membrane properties calculated according to the geometry of the gMμE (Table 19.1), we simulated the expected recordings of APs and subthreshold potentials using an equivalent electrical circuit. We found that the calculated values used for the junctional resistance were insufficient to generate the neuron–electrode coupling coefficient as that obtained experimentally. To reach the coupling level observed in the experiments we had to increase the junctional membrane conductance by at least an order of magnitude [64, 65]. The mechanism underlying the increased junctional membrane conductance is not known. It is conceivable that in association with the cytoskeleton restructuring around the gMμE, voltage-independent ion channels are recruited to the junctional membrane to sufficiently increase the junctional conductance without leading to noticeable effects on the passive and active membrane properties. An alternative explanation could be that the mechanical tension generated at the curvature of the gMμE generates non-specific membrane nanopores [58, 69]. An important question that was brought up in this relation is whether the growth of *Aplysia* neurons on a matrix of gMμEs is altering the neuron's physiological properties. We found that the growth of the neurons on gMμEs did not alter the biophysical properties of the neurons or their synaptic communication. It does however alter the typical growth patterns of the neurites [63].

Stimulation of cells by substrate-integrated electrodes (either passive metal electrodes or FETs) often involves undesired electrochemical reaction products at the electrode interface and consequently damage to the cells that goes beyond transient electroporation. Measurements of the neurons input resistance before and after stimulation by gMμEs showed no change, affirming that high enough charge transfer can be applied to evoke APs without damage to the cell [64].

Out of the five criteria to evaluate of the benefits of the approaches, the gMμE-based MEA provided multisite, simultaneous, intracellular recording and stimulation for periods of days (which is for as long as we carried out the recordings). The filtering properties of the gold electrodes and the a.c. amplifier used do not enable the resting potentials of the neurons to be recorded. Nevertheless, the configuration successfully monitored subthreshold synaptic potentials and APs (Fig. 19.5c, middle panel). The filtering nature of the recording system can be deconvoluted and thus unfiltered high-quality recordings of APs and synaptic potentials can be retrieved. A stable electrical coupling between gMμE and a neuron coincided with the formation of cytoskeletal actin rings surrounding the stalks of the mushroom-like structure [59, 63] (Fig. 19.5a, lower panel). Individual gMμE enables both voltage recordings and application of current [64]. So far, attempts to obtain in-cell recordings and stimulation from rat hippocampal neurons and primary cardiomyocytes were unsuccessful. It should be noted nevertheless that these attempts were limited to gMμE functionalized with poly-D-lysine rather than by the engulfment promoting peptide.

19.4 Nanopillars for Intracellular Recordings and Stimulation

Sharp glass electrodes (Fig. 19.4a) and patch electrodes (Fig. 19.4b) provide excellent intracellular recording by penetrating the plasma membrane and directly accessing the cytosol while generating effective R_{seal} with the plasma membrane (hundreds of MΩ to a few GΩ, respectively, Table 19.1). A recent study from Park's laboratory [71] used vertical nanowire electrode arrays (VNEAs) constructed from a doped silicon core encapsulated by silicon dioxide and tipped by Ti/Au, to generate an identical configuration to sharp intracellular electrodes (Figs. 19.4f and 19.5d). In the study, 3 × 3 arrays of 9 nanopillars, 150 nm in diameter, 3 μm in height at 2 μm pitch were grown on 16 sensing pads. Embryonic rat cortical neurons or HEK293 cells were then cultured on the VNEAs for a number of days. About 50% of the VNEAs spontaneously penetrated through the plasma membrane of the HEK293 cells as demonstrated by the fact that current injection through the VNEAs generated a voltage drop across the plasma membrane. In cases where spontaneous penetration of the membrane was not evident, an electroporating pulse (approximately ±6 V, 100 ms) was applied to penetrate the membrane of the neuron. The effect of the electroporating pulse on the integrity of the membrane was not shown. The seal resistance formed between the VNEAs and the plasma membrane was estimated to be 100–500 MΩ. The VNEAs were used in two regimes: in the Faradic regime, when a bias of ~−1.5 V was applied to the nanowire and the access resistance was reduced to 300 MΩ, and in the capacitive regime, when no bias was delivered and the access resistance was infinite. Accordingly, in the Faradic regime the electrical coupling for APs between the cell and the VNEA was about 10% whereas in the capacitive regime the attenuation was larger, reaching a lower coupling of ~0.1–0.3% (Table 19.1).

Table 19.1 Methods for recording and stimulation of electrogenic cells

	Seal (Ω)	Maximum AP (mV)	Maximum EPSP (mV)	Cell type studied	Coupling coefficient	Duration of stable recording	Electrode impedance (Ω)	Primary electrode material	Insertion method	Coating	Ref.
Sharp-glass electrodes	10^8	*	*	Multiple	1	Hours	10^6	Glass micro-pipette	Mechanical	None	4
Patch-clamp electrodes (whole cell)	$>10^9$	*	*	Multiple	1	Hours	10^6	Glass micro-pipette	Mechanical	None	3
Planar MEA and transistors arrays	10^6	<1	None	Multiple	0.001–0.01	>months	10^5 cm^{-2}	Metals	None	Multiple	45
Gold mushroom-shaped MEA (gMμE)	10^8	25	5	Neurons (*Aplysia*)	~0.5	>2 days	10^{11}	Au	Biological endocytosis	RGD motif peptide	64
Kinked silicon-nanowire FET	N/A	80**	None	Cardiomyocytes	~1	Seconds	Irrelevant (FET device)	Si	Mechanical or spontaneous	Phospholipids	77
Vertical nanowire electrode arrays (VNEA)	$>10^8$	4	None	Neurons (cortical rat)	~0.3	10 min	3×10^8 (with application of bias voltage = 1.5 V)	Si with Au tip	Bias voltage	Silanization	71
Branched intracellular nanotube-FET	N/A	75–100	None	Cardiomyocytes (embryonic chicken)	~1	1 h	Irrelevant (FET device)	Si	Mechanical or spontaneous	Phospholipids	78
Nanopillar electroporation	N/A	12	None	HL-1 cell line	~0.1	10 min per daily session, 3 days	6×10^6 to 18×10^6 (depending on the number of pillars)	Pt	Electroporating current	Fibronectin	57
gMμE electroporation	10^8	6	None	Neurons (*Aplysia*)	~0.1	5 min	10^{11}	Au	Electroporating current	Poly-L-lysine	58
Active silicon nanotube transistor	N/A	80	None	Cardiomyocytes (embryonic chicken)	~1	N/A	Irrelevant (FET device)	Si nanotube	Mechanical or spontaneous	Phospholipids	84

Note: Electrical and methodological specifications of the available modalities of cellular electrical recordings: from sharp-glass and patch-clamp electrodes to planar microelectrode and FET arrays and through to the recently developed intracellular-recording nano- and microdevices. *Glass micropipettes provide maximal (one-to-one) AP and EPSP amplitudes.

**Estimated from conductivity change.

Consistent with the intracellular positioning of the VNEA, all recorded APs were positive monophasic (as discussed in [18, 70]). However, the coupling coefficient and signal-to-noise ratio were insufficient to enable recordings of subthreshold synaptic potentials. In fact, in relation to the five examination criteria listed above, the recordings obtained by the VNEA do not provide significant advantages over extracellularly positioned gMμE in cultured hippocampal neurons [70]. The relatively low coupling coefficient of the VNEA-based sensor is most likely due to the high impedance of the VNEAs (Table 19.1). An advantage of recording by VNEAs over recordings by classical substrate-integrated planar electrodes (but not small electrodes [7]) is that a single pad records APs from an individual neuron. Although Robinson et al. [71] emphasize the usefulness of VNEAs to address individual neurons by electrical stimulation and record the ensuing synaptic potentials by a patch electrode, it should be noted that different types of extracellular micrometre-size planar (Fig. 19.4c) and protruding microelectrodes have been shown to effectively record or stimulate single cells [19, 56, 72, 73]. For example, Hofmann et al. [73] constructed a liquid-filled nanocavity (50 nm in height) that accesses a low impedance electrode by the fluid that fills the cavity. The functional contact of a cell with the electrode is through an aperture with a diameter smaller than the cell's. This configuration increases the spatial resolution of a MEA and increases the R_{seal} formed between the cell and the device. The cell–nanocavity configuration enables negative almost monophasic FPs with amplitudes >1 mV to be recorded. Using an array of 600 nm-thick tungsten protruding micronail-like electrodes Huys et al. [19] recorded biphasic or monophasic negative FPs of 50–100 μV. These micronails also enabled single cells to be stimulated effectively.

19.5 Intracellular Recordings by Membrane Electroporation

As pointed out by Robinson et al. [71], only a fraction of the nanowires spontaneously penetrate the plasma membrane.

Nevertheless, application of current through the pillars readily leads to their functional penetration.

Four recent studies demonstrated for the first time that localized membrane electroporation may lead to transient intracellular recordings of attenuated APs. Xie et al. demonstrated electroporation of cultured cardiomyocytes by vertical nanopillar electrodes [57], Hai and Spira [58], and Fendyur and Spira [59] demonstrated electroporation of cultured *Aplysia* neurons and primary cultured rat cardiomyocytes, respectively, by the gMμEs, and Breaken et al. demonstrated single-cell cardiomyocytes electroporation using micrometre-sized TiN protruding electrodes [60]. In all four studies the intracellular access was transient suggesting that electroporation activates repair mechanisms that seal off the electroporated nanopores [58], leaving the protruding nano- or microelectrodes out of the cell (Fig. 19.6). In these studies, the electroporating pulses reduced the junctional membrane resistance sufficiently to transform a typical biphasic cardiomyocyte FP to a 1–11 mV positive monophasic AP. The shape of the attenuated AP was typical of cardiomyocyte potentials.

The transient nature of the electrical coupling and the attenuation of the APs indicate that electroporation cannot be the method of choice to improve the quality of the interface formed between micro- or nano-based MEAs and the cell's membrane. Nevertheless, as it was demonstrated by the Melosh laboratory [74–76] and already applied by the Lieber laboratory [77, 78], it is conceivable that membrane–electrode fusion and GΩ seal formation can be facilitated by proper surface functionalization of the electrodes using lipid-based agents.

It is of interest to note that the nanopillar approach does not provide a significant advantage over planar extracellular electrodes as the recorded potentials are attenuated by at least an order of magnitude by the inherent high electrode impedance and the insufficient R_{seal}. It should also be noted that even if a single sensing pad carries multiple nanopillars, and a number of them penetrate the plasma membrane, the electrodes impedance is too high to enable recording of subthreshold potentials [57, 71] (Table 19.1). Theoretically, the impedance problem could be solved by increasing the density of the nanopillars over the sensing pad. Nevertheless,

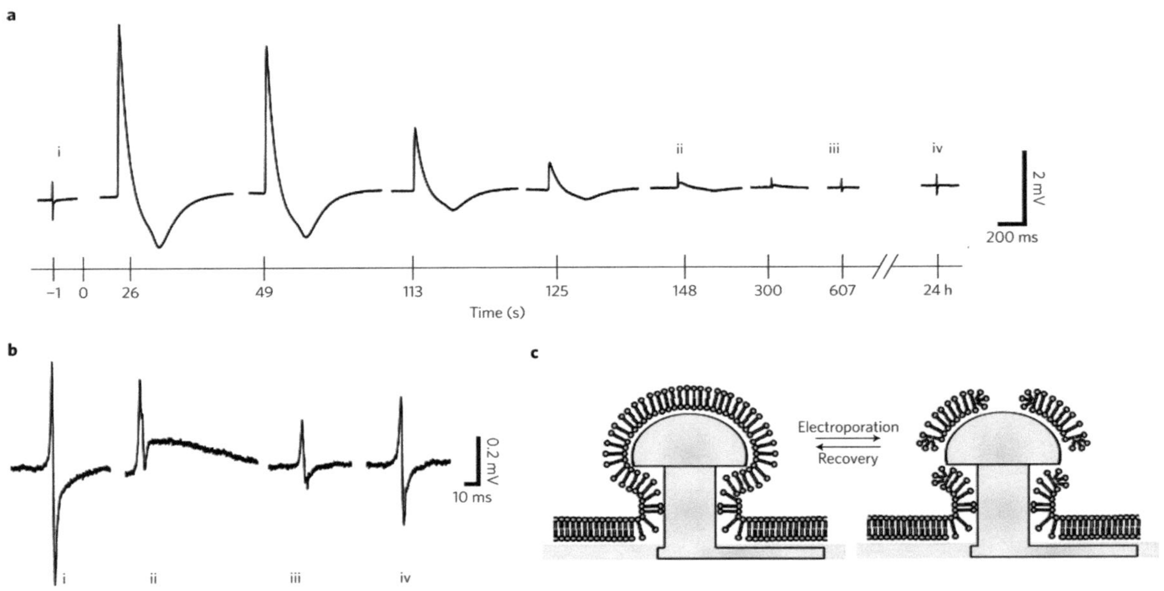

Figure 19.6 From extracellular field-potential recordings to intracellular recordings and the recovery process. (a) Before electroporation, a gMμE recorded a rat cardiomyocyte extracellular FP (a, (i); enlarged in b, (i)). After the delivery of an electroporating pulse (100 ms, 1 V) (at 0 s) the biphasic FP transformed into a monophasic 5 mV positive potential with a shape similar to that reported by intracellular recordings. The amplitude of the AP diminished over time 125–148 s, gradually resuming the shape of the extracellular FP after electroporation. Thereafter the shape of the FP gradually recovered (between 148–607 s, enlarged in b, i–iv) regaining the typical biphasic shape (a, (iv) and b, (iv)). (b) Enlargements of the FPs before electroporation (i) and 148 s (ii), 607 s (iii) and 24 h (iv) after electroporation. (c) Schematic drawing of the presumed reversible effects of an electroporating pulse on the plasma membrane facing a gMμE. Figure reproduced with permission from [59], © 2012 Frontiers Media.

when the density of protruding nanostructures exceeds a certain bound, the pillars do not penetrate the cell membrane, analogous to a dense 'bed of nails'. For example, Bruggermann et al. fabricated densely packed vertical (60 nm diameter, 300–400 nm-high) gold pillars on 15 μm-diameter pads [52]. In spite of the nanometric dimensions of the pillars, spontaneously firing HL-1 cells cultured on the MEA generated large ~1 mV negative monophasic FPs. This clearly pointed out that the pillar nanoelectrodes maintained an extracellular position. In a similar manner, densely packed (2 μm pitch) micronail-shaped structures record low amplitude (~100 μV) biphasic or negative-monophasic extracellular FPs from cultured rat cardiomyocytes [79]. In conclusion, although physical contact between a nanoparticle and the plasma membrane may be sufficient to induce particle penetration through the plasma membrane [80], higher-density nanocontacts (aiming at lowering the electrode impedance) may prevent membrane penetration from occurring. Thus, optimization of the pillar number, to reduce the impedance and pillar densities to promote internalization is critical.

19.6 Overcoming the Constraints of Nanopillar Impedance

Using advanced semiconductor-based nanotechnology and the classical concepts of mechanically penetrating the cell plasma membrane by sharp glass microelectrodes, Lieber's laboratory demonstrated intracellular recordings of 80–100 mV APs from beating cultured cardiomyocytes (Fig. 19.5g). This was done by either the so-called kinked nanowires [77] or pillar-shaped protruding silicon nanowires [78], nanofabricated as the sensing gate electrode of an FET.

In the work described by Tian and Cohen-Karni et al., an FET was generated at the tip of an ~80 nm kinked silicon nanowire by way of in situ doping [77]. In the work described by Duan et al. a protruding silicon nanowire was integrated onto the gate of the FET [78] (Fig. 19.5g). To facilitate the penetration of the electrodes into the cells the device's surface was modified by phospholipids

[77]. Spontaneous fusion of the applied phospholipids with the lipid membrane of the cells seems to underlie the formation of the GΩ seal. Using both types of nanosensors, full blown cardiac APs of 75–100 mV, ~200 ms were recorded (Fig. 19.5h,i). The high-quality measurements of the APs were made possible by three factors: (a) the nanoscale size of the sensors that enabled its insertion into the cytosol through the plasma membrane of the cells, (b) the formation of GΩ resistance between the plasma membrane and the nanostructures, (c) the fact that the size of the sensing area does not affect its sensitivity [78, 81, 82]. It should be noted that whereas the aspects of nano-dimensions and GΩ seal formation are essential components to enable high-quality recordings, the key to the success is the use of the gate electrode of an FET as the sensing electrode rather than passive metal or silicon-based micro- or nanoelectrodes. Thus, in contrast with passive conducting lines, where the signal is significantly attenuated due to stray capacitance, an FET effectively amplifies the signal in situ. It should be noted that FETs are more susceptible to failure due to leakage currents, whereas passive electrodes are not affected as dramatically by device imperfections.

The approach used by Lieber's laboratory to insert the nanopillars into the cells [77, 78, 83, 84] was to transfer a layer of 'mature' cardiomyocytes grown on a thin piece of polydimethylsiloxane, upside down, onto the device surface and apply gentle downward pressure onto the substrate. This manipulation led to the insertion of the nanoelectrodes into the cells within approximately 45 s. Whereas at the 'proof-of-concept' level the approach taken by Lieber's laboratory is sufficient, it should be further developed to enable experiments in which cardiomyocytes and neurons can be cultured and grown in continuous contact with the MEA substrate and maintain the cells–electrodes contact for long periods rather than be acutely manipulated.

When considering the branched intracellular nanotube-FET or kinked nanoelectrode devices as tools to map functional synaptic connectivity, a major hurdle is the signal-to-noise level of the device. Examination of some recordings (for example, see Fig. 4 in [78]) reveals noise levels of more than 20 mV. Whereas the noise is attributed to the nano-dimensions of the FETs and thus can be reduced by adjusting the FET size, the present device does not provide the resolution to enable the recording of miniature potentials, synaptic potentials and small membrane oscillations. Another unsolved problem that would need further study is to enable the accurate recordings of the resting potentials.

19.7 Planar Patch-Clamp MEA Technology

Another line of investigation to overcome the problems of high junctional membrane resistance, electrode impedance and of low R_{seal} revolves around microfluidic-based MEAs able to patch clamp neurons under in vitro conditions. The approach is extensively used to acutely patch cells in suspension relying on suction to draw individual cells to the aperture and to form a giga-seal resistance. Thus far the approach was not suitable for studies of long-term adhering cultured neuronal networks [85, 86]. Recently Martina et al. [87] began demonstrating that the approach has a potential to be adapted to neuronal networks. In this study a silicon oxide substrate or silicon oxide laminated by polyimide film containing 2 to 4 μm-sized apertures were used. Each aperture was connected to a microfluidic channel. Two isolated cell bodies of identifiable *Lymnaea* neurons were co-cultured over the apertures for 8–12 h. Within this time the cell bodies formed chemical synapses and adhered to the substrate to spontaneously form an unusually high GΩ seal resistance. Negative pressure pulse through the microfluidic system broke the junctional membrane establishing a classical whole-cell patch-clamp configuration. In a fraction of the experiments the patch configuration was stable for a number of hours and the properties of the synapses formed between the two cells could be investigated. The signal-to-noise ratio obtained by the planar patch-clamp device matches that of conventional patch-clamp recording. Although promising, it should be noted that the somata of the isolated neurons adhered to each other but did not extend neurites on the culture substrate. Thus the neuron–device configuration did not simulate the complex growth pattern of cultured mammalian neurons but rather is closer to the cell suspension mode of patch-clamp recordings.

When considering the potential use of planar patch-clamp MEAs as tools to map functional synaptic connectivity among cultured neurons, two major problems have to be dealt with: (a) the recording duration is expected to be limited by the perfusion of the neuron by the microfluidic solutions (Fig. 19.4h), (b) the sensor's density (apertures) is expected to be limited by the backside fluidic system.

19.8 Conclusions

On the basis of the results reviewed in the chapter and theoretical considerations, we estimate that nano- and micro-electrophysiological technologies enabling simultaneous, long-term, multisite, intracellular recording and stimulation from many neurons under in vitro and in vivo conditions will become available to the neuroscientist community within a number of years.

The approaches that at present reveal the best potential are: the bioinspired use of protruding electrodes that are engulfed by neurons, and the use of nanostructures that penetrate the plasma membrane in a similar way to classical sharp microelectrodes. Experimentations with passive nanopillar-based protruding structures functionalized by lipid layers revealed that whereas single or multiple nanopillars can penetrate the cell's plasma membrane forming relatively high R_{seal}, the high nano-electrode impedance/R_{seal} ratio value attenuates the recorded potential to a level that makes it impossible to record synaptic potentials or subthreshold membrane oscillations. Attempts to reduce the electrode impedance by fabricating multiple nanopillars on a single sensing pad did not solve the problem [71]. It seems that increasing the number and density of the nanopillars to reach low enough impedance (and improve the electrode impedance/R_{seal} ratio) is limited by cell biological properties as cells cultured on high-density nanopillars do not extend the plasma membrane into narrow spaces between the pillars [52, 79] and that physically inserting a dense population of nanopillars would be damaging to the cell. It is for these reasons that current passive nanopillar electrodes cannot be used as substitutes for traditional glass micropipette electrodes for intracellular recordings from neurons. Another severe limitation with the use of nanopillars is the instability of the intracellular configuration. So far, mechanisms to stabilize the intracellular positioning of the nanopillars have not been addressed.

On the other hand, the use of either kinked nanowires [77] or pillar-shaped protruding silicon nanowires [78], as the gate electrodes of an FET device, bypasses the problem of high electrode impedance (as FET recordings are independent of gate impedance [78, 81, 82]). Because of their nanoscale dimensions, the use of FET-based nanosensors can also enable the simultaneous recordings from subcellular compartments (dendrite, somata, axon, varicosities and others). The current limitations of the nanopillars–FET devices for intracellular recordings are: the inherent noise level of the nano-FETs, and the need to mechanically manipulate the cultured cells and the substrate on which they grow into physical contact with the electrodes. The problem of noise may be dealt with by modulating the transistor size and geometry [88] while taking into consideration the limitations of ambient thermal noise in electrolytic solutions [89]. The issue of having to mechanically press the cells and device to acutely form physical contact may be solved by merging some of the concepts developed by us and the Lieber laboratory.

So far, the gMμE-based MEA is the only device that enabled the recordings of both APs and subthreshold synaptic potentials, and that can also be used for effective intracellular stimulation [64]. Nevertheless, this coupling was demonstrated using large *Aplysia* neurons but as of yet has not successfully been applied to rat hippocampal neurons and primary rat cardiomyocytes.

It is conceivable that merging the cell-biological principles of evoking engulfment of the electrodes on the one hand and the use of FETs on the other may provide both a stable neuron–electrode configuration and intracellular access. Once achieved, such a device may be applied in arrays that make use of the well-established multiplexing capabilities of ultra-large-scale integrated transistor arrays.

Acknowledgements

Spira's laboratory is currently supported by EU FP7 MERIDIAN Grant agreement no. 280778., EU FP7 Marie Curie ITG, Grant agreement no. 264872., and the Charles E. Smith and Prof. Elkes Laboratory for Collaborative Research in Psychobiology. A. Hai was supported by a scholarship from The Israel Council for Higher Education.

Competing Financial Interests

The authors declare no competing financial interests.

References

1. Hodgkin, A. L., and Huxley, A. F. Action potentials recorded from inside a nerve fibre. *Nature* **144**, 710–711 (1939).
2. Grundfest, H. The mechanisms of discharge of the electric organs in relation to general and comparative electrophysiology. *Prog. Biophys. Biophys. Chem.* **7**, 1–85 (1957).
3. Sakmann, B., and Neher, E. Patch clamp techniques for studying ionic channels in excitable membranes. *Annu. Rev. Physiol.* **46**, 455–472 (1984).
4. Verkhratsky, A., Krishtal, O. A., and Petersen, O. H. From Galvani to patch clamp: the development of electrophysiology. *Pflugers Arch.* **453**, 233–247 (2006).
5. Hutzler, M. et al. High-resolution multitransistor array recording of electrical field potentials in cultured brain slices. *J. Neurophysiol.* **96**, 1638–1645 (2006).
6. Eytan, D., and Marom, S. Dynamics and effective topology underlying synchronization in networks of cortical neurons. *J. Neurosci.* **26**, 8465–8476 (2006).
7. Berdondini, L. et al. Active pixel sensor array for high spatio-temporal resolution electrophysiological recordings from single cell to large scale neuronal networks. *Lab Chip* **9**, 2644–2651 (2009).

8. Frey, U., Egert, U., Heer, F., Hafizovic, S., and Hierlemann, A. Microelectronic system for high-resolution mapping of extracellular electric fields applied to brain slices. *Biosens. Bioelectron.* **24**, 2191–2198 (2009).
9. Hochberg, L. R. et al. Neuronal ensemble control of prosthetic devices by a human with tetraplegia. *Nature* **442**, 164–171 (2006).
10. Blanche, T. J., Spacek, M. A., Hetke, J. F., and Swindale, N. V. Polytrodes: high-density silicon electrode arrays for large-scale multiunit recording. *J. Neurophysiol.* **93**, 2987–3000 (2005).
11. Buzsaki, G., Anastassiou, C. A., and Koch, C. The origin of extracellular fields and currents—EEG, ECoG, LFP and spikes. *Nature Rev. Neurosci.* **13**, 407–420 (2012).
12. Shoham, S., O'Connor, D. H., and Segev, R. How silent is the brain: is there a "dark matter" problem in neuroscience? *J. Comp. Physiol.* **192**, 777–784 (2006).
13. Mayford, M., Siegelbaum, S. A., and Kandel, E. R. Synapses and memory storage. *Cold Spring Harb. Perspect. Biol.* **4**, a005751 (2012).
14. Thomas, C. A. Jr, Springer, P. A., Loeb, G. E., Berwald-Netter, Y., and Okun, L. M. A miniature microelectrode array to monitor the bioelectric activity of cultured cells. *Exp. Cell Res.* **74**, 61–66 (1972).
15. Gross, G. W., Williams, A. N., and Lucas, J. H. Recording of spontaneous activity with photoetched microelectrode surfaces from mouse spinal neurons in culture. *J. Neurosci. Methods* **5**, 13–22 (1982).
16. Regehr, W. G., Pine, J., Cohan, C. S., Mischke, M. D., and Tank, D. W. Sealing cultured invertebrate neurons to embedded dish electrodes facilitates long-term stimulation and recording. *J. Neurosci. Methods* **30**, 91–106 (1989).
17. Connolly, P., Clark, P., Curtis, A. S., Dow, J. A., and Wilkinson, C. D. An extracellular microelectrode array for monitoring electrogenic cells in culture. *Biosens. Bioelectron.* **5**, 223–234 (1990).
18. Nam, Y., and Wheeler, B. C. In vitro microelectrode array technology and neural recordings. *Crit. Rev. Biomed. Eng.* **39**, 45–61 (2011).
19. Huys, R. et al. Single-cell recording and stimulation with a 16k micro-nail electrode array integrated on a 0.18 μm CMOS chip. *Lab Chip* **12**, 1274–1280 (2012).
20. Schwartz, A. B. Cortical neural prosthetics. *Annu. Rev. Neurosci.* **27**, 487–507 (2004).
21. Fee, M. S., Mitra, P. P., and Kleinfeld, D. Automatic sorting of multiple unit neuronal signals in the presence of anisotropic and non-Gaussian variability. *J. Neurosci. Methods* **69**, 175–188 (1996).

22. Brown, E. N., Kass, R. E., and Mitra, P. P. Multiple neural spike train data analysis: state-of-the-art and future challenges. *Nature Neurosci.* **7**, 456–461 (2004).
23. Kauer, J. S., Senseman, D. M., and Cohen, L. B. Odor-elicited activity monitored simultaneously from 124 regions of the salamander olfactory bulb using a voltage-sensitive dye. *Brain Res.* **418**, 255–261 (1987).
24. Siegel, M. S., and Isacoff, E. Y. A genetically encoded optical probe of membrane voltage. *Neuron* **19**, 735–741 (1997).
25. Shoham, D. et al. Imaging cortical dynamics at high spatial and temporal resolution with novel blue voltage-sensitive dyes. *Neuron* **24**, 791–802 (1999).
26. Kralj, J. M., Douglass, A. D., Hochbaum, D. R., Maclaurin, D., and Cohen, A. E. Optical recording of action potentials in mammalian neurons using a microbial rhodopsin. *Nature Methods* **9**, 90–95 (2012).
27. Loew, L. M., Cohen, L. B., Salzberg, B. M., Obaid, A. L., and Bezanilla, F. Charge-shift probes of membrane potential. Characterization of aminostyrylpyridinium dyes on the squid giant axon. *Biophys. J.* **47**, 71–77 (1985).
28. Stosiek, C., Garaschuk, O., Holthoff, K., and Konnerth, A. In vivo two-photon calcium imaging of neuronal networks. *Proc. Natl Acad. Sci. USA* **100**, 7319–7324 (2003).
29. Higley, M. J., and Sabatini, B. L. Calcium signaling in dendrites and spines: practical and functional considerations. *Neuron* **59**, 902–913 (2008).
30. Rothschild, G., Nelken, I., and Mizrahi, A. Functional organization and population dynamics in the mouse primary auditory cortex. *Nature Neurosci.* **13**, 353–360 (2010).
31. Grinvald, A., Lieke, E., Frostig, R. D., Gilbert, C. D., and Wiesel, T. N. Functional architecture of cortex revealed by optical imaging of intrinsic signals. *Nature* **324**, 361–364 (1986).
32. Nagel, G. et al. Channelrhodopsin-1: a light-gated proton channel in green algae. *Science* **296**, 2395–2398 (2002).
33. Bernstein, J. G., and Boyden, E. S. Optogenetic tools for analyzing the neural circuits of behavior. *Trends Cogn. Sci.* **15**, 592–600 (2011).
34. Homma, R. et al. Wide-field and two-photon imaging of brain activity with voltage- and calcium-sensitive dyes. *Phil. Trans. R. Soc. Lond. B* **364**, 2453–2467 (2009).
35. Spatz, J. P., and Geiger, B. Molecular engineering of cellular environments: cell adhesion to nano-digital surfaces. *Methods Cell Biol.* **83**, 89–111 (2007).

36. Rutten, W. L. Selective electrical interfaces with the nervous system. *Annu. Rev. Biomed. Eng.* **4**, 407–452 (2002).
37. Fromherz, P. Three levels of neuroelectronic interfacing: silicon chips with ion channels, nerve cells, and brain tissue. *Ann. NY Acad. Sci.* **1093**, 143–160 (2006).
38. Jones, I. L. et al. The potential of microelectrode arrays and microelectronics for biomedical research and diagnostics. *Anal. Bioanal. Chem.* **399**, 2313–2329 (2011).
39. Bershadsky, A. D., Balaban, N. Q., and Geiger, B. Adhesion-dependent cell mechanosensitivity. *Annu. Rev. Cell Dev. Biol.* **19**, 677–695 (2003).
40. Wrobel, G. et al. Transmission electron microscopy study of the cell-sensor interface. *J. R. Soc. Interface* **5**, 213–222 (2008).
41. Braun, D., and Fromherz, P. Fluorescence interference-contrast microscopy of cell adhesion on oxidized silicon. *Appl. Phys. A* **65**, 341–348 (1997).
42. Iwanaga, Y., Braun, D., and Fromherz, P. No correlation of focal contacts and close adhesion by comparing GFP-vinculin and fluorescence interference of DiI. *Eur. Biophysics J. Biophysics Lett.* **30**, 17–26 (2001).
43. Lambacher, A., and Fromherz, P. Luminescence of dye molecules on oxidized silicon and fluorescence interference contrast microscopy of biomembranes. *J. Opt. Soc. Am. B* **19**, 1435–1453 (2002).
44. Gleixner, R., and Fromherz, P. The extracellular electrical resistivity in cell adhesion. *Biophys. J.* **90**, 2600–2611 (2006).
45. Fromherz, P. in *Neuroelectronic Interfacing: Semiconductor Chips with Ion Channels, Nerve Cells, and Brain* (ed. Waser, P.) Ch. 32, 781–810 (Wiley, 2003).
46. Maccione, A. et al. Experimental investigation on spontaneously active hippocampal cultures recorded by means of high-density MEAs: analysis of the spatial resolution effects. *Front. Neuroeng.* **3**, 1–12 (2010).
47. Buitenweg, J. R., Rutten, W. L., and Marani, E. Geometry-based finite-element modeling of the electrical contact between a cultured neuron and a microelectrode. *IEEE Trans. Biomed. Eng.* **50**, 501–509 (2003).
48. Pine, J. Recording action potentials from cultured neurons with extracellular microcircuit electrodes. *J. Neurosci. Methods* **2**, 19–31 (1980).
49. Oka, H., Shimono, K., Ogawa, R., Sugihara, H., and Taketani, M. A new planar multielectrode array for extracellular recording: application to hippocampal acute slice. *J. Neurosci. Methods* **93**, 61–67 (1999).

50. Grumet, A. E., Wyatt, J. L. Jr, and Rizzo, J. F. 3rd. Multi-electrode stimulation and recording in the isolated retina. *J. Neurosci. Methods* **101**, 31–42 (2000).
51. Kim, J. H., Kang, G., Nam, Y., and Choi, Y. K. Surface-modified microelectrode array with flake nanostructure for neural recording and stimulation. *Nanotechnology* **21**, 85303 (2010).
52. Bruggemann, D. et al. Nanostructured gold microelectrodes for extracellular recording from electrogenic cells. *Nanotechnology* **22**, 265104 (2011).
53. Shein, M. et al. Engineered neuronal circuits shaped and interfaced with carbon nanotube microelectrode arrays. *Biomed. Microdevices* **11**, 495–501 (2009).
54. Keefer, E. W., Botterman, B. R., Romero, M. I., Rossi, A. F., and Gross, G. W. Carbon nanotube coating improves neuronal recordings. *Nature Nanotech.* **3**, 434–439 (2008).
55. Akaike, N., and Harata, N. Nystatin perforated patch recording and its applications to analyses of intracellular mechanisms. *Jpn. J. Physiol.* **44**, 433–473 (1994).
56. Braeken, D. et al. Single-cell stimulation and electroporation using a novel 0.18 μ CMOS chip with subcellular-sized electrodes. *Conf. Proc. IEEE Eng. Med. Biol. Soc.* **2010**, 6473–6476 (2010).
57. Xie, C., Lin, Z., Hanson, L., Cui, Y., and Cui, B. Intracellular recording of action potentials by nanopillar electroporation. *Nature Nanotech.* **7**, 185–190 (2012).
58. Hai, A., and Spira, M. E. On-chip electroporation, membrane repair dynamics and transient in-cell recordings by arrays of gold mushroom-shaped microelectrodes. *Lab Chip* **12**, 2865–2873 (2012).
59. Fendyur, A., and Spira, M. E. Toward on-chip, in-cell recordings from cultured cardiomyocytes by arrays of gold mushroom-shaped microelectrodes. *Front. Neuroeng.* **5**, 21 (2012).
60. Braeken, D. et al. Open-cell recording of action potentials using active electrode arrays. *Lab Chip* **12**, 4397–4402 (2012).
61. Spira, M. E. et al. Improved neuronal adhesion to the surface of electronic device by engulfment of protruding micro-nails fabricated on the chip surface. *Transducers '07 and Eurosensors XXI* 1247–1250 (IEEE, 2007).
62. Hai, A. et al. Spine-shaped gold protrusions improve the adherence and electrical coupling of neurons with the surface of micro-electronic devices. *J. R. Soc. Interface* **6**, 1153–1165 (2009).

63. Hai, A. et al. Changing gears from chemical adhesion of cells to flat substrata toward engulfment of micro-protrusions by active mechanisms. *J. Neural Eng.* **6**, 066009 (2009).

64. Hai, A., Shappir, J., and Spira, M. E. Long-term, multisite, parallel, in-cell recording and stimulation by an array of extracellular microelectrodes. *J. Neurophysiol.* **104**, 559–568 (2010).

65. Hai, A., Shappir, J., and Spira, M. E. In-cell recordings by extracellular microelectrodes. *Nature Methods* **7**, 200–202 (2010).

66. Aderem, A., and Underhill, D. M. Mechanisms of phagocytosis in macrophages. *Annu. Rev. Immunol.* **17**, 593–623 (1999).

67. Hayashi, Y., and Majewska, A. K. Dendritic spine geometry: functional implication and regulation. *Neuron* **46**, 529–532 (2005).

68. May, R. C., and Machesky, L. M. Phagocytosis and the actin cytoskeleton. *J. Cell Sci.* **114**, 1061–1077 (2001).

69. Cohen, A., Shappir, J., Yitzchaik, S., and Spira, M. E. Reversible transition of extracellular field potential recordings to intracellular recordings of action potentials generated by neurons grown on transistors. *Biosensors Bioelectronics* **23**, 811–819 (2008).

70. Fendyur, A., Mazurski, N., Shappir, J., and Spira, M. E. Formation of essential ultrastructural interface between cultured hippocampal cells and gold mushroom-shaped MEA—toward "IN-CELL" recordings from vertebrate neurons. *Front. Neuroeng.* **4**, 14 (2011).

71. Robinson, J. T. et al. Vertical nanowire electrode arrays as a scalable platform for intracellular interfacing to neuronal circuits. *Nature Nanotech.* **7**, 180–184 (2012).

72. Schoen, I., and Fromherz, P. Extracellular stimulation of mammalian neurons through repetitive activation of Na^+ channels by weak capacitive currents on a silicon chip. *J. Neurophysiol.* **100**, 346–357 (2008).

73. Hofmann, B., Katelhon, E., Schottdorf, M., Offenhausser, A., and Wolfrum, B. Nanocavity electrode array for recording from electrogenic cells. *Lab Chip* **11**, 1054–1058 (2011).

74. Almquist, B. D., and Melosh, N. A. Fusion of biomimetic stealth probes into lipid bilayer cores. *Proc. Natl Acad. Sci. USA* **107**, 5815–5820 (2010).

75. Almquist, B. D., and Melosh, N. A. Molecular structure influences the stability of membrane penetrating biointerfaces. *Nano Lett.* **11**, 2066–2070 (2011).

76. Almquist, B. D., Verma, P., Cai, W., and Melosh, N. A. Nanoscale patterning controls inorganic-membrane interface structure. *Nanoscale* **3**, 391–400 (2011).
77. Tian, B. et al. Three-dimensional, flexible nanoscale field-effect transistors as localized bioprobes. *Science* **329**, 830–834 (2010).
78. Duan, X. et al. Intracellular recordings of action potentials by an extracellular nanoscale field-effect transistor. *Nature Nanotech.* **7**, 174–179 (2012).
79. Huys, R. et al. A novel 16k micro-nail CMOS-chip for in-vitro single-cell recording, stimulation and impedance measurements. *Conf. Proc. IEEE Eng. Med. Biol. Soc.* **2010**, 2726–2729 (2010).
80. Doherty, G. J., and McMahon, H. T. Mechanisms of endocytosis. *Annu. Rev. Biochem.* **78**, 857–902 (2009).
81. Cohen-Karni, T., Timko, B. P., Weiss, L. E., and Lieber, C. M. Flexible electrical recording from cells using nanowire transistor arrays. *Proc. Natl Acad. Sci. USA* **106**, 7309–7313 (2009).
82. Sadiku, M. N. O. *Elements of Electromagnetics* (Oxford Univ. Press, 2000).
83. Tian, B., and Lieber, C. M. Design, synthesis, and characterization of novel nanowire structures for photovoltaics and intracellular probes. *Pure Appl. Chem.* **83**, 2153–2169 (2011).
84. Gao, R. et al. Outside looking in: nanotube transistor intracellular sensors. *Nano Lett.* **12**, 3329–3333 (2012).
85. Ionescu-Zanetti, C. et al. Mammalian electrophysiology on a microfluidic platform. *Proc. Natl Acad. Sci. USA* **102**, 9112–9117 (2005).
86. Lau, A. Y., Hung, P. J., Wu, A. R., and Lee, L. P. Open-access microfluidic patch-clamp array with raised lateral cell trapping sites. *Lab Chip* **6**, 1510–1515 (2006).
87. Martina, M. et al. Recordings of cultured neurons and synaptic activity using patch-clamp chips. *J. Neural Eng.* **8**, 034002 (2011).
88. Abidi, A. A. High-frequency noise measurements on FET's with small dimensions. *Electron. Dev., IEEE Trans. on* **33**, 1801–1805 (1986).
89. Voelker, M., and Fromherz, P. Nyquist noise of cell adhesion detected in a neuron-silicon transistor. *Phys. Rev. Lett.* **96**, 228102 (2006).

Index